PHYSICAL CONSTANTS

Speed of light in vacuum	c	2.998×10^8 m s^{-1}
Elementary charge	e	1.602×10^{-19} C
Avogadro constant	N_A, L	6.022×10^{23} mol^{-1}
Atomic mass unit	u	1.661×10^{-27} kg
Electron rest mass	m_e	9.109×10^{-31} kg
Proton rest mass	m_p	1.673×10^{-27} kg
Faraday constant	F	9.6485×10^4 C mol^{-1}
Planck constant	h	6.626×10^{-34} J s
Rydberg constant	R_∞	1.097×10^7 m^{-1}
Gas constant	R	8.314 J K^{-1} mol^{-1}
Boltzmann constant	k	1.381×10^{-23} J K^{-1}
Permittivity of vacuum	ϵ_0	8.854×10^{-12} C^2 N^{-1} m^2
	$1/4\pi\epsilon_0$	0.8988×10^{10} N m^2 C^{-2}

SI BASE UNITS

Physical quantity	Symbol for quantity	SI unit	Symbol for unit
Length	l	meter	m
Mass	m	kilogram	kg
Time	t	second	s
Thermodynamic temperature	T	Kelvin	K
Electric current	I	ampere	A
Luminous intensity	I_v	candela	cd
Amount of substance	n	mole	mol

SI PREFIXES

Fraction	Prefix	Symbol	Multiple	Prefix	Symbol
10^{-1}	deci	d	10	deka	da
10^{-1}	centi	c	10^2	hecto	h
10^{-3}	milli	m	10^3	kilo	k
10^{-6}	micro	μ	10^6	mega	M
10^{-9}	nano	n	10^9	giga	G
10^{-12}	pico	p	10^{12}	tera	T
10^{-15}	femto	f	10^{15}	peta	P
10^{-18}	atto	a	10^{18}	exa	E

CHEMICAL KINETICS

THIRD EDITION

Keith J. Laidler

University of Ottawa
Chairman, Commission on Chemical Kinetics,
International Union of Pure and Applied Chemistry

1817

HARPER & ROW, PUBLISHERS, New York
Cambridge, Philadelphia, San Francisco, Washington,
London, Mexico City, São Paulo, Singapore, Sydney

Dedicated to the memory of
Cyril Norman Hinshelwood,
Henry Eyring, and
Edgar William Richard Steacie
with respect and gratitude

Sponsoring Editor: Lisa S. Berger
Project Editor: Steven Pisano
Cover Design: Wanda Lubelska Design
Text Art: Volt Information Sciences, Inc.
Production: Willie Lane
Compositor: TAPSCO, Inc.
Printer and Binder: R. R. Donnelley & Sons Company

CHEMICAL KINETICS
Third Edition
Copyright © 1987 by Harper & Row, Publishers, Inc.

Library of Congress Cataloging-in-Publication Data

Laidler, Keith James, 1916–
    Chemical kinetics.

    Bibliography: p.
    1. Chemical reaction, Rate of.   I. Title.
QD501.L17   1987   541.3′94   86-18447
ISBN 0-06-043862-2

86 87 88 89 9 8 7 6 5 4 3 2 1

CONTENTS

Preface   xi

Chapter 1
BASIC KINETIC CONCEPTS   1

1.1   Scope of Chemical Kinetics   2
    1.1.1   Macroscopic and Microscopic Kinetics   3
1.2   Reaction Stoichiometry   4
1.3   Rates of Consumption and Formation   6
1.4   Extent of Reaction   7
1.5   Rate of Reaction   8
1.6   Volume Change During Reaction   9
1.7   Empirical Rate Equations   10
    1.7.1   Order of Reaction   10
    1.7.2   Rate Constants   11
    1.7.3   Reactions Having No Order   12
1.8   Elementary, Composite, and Chain Reactions   13
    1.8.1   Chain Reactions   14
1.9   Catalysis and Inhibition   14
    Problems   15
    References   16
    Bibliography   17

Chapter 2
ANALYSIS OF KINETIC RESULTS   18

2.1   Differential Method   18
2.2   Method of Integration   21
    2.2.1   First-Order Reactions   21
    2.2.2   Second-Order Reactions   23
    2.2.3   Reactions of the nth Order   25
2.3   Isolation Method   26
2.4   Half-Life   26
2.5   Comparison of Methods   28
2.6   Opposing Reactions   29
2.7   Reactions in Flow Systems   31

2.8   Techniques for Very Fast Reactions   33
        2.8.1   Stopped-Flow Method   34
        2.8.2   Relaxation Methods   35
        2.8.3   Shock-Tube Methods   39
        2.8.4   Flash Photolysis   39
2.9   Influence of Temperature on Reaction Rates   39
        2.9.1   Survey of Empirical Equations for Temperature Dependence   40
        2.9.2   Arrhenius Equation   44
        2.9.3   Improved Treatments of Temperature Dependence   45
        2.9.4   Temperature Dependence of the Pre-exponential Factor   46
Problems   48
References   50
Bibliography   51

Chapter 3
ENERGY OF ACTIVATION   54

3.1   Statistical Distribution of Molecular Energies   55
        3.1.1   Simple Statistical Expressions   55
        3.1.2   Tolman's Theorem   56
3.2   Potential-Energy Surfaces   59
3.3   Ab Initio Calculations of Potential-Energy Surfaces   64
        3.3.1   Treatments Based on the London Equation   64
        3.3.2   Variational Calculations   64
3.4   Semiempirical Calculations of Potential-Energy Surfaces   67
        3.4.1   London–Eyring–Polanyi (LEP) Method   68
        3.4.2   Sato Method   69
        3.4.3   Modified LEP Methods   70
        3.4.4   Bond-Energy–Bond-Order (BEBO) Method   70
3.5   Empirical Treatments of Activation Energy   71
Appendix: Statistical Distribution of Molecular Energies   74
    Motion in Two Dimensions   74
    Motion in Three Dimensions   75
    Motion in Many Degrees of Freedom   76
Problems   77
References   78
Bibliography   79

Chapter 4
THEORIES OF REACTION RATES   80

4.1   Kinetic Theory of Collisions   81
        4.1.1   Generalized Kinetic Theory   84
        4.1.2   Extensions of Collision Theory   87
4.2   Rate Theories Based on Thermodynamics   87
4.3   Rate Theories Based on Statistical Mechanics   88
4.4   Early Dynamical Theories of Rates   88

4.5   Conventional Transition-State Theory   89
    4.5.1   Equilibrium Hypothesis   90
    4.5.2   Statistical Mechanics and Chemical Equilibrium   93
    4.5.3   Derivations of the Rate Equation   94
    4.5.4   Symmetry Numbers and Statistical Factors   98
4.6   Some Applications of Conventional Transition-State Theory   106
    4.6.1   Reactions between Atoms   106
    4.6.2   Reactions between Molecules   107
    4.6.3   An Example: The Reaction $H + HBr \rightarrow H_2 + Br$ 109
4.7   Thermodynamic Formulation of Conventional Transition-State Theory   112
4.8   Assumptions and Limitations of Conventional Transition-State Theory   115
    4.8.1   Multiple Crossings and the Equilibrium Hypothesis   116
    4.8.2   Separability of the Reaction Coordinate   119
    4.8.3   Quantum Effects   121
4.9   Extensions of Transition-State Theory   123
    4.9.1   Variational Transition-State Theory   124
    4.9.2   Quantum-Mechanical Transition-State Theory   127
    4.9.3   Muonium Kinetics   129
4.10   Microscopic Reversibility and Detailed Balance   129
    Problems   131
    References   132
    Bibliography   135

Chapter 5
ELEMENTARY GAS-PHASE REACTIONS   137

5.1   Bimolecular Reactions   138
    5.1.1   Reactions of the Type $H + H_2$ 138
    5.1.2   Other Atom and Free-Radical Reactions   140
    5.1.3   Temperature Dependence of Pre-exponential Factors   145
    5.1.4   Ion–Molecule Reactions   146
5.2   Trimolecular Reactions   147
5.3   Unimolecular Reactions   150
    5.3.1   Lindemann–Christiansen Hypothesis   152
    5.3.2   Hinshelwood's Treatment   155
    5.3.3   Rice–Ramsperger–Kassel (RRK) Treatment   157
    5.3.4   Slater's Treatment   163
    5.3.5   Marcus's Extension of the RRK Treatment (RRKM)   164
    5.3.6   Influence of Foreign Gases   167
    5.3.7   Intermolecular Energy Transfer   168
    5.3.8   Intramolecular Energy Transfer   168
    5.3.9   Laser-Induced Unimolecular Reactions   170
    5.3.10   Decomposition of Ions   171
5.4   Combination and Disproportionation Reactions   171
    5.4.1   Mechanisms of Atom and Radical Combinations   173
    Problems   176
    References   177
    Bibliography   180

Chapter 6
ELEMENTARY REACTIONS IN SOLUTION   183

6.1   Solvent Effects on Reaction Rates   183
   6.1.1   Comparison between Gas-Phase and Solution Reactions   184
   6.1.2   Comparison between Different Solvents   184
6.2   Factors Determining Reaction Rates in Solution   185
   6.2.1   Collisions in Solution   186
   6.2.2   Transition-State Theory   188
   6.2.3   Influence of Internal Pressure   189
   6.2.4   Influence of Solvation   190
6.3   Reactions between Ions   191
   6.3.1   Influence of Solvent Dielectric Constant   191
   6.3.2   Pre-exponential Factors   194
   6.3.3   Single-Sphere Activated Complex   195
   6.3.4   Influence of Ionic Strength   197
   6.3.5   More Advanced Treatments   201
6.4   Ion–Dipole and Dipole–Dipole Reactions   203
   6.4.1   Pre-exponential Factors   204
   6.4.2   Influence of Ionic Strength   205
6.5   Influence of Hydrostatic Pressure   206
   6.5.1   Van't Hoff's Equation   206
   6.5.2   Volumes of Activation   207
6.6   Substituent and Correlation Effects   209
   6.6.1   Hammett Equation   209
   6.6.2   Compensation Effects   211
6.7   Diffusion-Controlled Reactions   212
   6.7.1   Full Microscopic Diffusion Control   212
   6.7.2   Partial Microscopic Diffusion Control   216
   6.7.3   Ionic Reactions   217
   Problems   222
   References   224
   Bibliography   226

Chapter 7
REACTIONS ON SURFACES   229

7.1   Adsorption   230
7.2   Adsorption Isotherms   230
   7.2.1   Simple Langmuir Isotherm   231
   7.2.2   Adsorption with Dissociation   232
   7.2.3   Competitive Adsorption   233
   7.2.4   Nonideal Adsorption   234
   7.2.5   Thermodynamics and Statistical Mechanics of Adsorption   235
7.3   Structures of Solid Surfaces and Adsorbed Layers   237
   7.3.1   Detailed Structural Studies   238
   7.3.2   Induced Heterogeneity   241
7.4   Mechanisms of Surface Reactions   241
   7.4.1   Kinetic Effects of Surface Heterogeneity   243
   7.4.2   Kinetic Effects of Interactions   244

7.5 Unimolecular Surface Reactions 244
    7.5.1 Inhibition 246
    7.5.2 Activation Energies 246
7.6 Bimolecular Surface Reactions 248
    7.6.1 Reaction between Two Adsorbed Molecules 249
    7.6.2 Reaction between a Gas Molecule and an Adsorbed Molecule 250
    7.6.3 Adsorption of Two Gases without Mutal Displacement 250
    7.6.4 Inhibition 251
    7.6.5 Activation Energies 251
    7.6.6 Parahydrogen Conversion 252
    7.6.7 Combination and Formation of Atoms at Surfaces 252
    7.6.8 Exchange Reactions 253
    7.6.9 Addition of Hydrogen to Ethylene 257
7.7 Transition-State Theory of Surface Reactions 258
    7.7.1 Rates of Chemisorption 259
    7.7.2 Rates of Desorption 261
    7.7.3 Unimolecular Surface Reactions 262
    7.7.4 Bimolecular Surface Reactions 264
    7.7.5 Comparison of Homogeneous and Heterogeneous Reaction Rates 266
Problems 268
References 270
Bibliography 273

Chapter 8
COMPOSITE REACTIONS 276

8.1 Types of Composite Mechanism 278
8.2 Rate Equations for Composite Mechanisms 278
    8.2.1 Simultaneous and Consecutive Reactions 279
    8.2.2 Steady-State Treatment 282
    8.2.3 Rate-Determining (Rate-Controlling) Steps 283
    8.2.4 Microscopic Reversibility and Detailed Balance 285
8.3 Chain Reactions 288
    8.3.1 Chain Initiation Processes 290
8.4 Some Inorganic Mechanisms 291
    8.4.1 Hydrogen–Bromine Reaction 291
    8.4.2 Hydrogen–Chlorine Reaction 295
    8.4.3 Hydrogen–Iodine Reaction 298
    8.4.4 Comparison of the Hydrogen–Halogen Reactions 300
    8.4.5 Formation and Decomposition of Phosgene 301
    8.4.6 Decomposition of Nitrogen Pentoxide 303
    8.4.7 Decomposition of Ozone 305
    8.4.8 Thermal Para–Ortho Hydrogen Conversion 307
8.5 Organic Decompositions 307
    8.5.1 Goldfinger–Letort–Niclause Rules 311
    8.5.2 Molecular Processes 314
    8.5.3 Decomposition of Ethane 316
    8.5.4 Decomposition of Acetaldehyde 319
    8.5.5 Inhibition Mechanisms 321

8.6   Gas-Phase Combustion   322
        8.6.1   Hydrogen–Oxygen Reaction   323
        8.6.2   Combustion of Hydrocarbons   328
8.7   Polymerization Reactions   330
        8.7.1   Molecular Mechanisms   330
        8.7.2   Free-Radical Mechanisms   332
        8.7.3   Cationic Polymerization   335
        8.7.4   Anionic Polymerization   338
        8.7.5   Emulsion Polymerization   339
Problems   340
References   342
Bibliography   346

Chapter 9
PHOTOCHEMICAL AND RADIATION-CHEMICAL REACTIONS   348

9.1   Photochemical Reactions   349
        9.1.1   Photochemical Primary Process   349
        9.1.2   Reactions of Electronically Excited Species   353
        9.1.3   Photochemical Thresholds   355
        9.1.4   Law of Photochemical Equivalence   355
        9.1.5   Rotating-Sector Technique   356
        9.1.6   Flash Photolysis   359
9.2   Laser Photochemistry   360
        9.2.1   Pulsed Lasers   361
        9.2.2   Multiphoton Excitation   362
9.3   Photosensitization   362
9.4   Radiation-Chemical Reactions   365
        9.4.1   Radiation-Chemical Primary Process   366
        9.4.2   Pulse Radiolysis   370
        9.4.3   Hydrated Electrons   370
9.5   Chemiluminescence   371
Problems   372
References   373
Bibliography   374

Chapter 10
HOMOGENEOUS CATALYSIS   377

10.1   General Catalytic Mechanisms   379
        10.1.1   Equilibrium Treatment: Arrhenius Intermediates   380
        10.1.2   Steady-State Treatment: van't Hoff Intermediates   382
        10.1.3   Activation Energies for Catalyzed Reactions   383
10.2   Acid–Base Catalysis   384
        10.2.1   General Acid–Base Catalysis   387
        10.2.2   Mechanisms of Acid–Base Catalysis   388
        10.2.3   Catalytic Activity and Acid–Base Strength   394

10.2.4   Salt Effects in Acid–Base Catalysis   396
10.2.5   Acidity Functions   397
10.3   Catalysis by Enzymes   399
10.3.1   Influence of Substrate Concentration   400
10.3.2   Influence of pH   406
10.3.3   Influence of Temperature   409
10.3.4   Transient-Phase Kinetics   410
10.3.5   Enzyme Mechanisms   412
10.4   Catalysis in Gaseous Systems   412
10.5   Chain Mechanisms   413
10.6   Catalysis by Ions of Variable Valency   414
10.7   Activation of Molecular Hydrogen   415
Problems   418
References   421
Bibliography   424

Chapter 11
ISOTOPE EFFECTS   427

11.1   Equilibrium Isotope Effects   428
11.1.1   Equilibria in Solution   432
11.2   Primary Kinetic Isotope Effects   433
11.2.1   Semiclassical Treatments   434
11.2.2   Quantum-Mechanical Tunneling   437
11.2.3   Reactions of the Type $H + H_2$ 438
11.2.4   Transfer of H^+, H, and H^- 442
11.2.5   Reactions of Muonium   442
11.2.6   Isotope Effects with Heavier Atoms   444
11.3   Secondary Kinetic Isotope Effects   446
References   447
Bibliography   448

Chapter 12
REACTION DYNAMICS   449

12.1   Molecular-Dynamical Calculations   451
12.1.1   The Reaction $H + H_2$ 454
12.1.2   The Reaction $Br + H_2$ 457
12.1.3   More Complex Reactions   458
12.2   Chemiluminescence   459
12.2.1   Highly Dilute Flames   459
12.2.2   Diffusion Flames   460
12.3   Features of Potential-Energy Surfaces   460
12.3.1   Attractive Surfaces for Exothermic Reactions   461
12.3.2   Repulsive Surfaces for Exothermic Reactions   462
12.3.3   Surfaces of Intermediate Types for Exothermic Reactions   462
12.3.4   Selective Enhancement of Reaction   466

12.3.5   Disposal of Excess Energy   469
12.3.6   Gradual and Sudden Surfaces   469
12.3.7   Influence of Rotational Energy   471
12.4   Molecular Beams   472
12.4.1   Stripping and Rebound Mechanisms   475
12.5   State-to-State Kinetics   476
12.5.1   Influence of Reactant Vibrational Energy   477
12.5.2   Influence of Reactant Rotational Energy   481
12.6   Spectroscopy of Transition Species   483
References   485
Bibliography   488

Biographical Sketches   491

Answers to Problems   523

Index   525

PREFACE

The object of this book is to present the more important experimental results and theories relating to the rates with which chemical reactions occur. Since the appearance of the second edition of this book in 1965, there have been many significant developments in both experimental and theoretical chemical kinetics. The subject is now so vast it is impossible to make any attempt at comprehensiveness. Instead, the present book attempts to describe the basic concepts as clearly as possible and confines its attention to those features of kinetics that seem to be most important for those who require a general knowledge of the subject. Those embarking on research in kinetics may find the book a useful initial guide but will need to read more detailed accounts of special aspects of the subject. At the end of each chapter there is a selected bibliography of some of the books and articles which I myself have found to be particularly helpful.

Some brief historical material was included in the first two editions of this book, and in the present edition these historical aspects are extended slightly. It seems to me that with the rapid advances being made it is more and more important to realize that some of the concepts we now take for granted are by no means self-evident and often were developed only after much struggle and controversy; examples in chemical kinetics are the temperature dependence of rates and the manner in which molecules acquire the energy needed for reaction. Improvements are still being made in the treatment of these topics and can only be made effectively by those who appreciate how the present points of view were attained. Brief biographical sketches of some of those who have made important contributions to kinetics are presented at the end of the book.

I am most grateful for suggestions that have been made by a number of experts, particularly the following: Professor Robert A. Alberty, Professor John T. Anderson, Professor Margaret H. Back, Professor Roy R. Baldwin, Professor G. C. Bond, Professor H. B. Dunford, Professor John T. Edward, Professor L. Peter Gold, Dr. Kenneth H. Holbrook, Professor Brian R. James, Dr. M. Christine King, Professor Charles A. McDowell, Professor John H. Meiser, Professor John M. Roscoe, Dr. C. R. Theocharis, Professor Donald G. Truhlar, and Professor J. S. Winn. Special thanks must go to Professor John T. Polanyi, who gave me great help with the difficult chapter on reaction dynamics, a field that is expanding so rapidly.

The book follows the recommendations of the International Union of Pure and Applied Chemistry as to terminology, symbols, and units.

KEITH J. LAIDLER

Chapter 1

Basic Kinetic Concepts

The birth of chemical kinetics often is taken to have occurred in 1850, when the German chemist Ludwig Ferdinand Wilhelmy (1812–1864) studied the rate of inversion of sucrose.[1] This pioneering work is of special significance as being the first in which a quantitative approach was made to reaction rates. Wilhelmy interpreted the course of the reaction by the use of a differential equation and also proposed an empirical equation to express the temperature dependence of the rate (see Section 2.9.1). Wilhelmy's work remained almost unnoticed for over 30 years, after which Friedrich Wilhelm Ostwald (1853–1932) called attention to it.[2]

Long before Wilhelmy carried out his work, even as early as the 18th century, some measurements had been made of rates of chemical reactions. For example, in 1777 C. F. Wenzel[3] described some measurements of the rate of solution of metals in acids but gave no details, merely saying that the rate increased with increasing concentration of acid. In 1818 the French chemist Louis Jacques Thénard (1777–1857) studied the rate of decomposition of hydrogen peroxide, a substance he had discovered.[4] However, none of these early studies made much of a quantitative attack on the problem of reaction rates. The early chemists were largely concerned with discovering new substances and not so much with interpreting chemical behavior. It was only in the second half of the 19th century that physical methods began to be applied to chemical problems and that investigations were carried out in the branch of science now known as physical chemistry.

Wilhelmy's work was followed by that of the French chemists Pierre Eugène Marcelin Berthelot (1827–1907) and Léon Pean de Saint-Gilles (1832–1863), who in 1862 published the results of a study of the reaction between ethanol and acetic acid to give ethyl acetate and water.[5] Their work was concerned mainly with the equilibrium that was established after a sufficient period of time had elapsed, but they obtained some results on the rate of combination of ethanol and acetic acid and found it to be proportional to the product of the reactant concentrations.

At about the same time, the Norwegian chemist Peter Waage (1833–1900) and his brother-in-law, the mathematician Cato Maximilian Guldberg (1836–1902), were carrying out investigations that led to what is now usually known, rather unsatisfactorily, as the *law of mass action*.[6] The significance of their work is by no means clearcut and has been the subject of some controversy.[7] Guldberg and Waage made rate measurements but did not investigate the kinetics of any reaction; instead, they assumed rate equations, and by equating such rate equations for a reaction in the forward and reverse directions they arrived at an equilibrium equation. It is known now that this procedure is not valid and that chemical equilibrium must be treated by the methods of thermodynamics or statistical mechanics. Thus, Guldberg and Waage did not solve satisfactorily a problem in either kinetics or thermodynamics; in spite of this, however, their work was of some significance in pointing the way to more reliable procedures.

Another collaboration between a chemist and a mathematician, carried out at about the same time and quite independently, was much more successful as far as the field of kinetics is concerned. In the years 1865 to 1867 Augustus George Vernon Harcourt (1834–1919) carried out very detailed experimental investigations on the reactions between hydrogen peroxide and hydrogen iodide and between potassium permanganate and oxalic acid, paying particular attention to the influence of the reactant concentrations on the rate. His results were analyzed mathematically, in terms of the integrated forms of differential equations, by William Esson (1839–1916), whose procedures were very similar to those that are used today.[8] Equations were obtained for the amount of product formed as a function of time, for "first-order" reactions, in which the rate is proportional to the concentration of a single reacting substance, and for "second-order" reactions, in which the rate is proportional to the product of two concentrations. Esson also developed a treatment for consecutive first-order reactions, in which the product of one reaction undergoes a subsequent reaction (see Section 8.2). Harcourt and Esson paid no attention to the then very popular but nebulous topic of "chemical affinity" and were not concerned with equilibrium states; this was probably fortunate, since at the time these questions tended to confuse the kinetic problem.

The investigations of Harcourt and Esson provided a great impetus to subsequent work in kinetics, a field that became very active by the turn of the century. One particularly important development, concerning temperature effects on rates, was to lead to greatly increased understanding of the molecular mechanisms of chemical reactions and is considered in Section 2.9.

1.1 SCOPE OF CHEMICAL KINETICS

Chemical kinetics deals with the rates of chemical reactions and with how the rates depend on factors such as concentration and temperature. Such studies are important in providing essential evidence as to the mechanisms of chemical processes. Valuable evidence about mechanisms also is provided by nonkinetic investigations—such as the detection of reaction intermediates and isotope exchange studies—but knowledge of a mechanism can be satisfactory only after a careful kinetic investigation has been carried out. Even then the mechanism cannot be deduced with certainty, since subsequent investigations may reveal unexpected complications. A kinetic study can disprove a mechanism but it cannot establish a mechanism with certainty.

Kinetic studies cover a very wide range, from several points of view. The half-life of a reactant, the time taken for half of the reactant to be consumed (Section 2.4), can range from a small fraction of a microsecond to a period longer than the age of the universe. For example, a stoichiometric mixture of hydrogen and oxygen gases at room temperature reacts so slowly that no change could be detected after hundreds of years; it is impossible to measure such a rate, but from reliable kinetic data it can be estimated that the half-life is greater than 10^{25} years. However, if a flame or a spark is applied to the mixture an explosion occurs with a half-life of less than 10^{-6} seconds—an increase in rate by a factor of over 10^{38}.

Chemical reactions occur in the gas phase, in solution in a variety of solvents, at gas–solid and other interfaces, in the liquid state, and in the solid state. Experimental methods, some of them very sophisticated, have been developed for studying the rates of these various types of reaction and even for following very rapid reactions such as explosions. Some of these methods are outlined in later sections. Theoretical treatments also have been worked out for the various types of reaction. We now have a reasonable understanding of the main factors influencing reaction rates; however, important problems remain unsolved and attention is directed to many of them throughout this book.

Some idea of the significance and wide scope of chemical kinetics is given by Table 1.1, which indicates some important branches of science to which the subject is relevant.

1.1.1 Macroscopic and Microscopic Kinetics

Most rate measurements are made on systems that are in thermal equilibrium, which means that energy is distributed among the molecules according to the Boltzmann distribution. Experiments of this kind can be referred to as "bulk" or "bulb" experiments, the latter expression relating to the glass bulbs commonly used for kinetic studies on gases. The term *macroscopic kinetics* describes this branch of kinetics, since the results relate to the behavior of a very large group of molecules in thermal equilibrium. Much of this book is concerned with macroscopic kinetics, since bulk experiments are by far the most common.

The macroscopic kinetic investigations that have been carried out have led to considerable insight into what is occurring at the molecular level during the course of

TABLE 1.1 SOME BRANCHES OF SCIENCE TO WHICH KINETICS IS RELEVANT

Branch	Applications of kinetics
Biology	Physiological processes (e.g., digestion and metabolism), bacterial growth
Chemical engineering	Reactor design
Electrochemistry	Electrode processes
Geology	Flow processes
Inorganic chemistry	Reaction mechanisms
Mechanical engineering	Physical metallurgy, crystal dislocation mobility
Organic chemistry	Reaction mechanisms
Pharmacology	Drug action
Physics	Viscosity, diffusion, nuclear processes
Psychology	Subjective time, memory

reaction. Information about this, however, must be arrived at by inference; hypotheses are proposed and then tested with respect to the behavior of bulk systems. There is a limit to what can be learned in this way. During recent years, further insight into chemical reactions has been provided by kinetic studies in which the reacting molecules are in well-defined states. Such studies are referred to as *microscopic*. A very important technique in this field is that of *crossed molecular beams*. If a bimolecular reaction $A + B \rightarrow Y + Z$ is to be studied, each reactant A and B is formed into a beam, characterized by well-defined velocities and internal energy states. The beams are made to cross one another so that some reaction takes place. The directions of motion of the reaction products and of the unreacted A and B molecules are then determined, together with their velocities and internal energy states. Investigations of this kind provide valuable information about the dynamics of both reactive and unreactive collisions, because the results of well-defined collisions are being observed. Further details of such studies are to be found in Chapter 12.

1.2 REACTION STOICHIOMETRY

A chemical reaction of known stoichiometry can be written in general as

$$aA + bB + \cdots \rightarrow \cdots + yY + zZ$$

It is convenient to use early letters of the alphabet for reactants and late letters for products; the letter X often is reserved for a reaction intermediate.

The symbol ν_1 is used to refer to the *stoichiometric coefficient* of a species in a balanced chemical equation and is negative for reactants, positive for products. In the equation written above, the letters y and z represent the stoichiometric coefficients for the products, while the stoichiometric coefficients for the reactants are $-a$ and $-b$. For the reaction

$$N_2 + 3H_2 \rightarrow 2NH_3$$

the stoichiometric coefficient for the product NH_3 is 2, that for N_2 is -1 and that for H_2 is -3.

Sometimes when substances react together an overall stoichiometric equation applies throughout the course of the reaction, and the reaction is said to have *time-independent stoichiometry*. An example of such a reaction is

$$H_2 + Br_2 \rightarrow 2HBr$$

In this reaction no intermediates are formed in significant amounts, so that the amount of HBr that has been formed at any time is twice the amount of H_2 consumed and twice the amount of Br_2 consumed. Suppose that a reaction is of stoichiometry

$$A + 3B \rightarrow 2Z$$

and that intermediates are not formed in significant amounts during reaction. If n_A^0, n_B^0, and n_Z^0 are the initial amounts of A, B, and Z, and n_A, n_B, and n_Z are the amounts at any later time, the equation

$$\frac{n_A - n_A^0}{-1} = \frac{n_B - n_B^0}{-3} = \frac{n_Z - n_Z^0}{2} \tag{1.1}$$

applies. This equation is obtained by dividing the change of each amount (final amount minus initial amount) by the corresponding stoichiometric coefficient. For the general reaction written earlier,

$$\frac{n_A - n_A^0}{-a} = \frac{n_B - n_B^0}{-b} = \cdots = \frac{n_Y - n_Y^0}{y} = \frac{n_Z - n_Z^0}{z} \tag{1.2}$$

Sometimes the stoichiometry changes throughout the course of the reaction, which is then said to show *time-dependent stoichiometry*. An example is the thermal decomposition of acetone, for which the overall reaction can be represented approximately by the equation

$$2CH_3COCH_3 \rightarrow 2CH_4 + C_2H_4 + 2CO$$

However, this equation does not represent the behavior very satisfactorily while the reaction is proceeding. Appreciable amounts of ketene, CH_2CO, are present during the course of the reaction, the following processes occurring:

$$CH_3COCH_3 \rightarrow CH_2CO + CH_4$$

$$2CH_2CO \rightarrow C_2H_4 + 2CO$$

Thus, the amounts of product present during the course of the reaction are not related by any stoichiometric equation to the amounts of reactants that have been consumed. Equation (1.2) therefore does not apply to this reaction. In such cases, the correct relationships between the amounts of reactant and product must include time-dependent terms and therefore must be based on a knowledge of the kinetics of the reaction.

The following recommendations have been made for writing chemical equations.[9] Two half-arrows, as in

$$H_2 + Br_2 \rightleftharpoons 2HBr$$

are suggested for use when the emphasis is on the equilibrium state only. A single full arrow, as in

$$H_2 + Br_2 \rightarrow 2HBr$$

can be used to refer to a reaction occurring in a single direction. If there is interest in the kinetics of a reaction in both directions, two full arrows are convenient, for example,

$$H_2 + Br_2 \rightleftarrows 2HBr$$

It is desirable to have a notation to indicate that a reaction is believed to be *elementary* (Section 1.8). A filled-in arrow has been suggested, as in

$$H + Br_2 \twoheadrightarrow HBr + Br$$

If this presents difficulties for printers, other devices may be used as long as they are identified clearly.

It is important to emphasize at the outset that the stoichiometric equation for a reaction should be used only to relate amounts of products formed to amounts of reactants consumed. Thus, the equation

$$N_2 + 3H_2 \rightarrow 2NH_3$$

tells us that when 1 mol of N_2 is consumed, 3 mol of H_2 are consumed and 2 mol of NH_3 are formed. In the absence of mechanistic information, the stoichiometric equation must not be used to make predictions about the kinetics of a reaction. The dependence of rate on reaction concentrations is determined by the reaction mechanism, and only in special cases is there a correspondence between the stoichiometry and the kinetics. This is usually true for elementary reactions (although even here there are some pitfalls, to be discussed in Section 5.3) and is sometimes true for reactions occurring in more than one stage. Frequently, however, the stoichiometry and the kinetic behavior are quite different. For example, the thermal decomposition of acetaldehyde can be represented to a good approximation by the equation

$$CH_3CHO \rightarrow CH_4 + CO$$

However, the rate of reaction frequently is proportional to the acetaldehyde concentration to the power of 1.5, for reasons to be considered in Section 8.5.4.

1.3 RATES OF CONSUMPTION AND FORMATION

An important part of any kinetic investigation is the measurement of rates of change of reactant and product concentrations. Consider, for example, the reaction

$$A + 3B \rightarrow 2Z$$

At first we assume that it occurs in a system for which there is no change in the total volume as reaction proceeds (this restriction is removed in Section 1.6). Figure 1.1

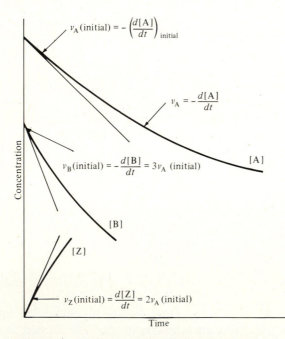

Figure 1.1 Variations with time of the concentrations of A, B, and Z for a reaction of the type $A + 3B \rightarrow 2Z$ occurring in a system at constant volume.

shows schematically the variations in concentrations of A, B, and Z if the kinetic experiment is begun with a mixture of A and B but no Z. At any time t the rate of consumption of A, v_A, is defined as the negative of the slope of the tangents to the curve for A at that time:

$$v_A \equiv -\frac{d[A]}{dt} \tag{1.3}$$

As a special case, we may draw a tangent at $t = 0$, corresponding to the beginning of the reaction; the negative of the slope is the initial rate of consumption of A. The rate of formation of Z, v_Z, is defined as

$$v_Z \equiv \frac{d[Z]}{dt} \tag{1.4}$$

In this particular reaction the stoichiometric coefficients are different for the three species, and the rates of change of their concentrations are correspondingly different; thus,

$$v_A = \tfrac{1}{3} v_B = \tfrac{1}{2} v_Z \tag{1.5}$$

Since a rate is a concentration divided by time its SI unit is $\text{mol m}^{-3}\,\text{s}^{-1}$. However, the unit used for rate is more commonly $\text{mol dm}^{-3}\,\text{s}^{-1}$ or $\text{mol cm}^{-3}\,\text{s}^{-1}$.

1.4 EXTENT OF REACTION

In dealing with certain thermodynamic and kinetic quantities it is convenient to make use of a concept introduced in 1922 by T. de Donder[10] and known as the *extent of reaction.* For any reactant or product in a reaction having time-independent stoichiometry the extent of reaction is defined by

$$\xi \equiv \frac{n - n^0}{\nu} \tag{1.6}$$

where n^0 is the initial amount of the substance, n is the amount at time t, and ν is the stoichiometric coefficient for that species in the particular equation that is written down for the reaction. What makes the extent of reaction so useful is that it is the same for every reactant and product. Thus, for the general equation

$$aA + bB + \cdots \rightarrow \cdots + yY + zZ$$

the extent of reaction is the amount Δn of any product formed divided by its stoichiometric coefficient:

$$\xi = \frac{\Delta n_Y}{y} = \frac{\Delta n_Z}{z} \tag{1.7}$$

It is also the change in the amount of any reactant (a negative quantity) divided by its stoichiometric coefficient, also a negative quantity. Thus, for the general reaction, the stoichiometric coefficients of A and B are $-a$ and $-b$, respectively, and therefore

$$\xi = \frac{\Delta n_A}{-a} = \frac{\Delta n_B}{-b} \tag{1.8}$$

The quantities in Eqs. (1.7) and (1.8) are all equal.

Suppose, for example, that 10.0 mol of N_2 and 20.0 mol of H_2 are caused to react, and that after a certain time 5.0 mol of ammonia have been produced. Before the extent of reaction can be calculated, the stoichiometric equation must be specified, and we first take it to be

$$N_2 + 3H_2 \rightarrow 2NH_3$$

The following table can then be set up:

	N_2	H_2	NH_3
Initial amount/mol	10.0	20.0	0
Final amount/mol	7.5	12.5	5.0
$(n - n^0)$/mol	−2.5	−7.5	5.0
Stoichiometric coefficient ν	−1	−3	2
ξ/mol	2.5	2.5	2.5

If, on the other hand, the stoichiometric equation is written as

$$\tfrac{1}{2}N_2 + \tfrac{3}{2}H_2 \rightarrow NH_3$$

the stoichiometric coefficients are halved and the extent of reaction is doubled, that is, 5.0 mol.

It is important to note that the extent of reaction has meaning only for time-independent stoichiometry and that the stoichiometric equation must be specified.

1.5 RATE OF REACTION

We have seen that rates of consumption and formation are not the same in general for all substances involved in a chemical reaction. It is convenient to define a rate that is the same for all species. For a reaction showing time-independent stoichiometry this can be done in terms of the extent of reaction.

The *rate of reaction,* independent of the reaction and product species, is defined as the time derivative of the extent of reaction divided by the volume:

$$v \equiv \frac{1}{V}\frac{d\xi}{dt} \equiv \frac{\dot{\xi}}{V} \tag{1.9}$$

For a species i the time derivative $\dot{\xi}$ is given by

$$\dot{\xi} = \frac{1}{\nu_i}\frac{dn_i}{dt} \tag{1.10}$$

where n_i is the amount of i and ν_i is its stoichiometric coefficient. The rate of reaction is therefore

$$v = \frac{1}{\nu_i V}\frac{dn_i}{dt} \tag{1.11}$$

Thus, for the general reaction,

$$v = -\frac{1}{aV}\frac{dn_A}{dt} = -\frac{1}{bV}\frac{dn_B}{dt} = \frac{1}{yV}\frac{dn_Y}{dt} = \frac{1}{zV}\frac{dn_Z}{dt} \tag{1.12}$$

A careful distinction should be made between v without a subscript, which is the rate of reaction, and v with a subscript [as in Eqs. (1.3)–(1.5)], which refers to a rate of consumption or formation.

If the volume does not change during the course of reaction, dn_i/V may be replaced by dc_i, where c_i is the concentration. Equation (1.11) then becomes

$$v = \frac{1}{v_i}\frac{dc_i}{dt} \tag{1.13}$$

Equation (1.12) becomes

$$v = -\frac{1}{a}\frac{d[A]}{dt} = -\frac{1}{b}\frac{d[B]}{dt} = \frac{1}{y}\frac{d[Y]}{dt} = \frac{1}{z}\frac{d[Z]}{dt} \tag{1.14}$$

Since the extent of reaction and the stoichiometric coefficients have no meaning except in relation to the equation for the reaction, that equation must be specified whenever a rate of reaction is quoted.

It is sometimes convenient to work with amounts of substances instead of with concentrations. The *rate of conversion* is defined as the time derivative of the extent of reaction; the rate of reaction is thus the rate of conversion divided by the volume. The rate of conversion is also independent of the species.

1.6 VOLUME CHANGE DURING REACTION

Equation (1.11) gives a general definition of the rate of reaction, but Eqs. (1.13)–(1.14) apply only if the volume remains constant during reaction. To remove the restriction of constant volume, we differentiate the relationship $n_B = [B]V$ for a species B and obtain

$$dn_B = V\,d[B] + [B]dV \tag{1.15}$$

Insertion into Eq. (1.11) gives

$$v = \frac{1}{v_B}\frac{d[B]}{dt} + \frac{[B]}{v_B V}\frac{dV}{dt} \tag{1.16}$$

This becomes equivalent to Eq. (1.13) when the volume remains constant. The final term in Eq. (1.16) is a correction for the change in concentration brought about by the change in volume. Even if no reaction were occurring, an increase in volume would bring about a decrease in concentration of the substances present. The rate of this decrease§ is equal to $([B]/V)dV/dt$, and therefore addition of the term $([B]/v_B V)dV/dt$ is required to make v equal to zero for this case of no reaction.

The above considerations are for rate of reaction, the stoichiometric coefficients

---

§ Suppose that the amount of B present in volume V is n_B, and that after time dt the volume is $V + dV$. The change in concentration is

$$\frac{n_B}{V + dV} - \frac{n_B}{V} = -\frac{n_B\,dV}{V(V + dV)} = -\frac{n_B\,dV}{V^2} = -\frac{[B]dV}{V}$$

The rate of change of the concentration is thus $-[B]dV/V\,dt$.

having been introduced. Rates of formation and consumption when there is a volume change are dealt with in a similar way. For a reactant, of amount n_B, the rate of consumption is

$$v_B = -\frac{1}{V}\frac{dn_B}{dt} \tag{1.17}$$

With Eq. (1.15) this becomes

$$v_B = -\frac{d[B]}{dt} - \frac{[B]}{V}\frac{dV}{dt} \tag{1.18}$$

When the volume is constant this reduces to Eq. (1.3). Similarly, for a product of reaction, of amount n_Z, the rate of formation is

$$v_Z = \frac{1}{V}\frac{dn_Z}{dt} = \frac{d[Z]}{dt} + \frac{[Z]}{V}\frac{dV}{dt} \tag{1.19}$$

These equations for varying volume usually are not needed for reactions in solution, where volume changes often are negligible. They may be needed, however, for liquid-phase reactions and for gas reactions in open systems—for example, for reactions in flames, where the gases are not confined in a vessel.

1.7 EMPIRICAL RATE EQUATIONS

For some reactions the rate of reaction v can be expressed by an equation of the form

$$v = k[A]^\alpha[B]^\beta \tag{1.20}$$

where k, α, and β are independent of concentration and time. When this is the case, the rate of consumption of a species A is expressible as

$$v_A = k_A[A]^\alpha[B]^\beta \tag{1.21}$$

The exponents α and β must be the same in the two equations, but since v in general differs from v_A (being v_A divided by the negative of the stoichiometric coefficient for A), k differs from k_A by the same factor. The same applies to a constant k_Z which relates to the product Z. Thus, if the stoichiometric equation is

$$A + 2B \rightarrow 3Z$$

the constants are related by

$$k = k_A = \tfrac{1}{2}k_B = \tfrac{1}{3}k_Z \tag{1.22}$$

1.7.1 Order of Reaction

The exponent α in the above equations is known as the *order of reaction*§ with respect to A. Similarly, the exponent β is the order with respect to B. These orders are known as *partial orders*. The sum $\alpha + \beta + \cdots$ is known as the overall order and usually is

§ The term "order," with its present meaning, was introduced by W. Ostwald [*Lehr. Allg. Chem.*, **2**, 634 (1887)]. Van't Hoff used "molecularity" for what is now called "order," and this terminology was used by kineticists until the 1920s and sometimes even later. Regrettably, some nonkineticists still use this confusing terminology.

given the symbol n. These orders α, β, $\cdots$ and n are purely experimental quantities and are not necessarily integral.

A very simple case is when the rate is proportional to the first power of the concentration of a single reactant:

$$v = k[A] \qquad (1.23)$$

Such a reaction is said to be of the *first order*. An example is the conversion of cyclopropane into propylene:

$$\underset{\underset{\displaystyle H_2C-CH_2}{\diagup \diagdown}}{CH_2} \rightarrow CH_3-CH=CH_2$$

The rate of this reaction is proportional to the first power of the cyclopropane concentration.

A *second-order* reaction is one for which the rate is proportional to the square of a reactant concentration,

$$v = k[A]^2 \qquad (1.24)$$

or to the product of two reactant concentrations,

$$v = k[A][B] \qquad (1.25)$$

In the latter case the partial orders are both unity ($\alpha = \beta = 1$) and the overall order is 2. For example, the reaction

$$H_2 + I_2 \rightleftarrows 2HI$$

is second order in both directions. If we study the reaction from left to right and remove the product HI as it is formed so that the back reaction cannot occur, the rate is given by

$$v = k_1[H_2][I_2] \qquad (1.26)$$

The partial orders are unity and the overall order is 2. If the hydrogen iodide decomposition is studied and the products (H_2 and I_2) removed, the rate is

$$v_{-1} = k_{-1}[HI]^2 \qquad (1.27)$$

Often, nonintegral orders of reaction are found. Thus, the acetaldehyde decomposition, which follows the stoichiometric equation

$$CH_3CHO \rightarrow CH_4 + CO$$

is of 3/2 order:

$$v = k[CH_3CHO]^{3/2} \qquad (1.28)$$

The reason for this order is considered in Section 8.5.4.

1.7.2 Rate Constants

The constant k which appears in the above equations is known as the *rate constant*. Its units depend on the order of the reaction. Suppose, for example, that a reaction is of the first order,

$$v = k[A] \qquad (1.29)$$

If the units of v are mol dm^{-3} s^{-1} and those of [A] are mol dm^{-3}, the unit of the rate constant is

$$\frac{\text{mol dm}^{-3}\,\text{s}^{-1}}{\text{mol dm}^{-3}} = \text{s}^{-1}$$

Similarly, for a second-order reaction, for which $v = k[A]^2$ or $v = k[A][B]$, the units of k are

$$\frac{\text{mol dm}^{-3}\,\text{s}^{-1}}{(\text{mol dm}^{-3})^2} = \text{dm}^3\,\text{mol}^{-1}\,\text{s}^{-1}$$

The units for other orders are worked out easily.

The rate of change of a concentration in general depends on the reactant or product with which we are concerned and the rate constant also reflects this dependence. For example, for the dissociation of ethane into methyl radicals,

$$C_2H_6 \rightarrow 2CH_3$$

the rate of formation of methyl radicals, v_{CH_3}, is twice the rate of consumption of ethane, $v_{C_2H_6}$:

$$v_{CH_3} = 2v_{C_2H_6} \tag{1.30}$$

Under certain conditions the reaction is first order from left to right, and the rate of ethane consumption can be expressed as

$$v_{C_2H_6} \equiv -\frac{d[C_2H_6]}{dt} = k_{C_2H_6}[C_2H_6] \tag{1.31}$$

The rate of formation of methyl radicals is

$$v_{CH_3} \equiv \frac{d[CH_3]}{dt} = k_{CH_3}[C_2H_6] \tag{1.32}$$

It follows that

$$k_{CH_3} = 2k_{C_2H_6} \tag{1.33}$$

In such cases it is important to specify the species to which the rate constant applies. This problem is avoided if rates of reaction are used instead of rates of consumption or formation; the corresponding rate constant k (without a subscript indicating a species) is now unambiguous. However, if this procedure is employed the stoichiometric equation must be specified.

1.7.3 Reactions Having No Order

Not all reactions behave in the manner described by Eq. (1.20), and the term "order" should not be used for those that do not follow such an equation. For example, reactions catalyzed by enzymes (Section 10.3) frequently obey an equation of the form

$$v = \frac{V[A]}{K_m + [A]} \tag{1.34}$$

where V and K_m are constants that are referred to in general as *kinetic parameters*. Reactions that do not have an order necessarily occur by composite mechanisms, and in later chapters (especially Chapters 8, 9, and 10) mechanisms are related to empirical rate equations.

1.8 ELEMENTARY, COMPOSITE, AND CHAIN REACTIONS

An *elementary reaction* is a reaction that occurs in a single step, with no experimentally detectable reaction intermediates. If no such intermediates can be detected or need to be postulated to interpret the behavior, a reaction is assumed tentatively to be elementary. However, it must always be borne in mind that further experimental work may reveal that a reaction originally believed to be elementary in reality occurs in more than one step. For example, the reaction

$$H_2 + I_2 \rightarrow 2HI$$

was long believed to be elementary. However, the reaction now is believed to occur in part by a mechanism involving the following elementary steps:

$$I_2 \rightleftarrows 2I$$

$$I + H_2 \rightarrow HI + H$$

$$H + I_2 \rightarrow HI + I$$

This matter is further discussed in Section 8.4.3.

The *molecularity* of an elementary reaction is the number of reactant particles (atoms, molecules, free radicals, or ions) that are involved in each individual chemical event. For example, the conversion of cyclopropane into propylene,

$$\begin{matrix} CH_2 \\ \\ H_2C-CH_2 \end{matrix} \rightarrow CH_3-CH=CH_2$$

is believed to be elementary, with individual cyclopropane molecules undergoing reaction in one stage; the molecularity is therefore unity, and the reaction is said to be *unimolecular*. For the elementary reaction

$$Br + H_2 \rightarrow HBr + H$$

the molecularity is two, and the reaction is said to be *bimolecular*. It is important to recognize that the molecularity is arrived at by inference from all the evidence available about the reaction; by contrast, the order of reaction simply indicates how the rate depends on concentration.

Composite reactions, which involve more than one elementary reaction, have also been called complex or stepwise reactions; the term "composite," recommended by IUPAC, is used in this book. The term "molecularity" has no meaning for a composite reaction.

It is convenient to number the elementary reactions that occur in a composite mechanism in such a way that reverse reactions are identified easily. For example, the reverse of reaction 1 is referred to as reaction -1, and the rates and rate constants are denoted by corresponding subscripts, for example, v_1, v_{-1}, k_1, and k_{-1}.

1.8.1 Chain Reactions

A composite reaction mechanism sometimes includes a cycle of reactions, such that certain reaction intermediates consumed in one step are regenerated in another. The intermediates may be atoms, free radicals, or ions. If such a cycle is repeated more than once the reaction is known as a *chain reaction.*

For example, if hydrogen and bromine react and the product HBr is removed as fast as it is formed, the process is believed to occur by the following steps:

(1) $Br_2 \xrightarrow{k_1} 2Br$ Initiation

(2) $Br + H_2 \xrightarrow{k_2} HBr + H$ ⎫

(3) $H + Br_2 \xrightarrow{k_3} HBr + Br$ ⎬ Chain propagation

(−1) $2Br \xrightarrow{k_{-1}} Br_2$ Termination

Reactions (2) and (3) constitute a cycle and are known as *chain-propagating steps;* in reaction (2) a bromine atom is consumed but is produced in reaction (3), while in reaction (3) a hydrogen atom is consumed but is produced in reaction (2). Under usual conditions this cycle occurs a number of times on the average, and the reaction is therefore a chain reaction. Reaction (1), which produces active intermediates (Br atoms), is known as an *initiation reaction,* and its reverse, reaction (−1), is called a *termination step* or a *chain-ending step.* Chain reactions always involve initiation and termination steps and two or more chain-propagating steps.

Chain reactions are dealt with further in Section 8.3, where a detailed discussion of the hydrogen–bromine reaction is given.

1.9 CATALYSIS AND INHIBITION

A *catalyst* is a substance that is both a reactant and a product of reaction; its concentration enters into the kinetic equation but not into the equilibrium constant for the reaction. For example, a catalyst might enter into a reaction as follows:

$$A + B + catalyst \rightleftarrows Y + Z + catalyst$$

and the rate from left to right might follow the kinetic equation

$$v_1 = k_1[A][B][catalyst] \tag{1.35}$$

The rate from right to left would then be given by

$$v_{-1} = k_{-1}[Y][Z][catalyst] \tag{1.36}$$

At equilibrium the rates are the same in the two directions,

$$k_1[A][B][catalyst] = k_{-1}[Y][Z][catalyst] \tag{1.37}$$

and the catalyst concentration cancels out in the equilibrium expression:

$$K = \frac{k_1}{k_{-1}} = \left(\frac{[Y][Z]}{[A][B]}\right)_{eq} \tag{1.38}$$

It follows that the catalyst does not modify the standard Gibbs energy change, $\Delta G° = -RT \ln K$, for the reaction. It also follows that a catalyst increases the rates to the same extent in forward and reverse directions.

Catalysis can be classified as *homogeneous* catalysis (Chapter 10), in which only one phase is involved, and *heterogeneous* or *surface* catalysis (Chapter 7), in which the reaction occurs at an interface between phases. Catalysis brought about by one of the products of a reaction is referred to as *autocatalysis.* Catalysis brought about by a group on a reactant molecule itself is called *intramolecular* catalysis.

Substances that act in some ways like catalysts but which are consumed in the course of reaction should not be called catalysts; they can be called *pseudocatalysts* or *activators,* the latter expression being used in particular in enzyme kinetics (Section 10.3).

An *inhibitor* is a substance that diminishes the rate of a chemical reaction. Inhibitors have sometimes been called "negative catalysts," but since their action is quite different from that of catalysts this usage is not recommended. In contrast to a catalyst, an inhibitor may be and frequently is consumed during the course of reaction.

If a reaction in the absence of an inhibitor proceeds with a rate v_0, and in the presence of inhibitor with rate v, the fraction

$$\epsilon_i \equiv \frac{v_0 - v}{v_0} = 1 - \frac{v}{v_0} \tag{1.39}$$

is known as the *degree of inhibition*.

PROBLEMS

1.1. A reaction having time-independent stoichiometry follows the equation

$$A + 2B \rightarrow 2Z$$

Suppose that initially there are present 0.28 mol of A, 0.39 mol of B, and 0.13 mol of Z.
(a) After a certain time 0.18 mol of A remains. What is the extent of reaction and what are the amounts of B and Z?
(b) What is the extent of reaction if the process goes to completion?

1.2. A reaction is of time-independent stoichiometry,

$$2A + B \rightarrow 3Z$$

If the initial rate of consumption of A is 3.36×10^{-4} mol dm^{-3} s^{-1}, what is the rate of consumption of B? What is the rate of formation of Z and what is the rate of reaction?

1.3. The stoichiometric equation for the oxidation of bromide ions by hydrogen peroxide in acid solution is

$$2Br^- + H_2O_2 + 2H^+ \rightarrow Br_2 + 2H_2O$$

$$v = k[H_2O_2][H^+][Br^-]$$

(a) If the concentration of H_2O_2 is increased by a factor of 3, by what factor is the rate of consumption of Br^- ions increased?

(b) If, under certain conditions, the rate of consumption of Br^- ions is 7.2×10^{-3} mol dm^{-3} s^{-1}, what is the rate of consumption of hydrogen peroxide? What is the rate of formation of bromine? What is the rate of reaction?

(c) If by the addition of water to the reaction mixture the total volume were doubled, what would be the effect on the rate of change of the concentration of Br^-? What would be the effect on the rate of reaction?

1.4. A reaction obeys the stoichiometric equation

$$A + 2B \rightarrow 2Z$$

Rates of formation of Z at various concentrations of A and B are given in the table below:

[A]/mol dm^{-3}	[B]/mol dm^{-3}	v_Z/mol dm^{-3} s^{-1}
3.5×10^{-2}	2.3×10^{-2}	5.0×10^{-7}
7.0×10^{-2}	4.6×10^{-2}	2.0×10^{-6}
7.0×10^{-2}	9.2×10^{-2}	4.0×10^{-6}

What are α and β in the rate equation

$$v_Z = k_Z[A]^\alpha[B]^\beta$$

and what is the rate constant k_Z? What is the rate constant k that relates to the rate of reaction?

1.5. A reaction has the stoichiometry

$$2A + 2B \rightarrow Y + 2Z$$

Some results for the rate of consumption of A are shown below:

[A]/mol dm^{-3}	[B]/mol dm^{-3}	v_A/mol dm^{-3} s^{-1}
1.4×10^{-2}	2.3×10^{-2}	7.4×10^{-9}
2.8×10^{-2}	4.6×10^{-2}	5.92×10^{-8}
2.8×10^{-1}	4.6×10^{-2}	5.92×10^{-6}

Deduce α and β and the rate constants k_A, k_B, k_Y, k_Z, and k.

REFERENCES

1. L. Wilhelmy, *Pogg. Ann.,* **81,** 413, 499 (1850).
2. F. W. Ostwald, *J. Prakt. Chem.,* **29,** 385 (1884).
3. C. F. Wenzel, *Lehre von der Verwandtschaft der Körper,* Dresden, 1777.
4. L. J. Thénard, *Ann. Chim. Phys.,* **9**(2), 314 (1818).
5. P. E. M. Berthelot and L. P. de Saint-Gilles, *Ann. Chim. Phys.* **63**(3), 385 (1862) and several later papers.
6. C. M. Guldberg and P. Waage, *Forh. Vid. Selsk. Christiania,* **35,** 92, 111 (1864); *Etudes sur les affinités chimiques,* Brøgger and Christie, Christiania (Oslo), 1867; *J. Prakt. Chem.,* **19,** 69 (1879).
7. See, for example, K. J. Mysels, *J. Chem. Educ.,* **33,** 178 (1956); E. A. Guggenheim, *J. Chem. Educ.,* **33,** 544 (1956); E. W. Lund, *J. Chem. Educ.,* **42,** 548 (1965).
8. A. G. V. Harcourt and W. Esson, *Proc. R. Soc. London,* **14,** 470 (1865); *Philos. Trans.,* **156,** 193 (1866); **157,** 117 (1867). Important excerpts from these papers are reproduced in

M. H. Back and K. J. Laidler (Eds.), *Selected Readings in Chemical Kinetics,* Pergamon, Oxford, 1967.

9. Symbols and Terminology in Chemical Kinetics, Appendix V of the IUPAC *Manual of Symbols and Terminology for Physicochemical Quantities and Units,* prepared by K. J. Laidler and published in *Pure Appl. Chem.,* **53,** 753 (1981).

10. T. de Donder, *Bull. Acad. Belg. Cl. Sci.,* **8,** 197 (1922).

BIBLIOGRAPHY

Historical aspects of chemical kinetics are covered in the following publications:

M. H. Back and K. J. Laidler (Eds.), *Selected Readings in Chemical Kinetics,* Pergamon, Oxford, 1967.

E. Farber, Early Studies Concerning Time in Chemical Reactions, *Chymia,* **7,** 135 (1961).

H. B. Hartley, *Studies in the History of Chemistry,* Clarendon, Oxford, 1971.

M. C. King, Experiments with Time: Progress and Problems in the Development of Chemical Kinetics, *Ambix,* **28,** 70 (1981); **29,** 49 (1982).

M. C. King, The Course of Chemical Change: The Life and Times of Augustus G. Vernon Harcourt (1834–1919), *Ambix,* **31,** 16 (1984).

K. J. Laidler, Chemical Kinetics and the Origins of Physical Chemistry, *Arch. Hist. Exact Sci.,* **32,** 43 (1985).

J. R. Partington, *A History of Chemistry,* Macmillan, London, 1961, especially vol. 4, chaps. 18 and 22.

J. Shorter, A. G. Vernon Harcourt: A Founder of Chemical Kinetics and a Friend of "Lewis Carroll," *J. Chem. Educ.,* **57,** 411 (1980).

For biographies of kineticists the reader should consult the following:

I. Asimov, *Biographical Encyclopedia of Science and Technology,* 2nd rev. ed., Doubleday, Garden City, NY, 1982.

C. C. Gillispie (Editor in Chief), *Dictionary of Scientific Biography* (in 14 vols.), Scribner, New York, 1970–1976.

Obituary Notices of Fellows of the Royal Society, Royal Society, London, 1932–1954.

Biographical Memoirs of Fellows of the Royal Society. Royal Society, London, 1955–present.

Basic definitions and symbols for kinetics are given in:

Symbolism and Terminology in Chemical Kinetics, Appendix V of the IUPAC *Manual of Symbols and Terminology for Physicochemical Quantities and Units,* prepared by K. J. Laidler and published in *Pure Appl. Chem.,* **53,** 753 (1981).

Extent of reaction has been treated in:

S. Cvitas and N. Kallay, *Chem. Br.,* **14,** 290 (1978).

M. L. McGlashan, *Chemical Thermodynamics,* Academic, New York, 1979.

Chapter 2

Analysis of Kinetic Results

An important problem in chemical kinetics is to determine how rates depend on the concentrations of reacting substances. It is also important to know how rates are affected by the products of reaction or by added substances, which may be catalysts or inhibitors (Section 1.9). Another problem is to analyze the effect of temperature on reaction rates. The present chapter is concerned with these aspects of chemical kinetics.

As far as reactant concentrations are concerned, there are two main procedures: the differential method and the method of integration. In the *differential method,* rates are measured directly, by determining the slopes of concentration–time curves, and an analysis is made of the way in which the slope depends on the reactant concentration. The *method of integration* involves first making a tentative decision as to what the order of the reaction might be. The differential equation corresponding to that order is integrated, resulting in an expression for concentration as a function of time. A test is then made as to whether the experimental concentration–time results fit the integrated equation; if they do, the right order has been chosen and the rate constant can be obtained readily. If the results do not fit the equation, another order must be chosen.

These two methods will now be considered in detail.

2.1 DIFFERENTIAL METHOD

In the differential method, which was first suggested in 1884 by van't Hoff,[1] the procedure is to determine rates by measuring the slopes of concentration–time curves (see Section 1.3). One does this at various concentrations c of a reactant, and if the reaction has an order n with respect to this particular reactant,

$$v = -\frac{dc}{dt} = kc^n \tag{2.1}$$

A double-logarithmic plot of $\ln v$ versus $\ln c$ gives a straight line of slope n; the intercept when $\ln c = 0$ is then $\ln k$. If a straight-line plot is not obtained, the rate cannot be represented by Eq. (2.1); that is, the reaction does not have an order with respect to that particular reactant.

The procedure may be applied in two different ways. In one of them, shown schematically in Fig. 2.1a, runs are carried out at different initial concentrations, and initial rates are determined by measuring initial slopes. A double-logarithmic plot then gives the order of reaction (Fig. 2.1b). This procedure, dealing with initial rates, avoids possible complications due to interference by products. Because of this, Letort[2] referred to the order determined in this way as the *order with respect to concentration,* or the *true order.* The symbol n_c is used to denote this order.

The second procedure involves considering a single run and measuring slopes at various times, corresponding to a number of values of the reactant concentration. This method is illustrated schematically in Fig. 2.2a, and again the logarithms of the rates are plotted against the logarithms of the corresponding reactant concentrations (Fig. 2.2b). The slope is the order; since time is now varying, Letort referred to this order as the *order with respect to time n_t.*

Figure 2.3 shows schematic plots in which the two procedures have been combined. The points at the extremities of the n_t plots correspond to initial conditions and give rise to the n_c plot.

The two orders are not always the same for a given reaction. In the thermal decomposition of acetaldehyde, for example, as discussed further in Section 8.5.4, Letort found that the order with respect to concentration (the true order) is 3/2, and that the order with respect to time is 2. The fact that the order with respect to time is greater than the order with respect to concentration means that as the reaction proceeds

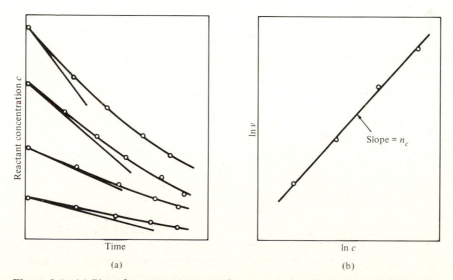

Figure 2.1 (a) Plot of reactant concentrations versus time for various initial concentrations. (b) A plot of $\ln v_i$ versus $\ln c_i$.

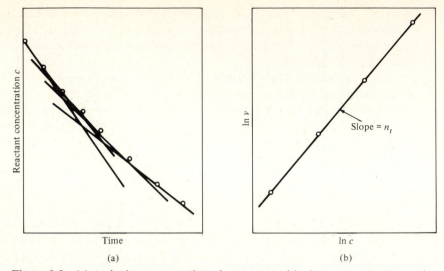

Figure 2.2   (a) A single concentration–time curve, with slopes measured at various reactant concentrations. (b) A plot of ln v versus ln c.

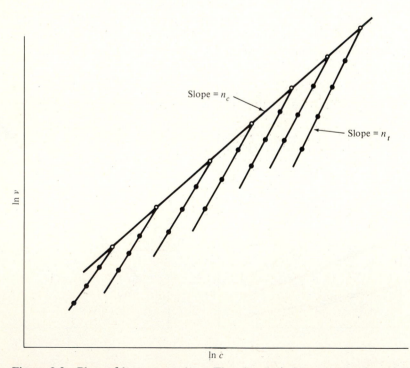

Figure 2.3   Plots of ln v versus ln c. The closed circles correspond to rates obtained in individual runs; the open circles are the initial rates.

the rate falls off more rapidly than if the true order applied to the time course of the reaction. This abnormally large falling off can only mean that some substance produced in the reaction is acting as an inhibitor. Conversely, if n_t is less than n_c, the rate is falling off less rapidly with time than expected on the basis of the true order. Therefore, some activation by the products of reaction exists, and the reaction is said to be *autocatalytic*.

2.2 METHOD OF INTEGRATION

The method of integration was first used in 1850 by Wilhelmy and some 15 years later was extended by Harcourt and Esson. In this section we discuss only the principles of the method, as applied to a few systems. Table 2.1 gives solutions for some of the simpler cases; many more cases have been treated by Capellos and Bielski.[3] Care must be taken with regard to the stoichiometry of the reaction and to the relationships between the different rate constants, as discussed in Section 1.7. The procedure adopted here, which is helpful in avoiding errors, is to focus attention on the consumption of a particular reactant, denoted as A.

2.2.1 First-Order Reactions

A reaction that is first-order with respect to a reactant A and is zero-order with respect to any other reactants may correspond to various stoichiometries, such as

$$A \rightarrow Z \qquad A \rightarrow 2Z \qquad 2A \rightarrow Z$$

$$A + B \rightarrow Z \qquad 2A + B \rightarrow Z$$

A procedure that covers all these possibilities is as follows. Suppose that at the beginning of the reaction ($t = 0$), with no product present, the concentration of A is a_0 and that at time t the amount of A that has been consumed per unit volume is x; the concentration of A is then $a_0 - x$, and the rate of consumption of A is $-d(a_0 - x)/dt = dx/dt$. Thus, we can write

$$\frac{dx}{dt} = k_A(a_0 - x) \tag{2.2}$$

where k_A is the first-order rate constant that relates to the consumption of A. This is converted into the rate constant k, which relates to rate of reaction, by use of the stoichiometric coefficients as discussed in Section 1.7. If there is a second reactant B, the rate constant k_B can be related in a similar way.

Separation of the variables x and t in Eq. (2.2) gives

$$\frac{dx}{a_0 - x} = k_A \, dt \tag{2.3}$$

and integration gives

$$-\ln (a_0 - x) = k_A t - I \tag{2.4}$$

TABLE 2.1 RATE EQUATIONS FOR IRREVERSIBLE PROCESSES

Stoichiometry	Order	Rate equation		Common unit for rate constant	Relationship between rate constants
		Differential form	Integrated form		
		A single reactant, or reactants at equal concentrations			
$A \rightarrow Z, A + B + \cdots \rightarrow Z$	0	$\dfrac{dx}{dt} = k_A$	$k_A = \dfrac{x}{t}$	mol dm^{-3} s^{-1}	a
	1/2	$\dfrac{dx}{dt} = k_A(a_0 - x)^{1/2}$	$k_A = \dfrac{2}{t}[a_0^{1/2} - (a_0 - x)^{1/2}]$	mol$^{1/2}$ dm$^{-3/2}$ s^{-1}	$k_A = k_B = k$
$A \rightarrow Z, A + B + \cdots \rightarrow Z$	1	$\dfrac{dx}{dt} = k_A(a_0 - x)$	$k_A = \dfrac{1}{t}\ln\left(\dfrac{a_0}{a_0 - x}\right)$	s^{-1}	$k_A = k_B = k$
	3/2	$\dfrac{dx}{dt} = k_A(a_0 - x)^{3/2}$	$k_A = \dfrac{2}{t}\left[\dfrac{1}{(a_0 - x)^{1/2}} - \dfrac{1}{a_0^{1/2}}\right]$	dm$^{3/2}$ mol$^{-1/2}$ s^{-1}	$k_A = k_B = k$
$A \rightarrow Z, A + B + \cdots \rightarrow Z$	2	$\dfrac{dx}{dt} = k_A(a_0 - x)^2$	$k_A = \dfrac{x}{t a_0(a_0 - x)}$	dm^3 mol^{-1} s^{-1}	$k_A = k_B = k$
$A + 2B \rightarrow Z$	2	$\dfrac{dx}{dt} = k_A(a_0 - x)(a_0 - 2x)$	$k_A = \dfrac{1}{t a_0}\ln\left(\dfrac{a_0 - x}{a_0 - 2x}\right)$	dm^3 mol^{-1} s^{-1}	$k_B = 2k_A = 2k$
$A \rightarrow Z, A + B + C + \cdots \rightarrow Z$	3	$\dfrac{dx}{dt} = k_A(a_0 - x)^3$	$k_A = \dfrac{2a_0 x - x^2}{2t a_0^2(a_0 - x)^2}$	dm^6 mol^{-2} s^{-1}	$k_A = k_B = k_C = k$
$A \rightarrow Z, A + B + \cdots \rightarrow Z$	n	$\dfrac{dx}{dt} = k_A(a_0 - x)^n$	$k_A = \dfrac{1}{t(n-1)}\left[\dfrac{1}{(a_0 - x)^{n-1}} - \dfrac{1}{a_0^{n-1}}\right]$	dm^{3n-3} mol^{1-n} s^{-1}	$k_A = k_B = k$
		Reactants at unequal initial concentrations			
$A + B \rightarrow Z$	2	$\dfrac{dx}{dt} = k_A(a_0 - x)(b_0 - x)$	$k_A = \dfrac{1}{t(a_0 - b_0)}\ln\left[\dfrac{b_0(a_0 - x)}{a_0(b_0 - x)}\right]$	dm^3 mol^{-1} s^{-1}	$k_A = k_B = k$
$A + 2B \rightarrow Z$	2	$\dfrac{dx}{dt} = k_A(a_0 - x)(b_0 - 2x)$	$k_A = \dfrac{1}{t(2a_0 - b_0)}\ln\left[\dfrac{b_0(a_0 - x)}{a_0(b_0 - 2x)}\right]$	dm^3 mol^{-1} s^{-1}	$2k_A = k_B = 2k$
$2A + B \rightarrow Z$	2	$\dfrac{dx}{dt} = k_A(a_0 - 2x)(b_0 - x)$	$k_A = \dfrac{1}{t(a_0 - 2b_0)}\ln\left[\dfrac{b_0(a_0 - x)}{a_0(b_0 - x)}\right]$	dm^3 mol^{-1} s^{-1}	$k_A = 2k_B = 2k$
$A \rightarrow X$ with autocatalysis[b]	1	$\dfrac{dx}{dt} = k_A(a_0 - x)(x_0 + x)$	$k_A = \dfrac{1}{t(a_0 - x_0)}\ln\left[\dfrac{a_0(x_0 + x)}{x_0(a_0 - x)}\right]$	dm^3 mol^{-1} s^{-1}	$k_A = k$

In all cases x is the amount of reactant A consumed per unit volume, and except in the autocatalytic case $x = 0$ when $t = 0$

[a] The equations for the zero-order reaction are true for any stoichiometry, but the relationship between the rate constants depends on the stoichiometry.

[b] In the autocatalytic case some product must be present initially (at concentration x_0) since otherwise the reaction will never start.

where I is the constant of integration. This constant may be evaluated using the boundary condition that $x = 0$ when $t = 0$; hence,

$$-\ln a_0 = I \tag{2.5}$$

and insertion of this into Eq. (2.4) leads to

$$\ln \left(\frac{a_0}{a_0 - x} \right) = k_A t \tag{2.6}$$

This equation can also be written as

$$x = a_0(1 - e^{-k_A t}) \tag{2.7}$$

and as

$$a_0 - x = a_0 e^{-k_A t} \tag{2.8}$$

This last equation shows that the concentration of reactant, $a_0 - x$, decreases exponentially with time, from an initial value of a_0 to a final value of zero.

The first-order equations can be tested and the constant evaluated using a graphical procedure. It follows from Eq. (2.6) that a plot of $\ln [a_0/(a_0 - x)]$ versus t will give a straight line if the reaction is first order; this is shown schematically in Fig. 2.4a. The rate constant is the slope of this plot. We may also plot $\ln (a_0 - x)$ versus t, as shown in Fig. 2.4b.

2.2.2 Second-Order Reactions

There are two possibilities for second-order reactions: the rate may be proportional to the product of two equal concentrations or to the product of two different ones. The first must occur when a single reactant is involved, as in the process

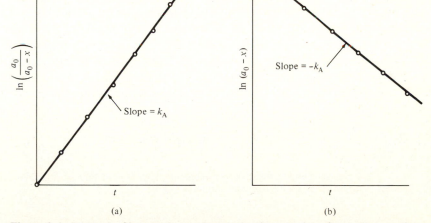

(a)   (b)

Figure 2.4 Method of integration and analysis of results for a first-order reaction: (a) plot of $\ln [a_0/(a_0 - x)]$ versus t and (b) plot of $\ln (a_0 - x)$ versus t.

$$2A \rightarrow Z$$

It is also found in second-order reactions having the stoichiometry

$$A + B \rightarrow Z$$

if the initial concentrations of A and B are the same.§

In such situations the rate may be expressed as

$$\frac{dx}{dt} = k_A(a_0 - x)^2 \tag{2.9}$$

where x is the amount of A that has reacted in unit volume at time t, and a_0 is the initial concentration of A. Separation of the variables leads to

$$\frac{dx}{(a_0 - x)^2} = k_A\, dt \tag{2.10}$$

which integrates to

$$\frac{1}{a_0 - x} = k_A t + I \tag{2.11}$$

where I is the constant of integration. The boundary condition is that $x = 0$ when $t = 0$; therefore,

$$I = \frac{1}{a_0} \tag{2.12}$$

Hence,
$$\frac{x}{a_0(a_0 - x)} = k_A t \tag{2.13}$$

The variation of x with t is no longer exponential.

Graphical methods can be employed to test this equation and to obtain the rate constant k_A. One procedure is to plot $x/a_0(a_0 - x)$ against t. If the equation is obeyed the points will lie on a straight line passing through the origin (see Fig. 2.5a), and the slope will be k_A. Alternatively, $x/(a_0 - x)$ may be plotted against t (Fig. 2.5b), in which case the slope is $a_0 k_A$.

If the rate is proportional to the product of the concentrations of two different reactants, and these concentrations are not initially the same, the integration proceeds differently. Suppose that the stoichiometry corresponds to $A + B \rightarrow Z$ and that the initial concentrations are a_0 and b_0; the rate after an amount x (per unit volume) of A has reacted is

$$\frac{dx}{dt} = k_A(a_0 - x)(b_0 - x) \tag{2.14}$$

---

§ The procedure must be modified for second-order reactions of different stoichiometries, for example, $A + 2B \rightarrow X$, for which Eq. (2.9) would become

$$\frac{dx}{dt} = k_A(a_0 - x)(a_0 - 2x)$$

See Table 2.1 for this and other cases.

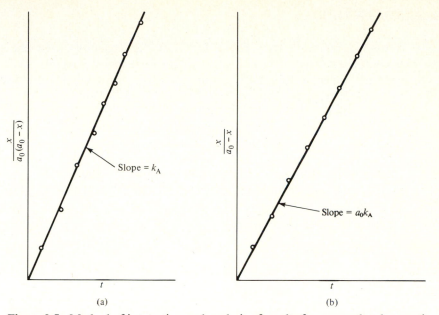

Figure 2.5 Method of integration and analysis of results for a second-order reaction involving a single reactant or two reactants of equal initial concentrations reacting according to the stoichiometry A + B → Z: (a) plot of $x/a_0(a_0 - x)$ versus t and (b) plot of $x/(a_0 - x)$ versus t.

The result of the integration, with the boundary condition $t = 0$ at $x = 0$, is

$$\frac{1}{a_0 - b_0} \ln \left[\frac{b_0(a_0 - x)}{a_0(b_0 - x)} \right] = k_A t \tag{2.15}$$

This equation can be tested by plotting the left-hand side versus t; if a straight line is obtained its slope is k_A.

2.2.3 Reactions of the *n*th Order

Suppose that a reaction is of the nth order and involves a single reactant of concentration a_0, or reactants of equal concentrations and with stoichiometry A + B + $\cdots$ → Z. If the concentration of A remaining after time t is $a_0 - x$, the rate of consumption of A is

$$v_A \equiv \frac{dx}{dt} = k_A(a_0 - x)^n \tag{2.16}$$

This must be integrated subject to the boundary condition that $x = 0$ when $t = 0$. If n is other than unity, the solution is

$$k_A = \frac{1}{t(n - 1)} \left[\frac{1}{(a_0 - x)^{n-1}} - \frac{1}{a_0^{n-1}} \right] \tag{2.17}$$

If n is unity the solution is given by Eq. (2.6).

2.3 ISOLATION METHOD

It was first pointed out by Esson[4] that if all the reactants except one are present in excess, the apparent order will be the order with respect to the one "isolated" reactant, since the concentrations of those in excess will not change very much during the course of the reaction. This method was generalized by Ostwald[5] as follows. If a reaction is of the αth order with respect to A, of the βth order with respect to B, and of the γth order with respect to C, and if B and C are in excess of A, the apparent order, determined by any of the methods described above, will correspond to α. The orders β and γ can be determined in a similar way. The method should always be employed in conjunction with other methods of determining the order, since in many cases it does not yield reliable results. Thus, if the reaction under investigation is composite, involving a number of stages, its mechanism may be influenced by the concentration changes, so that different mechanisms may be occurring under the different conditions of isolation.

2.4 HALF-LIFE

For a given reaction the *half-life* $t_{1/2}$ of a particular reactant is defined as the time required for its concentration to reach a value that is half-way between its initial and final values. The value of the half-life is always inversely proportional to the rate constant and in general depends on reactant concentrations. For a first-order reaction the rate equation is Eq. (2.6), and the half-life is obtained by putting x equal to $a_0/2$:

$$\ln \left(\frac{a_0}{a_0 - a_0/2} \right) = k_A t_{1/2} \tag{2.18}$$

and therefore

$$t_{1/2} = \frac{\ln 2}{k_A} \tag{2.19}$$

The half-life is independent of the initial concentration. Since there is only one reactant, the half-life of that reactant may be called the half-life of the *reaction.*

For a second-order reaction involving a single reactant, or two reactants of equal initial concentrations and reacting according to the stoichiometry

$$A + B \rightarrow Z$$

the rate equation is Eq. (2.13). Setting $x = a_0/2$ leads to

$$t_{1/2} = \frac{1}{a_0 k_A} \tag{2.20}$$

The half-life is now inversely proportional to the reactant concentration, a fact that may be understood as follows. Suppose that in one reaction system the reactant concentration is 10 times that in another. Ten times as much of it will be consumed when the half-life is reached, but the reaction is proceeding $10^2 = 100$ times as fast; it therefore takes 1/10 of the time.

So far, the emphasis has been on the half-life of a *reactant*. Sometimes it is legitimate to speak of the half-life of a *reaction,* but care must be taken. No difficulties arise if there is a single reacting substance: the half-life of that substance can be called the half-life of the reaction. The same is true if the stoichiometry is of the type A + B + $\cdots$ $\rightarrow$ products and the initial reactant concentrations are all the same.

If the stoichiometry is of the type A + B + $\cdots$ $\rightarrow$ products but the initial concentrations are different, the half-lives are different for the different reactants, and one should not speak of the half-life of the reaction. In this case, if one of the reactants has more than twice the concentration of another, half of it still will not have been consumed after infinite time.

Other stoichiometries lead to further complications. For example, a reaction may be of stoichiometry A + 2B $\rightarrow$ Z but be first order in both A and B. The rate equations are included in Table 2.1, and if the reactants are present in their stoichiometric ratios (i.e., if $b_0 = 2a_0$), the equations reduce to

$$\frac{dx}{dt} = k_A(a_0 - x)(2a_0 - 2x)$$

$$= 2k_A(a_0 - x)^2 = k_B(a_0 - x)^2 \tag{2.21}$$

Thus, the half-life of A is given by

$$t_{1/2,A} = \frac{1}{k_B a_0} = \frac{1}{k_A b_0} \tag{2.22}$$

and this is also the half-life of B. Note that this half-life is inversely proportional to the reactant concentrations.

This argument can be generalized to show that the concept of the half-life of a reaction is legitimate provided that the reactants are in their stoichiometric ratios. The above example, however, shows that care must be taken in relating the half-life to the

TABLE 2.2 EXPRESSIONS FOR REACTION HALF-LIVES

Order	Half-life $t_{1/2}$	Order	Half-life $t_{1/2}$
Single reacting substance A		Reactants in their stoichiometric ratios; A + $\cdots$ $\rightarrow$ Z	
0	$\dfrac{a_0}{2k_A}$	0	$\dfrac{a_0}{2k_A}$
1	$\dfrac{\ln 2}{k_A}$	1 ($v_A = k_A[A]$)	$\dfrac{\ln 2}{k_A}$
2	$\dfrac{1}{k_A a_0}$	2 ($v_A = k_A[A][B]$)	$\dfrac{1}{k_B a_0} = \dfrac{1}{k_A b_0}$
3	$\dfrac{3}{2k_A a_0^2}$	3 ($v_A = k_A[A][B][C]$)	$\dfrac{3}{2k_A b_0 c_0}$
n	$\dfrac{2^{n-1} - 1}{k_A(n - 1)a_0^{n-1}}$		

concentrations and rate constants. If the reactants are not in their stoichiometric ratios the concept of half-life of the reaction has no meaning; again, the reactant half-lives depend on the concentrations of all reactants. Some expressions for reaction half-lives are given in Table 2.2.

Since the half-lives of first-order reactions are independent of concentration, they may be quoted in place of rate constants. This is often done, for example, with radioactive decay. When the rate constant is used for a radionuclide, it is known as the *decay constant*. Unlike the rate constants for chemical processes, such decay constants are independent of temperature.

Even for reactions of different orders, half-lives provide a convenient way of making an approximate comparison of rates. For example, if two reactions are of different orders, their rate constants have different units, and it is not immediately obvious which reaction will have the higher rate under the experimental conditions. However, if the half-lives are given at the appropriate reactant concentrations, the rates can be compared.

2.5 COMPARISON OF METHODS

The differential method is the most reliable one for investigating a reaction about which there is not much previous information. The main reason is that it does not require a tentative decision as to what the order should be.§ Instead, it determines directly how the rate depends on concentration. If the double-logarithmic plots (Figs. 2.1–2.3) are linear, the reaction has an order, which is given by the slope. If they are not, other procedures are used to determine the rate–concentration relationships. Procedures used in enzyme kinetics will be explained in Section 10.3.

A second reason for the superiority of the differential method is that it distinguishes clearly between the two orders of reaction, n_c and n_t. This distinction is helpful in revealing information about the influence of products on the rates.

If the order of a reaction has been well established, the method of integration may be useful for providing accurate rate constants. Otherwise, the method has its dangers, for various reasons. In the first place, it creates a prejudice in favor of integral or half-integral orders, and deviations from such orders might escape notice. For example, a reaction having an order of 1.8 might fit the second-order integrated equation within the experimental error, and therefore the deviation from second-order kinetics would escape detection. The differential method would be more likely to indicate such a deviation.

Another difficulty with the method of integration is that the equations used, giving the variation of concentration with time, are often quite similar for different kinetic laws. For example, the time course of a simple second-order reaction is similar to that of a first-order reaction inhibited by products, and unless the experiments are done very carefully there is danger of confusion. The differential method, on the other hand, clearly reveals inhibition by products and avoids such confusion.

§ W. Lash Miller [*Trans. R. Soc. Can.,* **11,** Ser. 3, Sec. 3, 245 (1908)] referred to the method of integration as the "method of guess and try"; the differential and isolation method he called the "method of systematic exploration."

The method of integration leads in general to difficulties when the two orders n_c and n_t are different. The analysis of a given run by the method of integration involves n_t, the order with respect to time. If the work is repeated at different reactant concentrations, the rate constants obtained will be the same only if the order with respect to concentration, n_c, is the same as n_t. If they are different, the rate constants vary with the initial concentration, and a careful analysis of the variation allows n_c to be obtained.

Kinetic data have sometimes been analyzed by seeing how half-lives depend on concentration.§ This method, however, is valid only if the two orders, n_c and n_t, are the same. The half-life itself involves n_t, while its variation with concentration involves n_c in addition. Therefore, the order determined by this procedure will be a confusing mixture of n_c and n_t. The method is not recommended for reactions of previously unknown kinetics.

2.6 OPPOSING REACTIONS

A reaction may proceed to a state of equilibrium that differs appreciably from completion. The simplest case is when the reaction is of stoichiometry

$$A \underset{k_{-1}}{\overset{k_1}{\rightleftharpoons}} Z$$

and both forward and reverse reactions are first-order. If the experiment is started using pure A, of concentration a_0, and if after time t the concentration of Z is x, that of A is $a_0 - x$. The rate of reaction (1), if it occurred in isolation, is then equal to $k_1(a_0 - x)$, while the rate of the reverse reaction is $k_{-1}x$; the net rate of change of concentration of Z is

$$\frac{dx}{dt} = k_1(a_0 - x) - k_{-1}x \tag{2.23}$$

If x_e is the concentration of Z at equilibrium, when the net rate is zero,

$$k_1(a_0 - x_e) - k_{-1}x_e = 0 \tag{2.24}$$

Elimination of k_{-1} between Eqs. (2.23) and (2.24) gives rise to

$$\frac{dx}{dt} = \frac{k_1 a_0}{x_e}(x_e - x) \tag{2.25}$$

Integration of this equation, subject to the boundary condition that $x = 0$ when $t = 0$, gives

$$k_1 t = \frac{x_e}{a_0} \ln\left(\frac{x_e}{x_e - x}\right) \tag{2.26}$$

The amount of x present at equilibrium, x_e, can be measured directly. A procedure for obtaining k_1 is to obtain values of x at various values of t and to plot

$$\frac{x_e}{a_0} \ln\left(\frac{x_e}{x_e - x}\right)$$

§ This method is due to W. Ostwald [Z. *Phys. Chem.*, **2**, 127 (1888)].

versus t. The slope of the line then gives the value of k_1, and the constant k_{-1} for the reverse reaction can be obtained by use of the fact that the equilibrium constant is k_1/k_{-1}.

For certain purposes it is convenient to have this equation in a different form. Equation (2.24) rearranges to

$$x_e(k_1 + k_{-1}) = k_1 a_0 \tag{2.27}$$

so that

$$\frac{x_e}{a_0} = \frac{k_1}{k_1 + k_{-1}} \tag{2.28}$$

Therefore, Eq. (2.26) may be written as

$$k_1 = \frac{k_1}{k_1 + k_{-1}} \frac{1}{t} \ln\left(\frac{x_e}{x_e - x}\right) \tag{2.29}$$

or

$$k_1 + k_{-1} = \frac{1}{t} \ln\left(\frac{x_e}{x_e - x}\right) \tag{2.30}$$

Comparison of this equation with that for a simple first-order reaction [Eq. (2.6)] shows that the two are formally analogous, x_e replacing a_0 and $k_1 + k_{-1}$ replacing k_A.

In the foregoing derivation it has been assumed that the initial concentration of the product Z is zero. If this is not the case, and the initial concentrations of A and Z are a_0 and x_0, the net rate at time t is given by

TABLE 2.3 RATE EQUATIONS FOR OPPOSING REACTIONS

Stoichiometric equation	Rate equation[a]	
	Differential form	Integrated form
$A \rightleftharpoons Z$	$\dfrac{dx}{dt} = k_1(a_0 - x) - k_{-1}x$	
$A \rightleftharpoons Z$	$\dfrac{dx}{dt} = k_1(a_0 - x) - k_{-1}(x + x_0)$	$\dfrac{x_e}{a_0} \ln\left(\dfrac{x_e}{x_e - x}\right) = k_1 t$
$2A \rightleftharpoons Z$	$\dfrac{dx}{dt} = k_1(a_0 - x) - k_{-1}\left(\dfrac{x}{2}\right)$	
$A \rightleftharpoons 2Z$	$\dfrac{dx}{dt} = k_1\left(a_0 - \dfrac{x}{2}\right) - k_{-1}x$	
$A \rightleftharpoons Y + Z$	$\dfrac{dx}{dt} = k_1(a_0 - x) - k_{-1}x^2$	$\dfrac{x_e}{2a_0 - x_e} \ln\left[\dfrac{a_0 x_e + x(a_0 - x_e)}{a_0(x_e - x)}\right] = k_1 t$
$A + B \rightleftharpoons Z$	$\dfrac{dx}{dt} = k_1(a_0 - x)^2 - k_{-1}x$	$\dfrac{x_e}{a_0^2 - x_e^2} \ln\left[\dfrac{x_e(a_0^2 - xx_e)}{a_0^2(x_e - x)}\right] = k_1 t$
$A + B \rightleftharpoons Y + Z$	$\dfrac{dx}{dt} = k_1(a_0 - x)^2 - k_{-1}x^2$	$\dfrac{x_e}{2a_0(a_0 - x_e)} \ln\left[\dfrac{x(a_0 - 2x_e) + a_0 x_e}{a_0(x_e - x)}\right] = k_1 t$
$2A \rightleftharpoons Y + Z$	$\dfrac{dx}{dt} = k_1(a_0 - x)^2 - k_{-1}\left(\dfrac{x}{2}\right)^2$	

[a] In all cases x is the amount of A consumed per unit volume. The concentration of B is taken to be the same as that of A.

$$\frac{dx}{dt} = k_1(a_0 - x) - k_{-1}(x + x_0) \tag{2.31}$$

By methods similar to those used earlier it can be shown that this equation integrates to

$$k_1 + k_{-1} = \frac{1}{t} \ln\left[\frac{k_1 a_0 - k_{-1} x_0}{k_1 a_0 - k_{-1} x_0 - (k_1 + k_{-1})x}\right] \tag{2.32}$$

The concentration at equilibrium, x_e, is defined by

$$x_e = \frac{k_1 a_0 - k_{-1} x_0}{k_1 + k_{-1}} \tag{2.33}$$

Equation (2.32) reduces to

$$k_1 + k_{-1} = \frac{1}{t} \ln\left(\frac{x_e}{x_e - x}\right) \tag{2.34}$$

which is the same as Eq. (2.30).

The solutions for this and other cases are given in Table 2.3. The integration of equations of this type is facilitated by the use of an operator method, which is explained in the book by Capellos and Bielski.[6]

2.7 REACTIONS IN FLOW SYSTEMS

In kinetic investigations, particularly those concerned with obtaining fundamental information, it is common to employ a static system: the reactants are introduced into a reaction vessel and the concentration changes are followed. The equations developed so far in this chapter relate directly to such static systems. In some cases, however, it is more convenient to allow the reaction mixture to flow through a reaction vessel, known as a *reactor*. Such a flow system may be useful, for example, when it is desired to study a reaction at extremely low pressures or concentrations; to obtain enough product it may be necessary to pass a stream of reactants through the reactor for a considerable time. Flow systems also are useful in studies of very rapid reactions; a convenient technique is the "stopped-flow" method (Section 2.8.1), in which a rapid flow is stopped suddenly and an analysis is made of the change of concentration with time.

Flow systems are of two general types. In the first, there is no stirring in the reactor, and the flow through it is sometimes spoken of as *plug flow*. In the second, there is stirring which is sufficiently vigorous to effect complete mixing within the reactor. Intermediate situations are also possible, but they are difficult to analyze.

Plug flow is illustrated schematically in Fig. 2.6. The reaction mixture is passed through the reactor at a volume rate of flow (SI unit: $m^3\ s^{-1}$) equal to u. Consider an element of volume dV in the reactor and suppose for simplicity that the reaction rate depends on the concentration c of a single reactant. For a reaction of the nth order the rate of consumption of the substance is given by

$$v = kc^n \tag{2.35}$$

The rate of conversion of reactant in a volume dV is therefore $kc^n\ dV$.

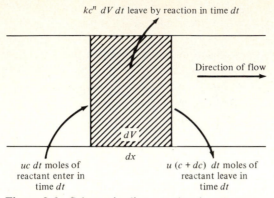

Figure 2.6 Schematic diagram showing the processes occurring in plug flow through a reaction vessel.

After the system has been operating for a sufficient time, a steady state is established; this means that there is no change, with time, in the concentration of reactant in the volume element. Three processes contribute to the steady state as follows:

1. Molecules of reactant enter the slab through the left face, the amount entering in time dt being $uc\,dt$.
2. Molecules leave the slab by the right face, the amount leaving in time dt being $u(c + dc)dt$.
3. Molecules disappear by chemical reaction; for a reaction of the nth order the amount $(-dn)$ consumed in time dt is $kc^n\,dV\,dt$.

The steady-state equation is obtained by equating the rate of entry of reactant into the slab (by process 1) to the sum of the rates of removal (by processes 2 and 3). The result is

$$uc\,dt = u(c + dc)dt + kc^n\,dV\,dt \qquad (2.36)$$

or
$$-\frac{dc}{c^n} = \frac{k}{u}\,dV \qquad (2.37)$$

This equation must be integrated over the volume V_0 of the reactor; at the entrance to the reactor $V = 0$ and $c = c_i$ (the initial concentration), while at the exit $V = V_0$ and $c = c_f$ (the final concentration of reactant). Therefore,

$$-\int_{c_i}^{c_f} \frac{dc}{c^n} = \frac{k}{u} \int_0^{V_0} dV \qquad (2.38)$$

For the particular case in which $n = 1$, integration gives

$$-\ln\left(\frac{c_f}{c_i}\right) = \frac{kV_0}{u} \qquad (2.39)$$

or
$$c_f = c_i e^{-kV_0/u} \qquad (2.40)$$

This equation may be compared with Eq. (2.8) for a static system, and the two are equivalent if V_0/u is replaced by t. This quantity V_0/u is known as the *contact time* for the reaction; it is the average time that a molecule takes to pass through the reactor. Equation (2.40) may be tested by varying V_0/u (by changing either the volume of the reactor or the flow rate), just as the time is varied in a static system. Reactions that are too rapid for convenient study in a static system may sometimes be studied in a flow system, the contact time being reduced by using a high flow rate and small volume.

The solution of Eq. (2.38) when n is other than unity is

$$\frac{1}{n-1}\left(\frac{1}{c_f^{n-1}} - \frac{1}{c_i^{n-1}}\right) = \frac{kV_0}{u} \tag{2.41}$$

This equation is to be compared with Eq. (2.17) with V_0/u equal to t.

In this derivation the tacit assumption was made that there is no volume change during the course of reaction; any such change will cause the volume flow rate through the reactor to vary. The inclusion of such volume changes complicates the handling of the rate equations and will not be considered here; reference may be made to treatments by Hougen and Watson and by Villermaux.[7]

In a *stirred-flow reactor,* in which the concentrations are maintained constant within the reactor, it is not necessary to consider a thin slab; instead, we consider the reactor as a whole. The rate of flow of reactants into the reactor is uc_i, and the rate of flow out is uc_f; the difference between these is the rate of reaction in the reactor, which is vV, where v is the rate per unit volume. Thus,

$$uc_i - uc_f = vV \tag{2.42}$$

or

$$v = \frac{u(c_i - c_f)}{V} \tag{2.43}$$

The measurement of c_i and c_f at a given flow rate allows the reaction rate to be calculated. The order of reaction and the rate constant can then be determined by working at different initial concentrations and rates of flow. The theory of stirred-flow reactors has been treated by Denbigh.[8]

2.8 TECHNIQUES FOR VERY FAST REACTIONS

Some reactions are so fast that special techniques have to be employed. Such techniques are of two main types: those of the first type employ essentially the same principles as are used for slow reactions, the procedures being modified to make them suitable for more rapid reactions; the second type are of a different character and involve special principles.

The reasons why conventional techniques lead to difficulties for very rapid reactions are:

1. The time that it takes to mix reactants or to bring them to a specified temperature may be significant in comparison to the half-life of the reaction. An

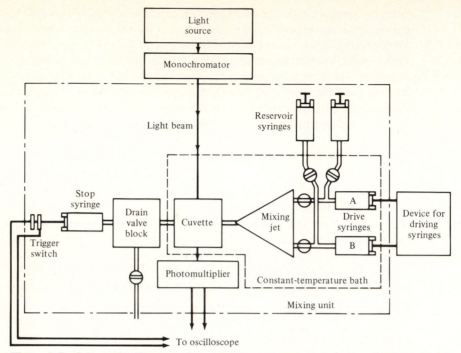

Figure 2.7 Schematic diagram of stopped-flow apparatus. Solutions in the two drive syringes A and B are forced rapidly through a mixing jet into a cuvette. When the flow is stopped, the oscilloscope is triggered and records light absorption as a function of time.

appreciable error therefore will be made, since the initial time cannot be determined accurately.

2. The time that it takes to make a measurement of concentration may be significant compared to the half-life.

2.8.1 Stopped-Flow Method

Sometimes the first difficulty can be surmounted by using special techniques for bringing the reactants very rapidly into the reaction vessel and for mixing them very rapidly.§ With the use of conventional techniques, it takes from several seconds to a minute to bring solutions into a reaction vessel and to have them mix completely at the temperature of the surroundings. This time can be reduced greatly by using a rapid flow, and flow techniques are often employed for rapid reactions. One method is the *stopped-flow* technique,¶ shown schematically in Fig. 2.7. This particular apparatus is

§ Flow methods were first used by E. Rutherford [*Philos. Mag.*, **44**, 422 (1897)] for gas reactions. The techniques were developed for solution reactions especially by H. Hartridge and F. J. W. Roughton [*Proc. R. Soc.*, **A 104**, 376 (1923)] and by G. A. Millikin [*Proc. R. Soc.*, **A155**, 277 (1936)].

¶ The stopped-flow method was introduced by B. Chance [*J. Franklin Inst.*, **229**, 455, 613, 727 (1940)], and the technique has been developed by many workers, for example by Q. H. Gibson and L. Milnes [*Biochem. J.*, **91**, 161 (1964)].

designed for the study of a reaction between two substances in solution. A solution of one of the substances is maintained initially in the syringe A, and a solution of the other is in syringe B. The plungers of the syringes can be forced in rapidly and a fast stream of the two solutions passes into the mixing system, which is designed in such a way that jets of the two solutions impinge on one another and give very rapid mixing. With a suitable design of the mixing chamber it is possible for mixing to be essentially complete in 0.001 s. From this mixing chamber the solution passes at once into the reaction cuvette. Sometimes the reaction cuvette is designed so as to be a mixing chamber also.

If a reaction is rapid it is necessary to employ analytical techniques that allow records to be obtained instantaneously. For reactions in solution, spectrophotometric methods may be employed. If the products absorb differently from the reactants at a particular wavelength, monochromatic light of this wavelength may be passed through the reaction vessel. By means of a photoelectric device with suitable electronic circuits the output can be displayed on a recorder or oscilloscope screen. If the reaction is not too fast a pen-and-ink recorder may respond sufficiently rapidly; otherwise, an oscilloscope can give a record of absorption versus time. Fluorescence, electrical conductivity, and optical rotation are also convenient properties to measure in such high-speed studies.

2.8.2 Relaxation Methods

The flow techniques just described are limited by the speed with which it is possible to mix solutions. There is no difficulty, using optical or other techniques, in following the course of a very rapid reaction, but for hydrodynamic reasons it is impossible to mix two solutions in less than about 10^{-3} s. If the half-life is less than this, the reaction is almost over before mixing has been achieved; any rate measurement will be of the rate of mixing, not of the rate of reaction. The neutralization in aqueous solution of strong acid by a strong base, that is, the reaction

$$H^+ + OH^- \rightarrow H_2O$$

under ordinary conditions has a half-life of 10^{-6} s or less (see Problem 2.5), and its rate therefore cannot be measured by any technique involving the mixing of solutions.

These technical problems were overcome by the development of a group of methods known as *relaxation methods*, the pioneer worker in this field being Manfred Eigen.[9] These methods differ fundamentally from conventional kinetic methods in that the system is initially at equilibrium under a given set of conditions. These conditions are then suddenly changed; the system is no longer at equilibrium, and it *relaxes* to a new state of equilibrium. The speed with which it relaxes can be measured, usually by spectrophotometry, and the rate constants can be obtained.

There are various ways in which the conditions are disturbed. One is by changing the hydrostatic pressure. Another technique is to increase the temperature suddenly, usually by the rapid discharge of a capacitor; this method is called the *temperature-jump* or *T-jump method*. It is possible to raise the temperature of a tiny cell containing a reaction mixture by a few degrees in less than 10^{-7} s, which is sufficiently rapid to permit the study of even the fastest purely chemical processes.

The principle of the method is illustrated in Fig. 2.8. Suppose that the reaction is of the simple type

$$A \underset{k_{-1}}{\overset{k_1}{\rightleftharpoons}} Z$$

the processes being first-order in both directions. At the initial state of equilibrium, the product Z is at a certain concentration. It stays at this concentration until the temperature jump occurs, at which time the concentration changes to another value which will be higher or lower than the initial value according to the sign of $\Delta H°$ for the reaction. From the shape of the curve during the relaxation phase, the sum of the rate constants, $k_1 + k_{-1}$, is obtained, as is shown by the following treatment.

Let a_0 be the sum of the concentrations of A and Z, and x the concentration of Z at any time; the concentration of A is $a_0 - x$. The kinetic equation is thus

$$\frac{dx}{dt} = k_1(a_0 - x) - k_{-1}x \tag{2.44}$$

If x_e is the concentration of Z at equilibrium,

$$k_1(a_0 - x_e) - k_{-1}x_e = 0 \tag{2.45}$$

The deviation of x from equilibrium, Δx, is equal to $x - x_e$, and therefore

$$\frac{d\,\Delta x}{dt} = \frac{dx}{dt} = k_1(a_0 - x) - k_{-1}x \tag{2.46}$$

Subtraction of the expression in Eq. (2.45) gives

$$\frac{d\,\Delta x}{dt} = k_1(x_e - x) + k_{-1}(x_e - x) \tag{2.47}$$

$$= -(k_1 + k_{-1})\Delta x \tag{2.48}$$

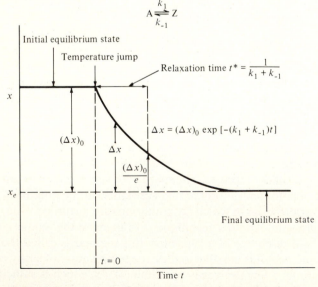

Figure 2.8 Principle of the temperature-jump technique.

Thus, the quantity Δx varies with time in the same manner as does the concentration of a reactant in a first-order reaction. Integration of Eq. (2.48), subject to the boundary condition that $\Delta x = (\Delta x)_0$ when $t = 0$, leads to

$$\ln\left[\frac{(\Delta x)_0}{\Delta x}\right] = (k_1 + k_{-1})t \tag{2.49}$$

The *relaxation time* τ is defined as the time corresponding to

$$\frac{(\Delta x)_0}{\Delta x} = e \tag{2.50}$$

or to

$$\ln\left[\frac{(\Delta x)_0}{\Delta x}\right] = 1 \tag{2.51}$$

The relaxation time is thus the time at which the distance from equilibrium is $1/e$ of the initial distance (see Fig. 2.8). From Eqs. (2.49) and (2.51) it follows that

$$\tau = \frac{1}{k_1 + k_{-1}} \tag{2.52}$$

Therefore, if τ is determined experimentally for such a system, $k_1 + k_{-1}$ can be calculated. However, the ratio k_1/k_{-1} is the equilibrium constant and can be determined directly; hence, the individual constants k_1 and k_{-1} can be obtained.

Relaxation curves for other types of reaction can be analyzed similarly to give rate constants. For example, consider a reaction of the type

$$A + B \underset{k_{-1}}{\overset{k_1}{\rightleftharpoons}} Z$$

which is second-order from left to right and first-order from right to left. If the concentrations of A, B, and Z at any time are $a_0 - x$, $b_0 - x$, and x,

$$\frac{dx}{dt} = k_1(a_0 - x)(b_0 - x) - k_{-1}x \tag{2.53}$$

As before, Δx is defined as $x - x_e$, and

$$\frac{d\,\Delta x}{dt} = \frac{dx}{dt} = k_1 a_0 b_0 - (k_1 a_0 + k_1 b_0 + k_{-1})x + k_1 x^2 \tag{2.54}$$

At equilibrium,

$$k_1(a_0 - x_e)(b_0 - x_e) = k_{-1}x_e \tag{2.55}$$

or

$$k_1 a_0 b_0 - (k_1 a_0 + k_1 b_0 + k_{-1})x_e + k_1 x_e^2 = 0 \tag{2.56}$$

Subtraction of Eq. (2.56) from (2.54) gives

$$\frac{d\,\Delta x}{dt} = -(k_1 a_0 + k_1 b_0 + k_{-1})(x - x_e) + k_1(x^2 - x_e^2) \tag{2.57}$$

$$= -(k_1 a_0 + k_1 b_0 + k_{-1})\Delta x + k_1(x + x_e)\Delta x \tag{2.58}$$

If the displacement from equilibrium is only slight, $x \approx x_e$, so that $x + x_e$ may be written as $2x_e$, Eq. (2.58) becomes

$$\frac{d \Delta x}{dt} = -(k_1 a_0 + k_1 b_0 + k_{-1} - 2k_1 x_e)\Delta x \tag{2.59}$$

Integration gives

$$\ln\left[\frac{(\Delta x)_0}{\Delta x}\right] = (k_1 a_0 + k_1 b_0 + k_{-1} - 2k_1 x_e)t \tag{2.60}$$

and the relaxation time, defined as before, is

$$\tau = \frac{1}{k_1 a_0 + k_1 b_0 + k_{-1} - 2k_1 x_e} \tag{2.61}$$

$$= \frac{1}{k_1(a_e + b_e) + k_{-1}} \tag{2.62}$$

where a_e and b_e, equal to $a_0 - x_e$ and $b_0 - x_e$, are the equilibrium concentrations of A and B. The constants k_1 and k_{-1} can now be separated by measuring τ at various values of $a_e + b_e$. In addition, use may be made of the fact that k_1/k_{-1} is the equilibrium constant.

The solutions for these and other cases are listed in Table 2.4. Note that the rate constants obtained refer to the second temperature, after the T-jump has occurred. Thus, if we want rate constants at 25°C, and the T-jump is 7°C, we should start at 18°C.

During recent years a considerable number of investigations have been made using this technique. An important reaction studied is the dissociation of water and combination of hydrogen and hydroxide ions:

$$2H_2O \rightleftharpoons H_3O^+ + OH^-$$

The reaction was followed by measuring the electrical conductivity of water; after the T-jump the conductivity increased with time, with a relaxation half-life of

TABLE 2.4 RELAXATION TIMES FOR SOME REACTION TYPES

System[a]	Rate equation	Relaxation time
$A \underset{k_{-1}}{\overset{k_1}{\rightleftharpoons}} Z$	$\frac{dx}{dt} = k_1(a_0 - x) - k_{-1}x$	$\frac{1}{k_1 + k_{-1}}$
$2A \underset{k_{-1}}{\overset{k_1}{\rightleftharpoons}} Z$	$\frac{dx}{dt} = k_1(a_0 - 2x)^2 - k_{-1}x$	$\frac{1}{4k_1 a_e + k_{-1}}$
$A + B \underset{k_{-1}}{\overset{k_1}{\rightleftharpoons}} Z$	$\frac{dx}{dt} = k_1(a_0 - x)(b_0 - x) - k_1 x$	$\frac{1}{k_1(a_e + b_e) + k_{-1}}$
$A \underset{k_{-1}}{\overset{k_1}{\rightleftharpoons}} Y + Z$	$\frac{dx}{dt} = k_1(a_0 - x) - k_{-1}(x_0 + x)(y_0 + x)$	$\frac{1}{k_1 + k_{-1}(x_e + y_e)}$
$A \underset{k_{-1}}{\overset{k_1}{\rightleftharpoons}} 2Z$	$\frac{dx}{dt} = k_1\left(a - \frac{x}{2}\right) - k_{-1}x^2$	$\frac{1}{\frac{1}{2}k_1 + 2k_{-1}x_e}$
$A + B \underset{k_{-1}}{\overset{k_1}{\rightleftharpoons}} Y + Z$	$\frac{dx}{dt} = k_1(a_0 - x)(b_0 - x) - k_{-1}(x_0 + x)(y_0 + x)$	$\frac{1}{k_1(a_e + b_e) + k_{-1}(x_e + y_e)}$

[a] In each case the rate constants k_1 and k_{-1} refer to the rate of change of the concentration of a substance on the right-hand side of the equation.

3.7×10^{-5} s at 23°C. From this value it can be calculated that the second-order rate constant for the combination of H_3O^+ and OH^- ions is 1.4×10^{11} dm^3 mol^{-1} s^{-1}, which is a remarkably high value (see also Section 6.7.3). The rate constant for the reverse dissociation, which is very small, can be calculated from this value and the equilibrium constant. A number of other hydrogen-ion transfer processes have also been studied using the T-jump technique.

2.8.3 Shock-Tube Methods

High-temperature gas-phase reactions that have half-lives between 10^{-3} and 10^{-6} s are studied conveniently by means of shock tubes. The principle of the method, first used by Vieille,[10] is as follows. A long tube is divided by a diaphragm into two compartments, one much smaller than the other. The smaller compartment contains a *driver gas,* which is usually helium or hydrogen, at a few atmospheres pressure. The larger compartment contains the reaction mixture at a low pressure, often about 10^{-3} atm. The process is started by rupturing the diaphragm, and a narrow shock front travels along the tube at supersonic speed. As the shock front passes through each element of volume there is a considerable temperature rise, the final temperature usually being 10^3–10^4 K. The heating process takes about 1 μs, which limits the applicability of the technique to reactions having half-lives in excess of this period of time. Reaction in each volume element commences as the shock wave passes: the driver gas lags behind and does not interfere with the process, but it pushes the reacted gas along the tube.

The course of the reaction is followed by means of an observation point at the end of the tube; concentrations are determined by spectroscopic methods. Measurements made as the shock front passes correspond to short reaction times. Gas that passes the observation point later has been undergoing reaction for longer periods. It is therefore possible to establish a profile of extent of reaction as a function of time. Some examples of reactions studied in shock tubes are discussed later.

2.8.4 Flash Photolysis

Another useful technique for investigating fast reactions is *flash photolysis,* a photochemical method that is considered further in Section 9.1.6. As originally employed by Porter and Norrish[11] the method could be used to study reactions with half-lives down to 10^{-6} s. By the use of pulsed laser flashes, however, Porter and Topp[12] have been able to investigate processes having half-lives as low as 10^{-12} s.

Figure 2.9 summarizes some of the techniques used for rapid reactions and indicates the approximate range of half-lives that is covered by each.

2.9 INFLUENCE OF TEMPERATURE ON REACTION RATES

Today, the influence of temperature on the rates of chemical reactions is usually interpreted in terms of what is known as the *Arrhenius equation.* According to this equation, a rate constant k is the product of a *pre-exponential factor A* and an exponential factor:

$$k = Ae^{-E_a/RT} \tag{2.63}$$

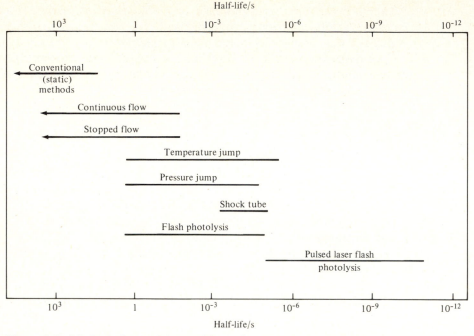

Figure 2.9 Methods for studying rapid reactions, and their approximate ranges of applicability.

The exponential factor involves the temperature T, the gas constant R, and an energy E_a which is known as the *activation energy* or the *energy of activation*. The pre-exponential factor A was known formerly as the "frequency factor," but since its dimensions are the same as those of the rate constant it is a frequency only in the case of a first-order reaction.

2.9.1 Survey of Empirical Equations for Temperature Dependence

Although the Arrhenius equation is usually used today to interpret kinetic data, the problem of temperature dependence was one of much uncertainty and controversy for over 60 years—from about 1850 to about 1910. For example, Mellor's influential monograph on equilibrium and kinetics,[13] published in 1904, quotes Ostwald as saying that the temperature-dependence of reaction rates is "one of the darkest chapters in chemical mechanics." It is instructive to consider briefly some of the proposed empirical relationships between the rate constant and temperature and to see how the final outcome was reached.[14]

In Wilhelmy's pioneering investigations[15] of 1850 he suggested, for the first time, an empirical equation relating the rate constant to the temperature; however, it did not satisfy the requirement that rate constants for reactions in forward and reverse directions must show a temperature dependence that is consistent with that of the equilibrium constant. In 1862 Berthelot[16] proposed the equation

$$k = Ae^{DT} \tag{2.64}$$

where A and D are constants. This equation, which was taken seriously at least until 1908, was supported by some very careful kinetic results published in 1885 by J. J. Hood.[17]

In 1884 van't Hoff published his pioneering textbook on chemical dynamics,[18] and it contained a discussion of the temperature dependence of equilibrium constants in terms of an equation that we can express as

$$\left(\frac{\partial \ln K_c}{\partial T}\right)_P = \frac{\Delta U^\circ}{RT^2} \tag{2.65}$$

where K_c is the concentration equilibrium constant and ΔU° the standard internal energy change. Consider the reaction

$$A + B \underset{k_{-1}}{\overset{k_1}{\rightleftharpoons}} Y + Z$$

which is elementary in both directions: the rate from left to right is $k_1[A][B]$ and that from right to left is $k_{-1}[Y][Z]$. At equilibrium the two rates are equal:

$$k_1[A][B] = k_{-1}[Y][Z] \tag{2.66}$$

The concentration equilibrium constant K_c is the ratio of rate constants,

$$\left(\frac{[Y][Z]}{[A][B]}\right)_{eq} = K_c = \frac{k_1}{k_{-1}} \tag{2.67}$$

and Eq. (2.65) can be written as

$$\frac{d \ln k_1}{dT} - \frac{d \ln k_{-1}}{dT} = \frac{\Delta U^\circ}{RT^2} \tag{2.68}$$

It was argued by van't Hoff that the rate constants k_1 and k_{-1} will be influenced by two different energy factors E_1 and E_{-1}, and he therefore split Eq. (2.68) into the two equations

$$\frac{d \ln k_1}{dT} = \frac{E_1}{RT^2} \tag{2.69}$$

and

$$\frac{d \ln k_{-1}}{dT} = \frac{E_{-1}}{RT^2} \tag{2.70}$$

The two energies E_1 and E_{-1} must be such that $E_1 - E_{-1} = \Delta U^\circ$.

van't Hoff was aware that ΔU°, which appears in the equilibrium equation [Eq. (2.65)] is not always temperature independent, and he therefore realized that the quantities E_1 and E_{-1} in Eqs. (2.69) and (2.70) also may be temperature dependent. In the event that E_1 is independent of temperature, Eq. (2.69) can be integrated to give

$$\ln k = \text{const} - \frac{E}{RT} \quad \text{or} \quad k = Ae^{-E/RT} \tag{2.71}$$

(we have now dropped the subscript). van't Hoff considered this possibility and, in addition, the possibility that E can be of the form

$$E = B + DT^2 \tag{2.72}$$

where B and D are temperature independent. The equation thus becomes

$$\frac{d \ln k}{dT} = \frac{B + DT^2}{RT^2} \tag{2.73}$$

and integration gives

$$\ln k = A' - \frac{B - DT^2}{RT} \tag{2.74}$$

where A' is the constant of integration. This equation also can be written as

$$k = Ae^{-(B-DT^2)/RT} \tag{2.75}$$

where $A = e^{A'}$. This equation received much experimental support.

A very important contribution to the problem was made in 1889 by Arrhenius,[19] who took van't Hoff's simpler equation, Eq. (2.71), as his starting point and on its basis proposed a general concept of how reactions occur. He pointed out that the magnitudes of the temperature effects on rates are usually much too large to be explicable on the basis of how temperature affects the molecular translational energies or—in the case of a reaction in solution—how the temperature affects the viscosity of the solvent. He concluded that an equilibrium is established between normal and active reactant molecules and that this equilibrium shifts in the manner predicted by van't Hoff's equation [Eq. (2.65)]. This is very much the position that is taken today, and the equation favored by Arrhenius, Eq. (2.71), is now usually known as the *Arrhenius equation.* We discuss its implications in more detail in Chapter 3.

Another empirical equation, proposed in 1893 by van't Hoff's student D. M. Kooij,[20] involves a temperature-dependent pre-exponential factor:

$$k = AT^m e^{-B/T} \tag{2.76}$$

From both the theoretical and empirical points of view, this equation is the most satisfactory; data sometimes show significant deviations from the simple Arrhenius equation [Eq. (2.71)] but usually they can be fitted very well by Eq. (2.76).

Some 30 years after Harcourt and Esson[21] made their pioneering contributions to chemical kinetics in 1865, by formulating and applying the method of integration, they collaborated again in suggesting an equation for temperature dependence. They gave very precise results for the reaction between hydrogen peroxide and hydrogen iodide, obtained from 0 to 50°C at intervals of about 5°C, and concluded that the results were represented best by the equation

$$k = AT^m \tag{2.77}$$

In 1912 they analyzed many other experimental results[22] and concluded that Eq. (2.77) was more satisfactory than any other equation that had been proposed. One particularly interesting feature of their work was that they were able to arrive at a "kinetic absolute zero," at which chemical reaction ceases. Their analysis led to the result that this absolute zero is −272.6°C, in remarkably good agreement with the modern value of −273.15°C.

In a book published in 1898, van't Hoff[23] pointed out that most of the equations that had been proposed for temperature dependence were special cases of the equation

$$k = AT^m e^{-(B-DT^2)/T} \tag{2.78}$$

This can be called a three-parameter equation, since the temperature dependence is determined by the three parameters B, m, and D. The three one-parameter equations obtained by dropping two of these parameters are listed in Table 2.5. Also included in the table is the two-parameter equation of Kooij.

It appears surprising at first sight that equations as widely different as those listed in Table 2.5 can give a reasonably good fit to the same experimental data. A fit to the one-parameter equations is confirmed if the following plots are linear:

Equation (2.71): $\ln k$ versus $1/T$.

Equation (2.77): $\ln k$ versus $\ln T$.

Equation (2.64): $\ln k$ versus T.

The reason that all these plots can give reasonable straight lines with the same data is that, over the narrow temperature ranges usually employed in kinetic studies, $1/T$, $\ln T$, and T are related linearly to each other, more or less. This is illustrated in Fig. 2.10 for the temperature range from 0 to 100°C. In most kinetic studies the range is even narrower, so that deviations from linearity are less obvious. It follows that if the temperature range is not too large and $\ln k$ is plotted against any one of the three functions $1/T$, $\ln T$, and T, and a straight line is obtained, the plot also will be almost linear if either of the other two functions is employed. This is illustrated in Fig. 2.11 for very reliable data of Harcourt and Esson[24] on the reaction between hydrogen peroxide and hydrogen iodide. Harcourt and Esson were correct in concluding *on purely empirical grounds* that the plot of $\ln k$ versus $\ln T$ is more linear than that of $\ln k$ versus $1/T$. We should note, however, that a double-logarithmic plot is intrinsically more likely to approach linearity than a semilogarithmic plot. This is well illustrated by the equation $xy = $ constant. A plot of y versus x is a hyperbola. A plot of $\ln y$ versus x or of y versus $\ln x$ shows much less curvature; while a plot of $\ln y$ versus $\ln x$ is a straight line.

TABLE 2.5 TEMPERATURE-DEPENDENCE EQUATIONS DERIVED FROM VAN'T HOFF'S THREE-PARAMETER EQUATION, $k = AT^m e^{-(B-DT^2)/T}$

Equation	Equation number in text	Suggested by
$k = Ae^{-B/T}$	(2.71)	van't Hoff, 1884
		Arrhenius, 1889
$k = AT^m$	(2.77)	Harcourt and Esson, 1895, 1912
$k = Ae^{DT}$	(2.64)	Berthelot, 1862
		Hood, 1885
$k = AT^m e^{-B/T}$	(2.76)	Kooij, 1893

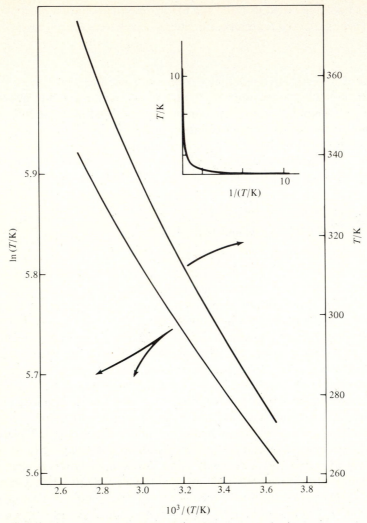

Figure 2.10 Plots of ln T (K) and T (K) versus $1/T$ (K) for the temperature range 0–100°C. The inset shows a plot of T (K) versus $1/T$ (K) for the range 0–10 K.

2.9.2 Arrhenius Equation

By about 1910 the only one-parameter temperature-dependence equation to survive was the Arrhenius equation. This was not because it was empirically the best; the Harcourt–Esson equation, Eq. (2.77), usually fits the data better. The Arrhenius equation was accepted because it provides an insight into how reactions proceed. We will see in Chapter 3 that the energy E_a that appears in the equation is related to the height of an energy barrier for the reaction. By contrast, the other two one-parameter equations are theoretically sterile. The parameter m that appears in the Harcourt–Esson equation,

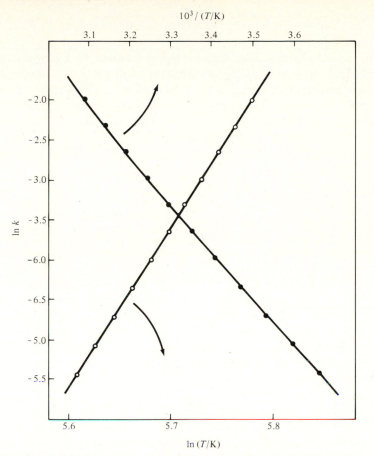

Figure 2.11 Plots for the data of Harcourt and Esson for the reaction between hydrogen peroxide and hydrogen iodide. The units of the rate constant are not stated but are probably $dm^3 mol^{-1} min^{-1}$. The natural logarithm of the rate constant is plotted against $\ln T$ (K) (open circles) and against $1/T$ (K) (filled circles).

Eq. (2.77), and the parameter D that appears in the Berthelot equation, Eq. (2.64), cannot be related simply to any physical quantity.

2.9.3 Improved Treatments of Temperature Dependence

The Arrhenius equation, Eq. (2.63), has been referred to as a one-parameter equation, since one parameter, E_a, expresses the temperature dependence. The three-parameter equation, Eq. (2.78), suggested by van't Hoff in 1898, will fit data better than the one-parameter equation, as will any of the two-parameter equations. In fact, the three-parameter equation was used by some of the earlier workers for fitting their kinetic results. For example, Bodenstein[25] used that equation to fit his very precise data on the gas-phase reaction between hydrogen and iodine and the reverse decomposition of hydrogen iodide. For the hydrogen–iodine reaction he obtained the formula

$$\ln k = -\frac{21\,832}{T} - 12.872 \ln T + 0.01751 T + 101.487 \qquad (2.79)$$

(note his use of more significant figures than is reasonable). Unfortunately, however, this procedure, in spite of giving an excellent fit to the data, is unsatisfactory. The difficulty is that identical data can be fitted almost equally well by the use of sets of widely different parameters, and a trifling change in the data may lead to a completely different empirical formula. In other words, the three-parameter equation is too sensitive even when applied to very precise data.

Of the possible two-parameter equations, the one that is most satisfactory is Eq. (2.76), which can be written

$$k = A T^m e^{-E_0/RT} \qquad (2.80)$$

where A, E_0, and m are temperature-independent constants. Gardiner[26] has analyzed data for a number of reactions and has shown that Eq. (2.80) applies satisfactorily, and better than the Arrhenius equation, to all of them. The procedure now commonly employed in compilations of kinetic data is to use the Arrhenius equation when it applies. However, if plots of $\ln k$ versus $1/T$ are not linear, the data are analyzed in terms of Eq. (2.80). It will be shown in Section 4.5 that there is a good theoretical basis for this equation, arising from the temperature dependence of partition functions which enter into the rate expression. In some cases, theory leads to a value of the exponent m. When this is so, the applicability of Eq. (2.80) can be tested by plotting $\ln(k/T^m)$ against $1/T$. If a straight line is obtained its slope is $-E_0/R$, so that E_0 can be calculated.

With the advent of modern experimental techniques in chemical kinetics it has become possible to study reaction rates over a much wider temperature range than was previously possible. For example, Whytock and coworkers[27] made accurate measurements of the rate constant of the elementary reaction

$$\text{Cl} + \text{CH}_4 \rightarrow \text{HCl} + \text{CH}_3$$

over the temperature range from 200 to 500 K. Figure 2.12e shows their data in an Arrhenius plot, which exhibits significant curvature. Whytock and coworkers carried out a statistical analysis of their results on the basis of Eq. (2.80) and found a good fit with $m = 2.5$. This is confirmed by the plot of $\ln(kT^{-2.5})$ versus $1/T$ which is shown in Fig. 2.12b; the curve is close to being linear. Figure 2.12c shows a Harcourt–Esson plot of $\ln k$ versus $\ln T$. There is somewhat less curvature than in the Arrhenius plot, but as we have discussed this is of no significance.

2.9.4 Temperature Dependence of the Pre-exponential Factor

It is convenient to *define* the experimental activation energy E_a by the equation§

$$\frac{d \ln k}{dT} \equiv \frac{E_a}{RT^2} \qquad (2.81)$$

§ The subscript a in E_a may be omitted if there is no danger of confusion with other types of energy. It is often convenient to omit it if other subscripts are needed, for example, E_1 and E_{-1}.

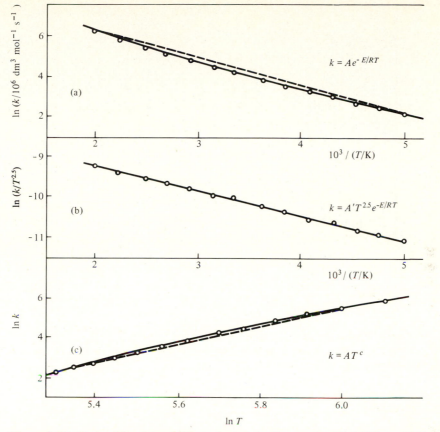

Figure 2.12 Plots of the data of Whytock et al. on the reaction $Cl + CH_4 \rightarrow HCl + CH_3$ at temperatures from 200 to 500 K: (a) simple Arrhenius plot; (b) plot of $\ln (k/T^{2.5})$ versus $1/T$; and (c) Harcourt–Esson plot of $\ln k$ versus $\ln T$.

This equation is equivalent to

$$\frac{d \ln k}{d(1/T)} \equiv -\frac{E_a}{R} \tag{2.82}$$

since $d(1/T) = -dT/T^2$. The significance of Eq. (2.82) is that if $\ln k$ is plotted against $1/T$, the slope at any point is equal to $-E_a/R$; thus, E_a is defined as the slope multiplied by $-R$. This definition applies whether or not the Arrhenius plot is linear; if it is not, the activation energy varies with the temperature.

This definition of the activation energy leads to a simple relationship between E_a and the temperature for a reaction obeying Eq. (2.80). In logarithmic form this equation is

$$\ln k = \ln A + m \ln T - \frac{E_0}{RT} \tag{2.83}$$

Differentiation gives

$$\frac{d \ln k}{dT} = \frac{m}{T} + \frac{E_0}{RT^2} = \frac{E_0 + mRT}{RT^2} \tag{2.84}$$

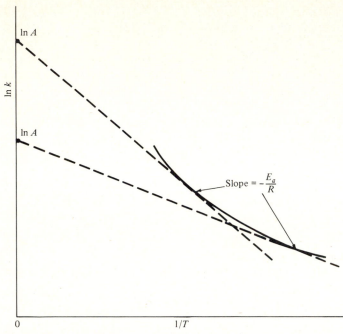

Figure 2.13   Schematic Arrhenius plot showing curvature; temperature dependence in E_a and in A must always go together, since $\ln A$ is $\ln k$ at $T = 0$.

and comparison of Eqs. (2.81) and (2.84) gives

$$E_a = E_0 + mRT \qquad (2.85)$$

It follows that the significance of E_0 is that it is the hypothetical activation energy at absolute zero.

    The above analysis shows that if the pre-exponential factor is temperature dependent, the activation energy is also temperature dependent. The converse is also true: a curved Arrhenius plot, meaning a temperature-dependent E_a, requires the pre-exponential factor to be temperature dependent. This is illustrated in Fig. 2.13.

PROBLEMS

2.1. A first-order reaction is 40% complete at the end of 1 h. What is the value of the rate constant? In how long will the reaction be 80% complete?

2.2. The half-life for the radioactive disintegration of radium is 1590 yr. Calculate the decay constant. In how many years will three-quarters of the radium have undergone decay?

2.3. A substance decomposes at 600 K with a rate constant of 3.72×10^{-5} s^{-1}. Calculate the half-life of the reaction. What fraction will remain undecomposed if the substance is heated for 3 h at 600 K?

2.4. How does the time required for a first-order reaction to go to 99% completion relate to the half-life of the reaction?

2.5. The rate constant for the reaction $H^+ + OH^- \rightarrow H_2O$ is 1.3×10^{11} dm³ mol⁻¹ s⁻¹. Calculate the half-life for the neutralization process if (a) $[H^+] = [OH^-] = 10^{-1}$ M and (b) $[H^+] = [OH^-] = 10^{-4}$ M.

2.6. The isotope ^{90}Sr emits radiation by a first-order process (as is always the case with radioactive decay) and has a half-life of 28.1 yr. When ingested by mammals it becomes incorporated permanently in bone tissue. If 1 μg is absorbed at birth, how much of this isotope remains after (a) 25 yr and (b) 70 yr?

2.7. The reaction $2NO(g) + Cl_2(g) \rightarrow 2NOCl(g)$ is second order in NO and first order in Cl_2. Five moles of nitric oxide and 2 mol of Cl_2 were brought together in a volume of 2 dm³ and the initial rate was 2.4×10^{-3} mol dm⁻³ s⁻¹. What will be the rate when one-half of the chlorine has reacted?

2.8. The following results were obtained by Letort[28] for the rate of decomposition of acetaldehyde corresponding to various degrees of decomposition:

Percent acetaldehyde decomposed	0	10	20	30	40	50
Rate of decomposition/ mmHg min⁻¹	8.53	6.74	5.14	4.31	3.11	2.29

Plot log (rate of decomposition) against log (percent acetaldehyde remaining) and deduce the order with respect to time.

2.9. The isotope $^{32}_{15}P$ emits β radiation and has a half-life of 14.3 d. Calculate the decay constant in s⁻¹. What percentage of the initial activity remains after (a) 10 d, (b) 20 d, and (c) 100 d?

2.10. The following counts per minute were recorded on a counter for the isotope $^{35}_{16}S$ at various times:

Time/days	Counts per minute (after correction)
0	4280
1	4245
2	4212
3	4179
4	4146
5	4113
10	3952
15	3798

Determine the half-life in days and the decay constant in s⁻¹. How many counts per minute would be expected after (a) 60 d and (b) 365 d?

2.11. A drug administered to a patient usually is consumed by a first-order process. Suppose that a drug is administered in equal amounts at regular intervals and that the interval between successive doses is equal to the $(1/n)$-life for the disappearance process (i.e., to the time that it takes for the fraction $1/n$ to be consumed). Prove that the maximum concentration of the drug in the patient's body is equal to n times the concentration produced by an individual dose.

2.12. Prove that for two simultaneous (parallel) reactions,

$$Z \xleftarrow{k_2} A \xrightarrow{k_1} Y$$

$[Y]/[Z] = k_1/k_2$ at all times.

2.13. The dissociation of a weak acid,

$$HA + H_2O \rightleftharpoons H_3O^+ + A^-$$

can be represented as

$$A \underset{k_{-1}}{\overset{k_1}{\rightleftharpoons}} Y + Z$$

The rate constants k_1 and k_{-1} cannot be measured by conventional methods but can be measured by the T-jump technique. Prove that the relaxation time is given by

$$\tau = \frac{1}{k_1 + 2k_{-1}x_e}$$

where x_e is the concentration of the ions (Y and Z) at equilibrium.

2.14. The reaction

$$\textit{cis}\text{-Cr(en)}_2\text{(OH)}_2^+ \rightleftharpoons \textit{trans}\text{-Cr(en)}_2\text{(OH)}_2^+$$

is first order in both directions. At 25°C the equilibrium constant is 0.16 and the rate constant $k_1 = 3.3 \times 10^{-4}$ s^{-1}. In the experiment starting with the pure *cis* form, how long would it take for half the equilibrium amount of the *trans* isomer to be formed?

2.15. A second-order reaction involving reactants both initially present at 0.1 M is 20% complete in 40 min. Calculate (a) the rate constant, (b) the half-life, and (c) the time it would take for the reaction to be 20% complete if the initial concentrations were 0.01 M.

2.16. The rate constant for a reaction at 30°C is exactly twice the value at 20°C. Calculate the activation energy.

2.17. The rate constant for a reaction at 230°C is exactly twice the value at 220°C. Calculate the activation energy.

2.18. Two second-order reactions have identical pre-exponential factors and activation energies differing by 20.0 kJ mol^{-1}. Calculate the ratio of their rate constants at (a) 0°C and (b) 1000°C. Assume the Arrhenius equation to apply; that is, the activation energies are temperature independent.

2.19. The water flea *Daphnia* performs a constant number of heart beats and then dies. On the average, the flea lives twice as long at 15°C as at 25°C. Estimate the activation energy for the reaction that controls the rate of its heart beat.

2.20. A sample of milk kept at 25°C is found to sour 40 times as rapidly as when it is kept at 4°C. Estimate the activation energy for the souring process.

2.21. The gas-phase reaction between nitric oxide and oxygen is third order. The following rate constants have been measured:

T/K	80.0	143.0	228.0	300.0	413.0	564.0
$k \times 10^{-9}$/cm^6 mol^{-2} s^{-1}	41.8	20.2	10.1	7.1	4.0	2.8

The behavior is interpreted in terms of a temperature-dependent pre-exponential factor, the rate equation being of the form

$$k = AT^m e^{-E/RT}$$

where A and m are constants. Assume the activation energy to be zero and determine m to the nearest half-integer.

REFERENCES

1. J. H. van't Hoff, *Etudes de dynamique chimique,* p. 84, Muller, Amsterdam, 1884.
2. M. Letort, Thesis, University of Paris, 1937; *J. Chim. Phys.,* **34,** 206 (1937); *Bull. Soc. Chim. Fr.,* **9,** 1 (1942).

3. C. Capellos and B. H. J. Bielski, *Kinetic Systems,* Wiley-Interscience, New York, 1972.

4. W. Esson, *Philos. Trans.,* **158,** 117 (1867).

5. W. Ostwald, *Lehr. Allg. Chem.,* **2,** pt. 2, 238 (1902).

6. C. Capellos and B. H. J. Bielski, *Kinetic Systems,* p. 32, Wiley-Interscience, New York, 1972.

7. O. A. Hougen and K. M. Watson, *Chemical Process Principles,* pt. 3, p. 834, Wiley, New York, 1947. J. Villermaux, *Génie de la Réaction Chimique. Conception et Fonctionement des Réacteurs,* Technique et Documentation, Paris, 1985.

8. K. G. Denbigh, *Trans. Faraday Soc.,* **40,** 352 (1944); *Discuss. Faraday Soc.,* **2,** 263 (1947).

9. M. Eigen, *Discuss. Faraday Soc.,* **17,** 194 (1954).

10. M. P. Vieille, *Comptes Rendus,* **129,** 1228 (1889).

11. G. Porter, *Proc. R. Soc. London,* **A200,** 284 (1950); R. G. W. Norrish and G. Porter, *Proc. R. Soc. London,* **A210,** 439 (1952).

12. G. Porter and M. R. Topp, *Proc. R. Soc. London,* **A315,** 163 (1970).

13. J. W. Mellor, *Chemical Statics and Dynamics,* Longmans, Green, London, 1904.

14. For a much more detailed account see K. J. Laidler, *J. Chem. Educ.,* **61,** 494 (1984).

15. L. Wilhelmy, *Pogg. Ann.,* **81,** 453, 499 (1850).

16. M. Berthelot, *Ann. Chim. Phys.,* **66**(3), 110 (1862); for recent applications and interpretations of the Berthelot equation see C. M. Hurd, *Phil. Mag. B50,* L29 (1984), *Solid State Phys.,* **18,** 6487 (1985).

17. J. J. Hood, *Philos. Mag.,* **20**(5), 323 (1885).

18. J. H. van't Hoff, *Etudes de dynamique chimique,* Muller, Amsterdam, 1884.

19. S. Arrhenius, *Z. Phys. Chem.,* **4,** 226 (1889); a translation of four pages of this paper which deal with the theory of temperature dependence is included in M. H. Back and K. J. Laidler, *Selected Readings in Chemical Kinetics,* pp. 31–35, Pergamon, Oxford, 1967.

20. D. M. Kooij, *Z. Phys. Chem.,* **12,** 155 (1893).

21. A. V. Harcourt and W. Esson, *Philos. Trans. A,* **186,** 817 (1895).

22. A. V. Harcourt (with a long Appendix by W. Esson), *Philos. Trans. A,* **212,** 187 (1912).

23. J. H. van't Hoff, *Lectures on Theoretical and Physical Chemistry,* pt. 1, *Chemical Dynamics,* Edward Arnold, London, 1898.

24. A. V. Harcourt and W. Esson, *Philos. Trans. A,* **186,** 817 (1895).

25. M. Bodenstein, *Z. Phys. Chem.,* **29,** 295 (1899).

26. W. C. Gardiner, *Acc. Chem. Res.,* **10,** 326 (1977).

27. D. A. Whytock, J. H. Lee, J. V. Michael, W. A. Payne, and L. J. Stief, *J. Chem. Phys.,* **66,** 2690 (1977).

28. M. Letort, Thesis, University of Paris, 1937.

BIBLIOGRAPHY

Solutions of the differential equations for many types of kinetic system are to be found in the following publications:

S. W. Benson, *The Foundations of Chemical Kinetics,* chaps. 2 and 3, McGraw-Hill, New York, 1960.

C. Capellos and B. H. J. Bielski, *Kinetic Systems,* Wiley-Interscience, New York, 1972.

A. A. Frost and R. G. Pearson, *Kinetics and Mechanism,* 2nd ed., chaps. 2 and 8, Wiley, New York, 1961.

D. Margerison, The Treatment of Experimental Data, in C. H. Bamford and C. F. H. Tipper (Eds.), *Comprehensive Chemical Kinetics,* vol. 1, chap. 5, pp. 343–421, Elsevier, Amsterdam, 1969.

Z. G. Szabó, Kinetic Characterization of Complex Reaction Systems, in C. H. Bamford and C. F. H. Tipper (Eds.), *Comprehensive Chemical Kinetics,* vol. 2, chap. 1, pp. 1–80, Elsevier Amsterdam, 1969.

For experimental methods see:

L. Batt, Experimental Methods for the Study of Slow Reactions, in C. H. Bamford and C. F. H. Tipper (Eds.), *Comprehensive Chemical Kinetics,* vol. 1, chap. 1, pp. 1–111, Elsevier, Amsterdam, 1969.

D. N. Hague, Experimental Methods for the Study of Fast Reactions, in C. H. Bamford and C. F. H. Tipper (Eds.), *Comprehensive Chemical Kinetics,* vol. 1, chap. 2, pp. 112–179, Elsevier, Amsterdam, 1969.

Methods for dealing with very rapid reactions are discussed in:

C. F. Bernasconi, *Relaxation Kinetics,* Academic, New York, 1976.

E. F. Caldin, *Fast Reactions in Solution,* Blackwell, Oxford, 1964.

M. Eigen and L. de Maeyer, in A. Weissberger (Ed.), *Techniques of Organic Chemistry,* 2nd ed., vol. 8, pt. 2, pp. 929–941, Interscience, New York, 1963.

D. N. Hague, Experimental Methods for the Study of Fast Reactions, in C. H. Bamford and C. F. H. Tipper (Eds.), *Comprehensive Chemical Kinetics,* vol. 1, chap. 2, pp. 112–179, Elsevier, Amsterdam, 1969.

Shock-tube methods for the study of chemical reactions are reviewed in:

J. N. Bradley, *Shock Waves in Chemistry and Physics,* Wiley, New York, 1962.

A. G. Gaydon and I. R. Hurle, *The Shock-Tube in High Temperature Chemical Physics,* Reinhold, New York, 1963.

E. F. Greene and J. P. Toennies, *Chemical Reactions in Shock Waves,* Academic, New York, 1964.

G. L. Schott and R. W. Getzinger, Shock-Tube Studies of the Hydrogen–Oxygen Reaction System, in B. P. Levitt (Ed.), *Physical Chemistry of Fast Reactions,* vol. 1, chap. 2, pp. 81–160, Plenum, London, 1972.

For accounts of early work on the temperature dependence of rates see:

K. J. Laidler, The Development of the Arrhenius Equation, *J. Chem. Educ.,* **61,** 494 (1984).

S. R. Logan, The Origin and Status of the Arrhenius Equation, *J. Chem. Educ.,* **59,** 279 (1982).

Other aspects of the activation energy are treated in:

J. R. Hulett, Deviations from the Arrhenius Equation, *Q. Rev. Chem. Soc.,* **18,** 277 (1964).

B. Perlmutter-Hayman, The Temperature Dependence of E_a, *Prog. Inorg. Chem.,* **20,** 229 (1976).

For more theoretical discussions of the activation energy see the bibliography at the end of Chapter 3. Statistical procedures for analyzing kinetic data are treated in:

E. S. Swinbourne, *Analysis of Kinetic Data,* Thomas Nelson, London, 1971.

R. de Levie, When, Why and How to use Weighted Least Squares, *J. Chem. Educ.,* **63,** 10 (1986) [this article includes a number of kinetic examples].

Chapter 3

Energy of Activation

The activation energy E_a and the pre-exponential factor A are two quantities of great importance in kinetics. Their interpretation proved to be a matter of difficulty, especially as far as exact treatments are concerned, and is still being actively studied.

The energy of activation can be looked at from several different points of view. van't Hoff's approach [Eqs. (2.65)–(2.70)] was from thermodynamics and can be represented in an approximate way by the schematic diagram shown in Fig. 3.1. The initial and final states differ by the energy $\Delta U°$, the standard internal energy change in the overall reaction. Between the initial and final states the energy usually passes through a maximum, and today we say that the system passes through a *transition state,* which corresponds to the top of the barrier. The molecular species at this maximum are now known as *activated complexes.* For reaction from left to right the barrier height is E_1 and for the reaction from right to left it is E_{-1}. These energies are related by the equation

$$E_1 - E_{-1} = \Delta U° \tag{3.1}$$

This way of looking at the situation is very useful, but from a modern point of view it is somewhat sketchy: nothing is said about zero-point levels or average energies. Later, we will see how this approach can be extended and made more precise (e.g., see Figs. 3.3 and 4.8).

Arrhenius accepted the approach that van't Hoff had suggested, but he added an alternative explanation (See Section 2.9.1). According to Arrhenius, an equilibrium between normal and active molecules shifts in the way predicted by van't Hoff's equation (2.65), and the rate is proportional to the concentration of active molecules. This point of view leads naturally to the idea of considering the statistical distribution of energies among the reacting molecules. As early as 1867 Pfaundler[1] had presented a qualitative discussion of reaction rates in terms of the recently formulated statistical

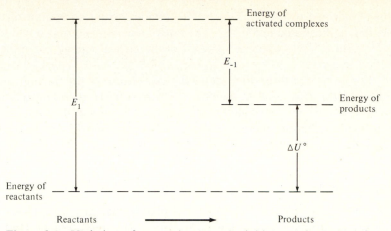

Figure 3.1 Variation of energy between the initial and final states for a reaction.

treatments of Maxwell and Boltzmann. In his 1884 book,[2] van't Hoff acknowledged having received inspiration from Pfaundler's work, which otherwise has been largely forgotten. Arrhenius himself made no explicit reference to the statistical approach, although his point of view was entirely consistent with it.

3.1 STATISTICAL DISTRIBUTION OF MOLECULAR ENERGIES

3.1.1 Simple Statistical Expressions

Some important statistical expressions, based on the distributions originally developed by Clerk Maxwell and Boltzmann, are listed in Table 3.1 and are discussed further in the appendix to this chapter. The simplest of these expressions, relating to the two-dimensional case, is that the fraction of molecules having energy in excess of a particular

TABLE 3.1 SOME STATISTICAL EXPRESSIONS

Number of degrees of freedom	Fraction of molecules having energy between ε and $\varepsilon + d\varepsilon$	Fraction of molecules having energy in excess of ε^*
1	$\dfrac{1}{\pi^{1/2}\varepsilon}\left(\dfrac{\varepsilon}{kT}\right)^{1/2}e^{-\varepsilon/kT}\,d\varepsilon$	See Appendix, Eq. (3.25)
2	$\dfrac{1}{\varepsilon}\left(\dfrac{\varepsilon}{kT}\right)e^{-\varepsilon/kT}\,d\varepsilon$	$e^{-\varepsilon^*/kT}$
3	$\dfrac{1}{\pi^{1/2}\varepsilon}\left(\dfrac{\varepsilon}{kT}\right)^{3/2}e^{-\varepsilon/kT}\,d\varepsilon$	See Appendix, Eq. (3.35)
n	$\dfrac{1}{(\frac{1}{2}n-1)!}\left(\dfrac{\varepsilon}{kT}\right)^{n/2-1}\dfrac{1}{kT}e^{-\varepsilon/kT}\,d\varepsilon$	$\dfrac{1}{(\frac{1}{2}n-1)!}\left(\dfrac{\varepsilon^*}{kT}\right)^{n/2-1}e^{-\varepsilon^*/kT}$ (if $\varepsilon^* \gg nkT$)

value ε^* is equal to $e^{-\varepsilon^*/kT}$. This result provides an interpretation of the Arrhenius equation, as illustrated in Fig. 3.2. The rate of reaction is proportional to this fraction $e^{-\varepsilon^*/kT}$ or to $e^{-E^*/RT}$, where E^* ($=L\varepsilon^*$) is the critical energy per mole required for reaction to occur.

For motion in three dimensions the expression for the fraction cannot be given in closed form; it is usual to employ $e^{-\varepsilon^*/kT}$ as a fairly good approximation for this case also. For energy distributed in a number of degrees of freedom, however, as in a unimolecular reaction, the expression $e^{-\varepsilon^*/kT}$ is in considerable error; this matter is discussed in the appendix to this chapter and in Section 5.3.

3.1.2 Tolman's Theorem

In 1920 Tolman[3] derived an important equation that gave significance to the experimental activation energy E_a defined by Eq. (2.81). According to Tolman's theorem the activation energy per molecule, which we write as ε_a, is given by

$$\varepsilon_a = \text{(average energy of molecules undergoing reaction)}$$
$$- \text{(average energy of colliding molecules)} + \tfrac{1}{2}\mathbf{k}T \qquad (3.2)$$

It was later shown by Fowler and Guggenheim[4] that the following very similar relationship is exact, whereas that of Tolman involves an approximation:

$$\varepsilon_a = \text{(average energy of molecules undergoing reaction)}$$
$$- \text{(average energy of reactant molecules)} \qquad (3.3)$$

The difference between the two equations arises from the fact that the average energy

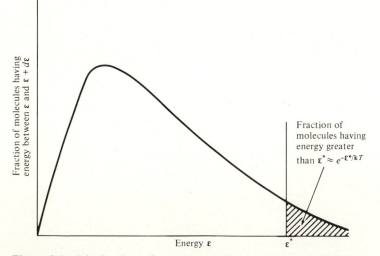

Figure 3.2 Distribution of energy according to the Maxwell–Boltzmann treatments. For molecules moving in two dimensions the fraction having energy in excess of ε^*, or of E^* per mole, is given by $e^{-\varepsilon^*/kT} = e^{-E^*/RT}$. This is also approximately correct for three dimensions.

of colliding molecules is greater by about $\frac{1}{2}kT$ than the average energy of all the reactant molecules. This is because molecules with higher relative energy are more likely to collide than molecules with less energy; the average translational energy of a reactant molecule is $\frac{3}{2}kT$, whereas the average translational energy of a molecule actually undergoing collision turns out to be about $2kT$.

The results expressed by Eqs. (3.2) and (3.3) are illustrated in Fig. 3.3. We will not give a rigorous proof of the relationship, which is complicated and has been given in a number of publications.[5] A simple proof, which takes into account translational energy only, is as follows. In Section 4.1 we discuss a collision theory equation, Eq. (4.23), which relates the rate constant to the translational energy ε_t:

$$k = \left[\frac{8}{\mu\pi(\mathbf{k}T)^3}\right]^{1/2} \int_0^\infty \varepsilon_t \sigma(\varepsilon_t) e^{-\varepsilon_t/\mathbf{k}T}\, d\varepsilon_t \tag{3.4}$$

Here μ is the reduced mass of the system and $\sigma(\varepsilon_t)$ is the reaction cross section, to which the reaction probability is proportional. These quantities are defined and further considered in Section 4.1. It follows from Eq. (2.82) that the activation energy ε_a per molecule is defined by

$$\varepsilon_a = -\mathbf{k}\frac{d\ln k}{d(1/T)} = \mathbf{k}T^2\frac{d\ln k}{dT} \tag{3.5}$$

Introduction of Eq. (3.4) into this expression gives

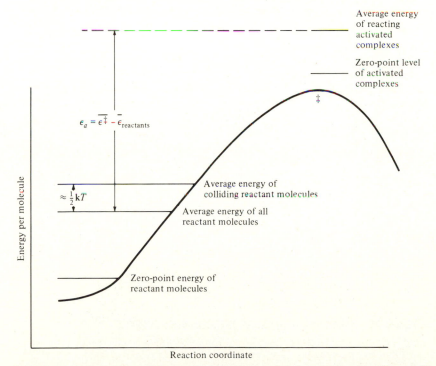

Figure 3.3 Illustration of the exact Fowler–Guggenheim relationship [Eq. (3.3)] and the approximate Tolman relationship [Eq. (3.2)].

$$\varepsilon_a = \frac{\int_0^\infty \varepsilon_t^2 \sigma(\varepsilon_t) e^{-\varepsilon_t/kT} \, d\varepsilon_t}{\int_0^\infty \varepsilon_t \sigma(\varepsilon_t) e^{-\varepsilon_t/kT} \, d\varepsilon_t} - \frac{3}{2} kT \qquad (3.6)$$

The final term in this expression, $\frac{3}{2}kT$, is the average translational energy per molecule; the first term, the ratio of the integrals, is the average energy for those collisions in which reaction takes place.

This relationship is demonstrated by some results calculated by Menzinger and Wolfgang[6] and illustrated in Fig. 3.4. The diagram relates to a temperature of 300 K and to a threshold energy E_0 of 33.5 kJ mol^{-1}. The lefthand curve is the Maxwell–Boltzmann distribution curve at 300 K, and the average energy ($\bar{E}_{reactants}$) is indicated. The cross section $\sigma(E)$ was assumed in these calculations to be proportional to $(E - E_0)^2$ when $E > E_0$ and to be zero when $E < E_0$. A plot of $\sigma(E)$ versus E, labeled "excitation function," is shown in the diagram. The curve designated "reaction function" represents the product

$$f(E)\sigma(E)$$

where $f(E)$ is the expression for the distribution of energies. This reaction function represents the energy distribution for those collisions that lead to reaction. The average energy $\overline{E^\ddagger}$ corresponding to this function is indicated on the diagram.

For the particular excitation function chosen, the experimental activation energy E_a is equal to 38 kJ mol^{-1} at 300 K and is substantially higher than the threshold energy E_0 (=33.5 kJ mol^{-1}). The activation energy E_a varies significantly with temperature; for example, at 900 K it is 49.4 kJ mol^{-1}.

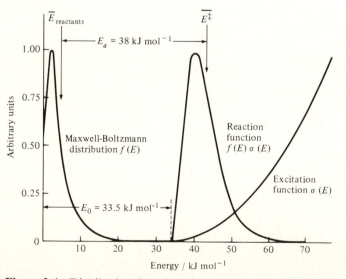

Figure 3.4 Distribution functions for a reaction at 300 K. The reaction is assumed to be sufficiently slow that equilibrium conditions are maintained. The excitation function is zero when $E < E_0$ and is proportional to $(E - E_0)^2$ when $E > E_0$.

3.2 POTENTIAL-ENERGY SURFACES

A very convenient way of considering the activation energy is to make use of a potential-energy surface, in which potential energy is plotted against bond distances and angles. The concept of a potential-energy surface springs from a suggestion made in 1914 by the French physical chemist René Marcelin (1885–1914).[7] He pointed out that the state of a molecular system can be represented in terms of Lagrange generalized coordinates: $q_1, q_2, q_3, \ldots, q_n$ for distance and $p_1, p_2, p_3, \ldots, p_n$ for momentum. He then regarded the course of reaction as the motion of a point in $2n$-dimensional phase space. He also made a statistical approach to this motion, as will be mentioned briefly in Section 4.3.

The procedure commonly used today is to plot potential energy against certain bond lengths. If the reaction occurs between two atoms only the distance between the nuclei is involved, and one can construct a potential-energy curve in which energy is plotted against this distance; this is a two-dimensional diagram. If three atoms are involved, as in a reaction of the type

$$A + B—C \rightarrow A—B + C$$

the system $A \cdots B \cdots C$ requires three parameters to describe it. These might be the A—B, A—C, and B—C distances or two distances and an angle. To plot energy against these three distances, a four-dimensional diagram would be necessary. Since such a diagram cannot be constructed or visualized, it is necessary to use a series of diagrams, in each of which one parameter has been fixed at a particular value. For example, the A—B—C angle might be fixed at 180° and a three-dimensional model constructed. Other three-dimensional surfaces could be constructed for other angles, and all these would be sections through the four-dimensional surface.

When this procedure of fixing the A—B—C angle is adopted, the resulting potential-energy surface has the form shown in Figs. 3.5, 3.6, and 3.7. On the left-hand face of Fig. 3.5 the B—C distance may be considered sufficiently great that one is dealing simply with the diatomic molecule A—B; the curve shown on that face is thus the potential-energy curve for the molecule A—B, the point R corresponding to the dissociation of the molecule and the point Q to the classical ground state. Similarly, on the right-hand face of the diagram there is a curve for the diatomic molecule B—C, the distance A—B now being sufficiently great that A has no effect on the energy of the system. The classical ground state of B—C is represented by the point P.

The course of the reaction is represented by a movement on the potential-energy surface from P to Q, and the system tends to travel along paths where the energy is not too high. The calculations have shown that the general form of a potential-energy surface is as represented by the perspective drawing shown as Fig. 3.6 and by the contour diagram shown as Fig. 3.7. The essential feature of such surfaces is that running in from the points P and Q there are two valleys, which meet at a *col* or *saddle point*. For the system to pass from P to Q, it travels up the first valley (the reactant valley) and passes over the col into the product valley. The paths of steepest descent from the col into the two valleys constitute a path that is known as the *minimum-energy path* (MEP). The actual *reaction paths,* or *trajectories,* followed by individual reaction

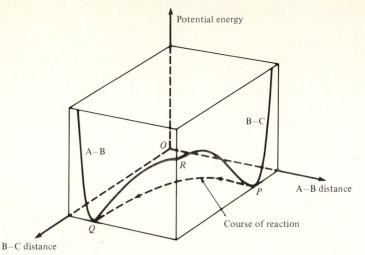

Figure 3.5 Variation of potential energy with the A—B and B—C distances, for the A···B···C system in which the A—B—C angle has been fixed.

systems are not exactly the same as this minimum-energy path and depend on the energy states and other features of the colliding molecules (for details see Chapter 12).

A section through the minimum-energy path, known as a *potential-energy profile,* is shown in Fig. 3.8. The maximum point in this profile, at the col in the potential-energy surface, is a point of particular significance. It is not only a position of maximum energy along the minimum-energy path, but it is a position of minimum energy with respect to motions at right angles to the path. Systems represented by a small region round this point are known as *activated complexes,* and their state is referred to as

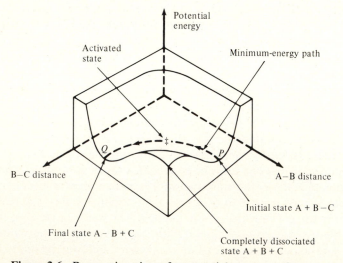

Figure 3.6 Perspective view of a potential-energy surface.

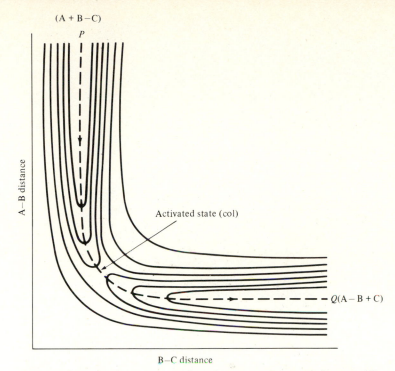

Figure 3.7   Potential-energy surface shown as a contour diagram. The dashed line shows the minimum-energy path (MEP), corresponding to the paths of steepest descent from the col into the two valleys.

the *transition state* for the reaction. Eyring introduced the symbol ‡ to represent activated complexes, and this is now used universally.§

Potential-energy surfaces for reactions involving more than three atoms are hypersurfaces in multidimensional space. Since we live in a three-dimensional world, these surfaces can be constructed only by holding constant all but two variables. Alternatively, as an approximation we may construct surfaces in which groups of atoms are treated as single particles; the reaction $CH_3 + H_2 \rightarrow CH_4 + H$, for example, can be treated as an A + BC system.

Most of the work on potential-energy surfaces has been done on the simplest of all bimolecular systems, the reaction

$$H^\alpha + H^\beta\!-\!H^\gamma \rightarrow H^\alpha\!-\!H^\beta + H^\gamma$$

where the three H atoms have been labeled with α, β, and γ. The rate of this reaction can be measured by measuring the rate of conversion of *ortho*-hydrogen to *para*-hydrogen. Also, the rates of various isotopically substituted reactions, such as

$$D + H_2 \rightarrow DH + H$$

§ In his handwritten manuscript Eyring had used an asterisk, but since the typewriter lacked this symbol his secretary typed a + sign with a − below. The printer set this as ‡, the Lorraine cross. Eyring was later amused to learn that ‡ resembles the Japanese word for "crazy."

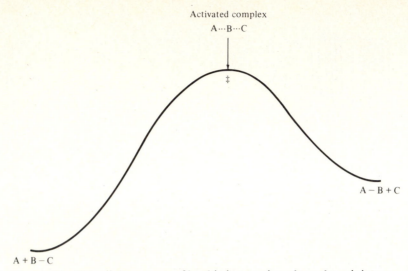

Activated complex

A···B···C

A + B − C

A − B + C

Figure 3.8   Potential-energy profile; this is a section along the minimum-energy path in a potential-energy surface.

can be measured (Chapter 11). In these reactions the most favorable line of approach for the H atom is along the axis of the H_2 molecule. Many of the calculations of potential-energy surfaces for this system therefore have been made for linear H · · · H · · · H complexes, and the surfaces have been shown as plots of energy against the H^α—H^β distance and the H^β—H^γ distance. Some calculations also have been made for nonlinear systems.

In 1928 Fritz London[8] discussed chemical reactions from the standpoint of molecular structure and suggested that quantum-mechanical methods could be used to calculate the energies of reaction intermediates such as the activated complexes themselves. A year previously he and Heitler[9] had given a simple quantum-mechanical treatment of the hydrogen molecule. According to a very approximate version of their treatment, the allowed energies for the hydrogen molecule are the sum and difference of two integrals which represent energies; that is, the energy may be expressed very approximately as

$$E = A \pm \alpha \tag{3.7}$$

The integral A is the *coulombic energy* and is roughly equivalent to the classical energy of the system. The integral α, the *exchange energy,* has no counterpart in classical theory but appears in quantum mechanics as a result of the fact that the electrons are indistinguishable and hence cannot be regarded as localized with respect to any particular nucleus; they can therefore be *exchanged*. At ordinary internuclear distances the energies A and α are negative in value, and the energy E in Eq. (3.7) is the energy relative to the separated atoms. Therefore, the lowest energy, corresponding to the most stable state, is obtained when the positive sign is taken.

For a triatomic system the energies corresponding to the diatomic pairs if they were isolated are indicated in Fig. 3.9. London[8] developed a very approximate

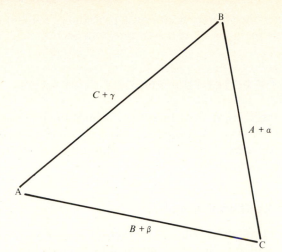

Figure 3.9 General triatomic complex ABC. The energy of the diatomic molecule B—C, with A removed, is $A + \alpha$, where A is the coulombic energy and α is the exchange energy. The energies of the other two diatomic molecules are as indicated. According to London's treatment, Eq. (3.8) gives the energy of the triatomic molecule.

quantum-mechanical treatment of the triatomic species and arrived at the following expression for the energy:

$$E = A + B + C \pm \left\{ \tfrac{1}{2}[(\alpha - \beta)^2 + (\beta - \gamma)^2 + (\gamma - \alpha)^2] \right\}^{1/2} \qquad (3.8)$$

This equation, known as the *London equation,* was used in 1931 by Eyring and Polanyi[10] in their pioneering semiempirical calculation of a potential-energy surface for the H$\cdots$H$\cdots$H system and has been used by a number of other workers for a variety of reaction systems. It should be mentioned that London§ himself was uneasy about the liberties taken with his equation, which he recognized as being very approximate.

In the original calculations of Eyring and Polanyi the quantities in the London equation were obtained empirically. Such calculations, which involve some quantum-mechanical theory (the approximate London equation) and some empiricism, are today classed as *semiempirical.* Before dealing with these and other semiempirical treatments (in Section 3.4), we will deal in Section 3.3 with calculations based solely on quantum mechanics. In these calculations, called *ab initio calculations,* no experimental data are employed other than the electronic and nuclear charges, the nuclear masses, and the value of the Planck constant. Besides these procedures, activation

§ For example, J. O. Hirschfelder [*J. Chem. Educ.,* **43**, 457 (1966)] says "London told me that he was appalled at the way chemists mangled his formula and still attached his name to the semiempirical results."

energies also have been obtained by purely empirical procedures, with no use of quantum mechanics, and these methods are reviewed briefly in Section 3.5.

3.3 AB INITIO CALCULATIONS OF POTENTIAL-ENERGY SURFACES

Two ab initio quantum-mechanical procedures to calculate potential-energy surfaces are discussed in this section. One is based on the London equation, which is a very approximate equation and incapable of giving accurate results. The second procedure involves the use of the variation method.

3.3.1 Treatments Based on the London Equation

The coulombic integrals A, B, and C and the exchange integrals α, β, and γ, which appear in the London equation (3.8), were computed by Sugiura[11] for the hydrogen molecule as functions of interatomic distance. Eyring and Polanyi,[12] in their pioneering calculation of a potential-energy surface for the $H + H_2$ reaction, considered introducing Sugiura's values into the London equation but decided that the errors were too great. Instead, they used a semiempirical procedure. Much more recently, J. C. Polanyi and coworkers[13] constructed a surface using Sugiura's values and found it to have a substantial energy well or basin of depth about 23 kJ mol^{-1}, corresponding to a symmetrical $H \cdots H \cdots H$ configuration. The heights of the two barriers on each side of the basin are 37 kJ mol^{-1} with respect to the energy of $H + H_2$. This value is close to the experimental barrier height. However, this agreement must be regarded as fortuitous in view of the very serious approximations involved in the London equation. Moreover, more reliable calculations make it unlikely that there is really a deep basin in the potential-energy surface for $H + H_2$.

3.3.2 Variational Calculations

Much more reliable calculations of molecular energies are made on the basis of the variational principle, and most of the more recent work has been done in this way.

The Hamiltonian operator $\hat{\mathscr{H}}$ for any system can be written down readily, but the difficulty lies in the solution of the differential equation

$$\hat{\mathscr{H}}\psi = E\psi \tag{3.9}$$

The point of the variational principle is that it can be shown that, if one selects any normalized function ϕ, the energy given by the integral

$$\int_{\text{all space}} \phi^* \hat{\mathscr{H}} \phi \, d\tau$$

(where ϕ^* is the complex conjugate of ϕ) cannot be below the true energy E for the ground state of the system. Thus, we can calculate the integral for a variety of functions ϕ and know that the lowest energy obtained will be the closest to the truth. More systematically, we can express ϕ as a function of one or more variables and calculate the integral as a function of these variables. The energy (i.e., the integral) can then be minimized with respect to the variables. The main difficulty with the method is that there is no simple way of deciding what type of function ϕ will be most satisfactory for a given problem.

Table 3.2 summarizes the results of some of the variational calculations that have been made on the $H + H_2$ system.[14] In this table two calculated activation energies are given, E_{cb} and E'_{cb}, and both are the classical energy§ at the col relative to the classical energy for $H + H_2$: therefore, both are classical energy barriers. The energy E_{cb} is the calculated energy barrier relative to the *experimental* value for $H + H_2$, while E'_{cb} is the barrier height relative to that for $H + H_2$ calculated using the same quantum-mechanical procedure. The calculated E'_{cb} values compare more favorably with experiment because there is some cancellation of errors when the same computational procedures are applied to H_3 as to $H + H_2$. However, the calculated E_{cb} provides a more demanding test of the reliability of the method.

The procedure for estimating the experimental classical barrier height from the kinetic results is not entirely straightforward. The results give the experimental activation energy E_a, as defined by Eq. (2.81), and the relationship between E_a and the classical barrier height depends on the theory used to interpret the results. It appears, however, that the classical barrier height must be about 40 kJ mol^{-1}.

The first ab initio calculations on the $H + H_2$ system, by Hirschfelder, Eyring, and Rosen, and many of the subsequent calculations, were done with simple calculators and were very laborious. As seen from Table 3.2, these early calculations did not yield barrier heights that were in good agreement with experiment. This early work, however, was of great value in leading to an understanding of the whole problem of activation energies. The calculations done in the 1950s and early 1960s made use of computers, and it was then possible to construct much more complicated trial wavefunctions than could be used in the earlier work. Better agreement with experiment could then be obtained, but the errors were still sufficiently large that the quantum-mechanical methods could not predict reliable activation energies even for relatively simple reactions.

In more recent years, computers have improved enormously in speed and capacity, allowing the calculations to enter an entirely new era. Trial wavefunctions having a considerable number of terms can be employed, and many parameters can be varied to obtain much more reliable energy values. In 1973 Liu published the results of calculations on the $H + H_2$ reaction, with the complex $H \cdots H \cdots H$ assumed to be linear, and he obtained a classical barrier height of 41.0 kJ mol^{-1}, which he estimated to be within 2–3 kJ mol^{-1} of the exact value. In 1978, Siegbahn and Liu published results for a complex H_3 that was not constrained to be linear; however, this did not lead to any lowering of the calculated energy. In 1984 Liu reported even more accurate calculations which yielded 40.2 ± 0.3 kJ mol^{-1}. For this reaction the latest calculations give values that are at least as reliable as the experimental ones.

Truhlar and Horowitz[15] made least-square fits to the values calculated by Liu and Siegbahn, and this enabled them to express the results in a functional form. They

---

§ In this context the word "classical" means that the zero-point energy is ignored. The classical energy for H_2, for example, is the energy at the minimum of the potential-energy curve.

TABLE 3.2 RESULTS OF SOME VARIATIONAL ACTIVATION
BARRIER CALCULATIONS

Year	Workers	Reference	E_{cb}/kJ mol^{-1}	E'_{cb}/kJ mol^{-1}	Remarks
1936	Hirschfelder, Eyring, and Rosen	1	236	80	Homopolar wavefunctions (Heitler–London)
			205	57	Homopolar and polar wavefunctions
			222	128	Homopolar wavefunctions; Z_{eff} varied[a]
			177	105	Homopolar and polar wavefunctions; Z_{eff} varied
1954	Barker and Eyring	2	155	84	Homopolar wavefunctions; two Z_{eff} varied
1957	Ransil	3	203	56	Molecular orbitals (MO) with configuration interaction
1958	Kimball and Trulio	4	155	83	MO composed of five atomic orbitals; configuration interaction; Z_{eff} varied
1959	Griffing, Jackson, and Ransil	5	172	100	MO composed of three atomic orbitals; configuration interaction; Z_{eff} varied
1959	Boys and Shavitt	6	122	64	MO composed of six atomic orbitals; configuration interaction; Z_{eff} varied
1965	Edmiston and Krauss	7	74	60	MO composed of 27 atomic orbitals; configuration interaction
1967	Conroy and Bruner	8	32	32	MO composed of six atomic orbitals; configuration interaction; extrapolation
1968	Shavitt, Stevens, Minn, and Karplus	9	48	46	MO composed of 15 atomic orbitals; configuration interaction
1973	Liu	10	41.0	41.0	Extensive configuration interaction
1978	Siegbahn and Liu	11	41.0	41.0	
1984	Liu	12	40.1	40.1	
1984	Mentch and Anderson	13	42.7	42.7	Random-walk importance sampling of wavefunctions
1984	Ceperlay and Alder	14	40.4	40.4	

[a] Z_{eff} is the effective nuclear charge used in generating the wavefunctions.

[1] J. O. Hirschfelder, H. Eyring, and N. Rosen, *J. Chem. Phys.,* **4,** 121 (1936).

[2] R. S. Barker and H. Eyring, *J. Chem. Phys.,* **22,** 1182 (1954).

[3] B. J. Ransil, *J. Chem. Phys.,* **26,** 971 (1957).

[4] G. E. Kimball and J. E. Trulio, *J. Chem. Phys.,* **28,** 493 (1958).

[5] V. Griffing, J. L. Jackson, and B. J. Ransil, *J. Chem. Phys.* **30,** 1066 (1959).

[6] S. F. Boys and I. Shavitt, Univ. Wisc. Naval Res. Lab. Tech. Rep., Wis-AF-13 (1959).

[7] C. Edmiston and M. Krauss, *J. Chem. Phys.,* **92,** 1119 (1965).

[8] H. Conroy and B. L. Bruner, *J. Chem. Phys.,* **47,** 971 (1967).

[9] I. Shavitt, R. M. Stevens, F. L. Minn, and M. Karplus, *J. Chem. Phys.,* **48,** 6 (1968).

[10] B. Liu, *J. Chem. Phys.,* **58,** 1925 (1973).

[11] P. Siegbahn and B. Liu, *J. Chem. Phys.,* **68,** 2457 (1978).

[12] B. Liu, *J. Chem. Phys.,* **80,** 581 (1984).

[13] F. Mentch and J. B. Anderson, *J. Chem. Phys.,* **80,** 2675 (1984).

[14] D. M. Ceperlay and B. J. Alder, *J. Chem. Phys.,* **81,** 5833 (1984).

presented potential-energy surfaces for H—Ĥ—H angles of 180°, 150°, 120°, 90°, and 60°, and Fig. 3.10 shows their diagrams for 180° and 90°. It is to be noted that whereas the E_{cb} value for 180° is 9.8 kcal mol^{-1} (41 kJ mol^{-1}), that for 90° is much higher, about 30 kcal mol^{-1} or 125 kJ mol^{-1}.

During recent years a number of activation-energy calculations have also been made on more complex reaction systems. The difficulties are considerably greater when more electrons are involved, but in spite of this it has proved possible to estimate energy barriers that in some cases can be relied on to predict which of two alternative processes involves a lower barrier and is therefore favored. Table 3.3 lists just a few of the hundreds of reactions that have been treated in this way.

3.4 SEMIEMPIRICAL CALCULATIONS OF POTENTIAL-ENERGY SURFACES

The calculations of potential-energy surfaces dealt with in the last section are purely quantum-mechanical ones, since they do not use any of the quantities that they are designed to interpret, such as bond-dissociation energies. No adjustments are made in the ab initio calculations with a view to obtaining a result that is closer to the experimental one.

By contrast, the semiempirical treatments, although based on quantum-mechanical theory, make use of experimental results of the kind they are interpreting, and adjustments are made to obtain more satisfactory results. The extent to which

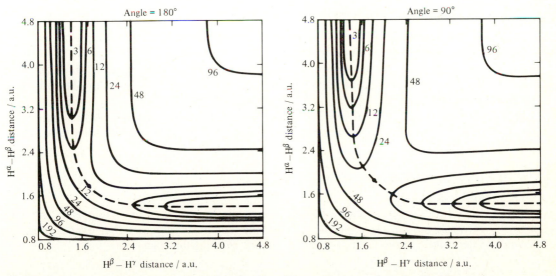

Figure 3.10 Potential-energy surfaces for the H + H$_2$ reaction, as given by Truhlar and Horowitz[15] on the basis of the ab initio calculations of Siegbahn and Liu. The numbers on the contours show the values of the energies in kcal mol^{-1} (1 kcal = 4.184 kJ); the unit for the bond distances is the atomic unit, or bohr (1 bohr = 52.92 pm).

TABLE 3.3   AB INITIO ENERGY BARRIER CALCULATIONS

Reaction	Reference
$H_2 + D_2 \rightarrow 2HD$	C. Woodrow Wilson and W. A. Goddard, *J. Chem. Phys.,* **56,** 5913 (1972)
HNCO isomerization	D. Poppinger, L. Radom, and S. A. Pople, *J. Am. Chem. Soc.,* **99,** 7806 (1977); D. Poppinger and L. Radom, *J. Am. Chem. Soc.,* **100,** 3674 (1978)
$O_3 \rightarrow O_2 + O$	J. S. Wright, S. K. Shih, and R. J. Buenker, *Chem. Phys. Lett.,* **75,** 513 (1980)
$CH_3NH_2^+ \rightarrow CH_2NH_3^+$	W. J. Bouma, J. M. Dawes, and L. Radom, *Org. Mass. Spectrosc,* **18,** 12 (1983)
$C_2H_3O^+ \rightarrow CH_3CO^+$	R. H. Nobes, W. J. Bouma, and L. Radom, *J. Am. Chem. Soc.,* **105,** 309 (1983)
$CH_3^+ + NH_3, CH_3^+ + H_2O$, etc.	R. H. Nobes and L. Radom, *Chem. Phys.,* **74,** 163 (1983)
$F + H_2 \rightarrow HF + H$	J. T. Muckerman, in *Theoretical Chemistry-Advances and Perspectives* (Ed. H. Eyring and D. Henderson), Academic Press, New York, 1981, Vol. 6A, pp. 1–77; for a discussion see D. M. Neumark, A. M. Wodtke, G. N. Robinson, C. C. Hayden, and Y. T. Lee, *J. Chem. Phys.,* **82,** 3045 (1985)
$H + CH_3 \rightarrow CH_4$	R. J. Duchovic, W. L. Hase, and H. B. Schlegel, *J. Phys. Chem.,* **88,** 1339 (1984)

this is done varies to a considerable degree; sometimes the adjustment is so extensive that the prefix "semi" seems to be an understatement.

3.4.1 London–Eyring–Polanyi Method

Potential-energy surfaces of the type calculated by Eyring and Polanyi[16] on the basis of the London equation (3.8) are known as London–Eyring–Polanyi (LEP) surfaces. The assumption is made that for a diatomic molecule the coulombic and exchange energies A and α are constant fractions of the total energy for all internuclear distances. The total energy E can be obtained readily as a function of internuclear distance r from analysis of spectroscopic data, and expressions such as that of Morse have been of great value in relating the potential energy to the distance. The Morse equation is

$$E = D(e^{-2\beta(r-r_0)} - 2e^{-\beta(r-r_0)}) \tag{3.10}$$

where r_0 is the equilibrium internuclear distance, D is the classical dissociation energy, and β is a constant.

From an inspection of Sugiura's calculations of the coulombic and exchange integrals for the H_2 molecule, Eyring and Polanyi concluded that over a range of interatomic distances (in particular, for $r > 80$ pm) the fraction

$$\rho = \frac{A}{A + \alpha} \tag{3.11}$$

is roughly constant at 10–15%. Therefore, for any triatomic configuration (see Fig. 3.9), it is possible to evaluate for each pair of atoms the coulombic and exchange energies, on the basis of the spectroscopic value for the total energy. The six energies A, B, C, α, β, and γ are thus known for the system, and insertion in the London equation (3.8) gives the required energy for the triatomic species.

Later workers tried taking different percentages of coulombic and exchange energies. Sometimes the procedure was to use the percentage as an empirical parameter to obtain satisfactory agreement for one particular reaction.

This method of Eyring and Polanyi led to an interpretation of experimental activation energies for very simple systems. Now that better procedures are available, however, the method is not much used. An unsatisfactory feature appears for the $H + H_2$ system: when the parameter ρ is adjusted to give a barrier height consistent with experiment, a substantial basin in the surface is found. Similar features are found with other reactions for which there is no reason to believe that a basin really exists.

3.4.2 Sato Method

Potential-energy basins appear to arise from the assumption of constant coulombic and exchange fractions. With a view to eliminating basins, Sato[17] proposed an alternative method in which ρ is treated as a function of the internuclear separation r. He obtained the dependence of ρ on r on the basis of the repulsive $^3\Sigma$ curve, the slope of which had been obtained theoretically (but, as we now know, not very accurately). According to the simplified Heitler–London treatment, the energy for the repulsive state of B—C is given by

$$E_r = A - \alpha \tag{3.12}$$

To obtain an analytic expression for E_r, Sato modified the Morse equation (3.10) by changing the sign between the two exponential terms from minus to plus. He also divided the expression by 2, since he found that this improved the agreement with the theoretical repulsive curve available to him. The resulting equation is

$$E_r = \frac{D}{2} \left(e^{-2\beta(r-r_0)} + 2e^{-\beta(r-r_0)} \right) \tag{3.13}$$

[compare Eq. (3.10)]. The energy values given by this equation can then be set equal to $A - \alpha$. Those for the ground state are set equal to $A + \alpha$ so that A and α can be obtained from the two equations. In this way, A and α values are calculated at a series of internuclear separations without any assumption of a constant ratio. Instead of basing calculations for the triatomic system A—B—C on the original London equation, Sato introduced an overlap integral S and used the form

$$E = \frac{1}{1 + S^2} \left(A + B + C \pm \left\{ \frac{1}{2} \left[(\alpha - \beta)^2 + (\beta - \gamma)^2 + (\gamma - \alpha)^2 \right] \right\}^{1/2} \right) \tag{3.14}$$

He used various values of S in calculations for different systems.

The Sato method is just as arbitrary and empirical as the LEP method, and in a comparison of the methods Weston[18] concluded that there is not much to choose between them. The Sato method is preferable to the LEP method in that it usually leads to surfaces free of basins. It is inferior, however, in giving barriers that are usually too thin. A consequence is that calculations for the Sato surface lead to considerably more tunneling than observed experimentally (see Section 4.8.3). This argument, however, is not a completely convincing one, since there is uncertainty in the theoretical treatment of tunneling itself (see Section 4.9.2).

3.4.3 Modified LEP Methods

A few methods have been developed that are basically of the LEP type but which avoid some of the approximations inherent in the London equation. They are essentially LEP treatments in which some of the integrals are adjusted on the basis of empirical evidence.

A useful calculation of this kind on the $H + H_2$ reaction was made by Porter and Karplus[19] who calculated some of the energy contributions theoretically and others semiempirically. As in the London equation, they expressed the coulombic bonding in the triatomic complex as the sum of the three contributions for the diatomic pairs. The exchange energies, however, were calculated in a less direct manner. In this way potential-energy surfaces were obtained for a variety of $H\cdots\hat{H}\cdots H$ angles. There are no basins and the classical barrier height is 35.9 kJ mol^{-1}: this is in very satisfactory agreement with experiment.

Various other extensions of the LEP method also have been proposed. For example, J. C. Polanyi and coworkers[20] have suggested an "extended LEPS method," which involves using a different Sato parameter for each kind of bond. This procedure has proved to be particularly useful.

3.4.4 Bond-Energy–Bond-Order (BEBO) Method

Johnston and coworkers[21] developed a bond-energy–bond-order (BEBO) treatment which can give activation energies in reasonable agreement with experiment. It has no adjustable parameters but involves a considerable amount of empiricism. However, it seems more satisfactory to classify the method as semiempirical rather than empirical since the empirical relationships it employs lie outside the field of kinetics.

One of the empirical relationships it uses is an equation proposed by Pauling to relate bond length r and bond order n:

$$r = r_s - 0.26 \ln n \tag{3.15}$$

where r_s is the single-bond length for a bond connecting two particular atoms. The bond energy D can be related to the single-bond energy D_s through the relationship

$$D = D_s n^p \tag{3.16}$$

where p is a constant. Johnston and Parr calculated the exponent from the equilibrium internuclear separation and bond energy in a noble-gas diatomic cluster ($n = 0$).

Johnston and coworkers postulated that for hydrogen-atom transfer the path of lowest energy from reactants to products is defined by

$$n_1 + n_2 = 1 \tag{3.17}$$

where n_1 and n_2 are the orders of the bonds being broken and formed. Through Eq. (3.15) this controls the changes in bond lengths, and through Eq. (3.16) the changes in energy of the A—B and B—C bonds. The assumption is that the bonds can be treated separately. A correction is made for A—C repulsion with the use of Sato's modified Morse function for the repulsive state [Eq. (3.13)].

Some idea of the usefulness of the method is given by the examples in Table 3.4. In some cases there are fairly serious discrepancies between theory and experiment, and a number of extended and improved procedures have been suggested.[22]

3.5 EMPIRICAL TREATMENTS OF ACTIVATION ENERGY

A number of empirical relationships have been proposed which relate activation energy to other quantities, such as heats of reaction. Such relationships are useful in allowing rough estimates of activation energies to be made quite easily. In addition, a few empirical procedures have been suggested for constructing complete potential-energy surfaces.

Evans and Polanyi[23] proposed a linear relationship between activation energies E_a and enthalpies of reaction, ΔH (or heats evolved, $Q = -\Delta H$) and discussed it in terms of potential-energy profiles. Their equation can be written as

$$E_a = \alpha \, \Delta H + c = -\alpha Q + c \qquad (3.18)$$

TABLE 3.4   ACTIVATION ENERGIES ESTIMATED BY THE BEBO METHOD

Reaction	Activation energy[a]/kJ mol^{-1}	
	Experimental	Estimated by BEBO method
$H + H_2 \rightarrow H_2 + H$	32	42
$H + CH_4 \rightarrow H_2 + CH_3$	50	46
$H + C_2H_6 \rightarrow H_2 + C_2H_5$	38	33
$H + H_2O \rightarrow H_2 + OH$	77	46
$H + HF \rightarrow H_2 + F$	146	138
$H + HCl \rightarrow H_2 + Cl$	15	21
$H + HBr \rightarrow H_2 + Br$	9	8
$H + HI \rightarrow H_2 + I$	3	4
$F + H_2 \rightarrow HF + H$	10	8
$F + CH_4 \rightarrow HF + CH_3$	8	13
$F + C_2H_6 \rightarrow HF + C_2H_5$	2	8
$Cl + H_2 \rightarrow HCl + H$	23	33
$Cl + CH_4 \rightarrow HCl + CH_3$	16	29
$Cl + C_2H_6 \rightarrow HCl + C_2H_5$	4	17
$Br + H_2 \rightarrow HBr + H$	82	88
$Br + CH_4 \rightarrow HBr + CH_3$	78	75
$Br + C_2H_6 \rightarrow HBr + C_2H_5$	57	59
$CH_3 + H_2 \rightarrow CH_4 + H$	51	63
$CH_3 + CH_4 \rightarrow CH_4 + CH_3$	62	54
$CH_3 + C_2H_6 \rightarrow CH_4 + C_2H_5$	51	42

[a] The experimental activation energies were taken from the compilations listed in the bibliography to Chapter 5. The BEBO estimates are given by H. S. Johnston, *Gas Phase Reaction Rate Theory*, Table 11.3, p. 212, Ronald Press, New York, 1966.

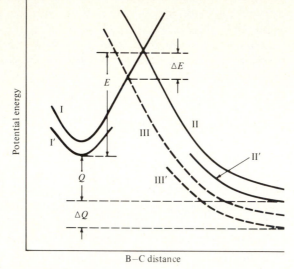

Figure 3.11 Potential-energy profiles for the reaction A + BC → AB + C, showing the relationship between the decrease ΔE_a in energy of activation and the increase ΔQ in heat evolved.

where the constant α is between zero and unity. The significance of this equation may be understood with reference to the potential-energy profiles shown in Fig. 3.11. Curve I relates to the system A + BC and shows the variation of potential energy with B—C distance, with the A—B distance held at the distance in the activated state. Curve II shows the corresponding variation for the system AB + C. Curve I′ is part of the potential-energy curve for BC when A is a significant distance away, and curve II′ is the curve for AB + C with the A—B distance corresponding to the normal molecule. The activation energy E_a is shown on the diagram; there will be some resonance splitting at the crossing point of curves I and II. The heat evolved in the reaction, $Q(= -\Delta H)$, is also shown.

The dashed curves III and III′ show the changes that occur if the reacting species A is replaced by something else. Curve I is hardly affected, but if the A—B bond is strengthened by the change, the curve is lowered and the heat evolved increases by an amount ΔQ. In ideal cases the change in E_a will parallel closely that in Q. If curves I and II are symmetrical where they intersect, the decrease in E_a will be one-half the increase in Q or one-half the decrease in ΔH (i.e., $\alpha = 0.5$). Similar arguments can be used to interpret the changes in activation energy that occur when B or C is changed.

An extension of the Evans–Polanyi equation, which includes terms for both bond forming and bond breaking, was proposed by Szabó.[24] His equation is

$$E_a = \sum_i D_i(\text{breaking}) - \alpha \sum_j D_j(\text{forming}) \tag{3.19}$$

where D_i and D_j are bond dissociation energies. Szabó applied this relationship to several series of homologous reactions and found it to be obeyed somewhat better than the simple Evans–Polanyi equation.

Similar types of relationship have been given by Semenov.[25] For a wide variety of abstraction reactions, Semenov plotted the energy of activation E_a against the heat evolved Q and found that for exothermic reactions (Q positive) almost all the points lie fairly close to the line represented by the equation

$$E_a/\text{kJ mol}^{-1} = 48.1 - 0.25Q/\text{kJ mol}^{-1} \tag{3.20}$$

As shown in Fig. 3.12, the activation energies rarely deviate from this relationship by more than 4 kJ mol^{-1}. For endothermic reactions, the best procedure is to estimate the energy of activation of the reverse reaction by using Eq. (3.20) and then to calculate that for the forward reaction. This is equivalent to saying that for a negative value of Q the relationship is

$$E_a/\text{kJ mol}^{-1} = 48.1 + 0.75Q/\text{kJ mol}^{-1} \tag{3.21}$$

These formulations are an improvement on an early suggestion by Hirschfelder[26] that for an exothermic reaction of the type

$$A + BC \rightarrow AB + C$$

the activation energy is given by

$$E_a = 0.055D_{\text{B–C}} \tag{3.22}$$

where $D_{\text{B–C}}$ is the dissociation energy of the bond that is being broken. This equation implies that E_a is independent of the reactant A, which is true only very approximately.

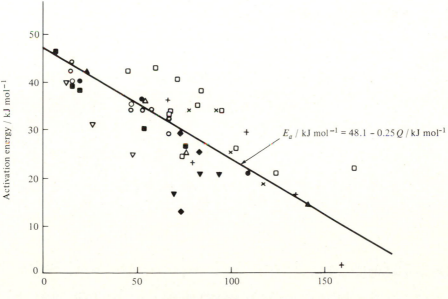

Figure 3.12 Activation energy E_a plotted against heat evolved Q for a number of exothermic reactions.

Relationships such as those mentioned above are reliable only for essentially homolytic mechanisms. They are apt to break down for reactions in which there is considerable charge development in the activated state.[27]

A number of suggestions have been made for the construction of complete potential-energy surfaces on a purely empirical basis. Sometimes the Morse function has been adapted to give a surface; this is a reasonable procedure since it ensures that a reliable potential-energy curve will be obtained if for the A + BC system one of the atoms is removed to infinity. This method was employed by Wall and Porter,[28] who generated a surface for a linear triatomic system by "rotating" a Morse curve through 90° about a fixed point corresponding to large A—B and B—C distances. As it rotated, the Morse curve was systematically distorted, undergoing a general rise and fall so as to sweep out the two valleys and lead to a saddle point. Various extensions of this procedure have been suggested.

A somewhat similar procedure was employed by Blais and Bunker,[29] who distorted the Morse function for B—C by reducing its depth as A approached. This method is more convenient for nonlinear and asymmetric systems. Karplus and Raff[30] improved this procedure by introducing a term that allows for some repulsion between A and BC.

APPENDIX: Statistical Distribution of Molecular Energies

An important statistical relationship, arrived at by Maxwell and Boltzmann in their treatment of the distributions of molecular speeds and energies, is that in one dimension the fraction of molecules having energies between ε_x and $\varepsilon_x + d\varepsilon_x$ is

$$\frac{dN_x}{N} = \frac{1}{(\pi\varepsilon_x \mathbf{k}T)^{1/2}} \, e^{-\varepsilon_x/\mathbf{k}T} \, d\varepsilon_x \tag{3.23}$$

This relationship applies to a molecule moving in the X direction, its energy ε_x being equal to $\frac{1}{2}mu_x^2$, where u_x is the speed in the X direction.

The fraction f^* of molecules having energies in excess of a specified quantity ε^* is relevant to kinetic problems. This fraction is

$$f^* = \frac{1}{(\pi\mathbf{k}T)^{1/2}} \int_{\varepsilon^*}^{\infty} \varepsilon^{-1/2} e^{-\varepsilon/\mathbf{k}T} \, d\varepsilon \tag{3.24}$$

The result of the integration, which cannot be expressed in closed form, is

$$f^* = 1 - \frac{2}{\pi^{1/2}} \sum_{j=0}^{\infty} \frac{(-1)^j (\varepsilon^*/\mathbf{k}T)^{j+1/2}}{j!(2j+1)} \tag{3.25}$$

Motion in Two Dimensions

Suppose that molecules are constrained to move only in two dimensions, X and Y. The probability that the molecule has translational energy between ε_x and $\varepsilon_x + d\varepsilon_x$ in the X direction is given by Eq. (3.23) and the corresponding probability in the Y direction is

$$\frac{dN_y}{N} = \frac{1}{(\pi\varepsilon_y kT)^{1/2}} \, e^{-\varepsilon_y/kT} \, d\varepsilon_y \tag{3.26}$$

The probability that the molecule has energy in the X direction between ε_x and $\varepsilon_x + d\varepsilon_x$ and simultaneously has energy between ε_y and $\varepsilon_y + d\varepsilon_y$ in the Y direction is

$$\frac{dN}{N} = \frac{1}{\pi kT(\varepsilon_x \varepsilon_y)^{1/2}} \, e^{-\varepsilon_x/kT} e^{-\varepsilon_y/kT} \, d\varepsilon_x \, d\varepsilon_y \tag{3.27}$$

What is needed, however, is the probability that the molecules have energy lying between $\varepsilon_x + \varepsilon_y$ and $\varepsilon_x + \varepsilon_y + d(\varepsilon_x + \varepsilon_y)$, without any restriction as to how the energy is distributed between the two dimensions. We write $\varepsilon = \varepsilon_x + \varepsilon_y$, and then ask the probability that the energy lies between ε and $\varepsilon + d\varepsilon$. We obtain this by putting $\varepsilon_y = \varepsilon - \varepsilon_x$ and integrating with respect to ε_x from 0 to ε:

$$\frac{dN}{N} = \frac{1}{\pi kT} \int_0^\varepsilon \varepsilon_x^{-1/2}(\varepsilon - \varepsilon_x)^{-1/2} e^{-\varepsilon_x/kT} e^{-(\varepsilon - \varepsilon_x)kT} \, d\varepsilon_x \, d\varepsilon \tag{3.28}$$

$$= \frac{e^{-\varepsilon/kT} \, d\varepsilon}{\pi kT} \int_0^\varepsilon \varepsilon_x^{-1/2}(\varepsilon - \varepsilon_x)^{-1/2} \, d\varepsilon_x \tag{3.29}$$

The value of the integral is π, and the result is therefore

$$\frac{dN}{N} = \frac{e^{-\varepsilon/kT} \, d\varepsilon}{kT} \tag{3.30}$$

Of special interest is the fraction f^* of molecules having energy in excess of a specified value ε^*. This is obtained by integrating Eq. (3.30) between the limits ε^* and infinity:

$$f^* = \frac{1}{kT} \int_{\varepsilon^*}^\infty e^{-\varepsilon/kT} \, d\varepsilon \tag{3.31}$$

$$= e^{-\varepsilon^*/kT} \tag{3.32}$$

This very simple result is illustrated in Fig. 3.2. It provides an interpretation of the Arrhenius equation: the rate of reaction is proportional to the fraction $e^{-\varepsilon^*/kT}$ of molecules having energy in excess of a critical energy ε^*.

Motion in Three Dimensions

For motion in three dimensions the fraction of molecules having energy between ε and $\varepsilon + d\varepsilon$ is

$$\frac{dN}{N} = \frac{2}{\pi^{1/2}(kT)^{3/2}} \, \varepsilon^{1/2} e^{-\varepsilon/kT} \, d\varepsilon \tag{3.33}$$

This is the usual form of the Maxwell–Boltzmann distribution law.

The fraction f^* of molecules having energy in excess of ε^*, that is,

$$f^* = \frac{2}{\pi^{1/2}(kT)^{3/2}} \int_{\varepsilon^*}^\infty \varepsilon^{1/2} e^{-\varepsilon/kT} \, d\varepsilon \tag{3.34}$$

cannot be given in closed form. The integration leads to

$$f^* = \frac{2}{\pi^{1/2}(kT)^{3/2}} \, e^{-1/2} e^{-\varepsilon^*/kT} + \frac{1}{kT}\left[1 - \mathrm{erf}\left(\frac{\varepsilon^*}{kT}\right)^{1/2}\right] \tag{3.35}$$

where the error function is defined by

$$\operatorname{erf} z = \frac{2}{\pi} \sum_{n=0}^{\infty} \frac{(-1)^n z^{2n-1}}{n!(2n+1)} \tag{3.36}$$

Because of the awkward form of this expression, the two-dimensional equation (3.32) is often used as an approximation for the three-dimensional case.

Motion in Many Degrees of Freedom

The equations developed so far have related specifically to translational motion in one, two, and three dimensions. It is not necessary to restrict the treatment to translational motion: the Maxwell–Boltzmann equation (3.23) applies equally well to any style of motion that is expressible as the square of a velocity. Translational motion in one dimension can be referred to as motion in one *square term;* translational motion in three dimensions as motion in three square terms. The same is true of rotational motion; each degree of rotational freedom also corresponds to a square term. The situation is a little different for vibrational motion, for which both kinetic and potential energies are involved. Each degree of vibrational freedom thus corresponds to *two* square terms. As an example, the following are the number of degrees of freedom and the number of square terms for the methane molecule:

	Degrees of freedom	Square terms
Translation	3	3
Rotation	3	3
Vibration	9	18
Total	15	24

When the energy is in n square terms, application of a treatment similar to that given earlier for two and three square terms leads to the following expression for the fraction of the molecules having energy between ε and $\varepsilon + d\varepsilon$, distributed in any way between the n square terms:

$$\frac{dN}{N} = \frac{1}{(\frac{1}{2}n - 1)!} \left(\frac{\varepsilon}{kT}\right)^{n/2-1} \frac{1}{kT} e^{-\varepsilon/kT} \, d\varepsilon \tag{3.37}$$

The fraction having energy in excess of ε^* is

$$f^* = \frac{1}{(\frac{1}{2}n - 1)!kT} \int_{\varepsilon^*}^{\infty} \left(\frac{\varepsilon}{kT}\right)^{n/2-1} e^{-\varepsilon/kT} \, d\varepsilon \tag{3.38}$$

The result of the integration is

$$f^* = e^{\varepsilon^*/kT}\left[\frac{1}{(\frac{1}{2}n - 1)!}\left(\frac{\varepsilon^*}{kT}\right)^{n/2-1} + \frac{1}{(\frac{1}{2}n - 2)!}\left(\frac{\varepsilon^*}{kT}\right)^{n/2-2} + \cdots + 1\right] \tag{3.39}$$

Provided that ε^*/nkT is much greater than unity, which is true in most kinetic situations, it is satisfactory to drop all except the first term of the expansion:

$$f^* = \frac{1}{(\frac{1}{2}n - 1)!}\left(\frac{\varepsilon^*}{kT}\right)^{n/2-1} e^{-\varepsilon^*/kT} \tag{3.40}$$

In dealing with unimolecular reactions (Section 5.3) we are concerned particularly with the distribution of energy among vibrational degrees of freedom. In that case we can replace $\frac{1}{2}n$ by the number s of vibrational degrees of freedom and write Eq. (3.40) in the form

$$f^* = \frac{1}{(s-1)!}\left(\frac{\varepsilon^*}{\mathbf{k}T}\right)^{s-1} e^{-\varepsilon^*/\mathbf{k}T} \tag{3.41}$$

PROBLEMS

3.1. Assuming the energy to be distributed between two degrees of freedom, calculate the fraction of collisions in which the energy is in excess of the following values at 500 K: (a) 40 kJ mol^{-1}, (b) 100 kJ mol^{-1}, (c) 200 kJ mol^{-1}, and (d) 300 kJ mol^{-1}. Repeat the calculations for 300 and 1000 K.

3.2. The potential energy of the hydrogen molecule is related to the internuclear distance r by the Morse equation

$$E = D(e^{-2\beta(r-r_0)} - 2e^{-\beta(r-r_0)})$$

with $D = 438.9$ kJ mol^{-1}, $r_0 = 74$ pm, and $\beta = 19.54$ nm^{-1}. Calculate E at the following r values:

$$r = 50, 75, 100, 150, 200, 300, 400 \text{ pm}$$

On the assumption that the coulombic energy is 14% of the total energy, calculate the coulombic and exchange energies for the above distances. Plot these two curves on the same graph.

Construct a potential-energy surface for linear $H \cdots H \cdots H$ as follows. First take $r_1 = r_2 = 50$ pm and calculate E using the London equation

$$E = A + B + C - \{\tfrac{1}{2}[(\alpha - \beta)^2 + (\beta - \gamma)^2 + (\gamma - \alpha)^2]\}^{1/2}$$

Repeat this with $r_1 = 50$ pm and r_2 having the values 75, 100, 150, 200, and 300 pm, and plot E versus r_2. Repeat this with r_1 values up to 200 pm, in each case going to $r_1 + r_2 = 400$ pm. Draw the four curves on the one graph.

On a separate graph prepare axes for r_1 and r_2, with the values going from 0 to 300 pm. Plot points corresponding to energies of

$$-50, -100, -150, -200, -250, -300, -400, -450, -500 \text{ kJ mol}^{-1}$$

and draw in the contour lines.

Then make calculations corresponding to additional points in the neighborhood of the saddle point to characterize the surface in more detail. Is there a basin in the region of the saddle point? If so, make an estimate of its depth E_d. Also, make an estimate of the height of the maximum point along the reaction path, with respect to the energy of $H + H_2$. This is the classical activation energy E_c.

3.3. An interesting exercise is to make calculations for different percentages of coulombic energy. The following percentages give a good range of results: 5, 7, 10, 14, 20, 25, and 30%. In each case the classical activation energy E_c and the depth at the col, E_d, should be calculated.

NOTE: Some interesting calculations of LEPS surfaces, to be done using a computer, are given by S. J. Moss and C. J. Coady, *J. Chem. Educ.*, **60**, 455 (1983).

REFERENCES

1. L. Pfaundler, *Ann. Phys. Chem.,* **131,** 55 (1867); *Ann. Phys. Chem.,* **144,** 428 (1872).

2. J. H. van't Hoff, *Etudes de dynamique chimique,* Muller, Amsterdam, 1884.

3. R. C. Tolman, *J. Am. Chem. Soc.,* **43,** 2506 (1920); his equation had been given previously by J. Rice [*Rep. Br. Assoc.,* 399 (1915)] but on the basis of obscure reasoning.

4. R. H. Fowler and E. A. Guggenheim, *Statistical Thermodynamics,* pp. 491–506, Cambridge University Press, Cambridge, U.K., 1939.

5. See, for example, Ref. 4; H. S. Johnston, *Gas Phase Reaction Rate Theory,* pp. 215–217, Ronald, New York, 1966; R. D. Levine and R. B. Bernstein, *Molecular Reaction Dynamics,* pp. 108–111, Oxford University Press, New York, 1974; D. G. Truhlar, *J. Chem. Educ.,* **55,** 309 (1978).

6. M. Menzinger and R. Wolfgang, *Angew. Chem.,* **81,** 483 (1969).

7. R. Marcelin, *Comptes Rendus,* **158,** 116, 407 (1914).

8. F. London, *Probleme der modernen Physik,* p. 104, Sommerfeld Festschrift, 1928.

9. W. Heitler and F. London, *Z. Phys.,* **44,** 455 (1927).

10. H. Eyring and M. Polanyi, *Z. Phys. Chem.,* **1312,** 279 (1931).

11. Y. Sugiura, *Z. Phys.,* **45,** 484 (1927).

12. H. Eyring and M. Polanyi, *Z. Phys. Chem.,* **B12,** 279 (1931).

13. P. J. Kuntz, E. N. Nemeth, and J. C. Polanyi, quoted in K. J. Laidler and J. C. Polanyi, in G. Porter (Ed.), *Progress in Reaction Kinetics,* Vol. 3, Pergamon, Oxford, 1965.

14. A much more extensive table has been given by D. G. Truhlar and R. E. Wyatt, in I. Prigogine and S. A. Rice (Eds), *Advances in Chemical Physics,* vol. 36, p. 141, Wiley-Interscience, New York, 1977.

15. D. G. Truhlar and C. J. Horowitz, *J. Chem. Phys.,* **65,** 2466 (1978).

16. H. Eyring and M. Polanyi, *Z. Phys. Chem. (Leipzig),* **B12,** 279 (1931); for a translation into English of the first part of this article, see M. H. Back and K. J. Laidler, *Selected Readings in Chemical Kinetics,* p. 41, Pergamon, Oxford, 1967.

17. S. Sato, *Bull. Chem. Soc. Jpn.,* **28,** 450 (1955); *J. Chem. Phys.,* **23,** 592, 2465 (1955); see also I. Yasumori, *Bull. Chem. Soc. Jpn.,* **32,** 1103, 1110 (1959).

18. R. E. Weston, *J. Chem. Phys.,* **31,** 892 (1959).

19. R. N. Porter and M. Karplus, *J. Chem. Phys.,* **44,** 1105 (1964).

20. P. J. Kuntz, E. M. Nemeth, J. C. Polanyi, S. D. Rosner, and C. E. Young, *J. Chem. Phys.,* **44,** 1168 (1966).

21. H. S. Johnston, *Adv. Chem. Phys.,* **3,** 131 (1960); H. S. Johnston and P. Goldfinger, *J. Chem. Phys.,* **37,** 700 (1962); H. S. Johnston and C. Parr, *J. Am. Chem. Soc.,* **85,** 2544 (1963); H. S. Johnston, *Gas Phase Reaction Rate Theory,* pp. 177–183, Ronald, New York, 1966.

22. For discussions of the BEBO method and proposals for its improvement see R. M. Jordan and F. Kaufman, *J. Chem. Phys.,* **63,** 1691 (1975); N. L. Arthur, K. F. Donchi, and J. A. McDonell, *J. Chem. Phys.,* **62,** 1585 (1975); A. A. Zavitsas and A. A. Melikian, *J. Am. Chem. Soc.,* **97,** 2757 (1975); R. D. Gilliom, *J. Chem. Phys.,* **65,** 5027 (1976); S. W. Mayer, L. Schieler, and H. S. Johnston, *Eleventh International Symposium on Combustion,* p. 837, Combustion Institute, Pittsburgh, 1967; B. C. Garrett and D. G. Truhlar, *J. Chem. Phys.,* **101,** 5207 (1979).

23. M. G. Evans and M. Polanyi, *Trans. Faraday Soc.,* **34,** 11 (1938); C. N. Hinshelwood, K. J. Laidler, and E. W. Timm, *J. Chem. Soc.,* (1938); S. Glasstone, K. J. Laidler, and H. Eyring, *The Theory of Rate Processes,* pp. 139–146, McGraw-Hill, New York, 1941.

24. Z. G. Szabó, *Chem. Soc. (London), Spec. Publ.,* **16,** 113 (1962); Z. G. Szabó and T. Bérces, *Z. Phys. Chem.,* **57,** 3 (1968).

25. N. N. Semenov, *Some Problems in Chemical Kinetics and Reactivity,* chap. 1, Pergamon, London, and Princeton University Press, Princeton, NJ, 1958.
26. J. O. Hirshfelder, *J. Chem. Phys.,* **9,** 645 (1941).
27. A. Maccoll, *Chem. Soc. (London), Spec. Publ.,* **16,** 116 (1962).
28. F. T. Wall and R. N. Porter, *J. Chem. Phys.,* **36,** 3256 (1962).
29. N. C. Blais and D. L. Bunker, *J. Chem. Phys.,* **37,** 2713 (1962); **39,** 315 (1963).
30. M. Karplus and L. M. Raff, *J. Chem. Phys.,* **41,** 1267 (1964).

BIBLIOGRAPHY

Various aspects of the energy of activation are discussed in:

W. C. Gardiner, Temperature Dependence of Bimolecular Gas Reaction Rates, *Acc. Chem. Res.,* **10,** 326 (1977).

P. D. Pacey, Changing Conceptions of Activation Energy, *J. Chem. Educ.,* **58,** 612 (1981).

Interesting reminiscences about the early calculations of potential-energy surfaces are to be found in:

J. O. Hirschfelder, A Forecast for Theoretical Chemistry, *J. Chem. Educ.,* **43,** 457 (1966).

J. O. Hirschfelder, My Fifty Years of Theoretical Chemistry. I. Chemical Kinetics, *Ber. Bunsenges. Phys. Chem.,* **86,** 349 (1982).

The earlier calculations of potential-energy surfaces are treated in more detail in:

K. J. Laidler, *Theories of Chemical Reaction Rates,* Chap. 2, McGraw-Hill, New York, 1969.

K. J. Laidler and J. C. Polanyi, Theories of the Kinetics of Bimolecular Reactions, in G. Porter (Ed.), *Progress in Reaction Kinetics,* vol. 3, p. 1, Pergamon, Oxford, 1965.

For later reviews of calculations of potential-energy surfaces see:

M. Baer, A Review of Quantum-Mechanical Approximate Treatments of Three-Body Systems, *Adv. Chem. Phys.,* **49,** 91 (1982).

K. R. Lawley (Ed.), *Potential-Energy Surfaces,* Advances in Chemical Physics, vol. 42, Wiley-Interscience, New York, 1980.

Potential-Energy Surfaces, *Faraday Discuss. Chem. Soc.,* **62,** 1–348 (1977).

D. G. Truhlar and R. E. Wyatt, H + H_2: Potential-Energy Surfaces and Elastic and Inelastic Scattering, *Adv. Chem. Phys.,* **36,** 141 (1977).

Chapter 4

Theories of Reaction Rates

Another important theoretical problem in chemical kinetics is to interpret the pre-exponential factor in the Arrhenius equation

$$k = Ae^{-E_a/RT} \tag{4.1}$$

Chapter 3 considered some of the approaches that have been made to the understanding and calculation of the activation energy E_a. When in addition a theory of the pre-exponential factor has been developed, a complete treatment of the rate constant of a chemical reaction has been achieved.

The formulation of a satisfactory treatment of the pre-exponential factor has presented considerable difficulty. Initially, this was due to the confusion that existed in the first quarter of the present century about the role played by molecular collisions. Today, we realize that the distribution of energy among molecules is brought about by the collisions that occur between them and that the rate of a reaction is determined by the frequency of these collisions and by factors related to energy distributions. For a period, however, it was believed by some that an important role was played by infrared radiation emitted by the walls of the reaction vessel and then absorbed by the reactant molecules. This so-called *radiation hypothesis*[1] appeared to receive support from the behavior of unimolecular gas reactions, since it was argued that the rates of such reactions could not depend on molecular collisions. However, as we will see in Section 5.3, such reactions can be explained in terms of collisions, and when this was realized, in the late 1920s, the radiation hypothesis was seen to be unnecessary and was discarded.

From that point on, progress was more rapid. There were four main lines of approach to the problem of the pre-exponential factor:[2]

1. Treatments based on the kinetic theory of collisions.
2. Treatments based on thermodynamics but with an additional hypothesis to explain reaction rates.
3. Treatments based on statistical mechanics, again with an additional hypothesis to explain rates.
4. Treatments based on molecular dynamics.

Many of the earlier treatments along these lines came together in 1935 when Henry Eyring[3] and, independently, M. G. Evans and Michael Polanyi[4] developed a simple and general formulation of rates that has come to be called *transition-state theory.* A good part of the present chapter is concerned with that theory. First, however, we consider some of the important theories that pre-dated transition-state theory and that contributed to its development. We also consider some more recent extensions of those earlier approaches.

4.1 KINETIC THEORY OF COLLISIONS

A treatment of reactions in terms of the kinetic theory of collisions, with the assumption that the molecules act as if they were hard spheres, was given by Max Trautz[5] in 1916 and by William Cudmore McCullagh Lewis[6] in 1918. Because of World War I neither was aware of the work of the other. Both Trautz and Lewis were firm supporters of the radiation hypothesis, and therefore their treatments were formulated in terms of that hypothesis. However, although they thought that radiation is concerned with the distribution of energy, they still regarded collisions as exerting important control over the rates of bimolecular reactions. As a result, their treatments were not affected when the radiation hypothesis was abandoned in the late 1920s.

The treatments of Trautz and Lewis were based on the expressions for the *collision number Z,* which is the total number of collisions per unit time per unit volume. For a gas containing only one type of molecule, A, the collision number is given by simple collision theory, in which molecules are treated as hard spheres, as

$$Z_{AA} = \tfrac{1}{2}\sqrt{2}\pi d^2 \bar{u} N_A^2 \qquad (4.2)$$

where N_A is the number of molecules per unit volume, d is the molecular diameter (equal to the distance between the centers when a collision occurs), and $\bar{u}$ is the mean molecular speed, given by kinetic theory to be

$$\bar{u} = \left(\frac{8\mathbf{k}T}{\pi m}\right)^{1/2} \qquad (4.3)$$

where m is the molecular mass. Introduction of this expression into Eq. (4.2) gives

$$Z_{AA} = 2d^2 N_A^2 \left(\frac{\pi \mathbf{k}T}{m}\right)^{1/2} \qquad (4.4)$$

This quantity is known as the *collision number,* and its SI unit is $m^{-3}\,s^{-1}$.

The corresponding expression for the collision number Z_{AB} for two unlike molecules A and B, of masses m_A and m_B, is

$$Z_{AB} = N_A N_B d^2 \left(8\pi k T \frac{m_A + m_B}{m_A m_B} \right)^{1/2} \tag{4.5}$$

The distance d is now the mean of the molecular diameters, or the sum of their radii. It is the distance between the centers of A and B when reaction occurs. A *reduced mass* μ can be defined by the equation

$$\mu \equiv \frac{m_A m_B}{m_A + m_B} \tag{4.6}$$

and Eq. (4.5) may then be written

$$Z_{AB} = N_A N_B d^2 \left(\frac{8\pi k T}{\mu} \right)^{1/2} \tag{4.7}$$

According to the hard-sphere collision theory, as formulated by Lewis and Trautz, the collision number multiplied by the Arrhenius factor $e^{-E/RT}$ gives the rate of formation of the products of reaction, in terms of the number of molecules formed per unit volume and per unit time. Thus, for reaction between two molecules A and B the rate is expressed as

$$v = N_A N_B d^2 \left(\frac{8\pi k T}{\mu} \right)^{1/2} e^{-E/RT} \tag{4.8}$$

Division by $N_A N_B$ gives a rate constant in molecular units (SI unit: $m^3 \ s^{-1}$). It can be put into molar units (SI unit: $m^3 \ mol^{-1} \ s^{-1}$) by multiplication by the Avogadro constant L:

$$k = L d^2 \left(\frac{8\pi k T}{\mu} \right)^{1/2} e^{-E/RT} \tag{4.9}$$

The pre-exponential factor in this expression is called the *collision frequency factor* and given the symbol z_{AB} (or z_{AA} if there is only one type of molecule):

$$z_{AB} = L d^2 \left(\frac{8\pi k T}{\mu} \right)^{1/2} \tag{4.10}$$

Thus, the rate constant is expressed as

$$k = z_{AB} e^{-E/RT} \tag{4.11}$$

Lewis applied this treatment to the reaction

$$2HI \rightarrow H_2 + I_2$$

and calculated a pre-exponential factor of $3.5 \times 10^{-7} \ dm^3 \ mol^{-1} \ s^{-1}$ at 556 K, in excellent agreement with the experimental value of $3.52 \times 10^{-7} \ dm^3 \ mol^{-1} \ s^{-1}$. Paradoxically, this agreement was somewhat unfortunate, since it led to undue confidence in the theory and delayed the development of the subject for some years. It became evident later that Lewis's choice of reaction was a lucky one for his treatment and that there are many reactions for which there are large discrepancies between the observed and calculated rate constants. For example, for gas reactions between molecules of any complexity the observed pre-exponential factors are often lower by several powers of 10 than the factors calculated by the simple collision theory (for examples see Section 5.1). Deviations from theory often are encountered with solution reactions between ions or dipolar substances.

These discrepancies show that simple hard-sphere collision theory is not adequate for chemical reactions. It may be fairly satisfactory for the treatment of the viscosity and diffusion of gases, but for a theory of chemical reactivity a more precise definition of a collision is required. If two molecules are to undergo a chemical reaction they must not merely collide with sufficient mutual energy: they must come together with such a mutual orientation that the necessary bonds can be broken and made. The simple kinetic theory counts every sufficiently energetic collision as an effective one: in reality, the molecules may not approach each other in the right way for reaction to occur, even if there is plenty of energy.

One way to remedy the situation, favored in the 1920s and 1930s, is to introduce into the pre-exponential factor a *steric factor P* that is supposed to represent the fraction of the total number of collisions that are effective from the orientation point of view. The rate constant can then be written

$$k = P z_{AB} e^{-E/RT} \tag{4.12}$$

This procedure is an improvement in some ways, but the evaluation of P cannot be done in a satisfactory manner. Moreover, factors other than orientation are involved, and again these cannot be estimated easily.

Aside from these difficulties, a more fundamental weakness of the simple hard-sphere collision theory of reactions is that it is not consistent with the fact that at equilibrium the ratio of rate constants in the forward and reverse directions is the equilibrium constant. Thus, if the rate constants in the two directions are formulated as

$$k_1 = z_1 e^{-E_1/RT} \quad \text{and} \quad k_{-1} = z_{-1} e^{-E_{-1}/RT} \tag{4.13}$$

the ratio is

$$K_c = \frac{k_1}{k_{-1}} = \frac{z_1}{z_{-1}} e^{-(E_1 - E_{-1})/RT} \tag{4.14}$$

This expression, however, lacks the entropy term that should appear in the expression for the equilibrium constant:

$$K_c = e^{\Delta S°/R} e^{-\Delta H°/RT} \tag{4.15}$$

In 1933 La Mer[7] tried to remedy this deficiency by introducing an entropy term into the kinetic theory expression:

$$k = z e^{\Delta^{\ddagger}S°/R} e^{-\Delta^{\ddagger}H°/RT} \tag{4.16}$$

Here $\Delta^{\ddagger}S°$ and $\Delta^{\ddagger}H°$ are the standard entropy and enthalpy changes in reaching the activated state. This procedure is not entirely satisfactory but was useful in paving the way to transition-state theory.

4.1.1 Generalized Kinetic Theory

During recent years there have been many important extensions of the original simple kinetic theory of collisions, and these have made significant contributions to rate theory. They are based on the concept of the reaction cross section σ for a reaction. In simple kinetic theory, the number of collisions per unit volume per unit time between N_A molecules of A and N_B molecules of B approaching one another with relative speed u is

$$Z_{AB} = \pi(r_A + r_B)^2 u N_A N_B \quad \text{(SI unit: m}^{-3}\text{ s}^{-1}\text{)} \tag{4.17}$$

This relationship is illustrated in Fig. 4.1. If every collision led to reaction, Eq. (4.17) would be the rate of reaction, in molecular units, for reactants approaching one another with speed u. In general, however, the rate is different from this and may be expressed by introducing a reaction probability P_r:

$$v = P_r Z_{AB} = P_r \pi d^2 u N_A N_B \tag{4.18}$$

Here we have written $r_A + r_B$ as d, the collision diameter. The quantity $P_r \pi d^2$ is called the *reaction cross section* and given the symbol σ. Thus, the reaction cross section is defined by the expression

$$\sigma \equiv P_r \pi d^2 \tag{4.19}$$

The rate of reaction $v(u)$, for reaction between molecules A and B moving with relative velocity u, is given by

$$v(u) = \sigma(u) u N_A N_B \tag{4.20}$$

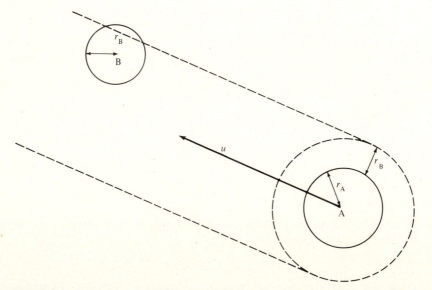

Figure 4.1 Frequency of collisions: molecule A is moving with speed u relative to B. In unit time the sphere of radius $r_A + r_B$ has swept out a volume $\pi(r_A + r_B)^2 u$ and has encountered $\pi(r_A + r_B)^2 u N_B$ molecules of B.

where $\sigma(u)$ is the particular reaction cross section that relates to that molecular velocity. The rate constant $k(u)$ for that velocity is

$$k(u) = \sigma(u)u \tag{4.21}$$

In reality, molecules are not all moving with the same speed; at a given temperature T there is a Maxwell–Boltzmann distribution of speeds. If the expression for that distribution is written as $f(T, u)$, the rate constant $k(u)$ has to be weighted by this function, with integration over speeds from zero to infinity, to give the overall average rate constant:

$$k(T) = \int_0^\infty f(T, u)\sigma(u)u \, du \tag{4.22}$$

This useful general expression is important in relating the macroscopic rate constant $k(T)$ to quantities that describe the situation on the molecular scale.

For a system at thermal equilibrium the function $f(T, u)$ is the Maxwell–Boltzmann distribution expression. It is convenient to integrate over the translational energy ε_t instead of the speed u; the change is made by putting ε_t equal to $\mu u^2/2$, where μ is the reduced mass of the system. The following expression is obtained:[8]

$$k(T) = \left[\frac{8}{\pi\mu(\mathbf{k}T)^3}\right]^{1/2} \int_0^\infty \varepsilon_t\sigma(\varepsilon_t)e^{-\varepsilon_t/\mathbf{k}T}d\varepsilon_t \tag{4.23}$$

This equation is useful for arriving at the macroscopic rate constant $k(T)$ from molecular properties. Note that this equation is concerned only with translation energy ε_t; internal energy is not taken into account.

We now consider what assumption about $\sigma(\varepsilon_t, T)$ needs to be made to obtain the simple collision theory expression of Lewis and Trautz [Eq. (4.9)]. At first sight it might appear that one should set $\sigma(\varepsilon_0, T)$ equal to zero when ε_t is less than a critical value ε_0 and that $\sigma(\varepsilon_t, T)$ would be constant when ε_t is greater than ε_0. In other words,

$$\sigma(\varepsilon_t, T) = 0 \qquad \text{when } \varepsilon_t < \varepsilon_0$$

$$\sigma(\varepsilon_t, T) = \pi d^2 \quad \text{when } \varepsilon_t > \varepsilon_0$$

where d is the sum of the radii (see Fig. 4.1). This approach to the problem represents an improvement, but it is more realistic to apply the conditions

$$\sigma(\varepsilon_t, T) = 0 \qquad\qquad \text{when } \varepsilon_t < \varepsilon_0$$

$$\sigma(\varepsilon_t, T) = \frac{\pi d^2(\varepsilon_t - \varepsilon_0)}{\varepsilon_t} \quad \text{when } \varepsilon_t > \varepsilon_0$$

The reason that the energy enters into the collision cross section can be seen qualitatively with reference to Fig. 4.2. The molecule A is not moving directly towards B, and the *impact parameter b* is defined as the closest distance between the two molecules if they were able to continue their motion without interacting with each other. If all collisions were head-on collisions ($b = 0$), it would be fairly satisfactory to treat $\sigma(\varepsilon_t, T)$ as a constant when $\varepsilon_t > \varepsilon_0$. Some collisions are not head-on collisions, however, and then reaction will occur only if ε_t is sufficiently great that the energy *along the*

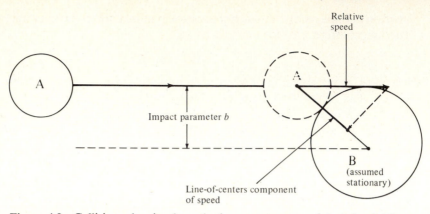

Figure 4.2 Collision, showing how the impact parameter b is defined. The energy along the line of centers on impact must exceed the critical energy ε_0.

line of centers is greater than ε_0. For $\varepsilon_t > \varepsilon_0$, energetic collisions are more likely to lead to reaction than less energetic ones. The cross section $\sigma(\varepsilon_t, T)$ thus increases with increasing ε_t and it turns out that multiplication by the factor $(\varepsilon_t - \varepsilon_0)/\varepsilon_t$ leads to the simple collision theory rate constant given by Eq. (4.9). Figure 4.3 shows schematically

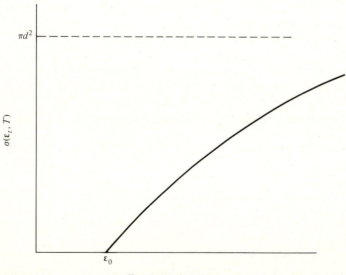

Figure 4.3 Variation of the reaction cross section $\sigma(\varepsilon_t, T)$ with ε_t for the condition

$$\sigma(\varepsilon_t, T) = 0 \qquad\qquad \text{when } \varepsilon_t < \varepsilon_0$$

$$\sigma(\varepsilon_t, T) = \frac{\pi d^2(\varepsilon_t - \varepsilon_0)}{\varepsilon_t} \quad \text{when } \varepsilon_t > \varepsilon_0.$$

the dependence of $\sigma(\varepsilon_t, T)$ on ε_t according to this model, which is often referred to as the *line-of-centers model*.

4.1.2 Extensions of Collision Theory

Evidence from recent experiments, of the kind to be described in Chapter 12, has suggested, however, that the behavior is more complicated than that represented in Fig. 4.3. During recent years progress has been made in measuring reaction cross sections when the reactant molecules are in prescribed vibrational and rotational states (Section 12.5). If full information of this kind were available for a reaction system it would be possible, by a summation procedure, to obtain a weighted average cross section, corresponding to various energies, and hence the rate constant.

4.2 RATE THEORIES BASED ON THERMODYNAMICS

Thermodynamics itself is concerned with systems at equilibrium and can give no complete treatment of reaction rates. However, if additional hypotheses are made, thermodynamics may make an important contribution to theories of rates.

The thermodynamic approaches to rates all stem from van't Hoff's contribution of 1884, which was considered in Section 2.9. The essential feature of van't Hoff's argument was that rates must have a temperature dependence similar to that for equilibrium constants. The next important contribution was that of René Marcelin,[9] who proposed an equation that in modern terminology may be written as

$$v = \text{const} \times e^{-\Delta^{\ddagger} G/RT} \tag{4.24}$$

where $\Delta^{\ddagger}G$ is the *Gibbs energy change* in going from the initial to the activated state. Similar but more explicit ideas were presented by Kohnstamm, Scheffer, and Brandsma.[10] Their arguments were somewhat similar to those of van't Hoff, but they split the overall *Gibbs energy change* into quantities relating to forward and reverse reactions. Thus, they split the thermodynamic equation

$$\ln K_c = -\frac{\Delta G^{\circ}}{RT} \tag{4.25}$$

into

$$\ln k_1 = -\frac{\Delta^{\ddagger} G_1^{\circ}}{RT} + \text{const} \tag{4.26}$$

for the forward reaction and

$$\ln k_{-1} = -\frac{\Delta^{\ddagger} G_{-1}^{\circ}}{RT} + \text{const} \tag{4.27}$$

for the reverse reaction. In general,

$$k = v^{\ddagger} e^{-\Delta^{\ddagger} G^{\circ}/RT} \tag{4.28}$$

where $v^{\ddagger}$ is a factor that is the same for all reactions. Kohnstamm and Scheffer[11] further split $\Delta^{\ddagger}G^{\circ}$ into entropy and enthalpy terms:

$$k = \nu^{\ddagger} e^{\Delta^{\ddagger} S^{\circ}/R} e^{-\Delta^{\ddagger} H^{\circ}/RT} \tag{4.29}$$

They were the first to introduce the concepts of the Gibbs energy of activation $\Delta^{\ddagger} G^{\circ}$, the entropy of activation $\Delta^{\ddagger} S^{\circ}$, and the enthalpy of activation $\Delta^{\ddagger} H^{\circ}$. However, they were unable to interpret the multiplying factor $\nu^{\ddagger}$. We shall see (Section 4.5) that according to transition-state theory this factor is equal to $\mathbf{k}T/h$, where $\mathbf{k}$ is the Boltzmann constant and h the Planck constant.

4.3 RATE THEORIES BASED ON STATISTICAL MECHANICS

Although the Maxwell–Boltzmann distributions were formulated in 1860, it was not until 1912 that the distributions first were applied to reaction rates; this was done by Berthoud.[12] Two years later, Marcelin[13] developed his idea that reaction is the motion of a point in phase space (Section 3.2) and applied the methods of statistical mechanics to obtain an expression for the concentration of species present at a critical surface in phase space.

In 1919 an important approach to reaction rates was made by Herzfeld,[14] who considered the dissociation of a diatomic molecule:

$$AB \rightleftharpoons A + B$$

He applied statistical mechanics to the equilibrium constant and collision theory to the reverse reaction. For the dissociation reaction he then obtained the expression

$$k_1 = \frac{\mathbf{k}T}{h} (1 - e^{-h\nu/\mathbf{k}T}) e^{-Q/RT} \tag{4.30}$$

where Q is the dissociation energy and ν the vibrational frequency. This equation is of particular interest because it is the first time that the expression $\mathbf{k}T/h$, so essential a feature of transition-state theory, appeared in a rate equation.

A number of other statistical-mechanical approaches to rates were made prior to the birth in 1935 of transition-state theory. In particular, mention may be made of the contributions of Rideal,[15] Dushman,[16] Rodebush,[17] Topley,[18] Tolman,[19] and La Mer.[20]

4.4 EARLY DYNAMICAL THEORIES OF RATES

Dynamical treatments of reaction rates have been developed, and these are to be distinguished from the collision theories based on kinetic theory (Section 4.1). Those theories lie in the area of *kinematics,* which deals with properties of motion that are independent of the forces bringing about the motion. The term *dynamics,* on the other hand, relates to the actual forces, and these are conveniently dealt with in terms of potential-energy surfaces which were considered in the last chapter.

In their classical paper on the construction of a potential-energy surface, Eyring and Polanyi[21] considered the possibility of making dynamical calculations but did not carry them out. During the next few years, Eyring and his co-workers[22] actually performed such calculations. However, in those early days such calculations required an enormous amount of labor. A trajectory over a surface had to be calculated point by point with the aid of simple mechanical calculators. A single trajectory took a considerable time, and a useful rate constant requires a large number of trajectories, corresponding to a range of starting conditions.

A very significant dynamical treatment of reaction rates was made in 1932 by Pelzer and Wigner,[23] and this was the first treatment to focus attention on the col or saddle point of a potential-energy surface. Pelzer and Wigner obtained an expression for the rate of reaction by considering the rate of passage of systems through the col. Their treatment is not as general as the transition-state theory later formulated by Eyring and by Evans and Polanyi, but it certainly contributed greatly to that theory.

More recent dynamical treatments of reaction rates are considered in Section 12.1.

4.5 CONVENTIONAL TRANSITION-STATE THEORY

The theory of reaction rates that was published almost simultaneously by Henry Eyring[24] and by M. G. Evans and M. Polanyi[25] in 1935 is referred to now as conventional transition-state theory (CTST).[26] There have been many improvements and extensions to the theory, and some of these are dealt with in Section 4.9; all the modified theories are considerably more complicated than CTST and cannot be expressed in as compact a form. The great value of CTST is that the resulting rate equation, although simple, provides a framework in terms of which even quite complicated reactions can be understood in a qualitative way.

It is possible to arrive at the CTST rate equation in a number of different ways, some of which are presented later in some detail. As in all scientific theories, various assumptions and approximations are involved, and these have come under considerable scrutiny. Here we simply state the main assumptions and approximations; they are considered in much more detail in Section 4.8. Section 4.9 gives a brief outline of treatments in which some of the assumptions and approximations have been removed.

The main assumptions of CTST are:

Assumption 1. Molecular systems that have surmounted the col in the direction of products *cannot turn back* and form reactant molecules again.

Assumption 2. The energy distribution among the reactant molecules is in accordance with the Maxwell–Boltzmann distribution. Furthermore, it is assumed that even when the whole system is not at equilibrium, the concentration of those activated complexes that are becoming products *can also be calculated using equilibrium theory.*

Assumption 3. It is permissible to *separate* the motion of the system over the col from the other motions associated with the activated complex.

Assumption 4. A chemical reaction can be satisfactorily treated in terms of *classical motion* over the barrier, quantum effects being ignored.

In addition, CTST involves the same assumptions and approximations that are made in the calculation of equilibrium constants using statistical mechanics. Usually, these are not serious, and corrections for them can be made.

4.5.1 Equilibrium Hypothesis

One of these assumptions, Assumption 2, will be considered here since it is the most troublesome at first sight. The assumption is that there is a certain type of equilibrium, sometimes called *quasiequilibrium,* between reactants and activated complexes. How can species that are undergoing transformation be in any sense at equilibrium?

To understand the apparent dilemma, we must be clear about what precisely is being assumed. Consider first the reaction

$$A + B \rightleftharpoons Y + Z$$

which has been allowed to go to complete equilibrium. This means that the process

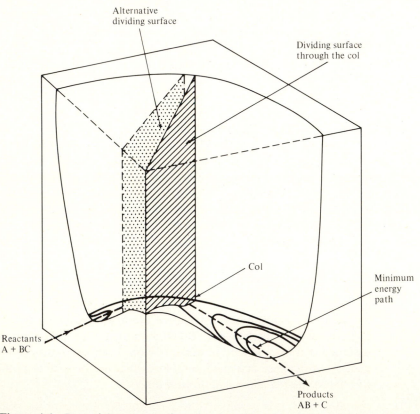

Figure 4.4 Potential-energy surface showing the col and the minimum-energy path, which is the path of steepest descent in each direction from the col. The dividing surface through the col is at right angles to the minimum-energy path, as is the alternative dividing surface.

from left to right is occurring at the same rate as that from right to left. In CTST, attention is confined to the situation at the col in the potential-energy surface, and we can imagine constructing a so-called *dividing surface* that passes through the col (Fig. 4.4). This dividing surface is constructed perpendicular to the minimum-energy path, which is the path of steepest descent from the lowest point of the col into the reactant and product valleys. Figure 4.4 shows both this dividing surface through the col and an alternative dividing surface that enters into some of the more modern formulations of transition-state theory (Section 4.9).

Conventional transition-state theory focuses attention entirely on the dividing surface that passes through the col and is concerned with the rate at which systems pass through that dividing surface. As shown in Fig. 4.5, it is convenient to consider two parallel dividing surfaces separated by a very small distance δ. Systems that are within this small distance are activated complexes by definition; those to the left are reactants and those to the right are products.

Following Assumption 1, we assume that systems entering the region of thickness δ from the left-hand side must exit into the product side and end up as products; similarly, those entering from the product side must become reactants. In Fig. 4.5

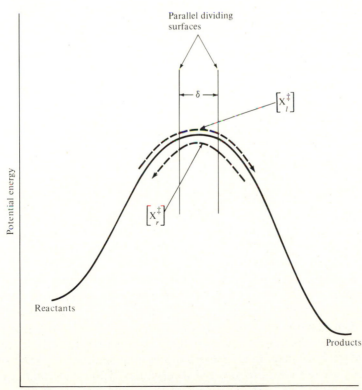

Figure 4.5 Profile through the minimum-energy path, showing two dividing surfaces at the col separated by a small distance δ.

activated complexes that have entered the region from the reactant side, and that are therefore moving from left to right, are designated $X_l^{\ddagger}$. Similarly, the species that have entered from the right-hand side are designated $X_r^{\ddagger}$. The concentrations $[X_l^{\ddagger}]$ and $[X_r^{\ddagger}]$ are defined as the amounts of those species divided by the volume of the system. Since δ is arbitrary, these concentrations are also arbitrary. At complete equilibrium in the system, the reaction rates in the two directions are the same, and it follows that $[X_l^{\ddagger}]$ and $[X_r^{\ddagger}]$ are equal to one another. Also, at complete equilibrium, the activated complexes are at equilibrium with both reactants and products. We therefore may write

$$[X_l^{\ddagger}] + [X_r^{\ddagger}] = [X^{\ddagger}] = K_c^{\ddagger}[A][B] \tag{4.31}$$

where $[X^{\ddagger}]$ is the total concentration of activated complexes and $K_c^{\ddagger}$ is the equilibrium constant. Since $[X_l^{\ddagger}] = [X_r^{\ddagger}]$, it follows that

$$[X_l^{\ddagger}] = \tfrac{1}{2}K_c^{\ddagger}[A][B] \tag{4.32}$$

It is possible therefore to obtain an expression for $[X_l^{\ddagger}]$ by applying statistical mechanics to the equilibrium constant $K_c^{\ddagger}$. If we also have an expression for the rate at which the species $X_l^{\ddagger}$ give rise to products, we have a complete treatment of the reaction rate for the system at equilibrium.

Usually, we are not interested in this equilibrium reaction rate but in the rate when the system is not at complete equilibrium.[27] Suppose that the system is first at complete equilibrium and that then the products of reaction Y + Z are removed. Obviously, the effect is to remove the species $X_r^{\ddagger}$, because they can be formed only from Y + Z. The concentration of the $X_l^{\ddagger}$ complexes, however, remains unchanged, since the formation of these complexes cannot depend in any way on whether or not the products are present. It is important to emphasize that in CTST when one states that the activated complexes are in equilibrium with the products, *one is referring only to those complexes $X_l^{\ddagger}$ that were reactant molecules in the immediate past.* The term *quasiequilibrium* is useful for emphasizing this point.

There is no assumption in CTST that there is a *classical* type of equilibrium between initial and activated states. If the equilibrium were classical, addition to the system of activated complexes $X_l^{\ddagger}$ would disturb the equilibrium, but this could not occur in the present situation. In CTST the activated complexes are transient species which do *not* linger at the activated state and make a decision as to whether to go either forward or backward: they are on their way to forming products and are bound to proceed in that direction.

Provided that Assumption 1 is satisfied, the equilibrium assumption is valid when the system A + B $\rightleftharpoons$ Y + Z is at equilibrium and introduces no error into the treatment of the equilibrium rate constant. The argument outlined above depends on the validity of Assumption 1, that systems pass through the dividing surface only once. If this is not true, the equilibrium hypothesis is no longer valid, because the concentration $[X_l^{\ddagger}]$, for example, then includes species that are not true activated complexes in that they are not about to give rise to products. This whole matter is considered further in Sections 4.8 and 4.9, where methods of dealing with the breakdown of Assumptions 1 and 2 are considered.

4.5.2 Statistical Mechanics and Chemical Equilibrium

An understanding of CTST requires some knowledge of how equilibrium constants are treated by the methods of statistical mechanics, which we now summarize very briefly. We do so without any proofs since such proofs can be found elsewhere (see the bibliography at the end of this chapter).

According to statistical mechanics, the molecular equilibrium constant for a reaction

$$aA + bB + \cdots \rightleftharpoons \cdots + yY + zZ$$

is given by

$$K_c = \frac{\cdots q_Y^y q_Z^z}{q_A^a q_B^b \cdots} e^{-E_0/RT} \tag{4.33}$$

where the q's are the partition functions per unit volume. The energy E_0 is the molar energy change at the absolute zero when a mol of A reacts with b mol of B, and so on, to form the products, all substances being in their standard states.

The manner of evaluating the partition functions is as follows. The total partition function q for a molecule is defined by

$$q \equiv \sum_i g_i e^{-\varepsilon_i/kT} \tag{4.34}$$

where the summation is taken over all energy levels. The energy ε_i is the energy of the ith state relative to the zero-point energy, and g_i is the degeneracy, that is, the number of energy states corresponding to the ith level. Usually, it is assumed that the various types of energy—electronic, vibrational, rotational, and translational—are independent of one another. The total energy corresponding to the ith energy state is thus expressed as the sum of the energies of the different types:

$$\varepsilon_i = e_i + v_i + r_i + t_i \tag{4.35}$$

The four energies on the right-hand side represent the four types of energy corresponding to the ith state. The partition function becomes

$$q = \sum_i g_{ei} e^{-e_i/kT} g_{vi} e^{-v_i/kT} g_{ri} e^{-r_i/kT} g_{ti} e^{-t_i/kT} \tag{4.36}$$

the g_i having factorized as well as the exponential terms. This equation may be written as

$$q = q_e q_v q_r q_t \tag{4.37}$$

where q_e, q_v, q_r, and q_t are separate partition functions, each referring to one type of energy. Thus, the partition function has been factorized, so that each term may be evaluated separately.

To calculate the partition function for the electronic states of an atom or molecule, one needs to know the electronic energy levels. The electronic partition function q_e is then calculated as

$$q_e = \sum_i g_{ei} e^{-e_i/kT} \tag{4.38}$$

At ordinary temperatures the excited electronic levels of an atom or molecule are usually too high to make a significant contribution to the partition function. If the lowest state is nondegenerate, the statistical weight g_e is unity, so that if the lowest state is taken as the zero level and all other levels are sufficiently high, the partition function q_e is approximately unity. As a rough guide, electronic levels may be neglected if their energy is more than $4\mathbf{k}T$. For oxygen, nitric oxide, and a few other molecules, the lowest level is not a singlet, so that the partition function q_e is no longer unity.

For the other types of energy it is possible to obtain compact expressions for the partition functions. These are listed in Table 4.1, which gives some typical values. The partition function for translational motion in three degrees of freedom involves only the mass of the molecule. The rotational partition function for a linear molecule involves the moment of inertia I. If the linear molecule is symmetrical (e.g., H_2 and CO_2) the partition function, when used to calculate an equilibrium constant [Eq. (4.33)], must be divided by 2; however, for rate constants a different procedure is recommended and discussed later (Section 4.5.4). For a nonlinear molecule the rotational partition function involves three moments of inertia—I_A, I_B, and I_C—which relate to the three principal axes of symmetry; these are perpendicular to one another.

The total number of degrees of freedom in a molecule containing N atoms is $3N$. For a molecule in the gas phase three relate to the translational motion. If the molecule is nonlinear, three degrees of freedom relate to rotational energy; the remainder, $3N - 6$, relate to vibrational energy. If the molecule is linear there are only two degrees of rotational freedom; there are therefore $3N - 5$ degrees of vibrational freedom. The expression given in Table 4.1 for the vibrational motion is for each degree of vibrational freedom, which can now be called a *normal mode of vibration*. For a diatomic molecule $3N - 5 = 1$, and the partition function is simply $(1 - e^{-h\nu/\mathbf{k}T})^{-1}$, where ν is the vibrational frequency. For a polyatomic molecule there will be the product of $3N - 5$ or $3N - 6$ such factors, each involving a characteristic ν corresponding to a normal mode of vibration.

4.5.3 Derivations of the Rate Equation

The rate equation for a bimolecular reaction, derived by the methods of conventional transition-state theory,[3,4] is

$$k = \frac{\mathbf{k}T}{h} \frac{q_{\ddagger}}{q_A q_B} e^{-E_0/RT} \tag{4.39}$$

The partition functions q_A and q_B relate to the two reactants A and B, and $q_{\ddagger}$ is a special type of partition function for the activated complex. In this partition function $q_{\ddagger}$, the factor relating to the actual motion over the col is omitted.

The rate equation has been derived in a considerable number of different ways, and to gain a deep insight into the theory it is useful to examine all of them. Here we outline only two of them but later we make reference to some of the others. The derivations are given for bimolecular gas reactions; modifications for reactions in solution (Chapter 6) or on surfaces (Chapter 7) are discussed later.

TABLE 4.1   PARTITION FUNCTIONS FOR DIFFERENT TYPES OF MOTION

Motion	Degrees of freedom	Partition function	Order of magnitude
Translation	3	$$\frac{(2\pi mkT)^{3/2}}{h^3}$$ (per unit volume)	10^{31}–10^{32} m^{-3}
Rotation (linear molecule)	2	$$\frac{8\pi^2 IkT}{\sigma h^2}$$	10–10^2
Rotation (nonlinear molecule)	3	$$\frac{8\pi^2(8\pi^3 I_A I_B I_C)^{1/2}(kT)^{3/2}}{\sigma h^3}$$	10^2–10^3
Vibration (per normal mode)	1	$$\frac{1}{1 - e^{-h\nu/kT}}$$	1–10
Free internal rotation	1	$$\frac{(8\pi^2 I'kT)^{1/2}}{h}$$	1–10

where m = mass of molecule
          I = moment of inertia for linear molecule
I_A, I_B, and I_C = moments of inertia for a nonlinear molecule about three axes at right angles
          to one another
          I' = moment of inertia for internal rotation
          ν = normal-mode vibrational frequency
          k = Boltzmann constant
          h = Planck constant
          T = absolute temperature
          σ = symmetry number[a]

It is useful to remember that the power to which h appears is equal to the number of degrees of freedom.

[a] As discussed in the text (Section 4.5.4), symmetry numbers are used in the calculation of equilibrium constants, but for rates an alternative procedure is recommended.

Derivation 1   The first derivation is that of Wynne-Jones and Eyring.[28] It considers the equilibrium between initial and activated states and then treats the motion through the col as a *very loose vibration*.

The equilibrium between two reactants A and B and the activated complexes $X^{\ddagger}$ may be expressed by the equation

$$\frac{[X^{\ddagger}]}{[A][B]} = K_c^{\ddagger} \tag{4.40}$$

where $X_c^{\ddagger}$ is the concentration equilibrium constant.§ This constant may be expressed in terms of the partition functions,

$$\frac{[X^{\ddagger}]}{[A][B]} = \frac{q^{\ddagger}}{q_A q_B} e^{-E_0/RT} \tag{4.41}$$

§ For gas reactions it is usually satisfactory to use concentrations instead of activities. For reactions in solution it may be necessary to include activity coefficients. (See Section 6.3.4.)

where E_0 is the difference between the molar zero-point energy of the activated complexes and that of the reactants; this energy is the hypothetical activation energy at the absolute zero. The partition functions in this expression must be evaluated with respect to the zero-point levels of the respective molecules.

If molecule A contains N_A atoms and B contains N_B atoms, the activated complex $X^{\ddagger}$ contains $N_A + N_B$ atoms. If the complex is nonlinear, it has three degrees of rotational freedom and $3(N_A + N_B) - 6$ degrees of vibrational freedom; there are $3(N_A + N_B) - 5$ degrees for a linear complex. One of these has a different character from the rest, since it corresponds to such a loose vibration that there is no restoring force: the complex can form products without any restraint. For this one degree of freedom we may employ in place of the ordinary factor $(1 - e^{-h\nu/kT})^{-1}$ the value of this function calculated in the limit at which ν approaches zero. This is obtained by expanding the exponential and taking only the first term:

$$\lim_{\nu \to 0} \frac{1}{1 - e^{-h\nu/kT}} = \frac{1}{1 - (1 - h\nu/kT)} = \frac{kT}{h\nu} \tag{4.42}$$

We therefore include this expression in $q^{\ddagger}$ and write the remaining product of factors as $q_{\ddagger}$, which now refers only to $3(N_A + N_B) - 7$ degrees of vibrational freedom $[3(N_A + N_B) - 6$ if it is linear]:

$$q^{\ddagger} = q_{\ddagger} \frac{kT}{h\nu} \tag{4.43}$$

Equation (4.41) thus becomes

$$\frac{[X^{\ddagger}]}{[A][B]} = \frac{q_{\ddagger}(kT/h\nu)}{q_A q_B} e^{-E_0/RT} \tag{4.44}$$

which rearranges to

$$\nu[X^{\ddagger}] = [A][B] \frac{kT}{h} \frac{q_{\ddagger}}{q_A q_B} e^{-E_0/RT} \tag{4.45}$$

The frequency ν is the frequency of vibration of the activated complexes in the degree of freedom corresponding to their conversion into products. The expression on the left-hand side of Eq. (4.45), the product of the concentration of complexes and the frequency of their conversion into products, is therefore the rate of reaction v:

$$v = [A][B] \frac{kT}{h} \frac{q_{\ddagger}}{q_A q_B} e^{-E_0/RT} \tag{4.46}$$

The rate constant, defined by $v \equiv k[A][B]$, is

$$k = \frac{kT}{h} \frac{q_{\ddagger}}{q_A q_B} e^{-E_0/RT} \tag{4.47}$$

which is the CTST formula.

It is to be emphasized that when one uses an equilibrium expression such as that in Eq. (4.41), the concentration $[X^{\ddagger}]$ is the concentration when the reactants and products are at equilibrium. If product molecules are absent the concentration is half of this, as previously discussed; then the complexes are all $X_l^{\ddagger}$ complexes (see Fig. 4.5). The rate of reaction, however, is given correctly by multiplying $[X^{\ddagger}]$ by the frequency

v, since a frequency is by definition the frequency of crossing *in one direction*. The frequency of crossing *in either direction* is $2v$, and an alternative procedure is to multiply $2v$ by $[X_i^{\ddagger}]$; the result is the same since $[X_i^{\ddagger}] = \frac{1}{2}[X^{\ddagger}]$. Confusion about this point has led sometimes to the belief that Eq. (4.47) is in error by a factor of two.[29]

Derivation 2   The second derivation we give is the one originally presented by Eyring[30] and by Evans and Polanyi.[3] In this derivation the motion over the col is not expressed as a vibrational motion, as in Derivation 1, but as a translational motion. Figure 4.5 shows a schematic representation of the top of the potential-energy barrier, all complexes lying within the arbitrary length δ being activated complexes. The translational partition function corresponding to the motion of a particle of mass $m_{\ddagger}$ in a one-dimensional box of length δ is

$$q_t = \frac{(2\pi m_{\ddagger} kT)^{1/2}}{h} \delta \tag{4.48}$$

Thus, the partition function $q^{\ddagger}$ for the activated complexes may be written as

$$q^{\ddagger} = \frac{(2\pi m_{\ddagger} kT)^{1/2} \delta}{h} q_{\ddagger} \tag{4.49}$$

where $q_{\ddagger}$ is for all the motions except that over the col. Introduction of this equation into Eq. (4.41) then gives

$$[X^{\ddagger}] = [A][B] \frac{(2\pi m_{\ddagger} kT)^{1/2}}{h} \delta \frac{q_{\ddagger}}{q_A q_B} e^{-E_0/RT} \tag{4.50}$$

This again includes activated complexes moving in both directions and is therefore twice $[X_i^{\ddagger}]$.

According to kinetic theory the average speed of particles *moving from left to right* over the barrier is

$$\bar{x} = \left(\frac{kT}{2\pi m_{\ddagger}}\right)^{1/2} \tag{4.51}$$

The frequency with which the complexes $X_i^{\ddagger}$ pass over the barrier is this speed divided by δ:

$$v = \left(\frac{kT}{2\pi m_{\ddagger}}\right)^{1/2} \frac{1}{\delta} \tag{4.52}$$

The rate of reaction is the concentration of complexes, given by Eq. (4.50), multiplied by this frequency:

$$v = [A][B] \frac{(2\pi m_{\ddagger} kT)^{1/2}}{h} \delta \left(\frac{kT}{2\pi m_{\ddagger}}\right)^{1/2} \frac{1}{\delta} \frac{q_{\ddagger}}{q_A q_B} e^{-E_0/RT} \tag{4.53}$$

$$= [A][B] \frac{kT}{h} \frac{q_{\ddagger}}{q_A q_B} e^{-E_0/RT} \tag{4.54}$$

The rate constant k is again that given by Eq. (4.47).

The distance δ cancels out in this derivation, and its magnitude is thus irrelevant.

An analogy is that if we know the number of cars on a bridge, the length of the bridge, and the average speed with which the cars are traveling, we can calculate the number of cars that arrive at the end of the bridge in unit time. If we repeat the calculation for half the length of the bridge, the result is the same; the precise length chosen is arbitrary. It follows that the concentration $[X^{\ddagger}]$ given by Eq. (4.50) is arbitrary; its value depends on the length δ, and this is a matter of choice.

Derivations 1 and 2 appear at first sight to be different, but they are basically the same. Derivation 1 regards the passage over the barrier as a very loose vibration, while Derivation 2 regards the passage as a free translation. If a particle is vibrating and the restoring force is reduced gradually to zero, the vibration ultimately becomes a translation. The partition function for a very loose vibration ($\mathbf{k}T/h\nu$) therefore should pass smoothly into that for a translation. That this is the case can be seen as follows. If the expression (4.52) for ν is substituted into $\mathbf{k}T/h\nu$, the result is the expression in Eq. (4.49), which is the partition function for one-dimensional motion in the length δ.

Other Derivations   Several alternative derivations of the basic equation of CTST [Eq. (4.47)] have been given, and a study of all of them is helpful in gaining a complete appreciation of the treatment. Mahan[32] started with a different formulation of the statistical mechanical treatment of equilibrium. He regarded reaction as involving the motion of a system in phase space and considered an element of volume $dp_1 dp_2 \cdots dp_{3N} dq_1 dq_2 \cdots dq_{3N}$ for a system involving N atoms. He then wrote down the standard statistical-mechanical expression for the fraction of molecules existing in a particular element of phase space; this expression involves the classical Hamiltonian for the system. He then extracted from this Hamiltonian the contribution associated with passage over the col and finally arrived at the CTST rate equation (4.47). This derivation differs from the others only in starting with a different statistical-mechanical equation; the assumptions made are exactly the same.

There have been a number of discussions of CTST which put more emphasis on the dynamical *approach* of the system to the activated-complex region of the potential-energy surface, rather than on the equilibrium between initial and activated states. The first of these discussions was that of Wigner,[33] who did not derive the equation. A derivation along the same lines, formulated by Bishop and Laidler,[34] does not make the equilibrium assumption *explicitly*. We shall see later (Section 4.8.1) that if systems pass over the col only once, equilibrium must be established.

4.5.4 Symmetry Numbers and Statistical Factors

In the evaluation of equilibrium constants by the use of partition functions it is customary to divide the rotational partition function for each of the molecules by its *symmetry number* σ. The symmetry number is obtained by imaging all identical atoms to be labeled and by counting the number of different but equivalent arrangements that can be made by rotating (but not reflecting) the molecule. Thus, if the atoms in the hydrogen molecule are labeled 1 and 2, the following two arrangements are possible:

$$H^1 - H^2 \qquad H^2 - H^1$$

The symmetry number is therefore 2, as it is in the water molecule:

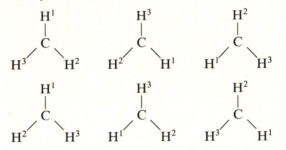

The symmetry number of ammonia (nonplanar) is 3:

That of the planar methyl radical is 6:

Note that the lower three are obtained by turning the flat molecules over, a procedure not possible with the nonplanar ammonia molecule. It is verified easily that the symmetry number of methane is 12, as is that of benzene.

The necessity of taking symmetry into account in this way may be understood by comparing the following two reactions:

$$(1) \quad H_2 + Cl \rightleftharpoons HCl + H$$

$$(2) \quad HD + Cl \rightleftharpoons HCl + D$$

The symmetry number of H_2 is 2, whereas all the other symmetry numbers are unity. If the partition functions *without symmetry numbers* are denoted by q^0, with the appropriate subscript indicating the species, the equilibrium constants for these two reactions would be

$$K_1 = \frac{2q^0_{HCl}q^0_H}{q^0_{H_2}q^0_{Cl}} e^{-E_0/RT} \tag{4.55}$$

and

$$K_2 = \frac{q^0_{HCl}q^0_D}{q^0_{HD}q^0_{Cl}} e^{-E_0/RT} \tag{4.56}$$

The reason for the 2 in the numerator of Eq. (4.55) is that the partition function for H_2 is $q^0_{H_2}$ divided by the symmetry number 2. If the q^0's are unaffected by replacing H by D and the E_0's are the same, as is true to a good approximation, K_1 is twice K_2. If we label the H atoms in H_2, reaction 1 can produce HCl in two different ways:

$$H^1{-}H^2 + Cl \rightarrow H^1{-}Cl + H^2$$

and

$$H^1{-}H^2 + Cl \rightarrow H^2{-}Cl + H^1$$

There is a double chance of getting HCl, which is not the case with reaction 2.

This standard procedure of dividing by symmetry numbers is always correct for equilibrium constants. For rate constants, however, unless certain special conventions are followed, the procedure sometimes produces anomalies. An example is provided by the reaction

$$(1) \quad H + H{-}H \rightarrow H\cdots H\cdots H^{\ddagger} \rightarrow H{-}H + H$$
$$\qquad\qquad \sigma = 2 \qquad\qquad \sigma = 2$$

The use of symmetry numbers with CTST gives

$$k_1 = \frac{kT}{h} \frac{q_{\ddagger}^0/2}{q_H^0 q_{H_2}^0/2} e^{-E_0/RT} \tag{4.57}$$

$$= \frac{kT}{h} \frac{q_{\ddagger}^0}{q_H^0 q_{H_2}^0} e^{-E_0/RT} \tag{4.58}$$

Since both H_2 and the activated complex have a symmetry number of 2, the factor of 2 cancels out.

Equation (4.58) must be incorrect, as is seen by comparing it with that obtained for the reaction

$$(2) \quad D + H{-}H \rightarrow D\cdots H\cdots H^{\ddagger} \rightarrow D{-}H + H$$
$$\qquad\qquad \sigma = 2 \qquad\qquad \sigma = 1$$

The rate equation using the same convention is now

$$k_2 = \frac{kT}{h} \frac{q_{\ddagger}^0}{q_D^0 q_{H_2}^0/2} e^{-E_0/RT} \tag{4.59}$$

$$= 2 \frac{kT}{h} \frac{q_{\ddagger}^0}{q_D^0 q_{H_2}^0} e^{-E_0/RT} \tag{4.60}$$

The conclusion is that reaction 2 is favored over reaction 1 by a factor of 2. This, however, violates common sense, since in both reactions a hydrogen atom is being abstracted from H_2; there cannot be a factor of 2 according to whether a hydrogen atom or some other atom is doing the abstracting. The factor of 2 should be in Eq. (4.58) as well as in Eq. (4.60).

Another anomaly is found if a species S (assumed to have $\sigma = 1$) abstracts a proton from the pyramidal ions H_3O^+, H_2DO^+, and HD_2O^+:

$$H_3O^+ + \quad S \quad \rightarrow \quad \overset{\overset{\textstyle H}{\displaystyle |}}{O}\!\!\overset{\cdot\cdot H}{\cdots}\!H\cdots S^{\ddagger} \rightarrow H_2O + HS^+$$
$$\sigma = 3 \quad \sigma = 1 \qquad\qquad \sigma = 1$$

$$H_2DO^+ + \quad S \quad \rightarrow \quad \overset{\overset{\textstyle D}{\displaystyle |}}{O}\!\!\overset{\cdot\cdot H}{\cdots}\!H\cdots S^{\ddagger} \rightarrow HDO + HS^+$$
$$\sigma = 1 \quad \sigma = 1 \qquad\qquad \sigma = 1$$

$$HD_2O^+ + \quad S \quad \rightarrow \quad \overset{\overset{\textstyle D}{\displaystyle |}}{O}\!\!\overset{\cdot\cdot D}{\cdots}\!H\cdots S^{\ddagger} \rightarrow D_2O + HS^+$$
$$\sigma = 1 \quad \sigma = 1 \qquad\qquad \sigma = 1$$

The conventional use of symmetry numbers now predicts factors in the ratio 3:1:1 for these three reactions; common sense, however, demands 3:2:1, since this is the ratio of the numbers of protons that are available for transfer.

There is a need for modification of the symmetry number procedure, or for the use of an alternative procedure. Some workers[35] have proposed certain rules that will make the conventional procedure work. It has been suggested that when, the products are identical with the reactants, as in the reaction $H + H_2$, the expression (4.58) should be multiplied by 2; the justification is that reaction can occur in both directions. Also, when the activated complex is an enantiomer, multiplication by 2 is required, since the two forms can be produced.§ This is true for the reaction

$$H_2DO^+ + \quad S \quad \rightarrow \quad \overset{\overset{\displaystyle D}{\overset{\displaystyle |}{}}}{O}\overset{\displaystyle \cdot\,\cdot H}{\underset{}{\cdots H \cdots S^\ddagger}} \rightarrow DHO + HS^+$$

$$(\sigma = 1) \quad \sigma = 1 \quad (\sigma = 1, \text{ but an enantiomer})$$

Multiplication by 2 gives the correct ratio 3:2:1 for the trio of reactions mentioned.

These procedures are perfectly correct, but it is too easy to make mistakes; in particular, the formation of an enantiomeric activated complex may well escape detection. The author prefers an alternative procedure, involving the use of statistical factors.[36] This procedure also has some pitfalls, but these are avoided more easily than when symmetry numbers are used.[37]

The statistical factor for a process is defined as the number of different sets of chemically plausible products that can be formed if all identical atoms in the reactant molecules are labeled. Thus, for the overall reaction

$$H + CH_4 \rightleftharpoons H_2 + CH_3 \text{(planar)}$$

the species on the left-hand side can be labeled as follows:

$$H^1 + CH^2H^3H^4H^5$$

Four sets of chemically plausible¶ products can be formed:

$$(1) \quad H^1H^2 + CH^3H^4H^5 \qquad (2) \quad H^1H^3 + CH^2H^4H^5$$

$$(3) \quad H^1H^4 + CH^2H^3H^5 \qquad (4) \quad H^1H^5 + CH^2H^3H^4$$

The statistical factor, designated l for a reaction from left to right, is therefore 4. If we now consider the reaction from right to left, the species $H_2 + CH_3$ can be labeled as follows:

$$H^1H^2 + CH^3H^4H^5$$

Again, we can obtain four sets of chemically plausible products:

---

§ V. Gold [*Trans. Faraday Soc.*, **60**, 739 (1964)] has suggested the alternative but equivalent procedure of setting $\sigma = \frac{1}{2}$ for an enantiomeric mixture.

¶ Products such as $H^2H^3 + CH^1H^4H^5$ are not considered to be chemically plausible. They can be included if desired, but there is no advantage. Bishop and Laidler [*Trans. Faraday Soc.*, **66**, 1685 (1970)] proved that Eq. (4.61) applies whether one confines oneself to chemically plausible products or includes the implausible ones.

$$(1) \quad H^1 + CH^2H^3H^4H^5 \qquad (2) \quad H^2 + CH^1H^3H^4H^5$$

$$(3) \quad H^1 + CH^3H^2H^4H^5 \qquad (4) \quad H^2 + CH^3H^1H^4H^5$$

[the methanes formed in (1) and (3) and in (2) and (4) are enantiomers]. The statistical factor for the reaction from right to left, now designated r, is thus 4.

For the overall reaction we have

$$H \;+\; CH_4 \underset{r=4}{\overset{l=4}{\rightleftharpoons}} H_2 \;+\; CH_3$$
$$\sigma = 1 \quad\; \sigma = 12 \quad\;\; \sigma = 2 \quad\; \sigma = 6$$

It is to be noted that the ratio l/r (=1) is equal to the ratio of symmetry numbers $\sigma_H\sigma_{CH_4}/\sigma_{H_2}\sigma_{CH_3}$ (=1 × 12/2 × 6 = 1). Bishop and Laidler have proved that for any reaction

$$A + B + \cdots \underset{r}{\overset{l}{\rightleftharpoons}} \cdots + Y + Z$$

the ratio of statistical factors is equal to the ratio of symmetry numbers:

$$\frac{l}{r} = \frac{\sigma_A\sigma_B\cdots}{\cdots\sigma_Y\sigma_Z} \tag{4.61}$$

It follows that the equilibrium constant can be formulated equally well by the use of symmetry numbers,

$$K = \frac{\cdots(q_Y^0/\sigma_Y)(q_Z^0/\sigma_Z)}{(q_A^0/\sigma_A)(q_B^0/\sigma_B)\cdots} e^{-E_0/RT} \tag{4.62}$$

and by the use of statistical factors,

$$K = \frac{l}{r}\frac{\cdots q_Y^0 q_Z^0}{q_A^0 q_B^0 \cdots} e^{-E_0/RT} \tag{4.63}$$

The latter formulation has advantages, since it gives a clear intuitive grasp of why these numerical factors are necessary. Also, it is a little easier to count statistical factors than symmetry numbers. Most important, it is easier to avoid errors with the statistical factor method. A particular source of error with the symmetry number method arises when the reaction introduces or removes an asymmetric center. A simple example is with the reaction

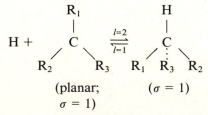

Since the H atom can be attached to either side of the plane, $l = 2$. The correct equilibrium constant is

$$K = 2\,\frac{q_{CHR_1R_2R_3}^0}{q_H^0 q_{CR_1R_2R_3}^0}\, e^{-E_0/RT} \tag{4.64}$$

In using the symmetry number method, it is easy to forget that an asymmetric center has been introduced and that therefore the factor of 2 is needed.

The treatment of the rate equation follows similar lines, but care must be taken with the choice of activated complex (this, of course, is equally true with the symmetry number procedure). The simplest example is the $H + H_2$ reaction. If all the hydrogen atoms are labeled, two equivalent linear activated complexes can be formed:

$$H^1 + H^2 - H^3 \begin{cases} \nearrow H^1 \cdots H^2 \cdots H^{3\ddagger} \rightarrow H^1 - H^2 + H^3 \\ \searrow H^1 \cdots H^3 \cdots H^{2\ddagger} \rightarrow H^1 - H^3 + H^2 \end{cases}$$

The statistical factor in forming the activated complex, which we now write as $l^{\pm}$, is therefore 2:

$$\underset{\sigma = 2}{H} + \underset{\sigma = 2}{H_2} \underset{r^{\ddagger}=2}{\overset{l^{\ddagger}=2}{\rightleftharpoons}} \underset{\sigma = 2}{H \cdots H \cdots H^{\ddagger}}$$

Note that the ratio $l^{\ddagger}/r^{\ddagger} = 1$ is the ratio of the symmetry numbers. The rate equation is

$$k = 2 \frac{kT}{h} \frac{q_{\ddagger}^0}{q_A q_B} e^{-E_0/RT} \tag{4.65}$$

We have avoided the error that might result from the symmetry number method, which does not give this value of 2 unless a special hypothesis is made.

A similar example is

$$OH + H_2 \rightarrow H_2O + H$$

The schematic potential-energy surface for this reaction, shown in Fig. 4.6, is helpful in understanding the statistical factors. The H^1—O radical can attack the H^2—H^3 molecule from either end to produce two activated complexes:

$$H^1 - O \atop \quad \cdots \atop H^2 \cdots H^{3\ddagger} \qquad \text{and} \qquad H^1 - O \atop \quad \cdots \atop H^3 \cdots H^{2\ddagger}$$

These are denoted by the points x_1 and x_2. The former activated complex can lead only to the products H^1—O—H^2 and H^3, and the latter can lead only to H^1—O—H^3 and H^2. The statistical factors are

$$\underset{\sigma = 2}{OH} + \underset{}{H_2} \underset{r^{\ddagger}=1}{\overset{l^{\ddagger}=2}{\rightleftharpoons}} X^{\ddagger} \underset{\sigma = 2}{\leftrightharpoons} H_2O + H$$

The correct rate equation is

$$k = 2 \frac{kT}{h} \frac{q_{\ddagger}^0}{q_{OH}^0 q_{H_2}^0} e^{-E_0/RT} \tag{4.66}$$

The rate equation obtained by either the symmetry number method or the statistical factor method depends on the symmetry chosen for the activated complex. The statistical factor procedure has been particularly useful in showing that there are

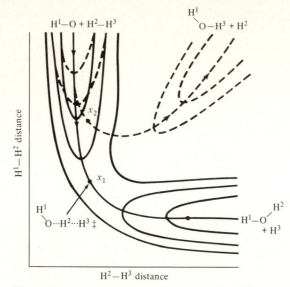

Figure 4.6 Schematic potential-energy surfaces for the reaction $OH + H_2 \rightarrow H_2O + H$, showing the two activated states at x_1 and x_2.

certain symmetry restrictions on activated complexes.[38] Consider, for example, an abstraction reaction such as

$$X + H_2 \rightarrow HX + H$$

where X may be a halogen atom. A schematic potential-energy surface for this reaction is shown in Fig. 4.7. The initial state is at the upper right, and the system has the choice of entering two valleys, one at the upper left leading to $X—H^2 + H^1$ and one at the lower right leading to $X—H^1 + H^2$. It is important to recognize that in such a situation there must be two alternative activated states, shown as points a and b in the diagram. The purely symmetrical activated state corresponding to point x in the diagram is excluded. The system will "cut the corners" and pass through either of the two activated states represented by the points a and b. The activated complex thus has a symmetry number of 1, and the statistical factors are as follows:

$$X + H_2 \underset{r^{\ddagger}=1}{\overset{l^{\ddagger}=2}{\rightleftharpoons}} \overset{X}{\underset{H}{\cdots \cdots}} \overset{H^{\ddagger}}{\cdots} \underset{r^{\ddagger}=1}{\overset{l^{\ddagger}=1}{\rightleftharpoons}} HX + H$$

$$\sigma = 2 \qquad \sigma = 1 \qquad \sigma = 1$$

The correct rate equation, given by either method, is

$$v = 2 \frac{kT}{h} \frac{q_{\ddagger}^0}{q_X^0 q_{H_2}^0} e^{-E_0/RT} \tag{4.67}$$

To summarize, the recommended procedure for dealing with symmetry is as follows. The rate equation first is written with the symmetry numbers omitted. Then it is multiplied by the statistical factor $l^{\ddagger}$ for the formation of the activated complex.

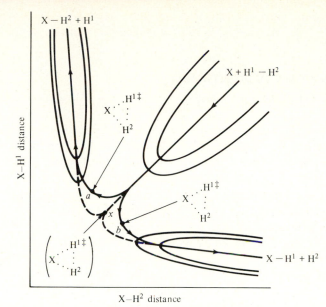

Figure 4.7 Schematic potential-energy surface for the reaction $X + H_2 \rightarrow HX + H$. The activated state is now at a confluence of two valleys, and there are alternative unsymmetrical activated complexes at points a and b.

Care must be taken when there is a confluence of more than two valleys at the activated state, as in the example illustrated in Fig. 4.7. It is then important to recognize that the system will "cut the corners" with a reduction of symmetry of the activated complex. In general, the symmetry of the reaction center in the activated complex is not greater than that of either the reactants or products.

In the language of group theory this conclusion may be expressed more precisely as follows: the point group of the activated complex can be no larger than the largest subgroup common to both reactant and product subgroups.[39] For example, suppose that a linear molecule A—B—B—A (point group $D_{\infty h}$) isomerizes into a molecule having the same symmetry as hydrogen peroxide (point group C_2):

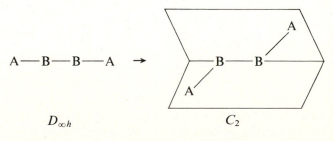

Intuitively, we would expect the activated complex to have C_2 symmetry, and this conclusion is confirmed by the group-theoretical argument. The linear form is of higher symmetry than the bent, and the $D_{\infty h}$ group includes C_2 (rotation through

180°) as a subgroup. Thus, the activated complex may have C_2 symmetry; a linear configuration ($D_{\infty h}$) is excluded, as is to be expected on intuitive grounds.

4.6 SOME APPLICATIONS OF CONVENTIONAL TRANSITION-STATE THEORY

Later in this book a number of detailed applications of transition-state theory will be made, with reference to particular reactions. Sometimes these specific applications are not entirely successful, and there are a number of reasons for this. For one thing, it is necessary to estimate the structure of the activated complex, with its vibrational frequencies, and this is difficult for systems of any complexity. Also, when there is the possibility of tunneling there is always a difficulty, since the theory of tunneling (Sections 4.8.3 and 4.9.2) is very complicated.

In spite of this, one can make a reasonable estimate of the partition function of an activated complex even when the potential-energy surface is not known. One is helped by the fact that at temperatures that are not too high the vibrational partition functions are not much greater than unity and are not very sensitive to the vibrational frequencies. Estimates of these frequencies therefore can be made without introducing much error. As a result, it has proved possible to make useful order-of-magnitude estimates of the magnitudes of pre-exponential factors even when the potential-energy surface is not known. Some examples are to be found in later chapters.

Aside from calculations for specific reactions, one of the great strengths of CTST is that it allows us to arrive at some broad and significant generalizations. One of these, first pointed out by Hinshelwood,[40] is that for bimolecular reactions between *atoms,* CTST and simple collision theory become identical. This is useful in putting both theories into some perspective. Transition-state theory also leads to the useful generalization that, other things being equal, in a bimolecular or trimolecular process the pre-exponential factor decreases as the reacting molecules become more complex. This conclusion is useful in throwing light on experimental results for a number of reactions.

4.6.1 Reactions between Atoms

Consider a reaction between two atoms A and B to form the diatomic molecule AB. This product molecule will retain the energy of bond formation and will dissociate on its first vibration. For present purposes, however, we assume that this energy is removed in some way and that each collision between A and B produces a product molecule AB, with zero activation energy.

The activated complex AB‡ has three degrees of translational freedom and two of rotational. If it were a normal diatomic molecule, it would have one degree of vibrational freedom, but because it is an activated complex this mode of vibration corresponds to decomposition, and the corresponding partition function therefore is omitted. The partition function $q_{\ddagger}$, with this omission, is

$$q_{\ddagger} = \frac{[2\pi(m_A + m_B)\mathbf{k}T]^{3/2}}{h^3} \frac{8\pi^2 I \mathbf{k}T}{h^2} \tag{4.68}$$

where m_A and m_B are the atomic masses, and I is the moment of inertia of the diatomic activated complex. The first term in this expression is the contribution from the three degrees of translational freedom, while the second is for rotation. If d_{AB} is the distance between the centers of the atoms in the activated complex, the moment of inertia is given by

$$I = d_{AB}^2 \frac{m_A m_B}{m_A + m_B} \tag{4.69}$$

The atoms A and B have only translational degrees of freedom, so that the respective partition functions are

$$q_A = \frac{(2\pi m_A \mathbf{k} T)^{3/2}}{h^3} \tag{4.70}$$

and

$$q_B = \frac{(2\pi m_B \mathbf{k} T)^{3/2}}{h^3} \tag{4.71}$$

With an activation energy of E_0 the rate constant is therefore

$$k_a = \frac{\mathbf{k} T}{h} \frac{q_{\ddagger}}{q_A q_B} e^{-E_0/RT} \tag{4.72}$$

Multiplication by L gives the value in molar units. Insertion of the expressions for the partition functions leads, after some reduction, to

$$k_a = L d_{AB}^2 \left(8\pi \mathbf{k} T \frac{m_A + m_B}{m_A m_B} \right)^{1/2} e^{-E_0/RT} \tag{4.73}$$

$$= L d_{AB}^2 \left(\frac{8\pi \mathbf{k} T}{\mu} \right)^{1/2} e^{-E_0/RT} \tag{4.74}$$

where μ is the reduced mass. This equation is identical with Eq. (4.9) which was obtained directly from simple collision theory.

When molecules rather than atoms are undergoing reaction, the result is different, as will now be seen.

4.6.2 Reactions between Molecules

The equations for reactions between molecules are more complicated, and a simplified treatment is useful; this was first given by Bawn.[41] We make the approximation that the partition function for a given type of energy has the same value for each degree of freedom. In reality, the values of the partition functions do not vary greatly with the molecular masses, moments of inertia, or vibrational frequencies, so that the procedure of treating them as equal does not involve a great error. Table 4.1 shows that partition functions per unit length for a single translational degree of freedom are of the order of 10^{10} or 10^{11} m^{-1}; those for rotation are often between 10 and 100 per degree of freedom, while those for vibration are not far from unity at ordinary temperatures.

The partition functions for simple translational, rotational, and vibrational degrees of freedom are written q_t, q_r, and q_v, respectively. For reaction between two atoms,

$$q_A = q_t^3 \qquad q_B = q_t^3 \qquad q_\ddagger = q_t^3 q_r^2 \tag{4.75}$$

since the complex must be linear. The rate constant is thus

$$k_a = \frac{\mathbf{k}T}{h} \frac{q_t^3 q_r^2}{q_t^3 q_t^3} e^{-E_0/RT} \tag{4.76}$$

$$= \frac{\mathbf{k}T}{h} \frac{q_r^2}{q_t^3} e^{-E_0/RT} \tag{4.77}$$

The significance of the ratio q_r^2/q_t^3 is that when two atoms form an activated complex, two degrees of rotational freedom are gained and three degrees of translational freedom are lost.

Suppose instead that A and B are complex nonlinear molecules; then

$$q_A = q_t^3 q_r^3 q_v^{3N_A-6} \qquad q_B = q_t^3 q_r^3 q_v^{3N_B-6} \tag{4.78}$$

where N_A and N_B are the numbers of atoms in the molecules. The activated complex contains $N_A + N_B$ atoms; if it were a normal species it would have $3(N_A + N_B) - 6$ degrees of vibrational freedom. In the activated complex, however, one of these corresponds to passage over the barrier, and only $3(N_A + N_B) - 7$ are to be included in the partition function, which is thus

$$q_\ddagger = q_t^3 q_r^3 q_v^{3(N_A+N_B)-7} \tag{4.79}$$

The rate constant is therefore

$$k_m = \frac{\mathbf{k}T}{h} \frac{q_t^3 q_r^3 q_v^{3(N_A+N_B)-7}}{q_t^3 q_r^3 q_v^{3N_A-6} q_t^3 q_r^3 q_v^{3N_B-6}} e^{-E_0/RT} \tag{4.80}$$

which reduces to

$$k_m = \frac{\mathbf{k}T}{h} \frac{q_v^5}{q_t^3 q_r^3} e^{-E_0/RT} \tag{4.81}$$

The significance of this is that when the molecules A and B form an activated complex there is a loss of three degrees of translational freedom and three degrees of rotational freedom; at the same time there is a gain of five degrees of vibrational freedom.

Comparison of Eqs. (4.77) and (4.81) shows that the pre-exponential factors for the two cases, A_m and A_a, are in the ratio

$$\frac{A_m}{A_a} = \frac{(q_v^5/q_t^3 q_r^3)}{q_r^2/q_t^3} = \frac{q_v^5}{q_r^5} \tag{4.82}$$

Since the collision theory and CTST give the same interpretation of A_a, the CTST pre-exponential factor A_m differs from the collision theory prediction by the factor q_v^5/q_r^5. Since q_v is usually close to unity, while q_r may be from 10 to 100 for a molecule, the discrepancy according to this treatment is 10^{-5}–10^{-10}. With less complex reacting species, the discrepancy is less: for an atom reacting with a linear molecule, forming a linear activated complex, the ratio is q_v/q_r which is 10^{-1}–10^{-2}.

The argument just given is helpful but should not be taken too seriously on the quantitative side. In particular, as Johnston[42] has pointed out, some of the vibrational motions in the activated state are very loose vibrations, so that the corresponding vibrational partition functions in the activated state may be much greater than unity. Consequently, the discrepancy between observed rates and those calculated from simple collision theory may be very much less than the factors of 10^{-5}–10^{-10} predicted in the above treatment. Even when this is taken into account, however, the difference is still very considerable.

For bimolecular gas reactions involving atoms or simple molecules the pre-exponential factors are in the range 10^9–$10^{11}\,dm^3\,mol^{-1}\,s^{-1}$. In such cases, both collision theory and CTST give values in reasonable agreement with experiment. For bimolecular gas reactions between complex molecules, the experimental pre-exponential factors are usually lower by several orders of magnitude; for some examples see Table 5.2. Simple collision theory can give no interpretation of these lower values; it predicts values of the same order of magnitude as for the simple reactions. CTST, however, does give a good interpretation of these lower values for the reactions involving more complex reactant molecules.

CTST gives a particularly good interpretation of the pre-exponential factors for trimolecular gas reactions, as will be seen in Section 5.2.

4.6.3 An Example: The Reaction $H + HBr \rightarrow H_2 + Br$

To illustrate the method used in making CTST calculations for individual reactions, the reaction between hydrogen atoms and the hydrogen bromide molecule is taken as an example. The following data are used:

> Barrier height E_0 (difference between zero-point levels of activated
>   complexes and reactants) = 5.0 kJ mol^{-1}
> HBr internuclear distance = 141.4 pm
> HBr vibrational frequency = 2650 cm^{-1}

The activated complex is taken to be linear, with the following distances:

> H—H distance = 150 pm
> H—Br distance = 142 pm

The real vibrational frequencies in the activated state are 2340 and 460 cm^{-1} (two degenerate bending vibrations). The fourth frequency corresponds to passage over the col.

First, we calculate the masses and moments of inertia of the reactant molecules and the activated complex.

H atom　　　Mass $m = \dfrac{1.008 \times 10^{-3}\ \text{kg mol}^{-1}}{6.022 \times 10^{23}\ \text{mol}^{-1}}$

$$= 1.674 \times 10^{-27}\ \text{kg}$$

HBr molecule   Mass $m = \dfrac{(1.008 + 79.92) \times 10^{-3} \text{ kg mol}^{-1}}{6.022 \times 10^{23} \text{ mol}^{-1}}$

$$= 1.338 \times 10^{-25} \text{ kg}$$

Moment of inertia $I = \dfrac{m_H m_{Br}}{m_H + m_{Br}} d_{HBr}^2$

$$= \dfrac{(1.674 \times 133.8) \times 10^{-27}}{1.674 + 133.8}$$

$$\times (141.4 \times 10^{-12})^2 \text{ kg m}^2$$

$$= 3.306 \times 10^{-47} \text{ kg m}^2$$

Activated complex $H \cdots H \cdots Br$: Let x pm be the distance between the center of mass and the end H atom:

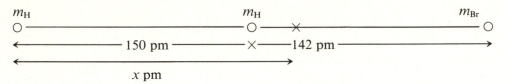

The following equation can then be set up to obtain the position of the center of mass:

$$m_H x + m_H(x - 150) = m_{Br}(292 - x)$$

Insertion of the masses leads to

$$x = 283.3 \text{ pm}$$

Thus, the moment of inertia of the activated complex is

$$I = m_H(283.3 \times 10^{-12} \text{ m})^2 + m_H(133.3 \times 10^{-12} \text{ m})^2 + m_{Br}(8.7 \times 10^{-12} \text{ m})^2$$

Insertion of the masses leads to

$$I = 1.74 \times 10^{-46} \text{ kg m}^2$$

The partition functions at 300 K are calculated as follows:

H Atom Here there is only translational motion and the partition function per unit volume is

$$q_H = \dfrac{(2\pi m_H \mathbf{k} T)^{3/2}}{h^3}$$

Insertion of the values gives

$$q_H = \dfrac{(2\pi \times 1.674 \times 10^{-27} \text{ kg} \times 1.381 \times 10^{-23} \text{ J K}^{-1} \times 300 \text{ K})^{3/2}}{(6.626 \times 10^{-34} \text{ J s})^3}$$

$$= 9.900 \times 10^{29} \text{ m}^{-3}$$

HBr Molecule The translational partition function is

$$q_{t,\mathrm{HBr}} = \frac{[2\pi(m_{\mathrm{H}} + m_{\mathrm{Br}})kT]^{3/2}}{h^3}$$

and insertion of the values gives

$$q_{t,\mathrm{HBr}} = 7.10 \times 10^{32} \text{ m}^{-3}$$

For the HBr molecule the rotational partition function is

$$q_{r,\mathrm{HBr}} = \frac{8\pi^2 I kT}{h^2}$$

and insertion of the values gives

$$q_{r,\mathrm{HBr}} = \frac{8\pi^2 \times 3.306 \times 10^{-47} \text{ kg m}^2 \times 1.381 \times 10^{-23} \text{ J K}^{-1} \times 300 \text{ K}}{(6.626 \times 10^{-34} \text{ J s})^2}$$

$$= 24.6$$

The vibrational partition function is

$$q_{v,\mathrm{HBr}} = \frac{1}{1 - e^{-h\nu/kT}}$$

where $\nu = 2650 \text{ cm}^{-1}$, which corresponds to a frequency of $7.94 \times 10^{13} \text{ s}^{-1}$. This leads to

$$\frac{h\nu}{kT} = \frac{6.626 \times 10^{-34} \text{ J s} \times 7.94 \times 10^{13} \text{ s}^{-1}}{1.381 \times 10^{-23} \text{ J K}^{-1} \times 300 \text{ K}}$$

$$= 12.7$$

This is sufficiently large to cause $q_{v,\mathrm{HBr}}$ to be very close to unity. Thus, the total partition function for HBr, per unit volume, is

$$q_{\mathrm{HBr}} = 7.10 \times 10^{32} \text{ m}^{-3} \times 24.6$$

$$= 1.75 \times 10^{34} \text{ m}^{-3}$$

Activated Complex The mass of the activated complex is $2m_{\mathrm{H}} + m_{\mathrm{Br}} = 1.37 \times 10^{-25}$ kg, and the translational contribution to its partition function is therefore

$$q_t^{\ddagger} = \frac{(2\pi \times 1.37 \times 10^{-25} \times 1.381 \times 10^{-23} \times 300)^{3/2}}{(6.626 \times 10^{-34})^3} \text{ m}^{-3}$$

$$= 7.32 \times 10^{32} \text{ m}^{-3}$$

The rotational partition function for the linear activated complex is

$$q_r^{\ddagger} = \frac{8\pi^2 \times 1.74 \times 10^{-46} \times 1.381 \times 10^{-23} \times 300}{(6.626 \times 10^{-34})^2}$$

$$= 129.7$$

The vibrational partition function corresponding to the real vibrational frequencies of the activated complex is

$$q_v^{\ddagger} = \frac{1}{(1 - e^{-h\nu_1/kT})(1 - e^{-h\nu_2/kT})^2}$$

Insertion of

$$\nu_1 = 2340 \text{ cm}^{-1} = 7.013 \times 10^{13} \text{ s}^{-1}$$

$$\nu_2 = 460 \text{ cm}^{-1} = 1.37 \times 10^{13} \text{ s}^{-1}$$

leads to

$$q_v^{\ddagger} = \frac{1}{(1 - e^{-11.2})(1 - e^{-2.19})^2}$$

$$= 1.27$$

Thus, the complete partition function for the activated complex is

$$q_{\ddagger} = 7.32 \times 10^{32} \times 129.7 \times 1.27 \text{ m}^{-3}$$

$$= 1.21 \times 10^{35} \text{ m}^{-3}$$

The *rate constant* is given by Eq. (4.47):

$$k = \frac{kT}{h} \frac{q_{\ddagger}}{q_H q_{HBr}} e^{-E_0/RT}$$

Insertion of the partition functions, with $E_0 = 5.0 \text{ kJ mol}^{-1}$, gives

$$k = \frac{1.381 \times 10^{-23} \times 300 \times 1.21 \times 10^{35} \text{ m}^{-3} \text{ s}^{-1}}{6.626 \times 10^{-34} \times 9.9 \times 10^{29} \times 1.75 \times 10^{34}} e^{-5000/8.31 \times 300}$$

$$= 5.88 \times 10^{-18} \text{ m}^3 \text{ s}^{-1}$$

for the rate constant at 300 K. These are molecular units; multiplication by L ($=6.022 \times 10^{23} \text{ mol}^{-1}$) and by 1000 dm^3 m^{-3} gives

$$k = 3.54 \times 10^9 \text{ dm}^3 \text{ mol}^{-1} \text{ s}^{-1}$$

4.7 THERMODYNAMIC FORMULATION OF CONVENTIONAL TRANSITION-STATE THEORY

For some purposes it is convenient to express rate constants in terms of thermodynamic quantities instead of partition functions. This is often done for reactions in solution, since partition functions for species in the liquid phase are hard to evaluate, whereas thermodynamic quantities may be easier to estimate. The thermodynamic formulation of CTST is due to Wynne-Jones and Eyring.[43]

The equilibrium constant for the formation of an activated complex $X^{\ddagger}$ from two reactants A and B may be written as

$$K_c^{\ddagger} = \left(\frac{[X^{\ddagger}]}{[A][B]}\right)_{eq} = \frac{q_{\ddagger}}{q_A q_B} e^{-E_0/RT} \tag{4.83}$$

Strictly speaking, the partition function for the activated complex that appears in this expression should include the motion over the col; that is, it should be $q^{\ddagger}$ and not $q_{\ddagger}$ [compare Eq. (4.43)]. The equilibrium constant of Eq. (4.83) is therefore a modified equilibrium constant. Comparison of Eqs. (4.83) and (4.47) leads to

$$k = \frac{\mathbf{k}T}{h} K_c^{\ddagger} \tag{4.84}$$

If $K_c^{\ddagger}$ is now expressed in terms of $\Delta^{\ddagger}G°$, the change in the standard Gibbs energy when the activated complexes are formed from the reactants, the result is

$$k = \frac{\mathbf{k}T}{h} e^{-\Delta^{\ddagger}G°/RT} \tag{4.85}$$

The quantity $\Delta^{\ddagger}G°$ is known as the *standard Gibbs energy of activation.* It can be split into $\Delta^{\ddagger}H° - T\,\Delta^{\ddagger}S°$ to give

$$k = \frac{\mathbf{k}T}{h} e^{\Delta^{\ddagger}S°/R} e^{-\Delta^{\ddagger}H°/RT} \tag{4.86}$$

where $\Delta^{\ddagger}S°$ is the *standard entropy of activation* and $\Delta^{\ddagger}H°$ is the *standard enthalpy of activation.*

It is useful to recast Eq. (4.86) into a form that involves the Arrhenius activation energy E_a rather than the enthalpy of activation $\Delta^{\ddagger}H°$; the two quantities are not quite the same. The Arrhenius activation energy is defined by Eq. (2.81):

$$\frac{d \ln k}{dT} \equiv \frac{E_a}{RT^2} \tag{4.87}$$

Since $K_c^{\ddagger}$ is a concentration equilibrium constant, its variation with temperature is given by the equation

$$\frac{d \ln K_c^{\ddagger}}{dT} = \frac{\Delta^{\ddagger}U°}{RT^2} \tag{4.88}$$

where $\Delta^{\ddagger}U°$ is the standard change in internal energy in passing from the initial to the activated state. Differentiation of the logarithmic form of Eq. (4.84) gives

$$\frac{d \ln k}{dT} = \frac{d}{dT}(\ln T + \ln K_c^{\ddagger}) \tag{4.89}$$

$$= \frac{1}{T} + \frac{d \ln K_c^{\ddagger}}{dT} \tag{4.90}$$

With Eq. (4.88) this gives

$$\frac{d \ln k}{dT} = \frac{1}{T} + \frac{\Delta^{\ddagger}U°}{RT^2} = \frac{RT + \Delta^{\ddagger}U°}{RT^2} \tag{4.91}$$

Comparison of this equation with Eq. (4.87) then gives

$$E_a = RT + \Delta^{\ddagger}U° \tag{4.92}$$

Since H is defined as $U + PV$, the general relationship between $\Delta^{\ddagger}H°$ and $\Delta^{\ddagger}U°$ is, for a process at constant pressure,

$$\Delta^{\ddagger}H° = \Delta^{\ddagger}U° + P\,\Delta^{\ddagger}V° \tag{4.93}$$

The quantity $\Delta^{\ddagger}V°$, the standard change in volume in going from the initial to the activated state, is known as the *standard volume of activation* (see Section 6.5). With Eq. (4.92) this equation gives

$$E_a = \Delta^{\ddagger}H° - P\,\Delta^{\ddagger}V° + RT \tag{4.94}$$

The quantity $P\,\Delta^{\ddagger}V°$ is different for different types of reaction. For a *unimolecular reaction* there is no change in the number of molecules, and therefore $\Delta^{\ddagger}V°$ is zero. In that case

$$E_a = \Delta^{\ddagger}H° + RT \tag{4.95}$$

The rate equation (4.86) therefore may be written as

$$k = \frac{\mathbf{k}T}{h}\,e^{\Delta^{\ddagger}S°/R}e^{-(E_a-RT)/RT} \tag{4.96}$$

$$= e\frac{\mathbf{k}T}{h}\,e^{\Delta^{\ddagger}S°/R}e^{-E_a/RT} \tag{4.97}$$

This equation also applies to a good approximation to any reaction in *solution* or in a *condensed* system, since then $\Delta^{\ddagger}V°$ is small.

The relationship, however, is different for a gas reaction other than a unimolecular one. For gas reactions, $P\,\Delta^{\ddagger}V°$ is given in general by

$$P\,\Delta^{\ddagger}V° = \Delta^{\ddagger}n\,RT \tag{4.98}$$

where $\Delta^{\ddagger}n$ is the change in the number of molecules when the activated complex is formed from the reactants. For a unimolecular reaction $\Delta^{\ddagger}n$ is zero and Eq. (4.97) applies. For a bimolecular gas reaction $\Delta^{\ddagger}n$ is -1, and Eq. (4.94) then becomes

$$E_a = \Delta^{\ddagger}H° + 2RT \tag{4.99}$$

Insertion of this into Eq. (4.86) leads to

$$k = e^2\frac{\mathbf{k}T}{h}\,e^{\Delta^{\ddagger}S°/R}e^{-E_a/RT} \tag{4.100}$$

Some of the relationships involving the thermodynamic parameters $\Delta^{\ddagger}U°$ and $\Delta^{\ddagger}H°$ are illustrated in Fig. 4.8, which applies to a bimolecular gas reaction. As previously discussed (Section 3.1.2), the experimental activation energy E_a, defined by Eq. (2.82), is the difference between the average energy of the activated complexes and the average energy of all the reactant molecules. For a bimolecular gas reaction, $\Delta^{\ddagger}H°$ is equal to $E_a - 2RT$ [Eq. (4.99)], while $\Delta^{\ddagger}U$ is $E_a - RT$ [Eq. (4.92)], and these relationships are shown in the figure. Pacey[44] has discussed this matter in some detail and has shown that the relationships are valid only if the transmission coefficient is independent of temperature.

If a rate constant for a bimolecular reaction is expressed in the units $dm^3\,mol^{-1}$ s^{-1}, the standard state to which the entropy of activation and the Gibbs energy of activation relate is $1\,mol\,dm^{-3}$. If the entropy of activation corresponding to this

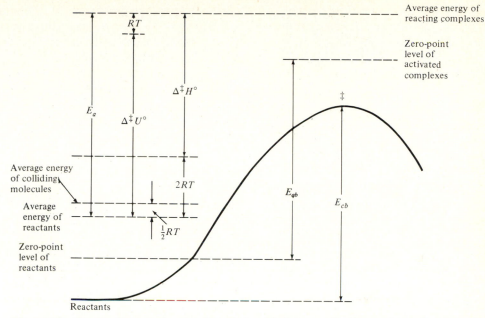

Figure 4.8 Energy diagram showing the relationship between the thermodynamic parameters and other energy changes: E_{cb} = classical energy barrier; E_{qb} = quantum energy barrier; and E_a = experimental activation energy, defined by Eq. (2.81).

standard state is about -50 J K^{-1} mol^{-1}, the pre-exponential factor is close to that given by the simple collision theory [Eq. (4.10)].

Equations such as (4.97) and (4.100) have sometimes been criticized as being dimensionally incorrect, and clumsy alternative formulations have been proposed to avoid this apparent difficulty. No problem arises if correct procedures are applied with respect to standard states, and this must always be done when rate constants or equilibrium constants are related to thermodynamic quantities. The IUPAC recommendation[45] is that equations such as (4.97) and (4.100) be used as they stand and that the more elaborate formulations are unnecessary.

4.8 ASSUMPTIONS AND LIMITATIONS OF CONVENTIONAL TRANSITION-STATE THEORY§

The derivations of the CTST equations that were given in Section 4.5 are somewhat deceptive, in that some important assumptions do not appear explicitly. These assumptions were listed briefly in Section 4.5, and in this section they are discussed in more detail. A few general points may be mentioned first.

A complete CTST calculation requires some knowledge of the potential-energy

§ The remainder of this chapter could be omitted by the reader who does not wish to explore these more advanced topics.

surface for the reaction, since one must know the barrier height, the interatomic distances and angles in the activated complex, and also the real vibrational frequencies in the activated complex. As seen in Chapter 3, such surfaces are difficult to calculate, and reliable surfaces are available only for reactions involving a relatively small number of atoms. However, if one is content with an order-of-magnitude estimate of a rate constant, empirical methods may provide adequate information about the activated state. The most difficult problem is with the barrier height, which enters exponentially into the rate equation and therefore has a strong influence on the rate. The interatomic distances and angles do not have as profound an effect on the rate and usually can be estimated satisfactorily. The vibrational frequencies are even less critical, since vibrational partition functions at ordinary temperatures are usually close to unity and are not strongly dependent on the frequencies. However, there are some exceptions.

Another possible source of error in CTST calculations is in the use of the standard expressions for partition functions listed in Table 4.1. Under ordinary conditions these are reliable, but modifications are necessary in some special situations. For example, the translational and rotational expressions in Table 4.1 were obtained by replacing summations by integrations and are therefore valid only if the spacing between the energy levels is very small compared to $\mathbf{k}T$. These expressions are therefore in error at low temperatures. The vibrational expression in Table 4.1 is a quantum expression, but it is based on the assumption that the vibrations are harmonic. This is no longer the case at high temperatures, and corrections are necessary. For details, the reader is referred to texts listed in the bibliography at the end of this chapter.

We now turn to a consideration of the assumptions and limitations of CTST itself.

4.8.1 Multiple Crossings and the Equilibrium Hypothesis

The first two assumptions of CTST (Section 4.5) are interrelated: there are no multiple crossings of the dividing surface at the col (Assumption 1), and the concentrations of activated complexes may be calculated using equilibrium theory (Assumption 2).

This matter may be considered more explicitly by following an argument first suggested by Anderson.[46] Figure 4.9 shows a number of different ways in which a system may cross a potential-energy surface. Six trajectories are shown, of which 1, 2, and 3 begin in the reactant valley and 4, 5, and 6 begin in the product valley. Trajectories 1 and 4 involve a simple crossing of the col, with no turning back. In trajectory 2 the system starts in the reactant valley, crosses the col at b, and is then bounced back so that it returns to the reactant valley. Trajectory 5 is similar but it starts in the product valley. Trajectories 3 and 6 start in one valley and end in the other, but in each the dividing surface is crossed three times.

If all trajectories were like trajectories 1 and 4 there would be no error in the quasiequilibrium hypothesis. In Fig. 4.9, however, there are six crossings from left to right, indicated by points a to f. Conventional transition-state theory counts of all these as contributing to reaction, but in reality it is only trajectories 1 and 3 that contribute to reaction from left to right. Thus, as a result of recrossing effects, conventional transition-state theory leads to rates that are too high.

This way of looking at the problem suggests how CTST might be improved. In

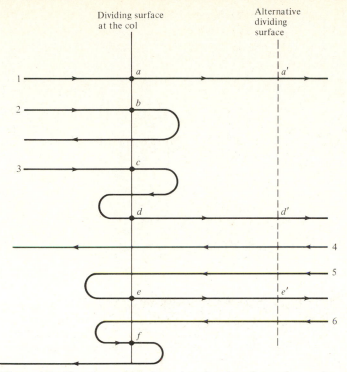

Figure 4.9 Six schematic trajectories, showing alternative modes of crossing the dividing surface at the col of a potential-energy surface. Crossings from the reactant valley into the product valley are at points *a–f*.

CTST the dividing surface is located at the col, as shown in Fig. 4.4, but it is possible to place it at other locations. The dashed line in Fig. 4.9 shows an alternative dividing surface, and there is now a crossing from left to right at only three points, *a'*, *d'*, and *e'*. Two of these, *a'* and *d'*, correspond to reaction from left to right, so that this alternative dividing surface has reduced the error due to recrossing effects. One therefore can consider a variety of positions for the dividing surface and calculate the rate corresponding to each. The minimum rate so obtained is the closest to the truth, and if quantum effects are negligible this rate is either equal to or greater than the true rate. This procedure, called *variational transition-state theory* (VTST), is considered in more detail in Section 4.9.1.

There are several reasons why activated complexes formed from reactants may not pass at once into products. One possibility is that there might be a basin at the junction of two valleys, as in Fig. 4.10. A system that enters such a basin might perform a number of vibrations before finding a way of escape from the basin. If this is so, there is a good possibility that the system will return to the reactant valley. The vibration of a system in the basin is represented schematically in Fig. 4.10, which shows the system returning to the reactant valley.

To take care of this situation, the CTST rate expression [Eq. (4.47)] can be

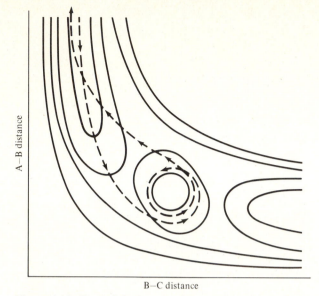

Figure 4.10 Potential-energy surface having a basin, showing a schematic trajectory.

multiplied by a fraction, which is known as the *transmission coefficient* and is represented by the symbol κ (Greek kappa). Thus, the rate equation is written as

$$k = \kappa \frac{\mathbf{k}T}{h} \frac{q_{\ddagger}}{q_A q_B} e^{-E_0/RT} \tag{4.101}$$

For a completely symmetric system having a basin, there is an equal chance of an exit into the reactant and product valleys, and $\kappa = \frac{1}{2}$.

The second reason for failure of activated complexes to form products arises if the shape of the potential-energy surface is such that some of the activated complexes encounter a potential-energy wall and are bounced back into the reactant valley. This situation has been treated a considerable number of times from a dynamical point of view.[47] A model potential-energy surface,[48] designed to illustrate in a qualitative way the results of more realistic calculations, is illustrated in Fig. 4.11. Included is a trajectory in which the system is bounced back into the reactant valley.

The dynamical calculations indicate that return into the entrance valley is most likely for collisions involving much more energy than is required to surmount the energy barrier. For chemical reactions under ordinary conditions, the majority of the collisions that lead to an activated complex are those in which there is little excess energy; this arises because of the application of the Boltzmann principle. As a result, for ordinary chemical reactions, the probability of a return into the entrance valley is not great, and little reduction in the transmission coefficient arises from this cause. This, however, is no longer the case for reactions having a low energy barrier or occurring at high temperatures.

A third situation in which the transmission coefficient is less than unity is for a bimolecular atom combination reaction in the gas phase:

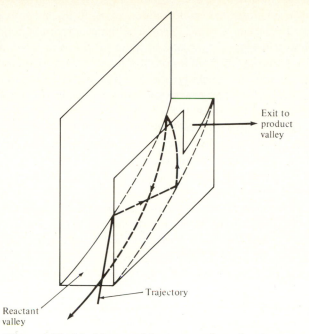

Figure 4.11   Model potential-energy surface, showing a trajectory in which the system is bounced back into the reactant valley.

$$A + A \rightarrow A_2$$

When two atoms collide there is no barrier to reaction and an activated complex is formed on every collision. However, the resulting molecule A_2 still contains the energy of bond formation and will decompose in the period of the first vibration. There is a small probability that the molecule will lose energy by radiation, but apart from this the transmission coefficient is zero. The situation is different if other molecules are present when the atoms come together, as will be discussed in Section 5.4.

The transmission coefficient may also be less than unity when there is a crossing of potential-energy surfaces, so that there are alternative reaction paths. This situation arises when excited electronic states are involved; this is the case in certain photochemical and chemiluminescent processes (Chapter 9).

4.8.2 Separability of the Reaction Coordinate

Assumption 3 of CTST is that the motion of the system over the col in the potential-energy surface can be separated out from all the other motions. First, this procedure is explained and then its validity is examined.

In both derivations of the CTST rate equation, it was noted that one of the vibrational factors in the partition function for the activated complex is of a different character from the remaining ones. This particular vibration was treated in Derivation 1 as a very loose vibration, and in Derivation 2 as a translation. The factor kT/h that appears in the final rate equation results from this free motion across the barrier.

   The situation is illustrated with reference to a linear activated complex A$\cdots$B$\cdots$C‡ for a reaction A + BC. Two schematic potential-energy surfaces for this system are shown in Fig. 4.12. In one of them (a) the valleys are arranged symmetrically. There are four normal modes of vibration for the activated complex; two, shown in the diagram, are for stretching, and two (not shown) are for bending. In the case illustrated in Fig. 4.12a, the passage over the col relates to the antisymmetric vibration, of frequency ν_3. Therefore, the CTST procedure for this case is to omit the vibration of frequency ν_3 from the partition function for the activated complex and to include only the vibration of frequency ν_1 and the two bending vibrations. Note that in this particular case the reaction coordinate does not correspond to the extension of a single bond, but rather to the simultaneous extension of the B—C bond and contraction of the A—B bond.

   A different situation is shown in Fig. 4.12b. Now the passage over the col corresponds to the stretching of the B—C bond; the reaction coordinate is thus the stretching of the B—C bond. For the triatomic case represented in Fig. 4.12b, one would consider the two bending motions and also the vibration of A with respect to B—C, the distance in the latter being fixed. The frequency of this stretching vibration is

$$\nu^* = \frac{1}{2\pi}\left(\frac{k_{AB}}{m^\ddagger}\right)^{1/2} \tag{4.102}$$

where k_{AB} is the force constant for the bond A—B, and $m^\ddagger$ is the reduced mass, given by

$$\frac{1}{m^\ddagger} = \frac{1}{m_B + m_C} + \frac{1}{m_A} \tag{4.103}$$

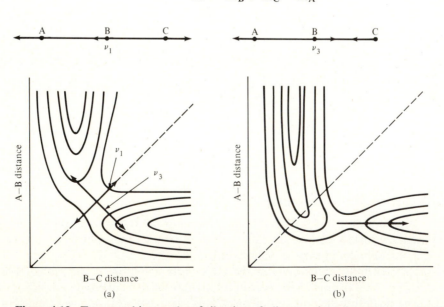

Figure 4.12 Two stretching modes of vibration of a linear activated complex A—B—C. In the contour diagram (a) the reaction coordinate corresponds to the antisymmetric vibration of frequency ν_2; in (b) it corresponds to a simple stretching of the B—C bond.

The procedure in more complicated cases is similar: if the reaction coordinate is a linear combination of two or more interatomic distances, that linear combination is held fixed, and a normal-mode analysis is carried out to determine the relevant vibrational frequencies.

If one is content with approximate estimates of rate constants, the exact choice of reaction coordinate is not of great importance. For exact treatments, however, such as those required for the prediction of kinetic-isotope effects (Chapter 11), the precise reaction coordinate is important and can be known only if one has a reliable potential-energy surface.

We now consider the question of the extent to which it is valid to separate out the motion along the reaction coordinate as the system passes over the col.

If the motion were classical there would be little problem. It is true that even in a classical theory there is some approximation in classifying motions as translational, rotational, and vibrational and regarding them as completely separate. There is some mixing of these types of energy and, if vibrations are anharmonic, there is a flow of energy between the normal modes of vibration. Usually these effects do not introduce much error.

However, the fact that motion is quantized raises serious questions about the validity of the separability assumption. A particularly clear discussion of this problem was first given by Johnston,[49] who proposed ways of circumventing the difficulty. These, however, have now been superseded largely by other treatments. Johnston considered the problem in terms of the relationship between the de Broglie wavelength corresponding to the motion over the col and the region at the col where there is essentially no curvature. Figure 4.5 showed a profile through the minimum-energy path, and a region of arbitrary length δ was regarded as comprising the transition state. In both derivations of the CTST rate equation it was considered that the system is moving freely through that region, with no restraining force, and this is valid only if δ is sufficiently small that the potential-energy curve is flat in that region. For the $H + H_2$ reaction, for example, Johnston and Rapp[49] estimated that δ must not be more than about 30 pm.

This, however, presents a serious difficulty. According to the de Broglie theory a system of mass m and velocity u has a wavelength λ equal to h/mu, where h is the Planck constant. For the transfer of a hydrogen atom, Johnston and Rapp[49] calculated de Broglie wavelengths of 101 pm at 300 K and 78 pm at 500 K. These values are very greatly in excess of 30 pm, the maximum permissible length of the flat region at the col. Therefore, it is not legitimate to regard the motion over the col as quite independent of the other motions of the activated complex.

4.8.3 Quantum Effects

Assumption 4 of CTST is that the motion over the col can be treated as classical motion. The argument just given shows one difficulty that arises when quantum effects are taken into account; we now consider other problems.

It is sometimes thought that the presence of the Planck constant h in the CTST rate equation (4.47) means that the theory is a quantum theory. When the partition functions in Eq. (4.47) are expressed, however, the Planck constant invariably cancels out, since there is always one more degree of freedom in the denominator. A simple example is the atom–atom reaction, treated in Eqs. (4.72)–(4.74).

Besides presenting a difficulty with separability, quantization of the motion introduces the possibility that the system may tunnel through the potential-energy surface. When this occurs, the system does not need to have enough energy to pass over the col; it may move part of the way up the entrance valley and tunnel into the exit valley, as shown schematically in Fig. 4.13. The activation energy is then less than if the system passed over the col, but the probability of tunneling and therefore the related transmission coefficient are now less than unity.

The theoretical treatment of tunneling has proved to be a matter of difficulty; some successful attacks on the problem are considered in Section 4.9.2. One important principle that has emerged is that tunneling is most probable for a thin energy barrier. Another is that tunneling is most important for the transfer of particles of small mass. Tunneling is therefore most probable for a reaction in which an electron is transferred from one species to another. It is also important in reactions in which a hydrogen atom, a proton H^+, or a hydride ion H^- is transferred or is otherwise involved at the reaction center (e.g., in $H + D_2 \rightarrow HD + D$). With any heavier species, even with deuterons, tunneling is negligible. Quantum-mechanical tunneling is an important factor as far as hydrogen–deuterium kinetic-isotope effects are concerned (see Chapter 11).

Because the theoretical treatment of tunneling has proved to be so difficult, there have been many experimental attacks on the problem. It is by no means a straightforward matter to obtain convincing experimental evidence for tunneling, but when tunneling occurs the following is to be expected:

1. Arrhenius plots (of $\ln k$ versus $1/T$) show deviations from linearity of the kind shown in Fig. 4.14. When tunneling occurs the species undergoing reaction are at a lower energy level than if they pass over the col, and the

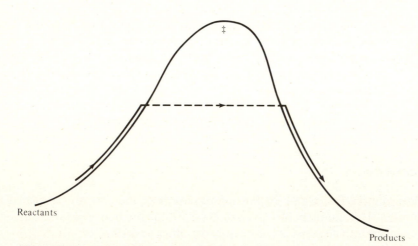

Figure 4.13 Schematic representation of quantum-mechanical tunneling through a potential-energy surface.

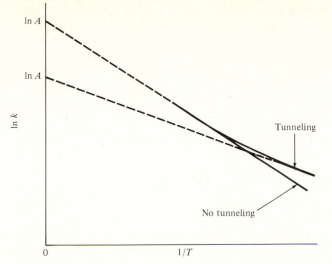

Figure 4.14 Schematic Arrhenius plot, showing the curvature that arises if quantum-mechanical tunneling is significant. The value of ln k extrapolated to $1/T = 0$ is ln A; this is abnormally low if tunneling occurs.

activation energy is reduced.[50] As the temperature is lowered, the rate of the reaction over the col becomes relatively less important than that of the tunneling process. As a result, the observed activation energy at lower temperatures is less than that at higher temperatures. Care must be taken in accepting this effect as evidence for tunneling, since there are other effects that can produce this type of curvature in Arrhenius plots. It can occur, for example, with certain types of composite mechanisms (see Problems 8.2 and 8.3).

2. It follows from this curvature of Arrhenius plots that tunneling will give rise to abnormally low pre-exponential factors, particularly at low temperatures (see Fig. 4.14). Evidence that in an elementary process the pre-exponential factor for a proton transfer is significantly less than that for the corresponding transfer of a deuteron (D^+) is a reliable indication that there is tunneling in the case of the proton.

3. When hydrogen–deuterium kinetic-isotope effects are studied, an abnormally high ratio of rates is expected if there is tunneling with hydrogen (for further details see Section 11.2). Abnormally large ratios thus provide some evidence for tunneling.

4.9 EXTENSIONS OF TRANSITION-STATE THEORY

After transition-state theory in its original form was put forward in 1935, it was recognized that it provided a very valuable insight into how chemical reactions occur. It was realized also that CTST has important limitations, some of which have been discussed in the previous section. These limitations led some workers to discard transition-state theory altogether, and a number of alternative treatments of rates were put forward; these include dynamical (trajectory) treatments, which are considered in Section 12.1.

Another approach to the problem was to make modifications to conventional transition-state theory. Some of these modifications have been extremely successful and have led to much closer agreement between theory and experiment. At the same time the modifications to the theory have given it a much wider applicability. The improved treatments usually have been referred to as *generalized transition-state theory* (GTST). At the present time, much reaction-rate theory is being developed along these lines, and a brief account of the main trends is given in the present section.

Some of the earlier modifications of CTST are referred to first. Discussions in the 1930s by Wigner[51] and by Horiuti[52] led Keck[53] in 1960 to the idea of varying the position of the dividing surface, instead of fixing it at the col as was done in CTST. As we have noted (Section 4.8.1) such a *variational transition-state theory* gives an upper bound to the rate, since it minimizes the effect of multiple crossings of the dividing surface.

Consistent with this idea was a suggestion[54] that the transition state should be taken to be at the maximum of a Gibbs energy profile along the reaction path, rather than at the maximum of a potential-energy profile as in CTST. Some actual calculations were made along these lines, and the rates were found to be significantly different from those given by CTST, especially at high temperatures. It was realized later that this approach to reaction rates is identical to one type of variational TST, namely, *canonical variational TST* (CVTST). The procedure used in CVTST, which is considered later, is to find the variationally best rate constant for a canonical ensemble by taking the structure of the activated complex as a variational parameter.

Another type of modification of CTST was the development of *adiabatic treatments,*[55] in which the reaction system was assumed to remain in quantized vibrational states during the course of reaction. As we have seen, in CTST the initial activated states are treated to some extent quantum mechanically, but the passage over the col is treated classically. The adiabatic treatments brought about some improvement by introducing quantization into the course of reaction. These approaches to the problem have now been superseded by variational treatments and by much improved quantum-mechanical versions of TST, developed in particular by W. H. Miller, D. G. Truhlar, B. C. Garrett, and their coworkers; these are considered later.

In the remaining two subsections generalized transition-state theories, including variational and quantum-mechanical treatments, are outlined. Space does not permit a more detailed discussion, and the reader is referred to the reviews listed in the bibliography to this chapter. Table 4.2 lists abbreviations commonly used in connection with these generalized theories and briefly summarizes their main features.

4.9.1 Variational Transition-State Theory

Variational transition-state theory was first presented in a useful form by Keck[56] in 1960, and it has been developed considerably during recent years, particularly by Truhlar and his co-workers.[57] The essence of the theory is that instead of concentrating

TABLE 4.2 SUMMARY OF CONCEPTS IN TRANSITION-STATE THEORY

Concept	Acronym	Description
Minimum-energy path	MEP	Path of steepest descent from the lowest point of the col into the reactant and product valleys
Dividing surface	DS	Plane surface perpendicular to the MEP
Transition state	TS	Section of potential-energy surface intersected by DS
Conventional transition-state theory	CTST	Conventional TST, with DS at the col
Variational transition-state theory	VTST	Position of DS varied to give minimum rate
(a) Microcanonical	μVTST	Deals with rates at fixed energy E; varies DS to minimize rate; equivalent to ATST with motion treated classically
(b) Canonical	CVTST	Variationally optimized rate at a given temperature; this is equivalent to locating TS at maximum Gibbs energy
(c) Improved canonical	ICVTST	CVTST with DS allowed to vary with E at low energies
Adiabatic transition-state theory	ATST	Reaction system remains in initial quantum states; equivalent to μVTST if motion is classical
Marcus–Coltrin path	MCP	Path corresponding to classical extremities of vibrations

on the situation at the col of the potential-energy surface, as is done in CTST, we can consider dividing surfaces at different positions along the reaction path; an alternative dividing surface was shown in Fig. 4.4. We then can calculate rates of crossing of these various dividing surfaces. As previously discussed, multiple crossings of the dividing surface tend to cause the calculated rates to be higher than the true ones, so that if several dividing surfaces are considered, the lowest rate is closest to the truth.

The way in which the rate equations in generalized transition-state theory are developed is as follows. We can define a *microcanonical*§ rate constant $k(E)$, which relates to reactant systems having a given total energy E. Integration over all energies, taking into account the Boltzmann distribution, then gives the *canonical* rate constant $k(T)$, which is the rate constant at a given temperature. The relationship between $k(T)$ and $k(E)$ is

$$k(T) = \frac{\int_0^\infty \phi_{\text{react}}(E)e^{-E/\mathbf{k}T}k(E)dE}{q_{\text{react}}(T)} \tag{4.104}$$

where $\phi_{\text{react}}(E)$ is the density of energy states for the reactants per unit energy and volume, and $q_{\text{react}}(E)$ is the partition function per unit volume for the reactants. According to CTST, the microcanonical rate constant is given by

$$k_{\text{CTST}}(E) = \frac{N^{\ddagger}(E)}{hq_{\text{react}}(E)} \tag{4.105}$$

where $N^{\ddagger}(E)$ is the number of energetically available internal states of the activated complex. The thermal rate constant in CTST is, as previously derived,

§ In statistical mechanics and rate theory the term "microcanonical" refers to molecular entities having a fixed energy E. The term "canonical" refers to a system in which the entities, such as the reactant molecules, are in thermal equilibrium at a given temperature.

$$k_{CTST}(T) = \frac{kT}{h} \frac{q_{\ddagger}(T)}{q_{react}(T)} e^{-E_0/kT} \qquad (4.106)$$

where E_0 is the energy at the col with reference to that of the reactants.

Variational transition-state theory can be developed in various ways, including the following:

1. *Microcanonical variational transition-state theory* ($\mu VTST$). This involves calculating microcanonical rate constants $k(E)$, corresponding to various dividing surfaces, and finding the minimum rate constant that is obtained when the position of the dividing surface is varied. This minimum $k(E)$ is then inserted into Eq. (4.104), and the thermal rate constant $k(T)$ is obtained by numerical integration.

2. *Canonical variational transition-state theory* (*CVTST*). Now the procedure is not to minimize the microcanonical rate constant $k(E)$ but to calculate $k(T)$ values by numerical integration of Eq. (4.104). The minimum value of this $k(T)$, when the dividing surface is varied, is then accepted as the best estimate of the rate constant.

3. *Improved canonical variational transition-state theory* (*ICVTST*). This is a modification of CVTST in which, for energies below the threshold energy, the position of the dividing surface is taken to be that of the microcanonical variational transition state. This forces the $k(E)$ values to be zero below the threshold energy. A compromise dividing surface is then chosen so as to minimize the contributions to the rate constant made by reactants having higher energies.

It has been shown by Garrett and Truhlar[58] that CVTST is equivalent to locating the transition state at the position of maximum Gibbs energy rather than maximum potential energy. Also, as long as motion through the dividing surface is treated classically, adiabatic transition-state theory is identical with μVTST.[59]

Some idea of the improvements obtained with the variational treatments is given by the values listed in Table 4.3. These values are the rate constants obtained from

TABLE 4.3 RATIOS OF GENERALIZED TRANSITION-STATE THEORY RATE CONSTANTS TO THOSE CALCULATED USING EXACT CLASSICAL DYNAMICS[a]

Type of transition-state theory	$H + H_2 \rightarrow H_2 + H$		$Br + H_2 \rightarrow HBr + H$	
	300 K	2400 K	300 K	2400 K
Conventional (CTST)	1.0	1.5	1.1	4.2
Microcanonical variational (μVTST)	1.0	1.2	1.1	3.7
Canonical variational (CVTST)	1.0	1.5	1.1	3.9
Improved canonical variational (ICVTST)	1.0	1.3	1.1	3.9

[a] These values are taken from D. G. Truhlar and B. C. Garrett, *Acc. Chem. Res.*, **13**, 440 (1980), where additional data are given; see also D. G. Truhlar, *J. Phys. Chem.*, **83**, 188 (1979).

the different versions of TST, compared with the results of exact *classical* dynamical calculations. In these calculations tunneling is neglected, although it is in fact important. For the two reactions $H + H_2$ and $Br + H_2$ at 300 K, all the treatments, including CTST, are equally satisfactory; this is because recrossing effects are not important at lower temperatures. At 2400 K, greater errors are found with conventional TST, and these are to some extent reduced by the variational treatments. Greater errors are found when, for example, a hydrogen atom is being transferred from one heavy atom to another; recrossing effects are then very important. Thus, for the reaction

$$C_4H_9 + H\text{—}C_4H_9 \rightarrow C_4H_9\text{—}H + C_4H_9$$

CTST at 2400 K overestimates the rate by a factor of 11.4. With the variational treatments there is considerable improvement, the ratios then being as follows:

μVTST: 1.7

CVTST: 2.0

IVTST: 1.9

4.9.2 Quantum-Mechanical Transition-State Theory

Various attempts have been made to introduce more quantum mechanics into the TST formulation. So far all such attempts, although helpful, have been somewhat piecemeal and involve serious approximations.

A number of important attacks on the problem have been made by W. H. Miller and his associates.[60] These workers have directed their efforts to the matter of the nonseparability of the reaction coordinate. As discussed in Section 4.8.2, the usual CTST assumption of separability is inconsistent with quantum theory, and problems arise in particular if the de Broglie wavelength for passage over the col is greater than the length of the flat region at the col. One procedure used by Miller and coworkers is to replace the classical expression for the rate of passage over the barrier by a quantum-mechanical expression: doing this properly presents formidable difficulties but semiclassical approximations have been suggested. Miller also made use of the variational procedure, locating the dividing surface at different positions. Some of the numerical results obtained by these procedures have been very impressive. For example, for the $H + H_2$ reaction, CTST without tunneling gives rates that are much too low, because tunneling is important for this system, while rates calculated by Miller's quantum TST are in complete agreement with those obtained by exact quantum-dynamical methods.

Different and very fruitful quantum-mechanical treatments of rates, based on transition-state theory, have been developed by Truhlar, Garrett, and their coworkers.[61] These treatments are extensions of canonical variational transition-state theory (CVTST), which preserves the convenient approximation of separability along the reaction coordinate. The procedure used by these workers has been to extend classical CVTST by adding suitable tunneling corrections.

In the earlier treatments along these lines, use was made of a convenient tunneling treatment proposed by Marcus and Coltrin.[62] A difficulty in dealing with tunneling is that systems move up the entrance valley along various paths, so that there is a wide variation in the probability of tunneling. Marcus and Coltrin concluded that a good approximation to the tunneling probability is obtained by confining attention to a path, shown in Fig. 4.15, that corresponds to the outermost vibrational turning points. This path is known as the *Marcus–Coltrin path*. This procedure has been found to be applicable particularly to potential-energy surfaces in which there is little curvature in the reaction path.

A more general procedure, which is satisfactory even if there is considerable curvature in the reaction path, was developed later by Garrett and Truhlar.[63] This is a variational method which employs a parameter that can be varied from a value appropriate for no curvature to one appropriate for large curvature. The procedure has been tested for a number of reactions[64] and is satisfactory even when tunneling leads to excited vibrational states of the product, such as in the reaction[65] $Cl + HBr \rightarrow HCl + Br$.

Table 4.4 illustrates, for the $H + H_2$ reaction, the type of improvement obtained by the use of refinements to conventional transition-state theory. The figures represent the ratios of calculated rates to those obtained from accurate quantum-dynamical calculations. The CTST values with tunneling neglected are all too low, especially at the lower temperatures where tunneling is relatively more important. Canonical variational theory (CVTST) gives no improvement for this reaction: for this symmetrical system, the best dividing surface is at the col and varying it does not help. Substantial improvement is obtained by making an allowance for tunneling, and the Marcus–Coltrin treatment is very satisfactory.

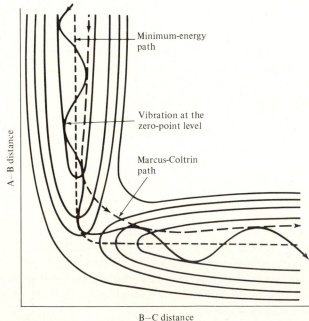

Figure 4.15 Schematic potential-energy surface showing the Marcus–Coltrin path.

TABLE 4.4   RATIOS OF CALCULATED RATES TO THOSE GIVEN BY EXACT QUANTUM-DYNAMICAL CALCULATIONS[a]

Method	200 K	300 K	400 K
Conventional TST (no tunneling)	0.034	0.30	0.50
Canonical variational TST	0.034	0.30	0.50
Canonical variational TST, with tunneling from the ground vibrational state along the Marcus–Coltrin path	1.14	0.88	0.83

[a] Values taken from D. G. Truhlar, A. D. Isaacson, R. T. Skodje, and B. C. Garrett, *J. Phys. Chem.,* **86,** 2252 (1982).

The extensions of conventional transition-state theory have now reached a stage beyond which little further improvement can be expected. The reliable treatment of tunneling is particularly noteworthy. Transition-state calculations can now be made as satisfactorily, and with less labor, as exact quantum-mechanical calculations.

4.9.3 Muonium Kinetics

During recent years tests of rate theory have been applied to some reactions of muonium (Mu). This species consists of a muon, a particle having a single positive charge, plus an electron. Its mass is 1/8.8 that of a hydrogen atom, and it has a lifetime of 2.2×10^{-8} s. In spite of this very short life, it has been possible to make experimental studies of the kinetics of reactions such as

$$Mu + H_2 \rightarrow MuH + H$$

$$Mu + D_2 \rightarrow MuD + D$$

$$Mu + Cl_2 \rightarrow MuCl + Cl$$

Some reviews of muonium chemistry, including its kinetics, are listed in the bibliography at the end of this chapter.

Since muonium is so much lighter than hydrogen, its reactions involve considerable quantum-mechanical tunneling. Experimental and theoretical studies of its reactions are therefore useful in providing tests for theories of tunneling and of rate theories in general. Blais, Truhlar, Garrett, and coworkers[66] for example, carried out trajectory calculations and variational transition-state theory calculations on the reaction between Mu and H_2, and were able to obtain very good agreement with experiment. The study of muonium reactions in relation to kinetic-isotope effects is considered in Section 11.2.5.

4.10 MICROSCOPIC REVERSIBILITY AND DETAILED BALANCE

An important principle that relates to the behavior of systems at equilibrium was first formulated by Tolman,[67] who referred to it as the *principle of microscopic reversibility at equilibrium.* This principle has important applications to kinetic problems, but it has not always been applied correctly; therefore, it is important to state clearly what the principle says and what consequences it has and does not have.

The principle may be expressed as follows:

> In a system at equilibrium, any molecular process and the reverse
> of that process occur, on the average, at the same rate.

The proof is as follows. Consider any molecular configuration, having certain bond angles and distances that define it. Such a species may be undergoing a particular motion; this motion can be described by expressing, as vectors, the velocities of the individual atoms. According to the principles of statistical mechanics, the probability of the existence of such a species, having the motion specified, depends only on the energy of the system, which in turn depends on the squares of the individual velocities. This being so, a species in which all of the motions are exactly reversed has the same energy, and therefore has the same probability of existence. Every molecular species in a system at equilibrium has an exact counterpart which is moving in the opposite direction and which, on the average, is present at the same concentration.

A closely related principle, the *principle of detailed balance at equilibrium* was first discussed and applied by Fowler.[68] This principle relates particularly to collisions between molecules, including reactive collisions. The principle states that in a system at equilibrium each collision has its exact counterpart in the reverse direction, and that the rate of every chemical process is exactly balanced by that of the reverse process.

In terms of transition-state theory it follows from the principles that when a system is at equilibrium the flow of systems over a potential-energy surface must be exactly balanced by a flow in the opposite direction. In particular, it can be concluded that the most probable reaction path in one direction (that is, the path along which most systems move) must also be the most probable path in the opposite direction. The activated state for a reaction in one direction must be the same as the activated state in the other direction. Such a conclusion is in any case necessary in terms of transition-state theory; indeed, if one is working within the framework of that theory, the principle of microscopic reversibility does not introduce anything new.

According to transition-state theory, the rate of a reaction in one direction depends on the increase in Gibbs energy in going from the initial to the activated state. Since the activated states must be the same in both directions, it follows that for an *elementary* reaction the Gibbs energy difference in the two directions must be equal to the overall Gibbs energy difference. It also follows that the ratio of rate constants must be equal to the equilibrium constant.

Wrong conclusions are often drawn from the principles by applying them to systems that are not elementary and are not at equilibrium. For example, it has been concluded that for any reaction the ratio of rate constants in forward and reverse directions must be equal to the equilibrium constant. However, rate constants are measured for systems not at equilibrium, and this conclusion is not necessarily correct for other than elementary reactions. This matter is further discussed in Section 8.2.4.

Sometimes even more unreasonable conclusions are falsely drawn on the basis of the principles. It has been stated, for example, that the rate constants in forward and reverse directions must both be affected in the same way by a change in pH, and by a change in dielectric constant. These conclusions are no more reasonable than the statement that the principles require the activation energies of reactions to be the same for a reaction and its reverse. Such statements would only be correct if pH, dielectric constant, and temperature could have no effect on equilibrium constants.

PROBLEMS

4.1. Two reactions of the same order have identical activation energies and entropies of activation differing by 50 J K^{-1} mol^{-1}. Calculate the ratio of their rate constants at any temperature.

4.2. The gas-phase reaction $H_2 + I_2 \rightarrow 2HI$ is second order. Its rate constant at 400.0°C is 2.34 × 10^{-2} dm^3 mol^{-1} s^{-1}, and its activation energy is 150 kJ mol^{-1}. Calculate, at 400°C, $\Delta^\ddagger H°$, $\Delta^\ddagger S°$, $\Delta^\ddagger G°$, and the pre-exponential factor.

4.3. A substance decomposes according to first-order kinetics, the rate constants at various temperatures being as follows:

Temperature/°C	Rate constant, k/s^{-1}
15.0	4.18 × 10^{-6}
20.0	7.62 × 10^{-6}
25.0	1.37 × 10^{-5}
30.0	2.41 × 10^{-5}
37.0	5.15 × 10^{-5}

Calculate the activation energy. Also, at 25.0°C, calculate $\Delta^\ddagger H°$, $\Delta^\ddagger G°$, $\Delta^\ddagger S°$, and the pre-exponential factor.

4.4. Using collision theory, estimate the collision number for 1 mol (6.022 × 10^{23} molecules) of hydrogen iodide present in a volume of 1 m^3 at 300 K. Take $d_{AA} = 0.35$ nm. If the activation energy for the decomposition of HI is 184 kJ mol^{-1}, what rate constant does kinetic theory predict at 300°C? To what pre-exponential factor and entropy of activation does this result correspond?

4.5. A second-order reaction in solution has rate constants of 5.7 × 10^{-5} dm^3 mol^{-1} s^{-1} at 25.0°C and 1.64 × 10^{-4} dm^3 mol^{-1} s^{-1} at 40.0°C. Calculate the activation energy and the pre-exponential factor, assuming the Arrhenius equation to apply. Also, at 25.0°C, calculate the Gibbs energy of activation $\Delta^\ddagger G°$, the entropy of activation $\Delta^\ddagger S°$, and the enthalpy of activation $\Delta^\ddagger H°$.

4.6. Give the statistical factors in each direction for the following reactions:
(a) $^{35}Cl-^{35}Cl + ^{37}Cl \rightleftharpoons ^{35}Cl-^{37}Cl + ^{35}Cl$
(b) $C\,^{35}Cl_4 + ^{37}Cl \rightleftharpoons C\,^{37}Cl\,^{35}Cl_3 + ^{35}Cl$
(c) $^{35}Cl_2O + ^{37}Cl \rightleftharpoons ^{37}Cl\,^{35}ClO + ^{35}Cl$
(d) $H_2 + Cl \rightleftharpoons HCl + H$
(e) $HD + Cl \rightleftharpoons HCl + D$
(f) $CH_3 + H_2 \rightleftharpoons CH_4 + H$
   (planar)
(g) $H_3O^+ + Cl^- \rightleftharpoons H_2O + HCl$
In each case, confirm that the ratio of the statistical factors in the two directions is equal to the ratio of the symmetry numbers for reactants and products.

4.7. One liter of oxygen is maintained at 400 mmHg pressure and a temperature of 300°C. Assume a collision diameter of 200 pm and calculate the following, on the basis of simple collision theory:
(a) The collision number Z_{AA}.
(b) The collision frequency factor z_{AA}.

(c) The number of collisions per second in the 1 liter of gas in which the excess energy in two degrees of freedom is 100 kJ mol^{-1}.

4.8. For the reaction $C_2H_5 + H_2 \rightarrow C_2H_6 + H$, the activation energy has been reported [J. R. Cao and M. H. Back, *Can. J. Chem.,* **60,** 3039 (1982)] to have the following values:

In the range 350–550 K: 50 kJ mol^{-1}.

In the range 800–1100 K: 71 kJ mol^{-1}.

From these values, deduce m and E_0 in the formula

$$k = AT^m e^{-E_0/RT}$$

4.9. Estimate, on the basis of CTST, the pre-exponential factors at 300 K for the following types of gas reaction:
- **(a)** A bimolecular reaction between an atom and a diatomic molecule, with the formation of a linear activated complex.
- **(b)** A bimolecular reaction between two diatomic molecules, the activated complex being nonlinear with no free rotation.
- **(c)** A trimolecular reaction between three diatomic molecules, the activated complex being linear with no free rotation.

Take the translation partition functions to be 10^{32} m^{-3}, the rotational functions for each degree of freedom to be 10, and the vibrational functions to be unity. Express the calculated pre-exponential factors in molecular units (e.g., m^3 s^{-1}) and in molar units (e.g., dm^3 mol^{-1} s^{-1}).

NOTE: Some interesting problems in conventional transition-state theory, involving the use of a computer, are to be found in S. J. Moss and C. J. Coady, *J. Chem. Educ.,* **60,** 455 (1983).

REFERENCES

1. For a historical account of the rise and fall of the radiation hypothesis see M. C. King and K. J. Laidler, *Arch. Hist. Exact Sci.,* **30,** 45 (1984).
2. For an account of these early struggles with the problem, leading to transition-state theory, see K. J. Laidler and M. C. King, *J. Phys. Chem.,* **87,** 2657 (1983).
3. H. Eyring, *J. Chem. Phys.,* **3,** 107 (1935). A more comprehensive treatment can be found in W. F. K. Wynne-Jones and H. Eyring, *J. Chem. Phys.,* **3,** 492 (1935). This article is reproduced in full in M. H. Back and K. J. Laidler (Eds.), *Selected Readings in Chemical Kinetics,* Pergamon, Oxford, 1967.
4. M. G. Evans and M. Polanyi, *Trans. Faraday Soc.,* **31,** 875 (1935); **33,** 448 (1937).
5. M. Trautz, *Z. Anorgan. Chem.,* **96,** 1 (1916).
6. W. C. McC. Lewis, *J. Chem. Soc. (London),* **113,** 471 (1918). An excerpt from this paper is included in M. H. Back and K. J. Laidler (Eds.), *Selected Readings in Chemical Kinetics,* Pergamon, Oxford, 1967.
7. V. K. Le Mer, *J. Chem. Phys.,* **1,** 289 (1933).
8. M. Eliason and J. O. Hirschfelder, *J. Chem. Phys.,* **30,** 426 (1959); M. Karplus, R. N. Porter, and R. D. Sharma, *J. Chem. Phys.,* **43,** 3259 (1965).
9. R. Marcelin, *Comptes Rendus,* **151,** 1052 (1920). Marcelin did not refer explicitly to Gibbs energy changes, but instead used the old concept of "affinity."
10. P. Kohnstamm and F. E. C. Scheffer, *Proc. Koninkl. Ned. Acad. Wetenschap.,* **13,** 789 (1911); W. F. Brandsma, *Chem. Weekblad,* **19,** 318 (1922); **47,** 94 (1928); **48,** 1205 (1929); F. E. C. Scheffer and W. F. Brandsma, *Rec. Trav. Chim.,* **45,** 5 (1926).

11. F. E. C. Kohnstamm and F. E. C. Scheffer, *Proc. Koninkl. Ned. Acad. Wetenschap.,* **13,** 789 (1911).

12. A. Berthoud, *J. Chim. Phys.,* **10,** 573 (1912); much earlier L. Pfaundler [*Ann. Phys. Chem.,* **131,** 55 (1867), **144,** 428 (1872)] had discussed the matter *qualitatively* in terms of statistical distributions.

13. R. Marcelin, *Comptes Rendus,* **158,** 116, 407 (1914).

14. K. F. Herzfeld, *Ann. Phys.,* **59,** 635 (1919).

15. E. K. Rideal, *Philos. Mag.,* **40,** 461 (1920).

16. S. Dushman, *J. Am. Chem. Soc.,* **43,** 397 (1921).

17. W. H. Rodebush, *J. Am. Chem. Soc.,* **45,** 606 (1923).

18. B. Topley, *Nature,* **128,** 115 (1931).

19. R. C. Tolman, *J. Am. Chem. Soc.,* **42,** 2506 (1920).

20. V. K. La Mer, *J. Chem. Phys.,* **1,** 289 (1933).

21. H. Eyring and M. Polanyi, *Z. Phys. Chem. B,* **12,** 279 (1931).

22. H. Eyring, H. Gerschinowitz, and C. E. Sun, *J. Chem. Phys.,* **3,** 786 (1935); J. O. Hirschfelder, H. Eyring, and B. Topley, *J. Chem. Phys.,* **4,** 170 (1936).

23. H. Pelzer and E. Wigner, *Z. Phys. Chem. B,* **15,** 445 (1932); E. Wigner, *Z. Phys. Chem. B,* **15,** 203 (1932).

24. H. Eyring, *J. Chem. Phys.,* **3,** 107 (1935).

25. M. G. Evans and M. Polanyi, *Trans. Faraday Soc.,* **31,** 875 (1935).

26. Transition-state theory is the expression now recommended by IUPAC [V. Gold, *Pure. Appl. Chem.,* **51,** 1725 (1979); K. J. Laidler, *Pure Appl. Chem.,* **53,** 753 (1981)]. The expressions "theory of absolute reaction rates" and "activated-complex theory" also have been used.

27. The argument presented in this paragraph was first given by M. Polanyi, *Trans. Faraday Soc.,* **34,** 75 (1938).

28. W. F. K. Wynne-Jones and H. Eyring, *J. Chem. Phys.,* **3,** 492 (1935).

29. This point of confusion also has been discussed by B. H. Mahan, *J. Chem. Educ.,* **51,** 709 (1974).

30. H. Eyring, *J. Chem. Phys.,* **3,** 107 (1935).

31. M. G. Evans and M. Polanyi, *Trans. Faraday Soc.,* **31,** 875 (1935).

32. B. H. Mahan, *J. Chem. Educ.,* **51,** 709 (1974). Essentially the same proof is given by I. W. M. Smith, *Kinetics and Dynamics of Elementary Gas Reactions,* Butterworth, London, 1980.

33. E. Wigner, *Trans. Faraday Soc.,* **34,** 29 (1938); *J. Chem. Phys.,* **5,** 720 (1937).

34. D. M. Bishop and K. J. Laidler, *J. Chem. Phys.,* **42,** 1688 (1965). This proof is repeated in K. J. Laidler, *Theories of Chemical Reaction Rates,* pp. 50–52, McGraw-Hill, New York, 1969.

35. E. Pollak and P. Pechukas, *J. Am. Chem. Soc.,* **100,** 2984 (1978); D. R. Coulson, *J. Am. Chem. Soc.,* **100,** 2992 (1978).

36. D. M. Bishop and K. J. Laidler, *J. Chem. Phys.,* **42,** 1688 (1965); *Trans. Faraday Soc.,* **66,** 1685 (1970).

37. J. N. Murrell and K. J. Laidler, *Trans. Faraday Soc.,* **64,** 371 (1968); J. N. Murrell and G. L. Pratt, *Trans. Faraday Soc.,* **66,** 1680 (1970); K. J. Laidler, in *Symposium on Etats de transition réactionnels,* p. 23, Société Chimique de France, Paris, 1970.

38. This was first pointed out by J. N. Murrell and K. J. Laidler, *Trans. Faraday Soc.,* **64,** 371 (1968), and is discussed in further detail by K. J. Laidler, *Symposium on Etats de transition réactionnels,* p. 23, Société Chimique de France, Paris, 1970. A rigorous group-theoretical treatment of the problem was given by P. Pechukas, *J. Chem. Phys.,* **64,** 1516 (1976).

39. P. Pechukas, *J. Chem. Phys.,* **64,** 1516 (1976).

40. C. N. Hinshelwood, *J. Chem. Soc.,* **5,** 635 (1937); *Trans. Faraday Soc.,* **34,** 74 (1938).
41. C. E. H. Bawn, *Trans. Faraday Soc.,* **31,** 1536 (1935); **32,** 178 (1936).
42. H. S. Johnston, *Gas Phase Reaction Rate Theory,* chap. 12, Ronald, New York, 1966.
43. W. F. K. Wynne-Jones and H. Eyring, *J. Chem. Phys.,* **3,** 492 (1935). Compare the treatment of Kohnstamm, Scheffer, and Brandsma (Section 4.2).
44. P. D. Pacey, *J. Chem. Educ.,* **58,** 612 (1981).
45. K. J. Laidler, *Pure Appl. Chem.,* **53,** 753 (1981).
46. J. B. Anderson, *J. Chem. Phys.,* **58,** 4684 (1973).
47. The first such treatment was that of J. O. Hirschfelder, H. Eyring, and B. Topley, *J. Chem. Phys.,* **4,** 170 (1936). For a more recent treatment see, for example, K. G. Tan, K. J. Laidler, and J. S. Wright, *J. Chem. Phys.,* **67,** 5883 (1977).
48. K. J. Laidler, K. G. Tan, and J. S. Wright, *Chem. Phys. Lett.,* **46,** 56 (1977).
49. H. S. Johnston, *Adv. Chem. Phys.,* **3,** 131 (1961); H. S. Johnston and D. Rapp, *J. Am. Chem. Soc.,* **83,** 1 (1961).
50. For a discussion see P. D. Pacey, *J. Chem. Phys.,* **71,** 2966 (1979) and H. Furue and P. D. Pacey, *J. Chem. Phys.,* **83,** 2878 (1985).
51. E. Wigner, *J. Chem. Phys.,* **5,** 729 (1937).
52. J. Horiuti, *Bull. Chem. Soc. Jpn.,* **13,** 210 (1938).
53. J. C. Keck, *J. Chem. Phys.,* **32,** 1035 (1960); *Adv. Chem. Phys.,* **13,** 85 (1967).
54. C. Steel and K. J. Laidler, *J. Chem. Phys.,* **34,** 1827 (1961); K. J. Laidler and J. C. Polanyi, *Prog. React. Kinet.,* **3,** 1 (1976); A. Tweedale and K. J. Laidler, *J. Chem. Phys.,* **53,** 2045 (1970).
55. D. G. Truhlar, *J. Chem. Phys.,* **53,** 2041 (1970); A. Tweedale and K. J. Laidler, *J. Chem. Phys.,* **53,** 2045 (1970); B. C. Garrett and D. G. Truhlar, *J. Chem. Phys.,* **83,** 1052, 1079, 3058 (1979); **84,** 682 (1980).
56. J. C. Keck, *J. Chem. Phys.,* **32,** 1035 (1960); *Adv. Chem. Phys.,* **13,** 85 (1967).
57. B. C. Garrett and D. G. Truhlar, *J. Phys. Chem.,* **83,** 1052 (1979); *J. Chem. Phys.,* **70,** 1593 (1979); and a number of later publications. For reviews see D. G. Truhlar and B. C. Garrett, *Acc. Chem. Res.,* **13,** 440 (1980); D. G. Truhlar, W. L. Hase, and J. T. Hynes, *J. Phys. Chem.,* **87,** 2664, 5523E (1983).
58. B. C. Garrett and D. G. Truhlar, *J. Am. Chem. Soc.,* **101,** 5207 (1979); **102,** 2559 (1980); D. G. Truhlar and B. C. Garrett, *Acc. Chem. Res.,* **13,** 440 (1980).
59. B. C. Garrett and D. G. Truhlar, *J. Phys. Chem.,* **83,** 1052 (1979).
60. W. H. Miller, *J. Chem. Phys.,* **55,** 3146 (1971); **62,** 1899 (1975) and many later papers. For reviews see W. H. Miller, *Acc. Chem. Res.,* **9,** 306 (1976); P. Pechukas, *Ber. Bunsenges. Phys. Chem.,* **86,** 372 (1982).
61. B. C. Garrett and D. G. Truhlar, *J. Phys. Chem.,* **83,** 200, 1079, 1915 (1979); *J. Am. Chem. Soc.,* **101,** 4534, 5207 (1979); *Proc. Natl. Acad. Sci. (USA),* **76,** 4755 (1979) and many later papers. For reviews see P. Pechukas, *Ber. Bunsenges. Phys. Chem.,* **86,** 372 (1982); *Ann. Rev. Phys. Chem.,* **32,** 159 (1981); D. G. Truhlar, W. L. Hase, and J. T. Hynes, *J. Chem. Phys.,* **87,** 2664 (1983); D. G. Truhlar and B. C. Garrett, *Ann. Rev. Phys. Chem.,* **35,** 159 (1984).
62. R. A. Marcus and M. E. Coltrin, *J. Chem. Phys.,* **67,** 2609 (1977).
63. B. C. Garrett and D. G. Truhlar, *J. Chem. Phys.,* **79,** 4931 (1983); **81,** 309 (1984).
64. See, for example, D. K. Bondi, J. N. L. Connor, B. C. Garrett, and D. G. Truhlar, *J. Chem. Phys.,* **78,** 5981 (1983); B. C. Garrett and D. G. Truhlar, *J. Chem. Phys.,* **81,** 309 (1984).
65. B. C. Garrett, N. Abusalbi, D. J. Kouri, and D. G. Truhlar, *J. Chem. Phys.,* **83,** 2252 (1985).
66. N. C. Blais, D. G. Truhlar, and B. C. Garrett, *J. Chem. Phys.,* **78,** 2363 (1983). See also D. K. Bondi, D. C. Clary, J. N. L. Connor, B. C. Garrett, and D. G. Truhlar, *J. Chem. Phys.,* **76,** 4986 (1982).

67. R. C. Tolman, *Phys. Rev.,* **23,** 699 (1924); *The Principles of Statistical Mechanics,* Clarendon, Oxford, 1938, p. 163.
68. R. H. Fowler, *Statistical Mechanics,* Cambridge University Press, Cambridge, 1936 (2nd ed.); especially pp. 659–660, 696–697, 716–719.

BIBLIOGRAPHY

For a historical account of early theories of reaction rates and of the development of conventional transition-state theory, see:

K. J. Laidler and M. C. King, The Development of Transition-State Theory, *J. Phys. Chem.,* **87,** 2657 (1983).

Conventional transition-state theory was first treated in detail in:

S. Glasstone, K. J. Laidler, and H. Eyring, *The Theory of Rate Processes,* McGraw-Hill, New York, 1941.

This book, long out of print, will be helpful as a starting point for those interested in how the theory originated. More recent accounts of the conventional theory are to be found in:

H. S. Johnston, *Gas Phase Reaction Rate Theory,* Ronald, New York, 1966.

K. J. Laidler, *Theories of Chemical Reaction Rates,* McGraw-Hill, New York, 1969.

R. P. Wayne, The Theory of the Kinetics of Elementary Gas Reactions, in C. H. Bamford and C. F. H. Tipper (Eds.), *Comprehensive Chemical Kinetics,* vol. 2, chap. 3, pp. 189–301, Elsevier, Amsterdam, 1969.

For accounts of limitations of conventional transition-state theory and extensions to it see:

S. H. Bauer, A Random Walk among Accessible States. *J. Chem. Educ.,* **63,** 377 (1986).

J. C. Keck, Variational Theory of Reaction Rates, *Adv. Chem. Phys.,* **13,** 85 (1967).

K. J. Laidler and A. Tweedale, The Current Status of Eyring's Rate Theory, in J. O. Hirschfelder and D. Henderson (Eds.), *Chemical Dynamics: Papers in Honor of Henry Eyring,* pp. 113–125, Wiley-Interscience, New York, 1971.

B. H. Mahan, Activated Complex Theory of Bimolecular Reactions, *J. Chem. Educ.,* **51,** 709 (1974).

W. H. Miller, Importance of Nonseparability in Quantum-Mechanical Transition-State Theory, *Acc. Chem. Res.,* **9,** 306 (1976).

D. G. Truhlar and R. E. Wyatt, History of H_3 Kinetics, *Ann. Rev. Phys. Chem.,* **27,** 1 (1976).

P. Pechukas, Statistical Approximations in Collision Theory, in W. H. Miller (Ed.), *Dynamics of Molecular Collisions, Part B,* p. 299, Plenum, New York, 1976.

B. C. Garrett and D. G. Truhlar, Generalized Transition-State Theory. Classical Mechanical Theory and Applications to Collinear Reactions of Hydrogen Molecules, *J. Phys. Chem.,* **83,** 1052 (1979).

D. G. Truhlar and B. C. Garrett, Variational Transition-State Theory, *Acc. Chem. Res.,* **13,** 440 (1980); *Ann. Rev. Phys. Chem.,* **35,** 159 (1984).

I. W. M. Smith, *Dynamics of Elementary Gas Reactions,* pp. 113–143, Butterworth, London, 1980.

P. Pechukas, Transition-State Theory, *Ann. Rev. Phys. Chem.,* **32,** 159 (1981).

P. Pechukas, Recent Developments in Transition-State Theory, *Ber. Bunsenges. Phys. Chem.,* **86,** 372 (1982).

D. G. Truhlar, W. L. Hase, and J. T. Hynes, The Current Status of Transition-State Theory, *J. Phys. Chem.,* **87,** 2664 (1983).

More detailed treatments of modern collision theory are to be found in:

M. A. Eliason and J. O. Hirschfelder, *J. Chem. Phys.,* **30,** 426 (1959).

E. F. Greene and A. Kuppermann, *J. Chem. Educ.,* **45,** 361 (1968).

W. C. Gardiner, *Rates and Mechanisms of Chemical Reactions,* pp. 75–95, Benjamin-Cummings, Menlo Park, CA, 1969.

J. Ross, J. C. Light, and K. E. Shuler, in A. R. Hochstim (Ed.), *Kinetic Processes in Gases and Plasmas,* Academic, New York, 1969.

For detailed treatments of statistical mechanics see:

T. L. Hill, *Introduction to Statistical Mechanics,* Addison-Wesley, Reading, MA, 1972.

L. K. Nash, *Elements of Statistical Mechanics,* Addison-Wesley, Reading, MA, 1972.

O. K. Rice, *Statistical Mechanics, Thermodynamics, and Kinetics,* Freeman, San Francisco, 1967.

G. S. Rushbrooke, *Introduction to Statistical Mechanics,* Clarendon, Oxford, 1962.

For treatments of quantum-mechanical tunneling see:

R. P. Bell, *The Tunnel Effect in Chemistry,* Chapman and Hall, London, 1980.

D. G. Truhlar and B. C. Garrett, Variational Transition-State Theory, *Ann. Rev. Phys. Chem.,* **35,** 159 (1984), especially pp. 178–182.

For reviews of muonium chemistry see:

H. J. Ache, *Positronium and Muonium Chemistry,* Advances in Chemistry No. 175, American Chemical Society, Washington, DC, 1979.

D. G. Fleming, D. M. Garner, and R. J. Mikula, *Hyperfine Interactions,* **8,** 337 (1981).

J. N. L. Connor, *Hyperfine Interactions,* **8,** 423 (1981).

D. C. Walker, *Muon and Muonium Chemistry,* Cambridge University Press, Cambridge, 1983.

Chapter 5

Elementary Gas-Phase Reactions

Most reactions involving organic compounds, whether they occur in the liquid or gas phase, take place in more than one step. For example, when ethane vapor is heated it decomposes into ethylene and hydrogen, and at first it was assumed that the process is§

$$C_2H_6 \rightarrow C_2H_4 + H_2$$

That is, ethane molecules were supposed to split off hydrogen molecules, in one step. Later work, however, has shown that the reaction occurs almost entirely by a chain mechanism, in which ethane molecules split initially into methyl radicals:

$$C_2H_6 \rightarrow 2CH_3$$

This process is followed by several reactions involving CH_3 radicals and also ethyl radicals, C_2H_5, and hydrogen atoms; the details are considered in Section 8.5.3.

Another reaction that was long thought to be elementary is the reaction between hydrogen and iodine,

$$H_2 + I_2 \rightarrow 2HI$$

For many years this reaction and the reverse dissociation of HI were quoted in textbooks as classical examples of elementary bimolecular reactions. It is now known, however, that there are complexities, in that processes such as $I + H_2 \rightarrow H + HI$ play a role (see Section 8.4.3).

This chapter is concerned with some of the overall gas reactions that are believed to be elementary and with some of the elementary reactions that occur as steps in overall reactions. An example is

$$CH_3 + C_2H_6 \rightarrow CH_4 + C_2H_5$$

§ Arrows with solid heads are used to emphasize that reactions are elementary.

which plays an important role in the thermal decomposition of ethane (Section 8.5.3.) This chapter also deals with some reactions which, while not known to occur as reaction steps, can be induced to occur in special ways, such as in a mass spectrometer.

5.1 BIMOLECULAR REACTIONS

Kinetic parameters for some bimolecular reactions between molecules are listed in Table 5.1. As a test of conventional transition-state theory (CTST), Herschbach, Johnston, Pitzer, and Powell[1] made calculations of the pre-exponential factors A for these reactions, using CTST and the kinetic theory of collisions. The CTST values are reasonably satisfactory. For all the reactions, simple kinetic theory leads to A values of 10^{10}–10^{11} dm^3 mol^{-1} s^{-1}, which is significantly too high in all cases. As discussed in Section 4.6.2, the kinetic theory values are too high for all except atom–atom reactions, because of neglect of the losses of rotational freedom when the activated complexes are formed.

5.1.1 Reactions of the Type H + H$_2$

A considerable amount of experimental and theoretical work has been carried out on the reaction

$$H + H_2 \rightarrow H_2 + H$$

This reaction is of special interest since it involves only three electrons, so that quantum-mechanical calculations of potential-energy surfaces, although difficult and time consuming, can be done more easily than for reactions involving more electrons.

The measurement of the rate of the reaction

$$H + H_2 \rightarrow H_2 + H$$

TABLE 5.1   KINETIC PARAMETERS FOR SOME BIMOLECULAR REACTIONS

| | | | log$_{10}$ (A/dm^3 mol^{-1} s^{-1}) | |
| | | | Calculated[a] | |
Reaction	Observed E_a/kJ mol^{-1}	Observed[a]	CTST	Simple collision theory
NO + O$_3$ → NO$_2$ + O$_2$	10.5	8.9	8.6	10.7
NO + O$_3$ → NO$_3$ + O	29.3	9.8	8.1	10.8
NO$_2$ + F$_2$ → NO$_2$F + F	43.5	9.2	8.1	10.8
NO$_2$ + CO → NO + CO$_2$	132	10.1	9.8	10.6
2NO$_2$ → 2NO + O$_2$	111	9.3	9.7	10.6
NO + NO$_2$Cl → NOCl + NO$_2$	28.9	8.9	8.9	10.9
2NOCl → 2NO + Cl$_2$	103	10.0	8.6	10.8
NO + Cl$_2$ → NOCl + Cl	84.9	9.6	9.1	11.0
F$_2$ + ClO$_2$ → FClO$_2$ + F	35.6	7.5	7.9	10.7
2ClO → Cl$_2$ + O$_2$	0	7.8	7.0	10.4

[a] The observed and calculated values are from D. R. Herschbach, H. S. Johnston, K. S. Pitzer, and R. E. Powell, *J. Chem. Phys.*, **25**, 736 (1956).

is made possible by the fact that H_2 exists in two forms, designated ortho (o) and para (p). The forms differ in the nuclear spins: in o-hydrogen the nuclear spins are in the same direction, while in p-hydrogen the spins are opposed. Pure p-hydrogen can be prepared by passing ordinary hydrogen through a tube containing a suitable catalyst (such as charcoal) and cooled in liquid nitrogen; at low temperatures p-H_2 is the predominant form. At room temperature and above, hydrogen is a mixture of three parts of o-H_2 and one part of p-H_2. Thus, when reaction with H atoms occurs at these higher temperatures, p-H_2 is converted into the equilibrium 3:1 mixture. This is due to the occurrence of the two reactions

$$H + p\text{-}H_2 \rightarrow p\text{-}H_2 + H$$

and
$$H + p\text{-}H_2 \rightarrow o\text{-}H_2 + H$$

the second taking place three times as rapidly as the first.

The rate of the reaction between atomic and molecular hydrogen was first measured by Geib and Harteck,[2] who caused atomic hydrogen to interact with hydrogen consisting largely of the para variety and determined the net rate of production of o-hydrogen in the mixture. Their experimental procedure involved passing a stream of hydrogen atoms and p-hydrogen through a reaction vessel and determining the concentrations of p- and o-hydrogen in the issuing gas by making use of the fact that there is a small difference in thermal conductivity between the two forms.

The atomic hydrogen was prepared by a method developed by R. W. Wood[3] and modified by Bonhoeffer.[4] This method depends on the fact that when a silent discharge is passed through hydrogen gas at low pressure, hydrogen atoms are produced and may be detected by their spectra. The apparatus used by Geib and Harteck[2] for obtaining a stream of atomic hydrogen, equivalent to that devised by Bonhoeffer, is illustrated in Fig. 5.1. A stream of hydrogen enters at the position indicated and passes through the tube between the two electrodes E_1 and E_2, the potentials at which differed by 3000 V; the current passing was 200 mA. The concentration of atomic hydrogen produced in the discharge tube was measured by means of a gauge devised by Harteck[5] and Wrede.[6]

These early investigations of the $H + H_2$ reaction did not lead to very precise results; in particular, reliable energies of activation could not be obtained. More recently, Le Roy and his coworkers[7] studied all four of the following reactions:

(1) $H + p\text{-}H_2 \rightarrow o\text{-}H_2 + H$

(2) $D + o\text{-}D_2 \rightarrow p\text{-}D_2 + D$

(3) $D + H_2 \rightarrow HD + H$

(4) $H + D_2 \rightarrow HD + D$

They employed a fast-flow technique for bringing the reactants together and used various modern methods of analysis. For reactions (3) and (4), in which there is isotopic exchange, they used electron spin resonance spectroscopy for measuring concentrations of H and D atoms. For reactions (1) and (2) they used probes that catalytically combined the atoms, the atom concentrations being determined from the heat evolved. The concentrations of the molecules were obtained by thermal conductivity methods. Measurements on reactions (3) and (4) were also made by Westenberg and de Haas.[8]

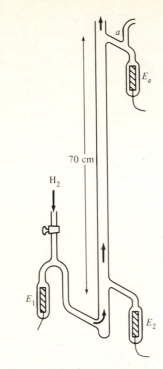

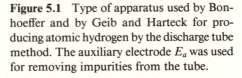

Figure 5.1   Type of apparatus used by Bonhoeffer and by Geib and Harteck for producing atomic hydrogen by the discharge tube method. The auxiliary electrode E_a was used for removing impurities from the tube.

Some results obtained by Le Roy and coworkers are shown as Arrhenius plots in Fig. 5.2. There is distinct curvature in the plots for $D + H_2$ and $H + H_2$, suggesting that there is quantum-mechanical tunneling (Section 4.8.3). Westenberg and de Haas carried out the measurements on $H + D_2$ and $D + H_2$ down to lower temperatures, as indicated by the dashed lines in Fig. 5.2, and also found curvature for these reactions.

Several attempts have been made to analyze these results in terms of conventional transition-state theory and generalized versions of transition-state theory. Figure 5.3 shows a comparison for the reaction $H + D_2 \rightarrow HD + H$. The experimental curve is a smooth curve based on the results of Schulz and Le Roy[7] and of Westenberg and de Haas.[8] The other curves are from rates calculated by Garrett and Truhlar[9] using the ab initio potential-energy surface given by Siegbahn and Liu.[10] The calculated rates given by CTST are too low, particularly at lower temperatures. Little improvement is given by varying the position of the dividing surface (e.g., by canonical variational theory) because for this potential-energy surface recrossing effects are not important. Inclusion of tunneling, however, by making use of the Marcus–Coltrin treatment (Section 4.9.2), brings about considerable improvement—the calculated values then being very close to the experimental ones.

The conclusions for the $H + H_2$ reaction are very similar, as indicated by the values given in Table 4.4.

5.1.2 Other Atom and Free-Radical Reactions

Kinetic results are available for other bimolecular reactions involving atoms and free radicals. Many of these are *abstraction* (also called *metathetical*) reactions, in which an atom or radical is abstracted. An example is

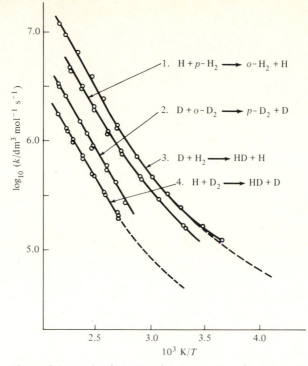

$$7.0$$

$$\log_{10}\,(k/\mathrm{dm^3\,mol^{-1}\,s^{-1}})$$

1. $H + p\text{-}H_2 \longrightarrow o\text{-}H_2 + H$

2. $D + o\text{-}D_2 \longrightarrow p\text{-}D_2 + D$

3. $D + H_2 \longrightarrow HD + H$

4. $H + D_2 \longrightarrow HD + D$

$$6.0$$

$$5.0$$

$$2.5 \qquad 3.0 \qquad 3.5 \qquad 4.0$$

$$10^3\ \mathrm{K}/T$$

Figure 5.2 Arrhenius plots for the results of Le Roy and coworkers on reactions of the type $H + H_2$. The dashed lines indicate the lower-temperature results of Westenberg and de Haas.

$$CH_3 + C_2H_6 \rightarrow CH_4 + C_2H_5$$

It is outside the scope of the present book to deal with experimental methods[11] in any detail, but some idea will be given of the general procedures used. There are two general methods of obtaining the rates and kinetic parameters of reactions involving free radicals,§ and they are used best in conjunction with each other. One method is to make a detailed study of the kinetics of a reaction that occurs in a number of steps and to deduce the characteristics of the individual steps from the results. For example, from a study of the decomposition of ethane it has been found possible to deduce the kinetic parameters for the reaction $CH_3 + C_2H_6 \rightarrow CH_4 + C_2H_5$, which is one of the reactions occurring in the mechanism. Some details of how reactions are analyzed in this way are given in Section 8.5.

The second general method is to isolate a free-radical reaction in some way. There are a number of methods of generating radicals and of measuring their concentrations, so that their reactions can be followed. As an example of this type of procedure, we may consider the abstraction by a methyl radical of a hydrogen atom from acetone,

§ From now on, for convenience and following modern usage, the term "radical" will be used to include an atom as a special case.

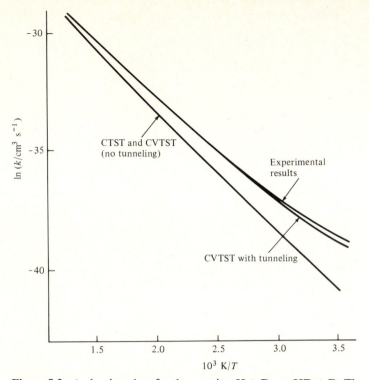

Figure 5.3   Arrhenius plots for the reaction H + D$_2$ → HD + D. The experimental curve is a smooth curve through the results of Le Roy and coworkers and of Westenberg and de Haas. Conventional transition-state theory (CTST) gives results that are too low and canonical variational theory (CVTST) with tunneling neglected gives little improvement. Canonical variational theory with tunneling included gives results close to the experimental values.

$$CH_3 + CH_3COCH_3 \xrightarrow{k_1} CH_4 + CH_2COCH_3$$

The rate of this reaction is

$$v_1 = \frac{d[CH_4]}{dt} = k_1[CH_3][CH_3COCH_3] \tag{5.1}$$

There is no difficulty about measuring v_1, the rate of production of methane, in such a system, but to obtain the rate constant k_1 it is necessary for the concentration of the free methyl radicals to be known. This presents a problem, because the concentration of methyl radicals is necessarily very small and difficult to measure directly.

A method that has been used frequently to overcome this difficulty involves comparing the rate of the hydrogen-atom abstraction reaction with that of the radical combination reaction

$$CH_3 + CH_3 \xrightarrow{k_2} C_2H_6$$

The kinetics of such reactions are considered in Section 5.4, and it will be seen that at sufficiently high pressures the reactions are second order. Therefore, the rate of formation of ethane is

$$v_2 = v_{C_2H_6} = k_2[CH_3]^2 \tag{5.2}$$

From Eqs. (5.1) and (5.2) it follows that

$$\frac{v_1}{v_2^{1/2}} = \frac{k_1}{k_2^{1/2}} [CH_3COCH_3] \tag{5.3}$$

Therefore, if one can measure the rates of formation of methane (v_1) and of ethane (v_2) at a given acetone concentration, one can calculate $k_1/k_2^{1/2}$. Also, if k_2 is known, one has obtained k_1, and by doing the work at different temperatures the activation energy and the pre-exponential factor are obtained.

The determination of k_2 has been carried out by a special photochemical technique, to be described in Section 9.1.5. Once this value is known, the method can be applied to a large variety of molecules and the kinetic parameters can be obtained for the abstraction reactions. It is possible, for example, to produce methyl radicals by the action of ultraviolet light on acetone and to do this in the presence of an additional organic substance which may be represented as RH. The methyl radicals then undergo the following reactions:

$$CH_3 + CH_3COCH_3 \xrightarrow{k_1} CH_4 + CH_2COCH_3$$

$$CH_3 + CH_3 \xrightarrow{k_2} C_2H_6$$

$$CH_3 + RH \xrightarrow{k_3} CH_4 + R$$

The rates are given by Eq. (5.2) and by

$$v_{CH_4} = k_1[CH_3][CH_3COCH_3] + k_3[CH_3][RH] \tag{5.4}$$

and combination of the two equations gives

$$\frac{v_{CH_4}}{v_{C_2H_6}^{1/2}[CH_3COCH_3]} = \frac{k_1}{k_2^{1/2}} + \frac{k_3[RH]}{k_2^{1/2}[CH_3COCH_3]} \tag{5.5}$$

The rates v_{CH_4} and $v_{C_2H_6}$ are measured, k_2 is known, and k_1 is determined by studies in the absence of acetone. Therefore, the rate constant k_3 can be calculated.

This procedure for obtaining the rate constant of an abstraction reaction is an example of the use of what has been called a *pilot reaction*. Advantage is often taken of isotopic labeling.[12] Thus, CD$_3$ produced photochemically may be caused to undergo the reactions

$$(1) \quad CD_3 + CD_3COCD_3 \xrightarrow{k_1} CD_4 + CH_2COCD_3$$

and $\quad (2) \quad CD_3 + RH \xrightarrow{k_2} CD_3H + R$

If the kinetic parameters of the pilot reaction (1) have been determined, those for reaction (2) can be calculated from the ratio of the concentrations of CD$_4$ and CD$_3$H that are produced.

Table 5.2 gives a selection of kinetic parameters for a number of abstraction reactions involving atoms and free radicals. Pre-exponential factors for some of these reactions have been calculated using conventional transition-state theory, on the basis

TABLE 5.2   KINETIC PARAMETERS FOR SOME ABSTRACTION REACTIONS[a]

Reaction	Temperature range/K	E_a/kJ mol^{-1}	$\log_{10}$ (A/dm^3 mol^{-1} s^{-1})
$H + H_2 \rightarrow H_2 + H$	450–750	31.8	10.6
$H + D_2 \rightarrow HD + H$	450–750	39.2	10.7
$H + CH_4 \rightarrow H_2 + CH_3$	372–1290	49.8	11.1
$H + C_2H_6 \rightarrow H_2 + C_2H_5$	290–509	38.1	10.9
$H + O_2 \rightarrow HO + O$	700–2500	70.3	11.4
$H + Br_2 \rightarrow HBr + Br$	973–1673	15.5	12.0
$CH_3 + H_2 \rightarrow CH_4 + H$	450–750	51.0	9.5
$CH_3 + D_2 \rightarrow CH_3D + D$	300–626	53.1	9.3
$CH_3 + C_2H_6 \rightarrow CH_4 + C_2H_5$	533–763	50.6	9.3
$CH_3 + CH(CH_3)_3 \rightarrow CH_4 + C(CH_3)_3$	573–733	34.3	8.5

[a] The data are collected from various sources listed in the bibliography at the end of this chapter.

of estimated partition functions. In most cases, as seen in Table 5.3, the agreement is fairly satisfactory. It is much better than with simple collision theory, which leads to about 10^{11} dm^3 mol^{-1} s^{-1} for all the reactions. The results in Tables 5.2 and 5.3 illustrate the point, mentioned in Section 4.6.2, that pre-exponential factors tend to decrease as the complexity of the reactants increases.

An alternative method of estimating pre-exponential factors, suggested by Benson,[13] makes use of the thermodynamic formulation of conventional transition-state theory. According to that formulation [Eq. (4.100)], the pre-exponential factor for a bimolecular reaction is given by

$$A = e^2 \frac{kT}{h} e^{\Delta^{\ddagger}S^0/R} \tag{5.6}$$

Reliable values are now available for the entropies of many molecules and radicals,

TABLE 5.3   OBSERVED AND CALCULATED KINETIC PARAMETERS FOR SOME ABSTRACTION REACTIONS

Reaction	E_a/kJ mol^{-1}	$\log_{10}$ (A/dm^3 mol^{-1} s^{-1}) Observed[a]	$\log_{10}$ (A/dm^3 mol^{-1} s^{-1}) Calculated using CTST
$H + H_2 \rightarrow H_2 + H$	31.8	10.6	10.7,[b] 10.8,[c] 10.7[d]
$H + CH_4 \rightarrow H_2 + CH_3$	49.8	11.1	10.3[c]
$H + C_2H_6 \rightarrow H_2 + C_2H_5$	38.1	10.9	10.1[b]
$CH_3 + H_2 \rightarrow CH_4 + H$	51.0	9.5	9.0,[b] 9.4,[c] 9.0[d]
$CH_3 + C_2H_6 \rightarrow CH_4 + C_2H_5$	50.6	9.3	8.0[b]
$CH_3 + i\text{-}C_4H_{10} \rightarrow CH_4 + C_4H_9$	28.0	8.5	6.8[b]
$Br + H_2 \rightarrow HBr + H$	82.4	11.4	11.1[d]
$CH_3 + CH_3COCH_3 \rightarrow CH_4 + CH_2COCH_3$	40.2	8.5	8.0[e]

[a] The observed parameters were obtained from sources listed in the bibliography at the end of this chapter.
[b] S. Bywater and R. Roberts, *Can. J. Chem.*, **30**, 773 (1952).
[c] J. C. Polanyi, *J. Chem. Phys.*, **23**, 1505 (1955); **24**, 493 (1956).
[d] D. J. Wilson and H. S. Johnston, *J. Am. Chem. Soc.*, **79**, 29 (1957).
[e] T. L. Hill, *J. Chem. Phys.*, **17**, 503 (1949).

and Benson's procedure involves making estimates of the entropies of activated complexes on the basis of the known values for stable molecules of similar structure. For example, the activated complex for the reaction

$$CH_3 + C_2H_6 \rightarrow CH_4 + C_2H_5$$

has the structure

A lower limit for the entropy of this complex is provided by the entropy of propane, which has a tighter structure. Contributions to the entropy also arise from symmetry changes; details are to be found in Benson's book.[13] This method is a useful one, especially in view of its simplicity.

One conclusion that has resulted from the experimental study of abstraction reactions is that reactions involving the abstraction of an atom occur with a much lower activation energy and with a much higher rate than those in which a radical is abstracted.

5.1.3 Temperature Dependence of Pre-exponential Factors

A number of abstraction reactions have been studied over a considerable range of temperature, and there is then usually a significant deviation from the Arrhenius equation. For example, for the reaction

$$C_2H_5 + H_2 \rightarrow C_2H_6 + H$$

the activation energy in the range 355–595 K was found to be 54.4 kJ mol^{-1}.[14] In the 1111–1200 K range, however, Cao and Back[15] obtained a value of 96 kJ mol^{-1}.

As was discussed in Section 4.7, deviations from Arrhenius behavior is treated conveniently in terms of the temperature dependence of the pre-exponential factor. A summary of the predictions of conventional transition-state theory for bimolecular and trimolecular reactions involving atoms, linear molecules, and nonlinear molecules is given in Table 5.4.

Unfortunately, it is difficult to test those predictions for reactions having a substantial activation energy, since the temperature dependence of the rate then arises largely from the exponential factor, which dominates the effect of the pre-exponential factor. The few tests that have been made with such reactions[16] do not lead to particularly good agreement with the predictions of CTST. The reason for this appears to be that for reactions having an energy barrier there is a significant temperature dependence of the transmission coefficient.[17] This effect is ignored in CTST but has been taken into account in some of the generalized versions of TST.

The situation, however, is different for reactions in which there is no energy barrier, that is, reactions having zero activation energy. Here the temperature dependence of the transmission coefficient is not involved. There are few examples for bimolecular reactions between neutral species, but later we shall see some examples for trimolecular reactions (Section 5.2).

TABLE 5.4 TEMPERATURE DEPENDENCES OF PARTITION FUNCTIONS AND OF PRE-EXPONENTIAL FACTORS ACCORDING TO CTST

Type of species	Temperature dependence of partition function		
	Translation	Rotation	Total[a]
Monatomic (A)	$T^{1.5}$	—	$T^{1.5}$
Linear (L)	$T^{1.5}$	T	$T^{2.5}$
Nonlinear (N)	$T^{1.5}$	$T^{1.5}$	T^{3}

Type of reaction	Temperature dependence of A according to CTST[b]	
	Linear activated complex	Nonlinear activated complex
	Bimolecular	
A + A	$T^{0.5}$	—
A + L	$T^{-0.5}$	T^{0}
A + N	T^{-1}	$T^{-0.5}$
L + L	$T^{-1.5}$	T^{-1}
L + N	T^{-2}	$T^{-1.5}$
N + N	$T^{-2.5}$	T^{-2}
	Trimolecular	
A + A + A	T^{-1}	$T^{-0.5}$
A + A + L	T^{-2}	$T^{-1.5}$
A + L + L	T^{-3}	$T^{-2.5}$
L + L + L	T^{-4}	$T^{-3.5}$
A + A + N	$T^{-2.5}$	T^{-2}
A + L + N	$T^{-3.5}$	T^{-3}
L + L + N	$T^{-4.5}$	T^{-4}
L + N + N	T^{-5}	$T^{-4.5}$
N + N + N	$T^{-5.5}$	T^{-5}

[a] Neglecting the temperature dependence of the vibrational partition function and assuming no free rotation. Each degree of free rotation contributes $T^{0.5}$ (Table 4.1).

[b] Neglecting the temperature dependence of the vibrational partition function and assuming no change in free rotation when the activated complex is formed. An additional $T^{-0.5}$ appears for each degree of freedom lost in forming the activated complex.

5.1.4 Ion–Molecule Reactions

During recent years, with the development of the techniques of mass spectrometry, some reactions between ions and molecules have been investigated over a range of temperatures. For example, the reaction

$$t\text{-}C_4H_9^+ + i\text{-}C_5H_{12} \rightarrow i\text{-}C_4H_{10} + t\text{-}C_5H_{11}^+$$

has been found[18] to have a pre-exponential factor proportional to $T^{-3.0}$. Such a reaction certainly will have a nonlinear activated complex, and if there is no loss of free rotation when the complex is formed CTST predicts a dependence on $T^{-2.0}$ (Table 5.4). To accommodate the $T^{-3.0}$ dependence, it is necessary to postulate that two free rotations of methyl radicals become inactive in the activated state; this is entirely reasonable.

A more complex type of behavior is found with the reaction

$$O^+ + N_2 \rightarrow NO^+ + N$$

which has been studied[19] over a wide temperature range, from about 10^2 to 10^5 K. The rate was found to pass through a minimum at about 800 K. The activated complex $(N_2O^+)^{\ddagger}$ is expected to be linear, and therefore the pre-exponential factor, according to Table 5.4 $(A + L \rightarrow L^{\ddagger})$, would be proportional to $T^{-0.5}$. If, as is likely, the activation energy is zero, the rate should vary as $T^{-0.5}$. This is approximately the case in the low-temperature range. To explain the increase in rate at the higher temperatures it has been pointed out that the effect of the vibrational partition functions can no longer be neglected. By use of estimated vibration frequencies of the activated complex, it has been shown[20] that the behavior at higher temperatures can be interpreted satisfactorily. This explanation seems preferable to the suggestion[21] that there is a change of mechanism at the higher temperatures.

5.2 TRIMOLECULAR REACTIONS

The first trimolecular gas reaction was discovered and investigated in 1914 by Trautz:[22] it was the reaction

$$2NO + Cl_2 \rightarrow 2NOCl$$

and he investigated it from 8 to 283°C. Later, Krauss and Saracini[23] confirmed that the reaction is homogeneous and third order over a wide pressure range. A number of other reactions have been found to be third order:[24] they include

$$2NO + Br_2 \rightarrow 2NOBr$$

$$2NO + O_2 \rightarrow 2NO_2$$

Subsequent work has indicated that each of these reactions occurs in a single chemical step, so that they are regarded as elementary. This classification does not exclude the possibility that there is first a loose association between two molecules, followed by a reaction with a third molecule:

$$A + B \rightleftharpoons AB$$

$$AB + C \rightarrow Y + Z$$

In 1922 Bodenstein[25] explained trimolecular reactions on this basis. He regarded the two molecules A and B as undergoing a "sticky collision" in which they remained together for a longer time than in an ordinary collision. He then made an estimate of the frequency of collisions between AB and C, and obtained the frequency with which ABC was formed from A + B + C. This procedure, however, led to a serious overestimate, by a factor of about 10^5, of the pre-exponential factor for these trimolecular reactions. In the light of transition-state theory, this is not surprising: the simple collision theory that Bodenstein used takes no proper account of the substantial loss of entropy that occurs when three molecules unite to form an activated complex.

Transition-state theory, on the other hand, has the virtue of allowing for this effect and also of circumventing the problem of deciding how the ternary complex is formed. As long as it can be assumed that the ternary activated complex is in equilibrium with the reactants, the manner of its formation is irrelevant to the calculation of the rate.

The detailed application of CTST was worked out by Gershinowitz and Eyring.[26] For the reaction

$$2NO + X_2 \rightarrow 2NOX$$

the rate constant is

$$k = \frac{\mathbf{k}T}{h} \frac{q_{\ddagger}}{q_{NO}^2 q_{X_2}} e^{-E_0/RT} \tag{5.7}$$

the statistical factor being unity. If there is no free rotation in the activated complex, this equation becomes

$$k = \frac{\mathbf{k}T}{h} \frac{\dfrac{(2\pi m_{\ddagger}\mathbf{k}T)^{3/2}}{h^3} \dfrac{8\pi^2(8\pi^3 I_A I_B I_C)^{1/2}(\mathbf{k}T)^{3/2}}{h^3} \prod^{11} q_{\ddagger}(\nu)}{\prod_i^3 \dfrac{(2\pi m_i \mathbf{k}T)^{3/2}}{h^3} \prod_i^3 \dfrac{8\pi^2 I_i \mathbf{k}T}{h^2} \prod_i^3 q_i(\nu)} e^{-E_0/RT} \tag{5.8}$$

where the subscript i refers to the initial state. The activated complex contains six atoms, so that there are $3 \times 6 - 7 = 11$ vibrational factors. Since these vary only slightly with temperature, the temperature dependence of the rate constant, to a good approximation, is given by

$$k \propto T^{-3.5} e^{-E_0/RT} \tag{5.9}$$

The pre-exponential factor for this reaction, of the type $L + L + L$ (compare Table 5.4) thus varies as $T^{-3.5}$. In the case of the $2NO + O_2$ reaction, E_0 is zero, and the rate constant itself, according to CTST, should decrease with increasing temperature, being proportional to $T^{-3.5}$.

Alternatively, if the activated complex has the structure

$$\begin{matrix} & O \cdots O & \\ O{=}N \cdots\quad\quad\cdots N{=}O & \end{matrix}$$

one of the vibrational partition functions in Eq. (5.8) must be replaced by the expression given in Table 4.1 for free rotation. The effect of this is to replace the term $T^{-3.5}$ in Eq. (5.9) by $T^{-3.0}$. The resulting rate expression can be written as

$$\ln k = \ln (\text{constant}) - 3 \ln T + f(T) \tag{5.10}$$

where the constant can be evaluated and $f(T)$ is the contribution from the vibrational partition functions. It follows that a plot of

$$\ln k + 3 \ln T - f(T) \tag{5.11}$$

versus $1/T$ should be a straight line of slope equal to $-E_0/R$. Such a plot is shown in Fig. 5.4, and it is seen that E_0 is extremely close to zero. This result may be associated

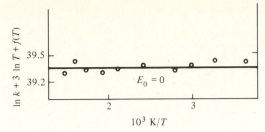

Figure 5.4 Plot of $\ln k + 3 \ln T - f(T)$ [see Eq. (5.11)] versus $1/T$ for the reaction between nitric oxide and oxygen.

with the fact that both nitric oxide and oxygen are paramagnetic, the former molecule having one unpaired electron and the latter having two unpaired electrons.

The rate constant therefore is calculated with E_0 taken as zero. The moments of inertia are derived from spectroscopic measurements, but the evaluation of those of the activated complex requires a knowledge of its configurations and dimensions. The relevant values are not very critical, and estimates were made on the basis of the normal interatomic distances in other molecules.

The observed results and those calculated by the methods outlined above are shown in Table 5.5. The agreement is very satisfactory, especially in that both theory and experiment show a minimum rate at about the same temperature. The contributions of the vibrational partition functions become greater the higher the temperature and eventually overcome the decrease due to the factor $T^{-3.0}$.

The reaction between nitric oxide and chlorine has been treated in a similar manner, the activated complex being assumed to have the configuration shown for $N_2O_4^{\ddagger}$, with a Cl_2 molecule replacing the O_2 molecule. A plot of the same function against $1/T$ again gave a straight line, but the slope corresponded to an energy of activation of 20.1 kJ mol^{-1}. Using this value, one can calculate the rate constants given in Table 5.6. The agreement between calculated and observed values is again quite satisfactory, especially in view of the assumptions made in the theory and the fact that the data are not very accurate.

TABLE 5.5 CALCULATED AND OBSERVED RATE CONSTANTS FOR THE NITRIC OXIDE–OXYGEN REACTION

	$k/10^3$ dm^6 mol^{-2} s^{-1}	
T/K	Calculated	Observed
80	86.0	41.8
143	16.2	20.2
228	5.3	10.1
300	3.3	7.1
413	2.2	4.0
564	2.0	2.8
613	2.1	2.8
662	2.0	2.9

TABLE 5.6   CALCULATED AND OBSERVED
RATES FOR THE NITRIC
OXIDE–CHLORINE REACTION

	$k/dm^6 \, mol^{-2} \, s^{-1}$	
T/K	Calculated	Observed
273	1.4	5.5
333	2.2	9.5
355	8.6	27.2
401	18.3	72.2
451	25.4	182
506	64.5	453
566	120.2	1130

The treatment of trimolecular gas reactions is one of the outstanding successes of conventional transition-state theory. Using very straightforward procedures, we can account for the unusual temperature dependences of these reactions in a simple way, and we can calculate reliable values for the pre-exponential factors. As was noted, a simple version of collision theory gave results in error by several orders of magnitude, and it gave no interpretation of the temperature dependence. Other theories can give satisfactory results only at the cost of considerable labor.

Kinetic data for some trimolecular reactions studied in the mass spectrometer have been compiled and discussed[27] and are listed in Table 5.7. All these reactions appear to occur with zero activation energy, so that the temperature dependence of the rates is attributed to that of the pre-exponential factors. The predicted temperature dependences for linear and nonlinear complexes are shown in the table (compare Table 5.4). In view of the simplicity of the model and the possible experimental error, the agreement between theory and experiment is very satisfactory. These trimolecular reactions occur much more slowly than bimolecular reactions having zero activation energy, because of the considerable entropy loss when the activated complex is formed.

5.3 UNIMOLECULAR REACTIONS

The early history of unimolecular gas reactions was one of considerable confusion, on both the experimental and theoretical sides. In the last century, van't Hoff was aware of two first-order gas reactions—the decompositions of arsine and of phosphine[28]—but he realized that they occurred to a considerable extent on the surface of the reaction vessel. In 1921, Daniels and Johnston[29] investigated the decomposition of nitrogen pentoxide,

$$2N_2O_5 \rightarrow 2N_2O_4 + O_2$$

and believed it to be unimolecular. Much work was done subsequently on this reaction, but eventually it was realized that it occurs by a composite mechanism, to be considered in Section 8.5.

In 1922 Trautz and Winkler[30] discovered and investigated the first true unimolecular gas reaction, the isomerization of cyclopropane into propylene,

TABLE 5.7   TEMPERATURE DEPENDENCES OF RATES OF TRIMOLECULAR
  ION–MOLECULE REACTIONS

Reaction	Experimental temperature dependence	Theoretical temperature dependence[a]		Reference to experimental work
		Linear complex	Nonlinear complex	
$He^+ + He + He \rightarrow He_2^+ + He$	T^{-1}	T^{-1}	$T^{-0.5}$	b
$N^+ + N_2 + He \rightarrow N_3^+ + He$	$T^{-1.7}$	T^{-2}	$T^{-1.5}$	c
$N_2^+ + N_2 + He \rightarrow N_4^+ + He$	$T^{-1.6}$	T^{-3}	$T^{-2.5}$	c
$N^+ + N_2 + N_2 \rightarrow N_3^+ + N_2$	$T^{-2.5}$	T^{-3}	$T^{-2.5}$	d
$O_2^+ + O_2 + O_2 \rightarrow O_4^+ + O_2$	T^{-3}	T^{-4}	$T^{-3.5}$	e
$N_2^+ + N_2 + N_2 \rightarrow N_4^+ + N_2$	$T^{-4.4}$	T^{-4}	$T^{-3.5}$	d
$H_3^+ + H_2 + H_2 \rightarrow H_5^+ + H_2$	$T^{-4.6}$	$T^{-4.5}$	T^{-4}	f
$H_3O^+ + H_2O + CH_4 \rightarrow H_5O_2^+ + CH_4$	$T^{-4.2}$	$T^{-5.5}$	T^{-5}	g

[a] M. Meot-Ner, J. J. Solomon, F. H. Field, and H. Gerschinowitz, *J. Phys. Chem.,* **78,** 1773 (1974).

[b] F. E. Niles and W. W. Robertson, *J. Chem. Phys.,* **42,** 3277 (1965).

[c] D. K. Bohme, D. B. Dunkin, F. C. Fehsenfeld, and E. E. Ferguson, *J. Chem. Phys.,* **51,** 863 (1969).

[d] A. Good, D. A. Durden, and P. Kebarle, *J. Chem. Phys.,* **52,** 212 (1970).

[e] D. A. Durden, P. Kebarle, and A. Good, *J. Chem. Phys.,* **50,** 805 (1969).

[f] R. C. Pierce and R. F. Porter, *Chem. Phys. Lett.,* **23,** 608 (1973).

[g] A. J. Cunningham, J. D. Paysant, and P. Kebarle, *J. Am. Chem. Soc.,* **94,** 7627 (1972).

$$\underset{H_2C-CH_2}{\overset{CH_2}{\diagup \diagdown}} \rightarrow CH_3CH=CH_2$$

Subsequent work on this reaction has given no evidence that it is other than elementary, and it appears to occur entirely in the gas phase and not at all on the surface of the vessel. Other truly unimolecular reactions discovered in the 1920s are the dissociation of molecular bromine[31]

$$Br_2 \rightarrow 2Br$$

the decomposition of sulfuryl chloride,[32]

$$SO_2Cl_2 \rightarrow SO_2 + Cl_2$$

and the conversion of pinene into *dl*-limonine.[33]

In 1919 Perrin[34] concluded that unimolecular gas reactions could be understood only on the basis of the radiation hypothesis,[35] according to which the energy required for reaction to occur is provided by radiation from the walls of the vessel. In 1919, however, there was no known unimolecular reaction, and there was not even any real reason to believe that such reactions existed. Perrin did not mention any example of a unimolecular reaction: he took it for granted that they existed and, most important, that they were first-order reactions at all pressures. His argument was that one could imagine expanding the reaction vessel to an infinite volume, when the molecules would be so far apart that no collisions took place. First-order kinetics implies that the probability that a given molecule A reacts, which is proportional to $(-d[A]/dt)/[A]$, is independent of [A]. According to Perrin, this probability remains the same even if [A] approaches zero, when there are no collisions. He therefore concluded that collisions could play no part in unimolecular processes, and the radiation hypothesis appeared to offer the only plausible alternative. The argument can be expressed in another way by saying that the frequency of collisions depends on the square of the pressure; dependence of the rate on the first power therefore appeared to be impossible.

To some[36] this argument appeared compelling, while others[37] were skeptical. The controversy raged for a number of years until finally an alternative explanation for unimolecular reactions was accepted. It is now easy to see the fallacy in Perrin's argument: there were no known experimental examples, and Perrin assumed behavior that eventually turned out to be incorrect. Unimolecular reactions do not remain first order at very low pressures.

5.3.1 Lindemann–Christiansen Hypothesis

In 1921 the Faraday Society held a discussion on "The Radiation Theory of Chemical Action." At this meeting, the whole problem, including that of unimolecular reactions, was hotly debated. One who expressed strong opposition to the radiation hypothesis was Lindemann (later Lord Cherwell). Realizing that if he destroyed the hypothesis he should suggest an alternative explanation for unimolecular reactions, he did so in a very brief communication which was later published.[38] A few days after Lindemann made his oral presentation, Christiansen[39] published his Ph.D. thesis in which the same treatment was included, in a more detailed and explicit form than appeared in Lindemann's written submission. This Lindemann–Christiansen hypothesis is the basis of all modern theories of unimolecular reactions, although a number of important modifications had to be made to it later.

The hypothesis provided a way in which, under some conditions, activation by collisions can give rise to first-order kinetics. It also led to the important prediction that at sufficiently low pressures all unimolecular reactions must become second order. Much later work has shown this to be the case.

The Lindemann–Christiansen hypothesis, as later interpreted and expanded by Hinshelwood[40] and others, can be formulated as follows. The process of forming a molecule A* having sufficient energy to undergo reaction, the energization process, involves collision between two molecules of A. The energized molecule A* may undergo de-energization by collision with a normal molecule, or it may undergo a unimolecular reaction to form products. These three processes are quite distinct, and the situation may be represented as

$$A + A \underset{k_{-1}}{\overset{k_1}{\rightleftharpoons}} A^* + A$$

$$A^* \overset{k_2}{\rightarrow} Y + Z$$

The rate of energization is $k_1[A]^2$. The energized species can be de-energized, at a rate $k_{-1}[A^*][A]$, or it can undergo conversion into products $Y + Z$ at a rate $k_2[A^*]$.

If the pressure is high enough, the rate $k_{-1}[A^*][A]$ with which A* is de-energized will be greater than the rate of product formation, $k_2[A^*]$. When this is the case, the energized molecules A* are essentially in equilibrium with normal molecules, and the concentration [A*] is thus proportional to the first power of [A], being given by

$$[A^*] = \frac{k_1}{k_{-1}} [A] \qquad (5.12)$$

Thus, the rate of reaction, equal to $k_2[A^*]$, is also proportional to [A]:

$$v = k_2[A^*] = k_2 \frac{k_1}{k_{-1}} [A] \qquad (5.13)$$

The kinetics are first order.

The situation, however, is different if the concentration of A is so small that de-energization is much less rapid than the rate at which A is converted into products. There is no longer an equilibrium involving A*. Instead, the rate of product formation is almost equal to the rate at which A* molecules are formed, since now these molecules nearly always become products. The rate of reaction therefore equals the rate of energization, which is $k_1[A]^2$; the kinetics are second order.

After Lindemann and Christiansen put forward these suggestions, some experiments were done with the object of seeing whether the kinetics did become second order at sufficiently low pressures. At first there was considerable confusion, since many of the reactions studied were not truly unimolecular; some were composite reactions and others occurred on the surface of the vessel. Much early work was done on the nitrogen pentoxide decomposition (see Section 8.4.6), which did not show the change of kinetics, but which eventually turned out to occur by a composite mechanism.

An important aspect of the Lindemann–Christiansen hypothesis is the concentration of A at which there should be an observable transition from first-order towards second-order kinetics. According to the hypothesis, the kinetics are first order only when the rate of de-energization, $k_{-1}[A^*][A]$, is considerably greater than the rate of conversion of A* into products, $k_2[A^*]$. When this is so, the rate of reaction is given by Eq. (5.13). A first-order rate constant under these conditions can be defined by

$$v \equiv k_\infty^1 [A] \qquad (5.14)$$

and thus is given by

$$k_\infty^1 = \frac{k_1 k_2}{k_{-1}} \qquad (5.15)$$

We now consider in particular the situation when the rates of de-energization and of product formation are equal: the condition is

$$k_{-1}[A^*][A]_{1/2} = k_2[A^*] \qquad (5.16)$$

where $[A]_{1/2}$ is the concentration at which this is satisfied. Now A* has only a 50% chance of becoming products, and the rate is just half of the rate given by Eq. (5.13). Thus,

$$v_{1/2} = \tfrac{1}{2} k_2 \frac{k_1}{k_{-1}} [A]_{1/2} \qquad (5.17)$$

$$= \tfrac{1}{2} k_\infty^1 [A]_{1/2} \qquad (5.18)$$

The first-order rate coefficient k^1, defined under any condition by

$$v \equiv k^1[A] \tag{5.19}$$

has fallen to one-half of its limiting (high-pressure) value k^1_∞.

Subsequent discussions of this problem, beginning with that of Hinshelwood,[40] have made use of the steady-state hypothesis, which we consider later (Section 8.2.2). The arguments so far are independent of that hypothesis, which provides a useful general equation for the rate. According to the steady-state hypothesis, the concentration $[A^*]$ can be regarded as essentially constant with time, so that $d[A^*]/dt = 0$. It follows from the mechanism that

$$\frac{d[A^*]}{dt} = k_1[A]^2 - k_{-1}[A^*][A] - k_2[A^*] = 0 \tag{5.20}$$

and therefore that

$$[A^*] = \frac{k_1[A]^2}{k_{-1}[A] + k_2} \tag{5.21}$$

The rate of formation of products is

$$v = k_2[A^*] = \frac{k_1 k_2[A]^2}{k_{-1}[A] + k_2} \tag{5.22}$$

At sufficiently high concentrations $k_{-1}[A] \gg k_2$, and the rate is

$$v = \frac{k_1 k_2}{k_{-1}}[A] = k^1_\infty[A] \tag{5.23}$$

as found previously [Eq. (5.13)]. At low concentrations, on the other hand, $k_2 \gg k_{-1}[A]$ and

$$v = k_1[A]^2 \tag{5.24}$$

It follows from Eq. (5.22) that the first-order rate coefficient, defined by Eq. (5.19), is given by

$$k^1 = \frac{k_1 k_2[A]}{k_{-1}[A] + k_2} \tag{5.25}$$

which for some purposes is written more conveniently as

$$k^1 = \frac{k_2(k_1/k_{-1})}{1 + k_2/k_{-1}[A]} \tag{5.26}$$

According to these equations, a plot of k^1 versus $[A]$ is a hyperbola, as shown in Fig. 5.5. The first-order coefficient k^1 is constant in the higher concentration range but falls off at lower concentrations. As previously shown, $k^1 = k^1_\infty/2$ when $k_{-1}[A] = k_2$, and we write the concentration at which this is true as $[A]_{1/2}$. Thus,

$$[A]_{1/2} = \frac{k_2}{k_{-1}} = \frac{k^1_\infty}{k_1} \tag{5.27}$$

using Eq. (5.23).

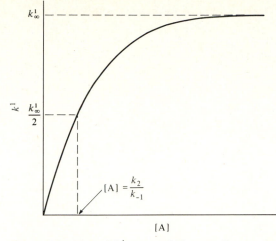

Figure 5.5 Plot of k^1 versus [A], according to Eq. (5.26).

However, when this relationship was applied to experimental data, some difficulties arose. The value of k^1_∞ is an experimental quantity, and it was thought that k_1 could be estimated easily on the basis of simple collision theory; it was originally set as $z_1 e^{-E_1/RT}$, where z_1 is the collision frequency factor and E_1 is the activation energy. However, the experimental values of $[A]_{1/2}$ were always much smaller than those estimated in this way. Since there can be no doubt about k^1_∞, which is an experimental quantity, the error must be in the estimation of k_1. Thus, a modification must be made to allow k_1 to be much larger than $z_1 e^{-E_1/RT}$.

A second difficulty with the Lindemann–Christiansen hypothesis becomes apparent when experimental results are plotted in another way. Equation (5.25) may be written as

$$\frac{1}{k^1} = \frac{k_{-1}}{k_1 k_2} + \frac{1}{k_1 [A]} \tag{5.28}$$

and a plot of $1/k^1$ against the reciprocal of the concentration should give a straight line. However, deviations from linearity have been found of the kind shown schematically in Fig. 5.6.

5.3.2 Hinshelwood's Treatment

The first difficulty with the original Lindemann–Christiansen hypothesis, that the first-order rates are maintained down to lower concentrations than the theory appeared to permit, was overcome successfully by Hinshelwood.[41] He pointed out that the expression $z e^{-E/RT}$ only applies if the energy is distributed among two degrees of freedom (see Section 3.1 and the appendix to Chapter 3). For some unimolecular reactions, however, the number of degrees of freedom s is considerable. The activation energy is distributed initially among these degrees of freedom, and there are many ways in

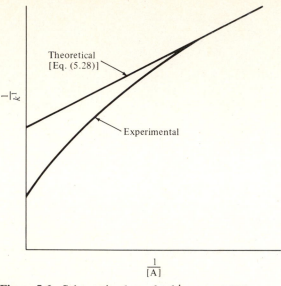

Figure 5.6   Schematic plots of $1/k^1$ versus $1/[A]$.

which this distribution may be effected. Once the energy is in the molecule, distributed in any way among s vibrational degrees of freedom, the molecule is in a position eventually to react. After a number of vibrations of the energized molecule A*—which may be a very considerable number—the energy may find its way into appropriate degrees of freedom so that A* can pass at once into products.

Hinshelwood derived the statistical expression (3.41) that was given in the appendix to Chapter 3. This formula gives the fraction of molecules having energy in excess of ε_0^* as

$$f^* = \frac{1}{(s-1)!} \left(\frac{\varepsilon_0^*}{kT}\right)^{s-1} e^{-\varepsilon_0^*/kT} \tag{5.29}$$

where s is the number of degrees of freedom. He then expressed the rate constant k_1 as

$$k_1 = z_1 \frac{1}{(s-1)!} \left(\frac{\varepsilon_0^*}{kT}\right)^{s-1} e^{-\varepsilon_0^*/kT} \tag{5.30}$$

instead of simply as $z_1 e^{-\varepsilon_0^*/kT}$. Thus, an additional factor has appeared,

$$\frac{1}{(s-1)!} \left(\frac{\varepsilon_0^*}{kT}\right)^{s-1}$$

and if s is sufficiently great this may be greater than unity by many powers of 10.

The activation energy for k_1 predicted in the Hinshelwood formulation differs from that predicted by the simple collision theory. Equation (5.30) can be written

$$k_1 \propto T^{1/2} \left(\frac{1}{T}\right)^{s-1} e^{-\varepsilon_0^*/kT} \tag{5.31}$$

This equation leads to

$$\frac{d \ln k_1}{dT} = \frac{(\frac{3}{2} - s)\mathbf{k}T + \varepsilon_0^*}{\mathbf{k}T^2} \tag{5.32}$$

Since by the definition of the experimental activation energy [Eq. (2.81)]

$$\frac{d \ln k_1}{dT} \equiv \frac{E_a}{RT^2} = \frac{\varepsilon_a}{\mathbf{k}T^2} \tag{5.33}$$

where ε_a is the experimental activation energy per molecule, it follows that

$$\varepsilon_0^* = \varepsilon_a + (s - \tfrac{3}{2})\mathbf{k}T \tag{5.34}$$

Equation (5.30) can give much higher rates of activation and therefore much higher k_1/k_{-1} values than simple collision theory. For example, if the experimental activation energy is 170 kJ mol^{-1} and s is 12, Eq. (5.30) gives a value of k_1/k_{-1} that is larger by more than 10^6 than that given by the older theory.

A number of difficulties still remain, however, particularly the following:

1. The number s of degrees of freedom required to give agreement with experiment on the basis of Hinshelwood's proposal usually turns out to be only about one-half of the total number of vibrational modes. No satisfactory explanation has been given for this.
2. According to Hinshelwood's treatment, k_∞ is given by

$$k_\infty^1 = \frac{k_1 k_2}{k_{-1}} = k_2 \frac{1}{(s-1)!} \left(\frac{\varepsilon_0^*}{\mathbf{k}T}\right)^{s-1} e^{-\varepsilon^*/\mathbf{k}T} \tag{5.35}$$

   Therefore, one would expect a strong temperature dependence of the pre-exponential factor, especially for large values of s. There is no experimental evidence for this.
3. Hinshelwood's treatment does not affect the argument that a plot of $1/k^1$ versus $1/[A]$ should be linear (Fig. 5.6). The lack of linearity found experimentally still calls for an explanation.

5.3.3 Rice–Ramsperger–Kassel (RRK) Treatment

The treatment of Hinshelwood and the further developments of the theory of unimolecular reactions are considered best in terms of the following scheme:

$$A + M \rightleftharpoons A^* + M$$

$$A^* \rightarrow A^\ddagger \rightarrow \text{products}$$

Here M is any molecule, including another A, that can transfer energy to A when a collision occurs. A distinction is made between an *activated* molecule, represented by the symbol $A^\ddagger$, and an *energized* molecule, represented as A^*. An *activated* molecule $A^\ddagger$ is by definition one that is passing directly through the dividing surface of the potential-energy surface. An *energized* molecule A^*, on the other hand, is one that has acquired all the energy it needs to become an activated molecule $A^{\ddagger}$; however, it must undergo vibrations before it does so. Figure 5.7 shows an energy diagram that relates to the Hinshelwood modification and also to the RRK mechanism now to be considered. To become an activated molecule $A^{\ddagger}$, the molecule A must acquire at least the energy ε_0^*. Hinshelwood's modification allowed A to acquire an amount of energy ε_0^* at an enhanced rate but regarded the rate at which A^* becomes $A^{\ddagger}$ to be independent of that energy. In the further developments of the theory, the rate constant k_2 was not treated as constant but was considered to be larger the greater the value of ε^*.

We may approach this problem by starting with Eq. (5.26), which now can be written in a more general form:

$$k^1 = \frac{k_2(k_1/k_{-1})}{1 + k_2/k_{-1}[M]} \tag{5.36}$$

In Hinshelwood's modification both k_2 and k_1/k_{-1} are treated as independent of ε^*, the amount of energy in the energized molecule. Hinshelwood's formula, Eq. (5.30), involves the critical energy ε_0^* but not ε^*, the amount of energy in an individual energized molecule.

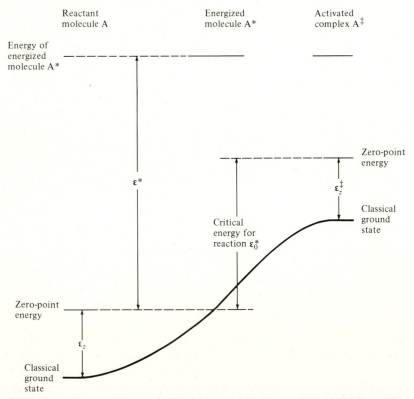

Figure 5.7 Energy scheme for Lindemann, Hinshelwood, and RRK mechanisms.

Very shortly after Hinshelwood suggested his modified formula, O. K. Rice and Ramsperger[42] and independently Kassel[43] proposed theories in which both k_2 and k_1/k_{-1} were treated as dependent on the energy ε^* contained within an individual energized molecule. These theories are referred to jointly as RRK theory. Equation (5.36) must now be modified: k_2 must be replaced by $k_2(\varepsilon^*)$ and k_1/k_{-1} (now written as f) by $f(\varepsilon^*)$, which is a distribution function. The expression

$$dk^1 = \frac{k_2(\varepsilon^*)\,f(\varepsilon^*)}{1 + k_2(\varepsilon^*)/k_{-1}[M]}\,d\varepsilon^* \tag{5.37}$$

is a *microcanonical* rate constant, relating to energized molecules having energy between ε^* and $\varepsilon^* + d\varepsilon^*$. To obtain the ordinary rate coefficient k^1, the expression must be integrated from ε_0^*—the minimum energy that can lead to reaction—to infinity:

$$k^1 = \int_{\varepsilon_0^*}^{\infty} \frac{k_2(\varepsilon^*)\,f(\varepsilon^*)}{1 + k_2(\varepsilon^*)/k_{-1}[M]}\,d\varepsilon^* \tag{5.38}$$

Figure 5.8 summarizes the way in which the theories arise from this general equation.

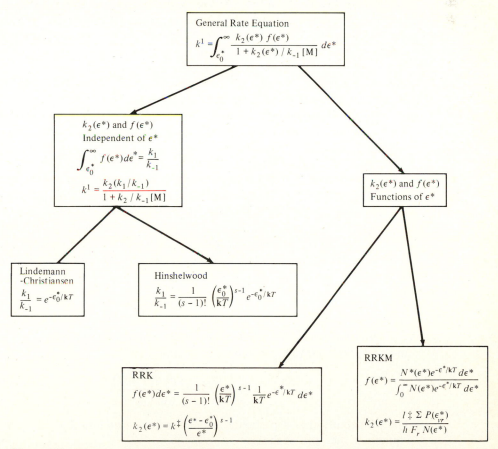

Figure 5.8 Diagram showing the relationship between some of the theories of unimolecular gas reactions. The symbols are defined in the text.

Although it is convenient to consider the Rice–Ramsperger and Kassel theories together as RRK theory, there are two important differences between them. First, Rice and Ramsperger used classical statistical mechanics, whereas Kassel introduced a quantum treatment as well as a statistical one. Second, Kassel proposed that reactions occur only when the critical energy ε_0^* is concentrated in one normal mode of vibration, that is, in two "square terms." In the Rice–Ramsperger treatment, the energy goes into one "square term," which is not realistic. Most later RRK treatments are essentially Kassel treatments, and the present account relates more closely to the Kassel version.

In RRK theory a molecule is regarded as a system of loosely coupled oscillators. In the energized molecule A* an amount of energy ε^* is distributed among the normal modes of vibration. Because the normal modes are coupled loosely, the energy can flow between them, and after a sufficient number of vibrations the critical amount of energy ε_0^* may be in a particular normal mode and reaction can occur. For example, energy in an energized molecule of ethane, $C_2H_6^*$, may pass into a normal mode which corresponds to an extension of the C−C bond. If the energy is sufficiently great the bond will break with the formation of two methyl radicals. In RRK theory, the oscillators are regarded as all having the same frequency of vibration.

An essential feature of RRK theory is that the energized molecules A* have *random lifetimes*. By this is meant that the energy is distributed randomly among the various normal modes, so that the process A* → A‡ depends entirely on statistical factors. In other words, the process by which A* is formed does not lead to a species in which the energy is more likely to be in certain normal modes, in which case there might be a higher or lower probability of forming A‡.

Another important feature of the theory is that the collisions that produce A* molecules and also those that deactivate them are *strong collisions*. By this is meant that in the collision processes large amounts ($\gg kT$) of energy are transferred. In other words, activation more often occurs in a single collision than as a result of a number of successive collisions. When an energized molecule A* undergoes a collision, it is assumed in the strong-collision theories that de-energization is almost inevitable.

The way in which RRK theory treats the dependence of k_2 on ε^* is as follows. The statistical weight of a system of s degrees of vibrational freedom containing j quanta of vibrational energy is equal to the number of ways in which j objects can be divided among s boxes, each of which can contain any number; the number of such ways is

$$w = \frac{(j + s - 1)!}{j!(s - 1)} \tag{5.39}$$

The statistical weight for states in which the s oscillators have j quanta among them, and one particular one has m quanta, is similarly

$$w' = \frac{(j - m + s - 1)!}{(j - m)!\,(s - 1)!} \tag{5.40}$$

The probability that a particular oscillator has m quanta and all s oscillators have j quanta is the ratio of these:

$$\frac{w'}{w} \equiv r = \frac{(j - m + s - 1)!\,j!}{(j - m)!\,(j + s - 1)!} \tag{5.41}$$

If the Sterling approximation ($n! = n^n/e^n$) is applied to this expression, the terms in e^n cancel out, and the result is

$$r = \frac{(j - m + s - 1)^{j-m+s-1}j^j}{(j - m)^{j-m}(j + s - 1)^{j+s-1}} \tag{5.42}$$

Provided that $j - m \gg s - 1$, this reduces to

$$r = \frac{(j - m)^{j-m+s-1}j^j}{(j - m)^{j-m}j^{j+s-1}} \tag{5.43}$$

$$= \left(\frac{j - m}{j}\right)^{s-1} \tag{5.44}$$

The total number of quanta j may be taken as proportional to ε^*, the total energy of the molecule, while m is proportional to ε_0^*, the minimum energy that a molecule must have for decomposition to take place.§ The expression given above therefore is equal to

$$r = \left(\frac{\varepsilon^* - \varepsilon_0^*}{\varepsilon^*}\right)^{s-1} \tag{5.45}$$

The rate with which the required energy ε_0^* passes into this particular oscillator is proportional to this quantity, so that

$$k_2 = k^{\ddagger}\left(\frac{\varepsilon^* - \varepsilon_0^*}{\varepsilon^*}\right)^{s-1} \tag{5.46}$$

In this expression $k^{\ddagger}$ is the rate constant corresponding to the free passage of the system through the dividing surface; when ε^* is sufficiently large, the energized molecule is essentially an activated molecule and therefore can pass immediately into the final state. The variation of k_2 with $\varepsilon^*/\varepsilon_0^*$, according to Eq. (5.46), is shown in Fig. 5.9.

RRK theory also treats the distribution function $f(\varepsilon^*)$ as dependent on ε^*. If the energy ε^* is distributed among s normal modes of vibration we have from Eq. (3.37) that

$$f(\varepsilon^*) = \frac{1}{(s - 1)!}\left(\frac{\varepsilon^*}{kT}\right)^{s-1}\frac{1}{kT}e^{-\varepsilon^*/kT}\,d\varepsilon^* \tag{5.47}$$

Introduction of this expression and Eq. (5.46) into the general rate equation (5.38) gives

$$k^1 = \int_{\varepsilon_0^*}^{\infty}\frac{k^{\ddagger}\left(\dfrac{\varepsilon^* - \varepsilon_0^*}{\varepsilon^*}\right)^{s-1}\dfrac{1}{(s - 1)!}\left(\dfrac{\varepsilon^*}{kT}\right)^{s-1}\dfrac{1}{kT}e^{-\varepsilon^*/kT}}{1 + \dfrac{k^{\ddagger}}{k_{-1}[M]}\left(\dfrac{\varepsilon^* - \varepsilon_0^*}{\varepsilon^*}\right)^{s-1}}\,d\varepsilon^* \tag{5.48}$$

This equation may be reduced by making the following substitutions:

$$x = \frac{\varepsilon^* - \varepsilon_0^*}{kT} \qquad b = \frac{\varepsilon_0^*}{kT} \tag{5.49}$$

§ This depends on the assumption that the oscillators all have the same frequency, so that all the quanta are equal in size.

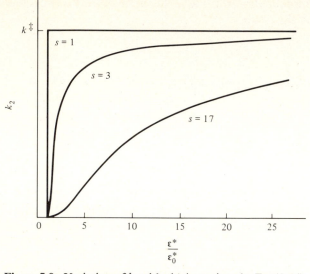

Figure 5.9 Variation of k_2 with $\varepsilon^*/\varepsilon_0^*$, as given by Eq. (5.46), for different values of s, the number of degrees of freedom.

Changing $d\varepsilon^*/\mathbf{k}T$ to dx and consequently the limits of integration, we obtain

$$k^1 = \int_0^\infty \frac{1/[(s-1)!]e^{-(x+b)}k^\ddagger x^{s-1}\,dx}{1 + (k^\ddagger/k_{-1}[\mathbf{M}])[x/(x+b)]^{s-1}} \tag{5.50}$$

$$= \frac{k^\ddagger e^{-b}}{(s-1)!} \int_0^\infty \frac{x^{s-1}e^{-x}\,dx}{1 + (k^\ddagger/k_{-1}[\mathbf{M}])[x/(x+b)]^{s-1}} \tag{5.51}$$

or
$$k^1 = \frac{k^\ddagger e^{-\varepsilon_0^*/\mathbf{k}T}}{(s-1)!} \int_0^\infty \frac{x^{s-1}e^{-x}\,dx}{1 + (k^\ddagger/k_{-1}[\mathbf{M}])[x/(x+b)]^{s-1}} \tag{5.52}$$

When [M] is very large, that is, at the high-pressure first-order limit, this expression reduces approximately to

$$k_\infty^1 = k^\ddagger e^{-\varepsilon_0^*/\mathbf{k}T} \tag{5.53}$$

For given values of [M], ε_0^*, and s, the integral in Eq. (5.52) can be obtained by numerical methods.[44] Usually, Eq. (5.52) gives reasonable agreement with experiment if s is taken to be about one-half the total number of normal modes in the molecule. This is one unsatisfactory feature of the theory: it is impossible to predict what value of s should be taken for a given reaction.

The significance of $k^\ddagger$ in RRK theory is also somewhat unsatisfactory. If there were a complete redistribution of energy among the normal modes on every vibration, the theory would predict that the pre-exponential factor for the reaction in its high-pressure limit [Eq. (5.53)] would be the average of the various normal-mode frequencies. It therefore would predict a pre-exponential factor of about 10^{13} s^{-1} for all unimolecular reactions. This is true for some reactions, but for many the values are very much greater than this. RRK theory provides no explanation for such high values. An interpretation[45] is provided by conventional transition-state theory, according to which $k^{\ddagger}$ is given by

$$k^{\ddagger} = \frac{\mathbf{k}T}{h} \frac{q_{\ddagger}}{q_r} \tag{5.54}$$

where $q_{\ddagger}$ is the partition function for the activated complexes and q_r for the reactants. If, as is plausible, the activated complexes have much looser structures than the reactants, $q_{\ddagger}$ may be much larger than q_r, and the large pre-exponential factors are explained.

5.3.4 Slater's Treatment

The RRK theory is a statistical theory, and brief mention should be made here of a very different approach, a dynamical one, which although ultimately unsuccessful did contribute considerably to the development of our understanding of unimolecular reactions. This was the theory of N. B. Slater, first proposed[46] in 1939 and elaborated by him in a book[47] and a number of subsequent papers.[48]

As an example consider the dissociation of ethane into two methyl radicals:

$$C_2H_6 \rightarrow 2CH_3$$

When the ethane molecule is energized as a result of collisions, the energy is distributed among the 18 normal modes of vibration. As it vibrates the C—C bond expands and contracts in a somewhat complicated way, because many of the 18 normal modes contribute to the extension and contraction of that bond. The essence of Slater's theory was that the bond breaks when, as a result of a number of normal-mode vibrations coming into phase, the C—C bond becomes extended by a critical amount. In the earlier versions of his theory, it was assumed that energy remains trapped in normal modes: this is in contrast to RRK theory, which assumes free exchange of energy between oscillators that have random lifetimes.

Some of the earlier quantitative treatments of individual reactions on the basis of Slater's theory were very encouraging. In particular, the theory seemed to be successful in avoiding the difficulty with RRK theory that for unknown reasons not all normal modes contribute to reaction. For example, for the isomerization of cyclopropane (see also Section 5.3.5), Slater was able to obtain good agreement with experiment when all 21 normal modes of vibration were allowed to contribute to reaction; for agreement with RRK theory only 12 normal modes could be used.

However, serious difficulties arose for certain reactions, such as the dissociation of hydrogen peroxide[49] and the decomposition of cyclobutane.[50] For these reactions the Slater theory gave low-pressure second-order rate constants that were much lower than the experimental values. The problem arises from Slater's assumption of no energy flow between normal modes. On that basis, certain modes cannot contribute to reaction at all. For example, in the hydrogen peroxide dissociation, torsional motions about the O—O bond bring about no extension of that bond and cannot lead to its dissociation. Energy in such modes thus makes no contribution, and therefore Slater's condition for energization is more stringent than the RRK condition.

Further evidence that there is a rapid exchange of energy between normal modes is provided by experiments with molecules that have been energized in different ways; this will be considered later (Section 5.3.8).

Attempts have been made[51] to modify Slater's treatment with a view to permitting flow between normal modes. However, in the meantime, Marcus had proposed his extension of RRK theory, and since this proved successful little further work was done on the basis of the Slater treatment.

In spite of this ultimate failure of Slater's theory, his work played an important role in the development of the theories of unimolecular reactions, particularly by suggesting crucial experiments. Aside from this, in the course of his work Slater gave important classical treatments of molecular vibrations and also suggested a quantum-mechanical treatment of unimolecular reactions.[52]

5.3.5 Marcus's Extension of the RRK Treatment (RRKM)

Marcus[53] has extended the Rice–Ramsperger–Kassel theory in a direction that brings it into line with transition-state theory. The essence of his theory, now generally known as the Rice–Ramsperger–Kassel–Marcus (RRKM) formulation, is that the individual vibrational frequencies of the energized species and activated complexes are considered explicitly: account is taken of the way the various normal-mode vibrations and rotations contribute to reaction, and allowance is made for the zero-point energies.

Since RRKM theory is described in detail in two excellent books[54] by Robinson and Holbrook and by Forst, only a brief general account of the main assumptions and procedures is given here. Figure 5.8 shows the relationship of RRKM theory to the other theories, and Fig. 5.10 shows the energy scheme. The total energy contained in the energized molecule is classified as either *active* or *inactive* (also referred to as *adiabatic*). The inactive energy is energy that remains in the same quantum state during the course of reaction and that therefore cannot contribute to the breaking of bonds. The zero-point energy is inactive, as is the energy of overall translation and rotation, since this energy is preserved as such when the activated molecule $A^{\ddagger}$ is formed. Vibrational energy and the energy of internal rotations are active.

In RRKM theory, the distribution function $f(\varepsilon^*)$ is expressed as

$$f(\varepsilon^*)d\varepsilon^* = \frac{N(\varepsilon^*)e^{-\varepsilon^*/kT}\,d\varepsilon^*}{\int_0^\infty N(\varepsilon^*)e^{-\varepsilon^*/kT}\,d\varepsilon^*} \qquad (5.55)$$

where $N(\varepsilon^*)$ is the density of states having energy between ε^* and $\varepsilon^* + d\varepsilon^*$. (The *density of states* is defined as the number of states per unit energy range.) The denominator in Eq. (5.55) is the partition function relating to the active-energy contributions.

The expression for $k_2(\varepsilon^*)$ in RRKM theory is

$$k_2(\varepsilon^*) = \frac{l^{\ddagger}\sum P(\varepsilon^*_{\text{active}})}{hN(\varepsilon^*)F_r} \qquad (5.56)$$

where $l^{\ddagger}$ is the statistical factor (Section 4.5) and $\sum P(\varepsilon^{\ddagger}_{\text{active}})$ is the number of vibration–rotational quantum states of the activated molecule, corresponding to all energies up to and including $\varepsilon^{\ddagger}_{\text{active}}$. The factor F_r is introduced to correct for the fact that the rotations may not be the same in the activated molecule as in the energized molecule.

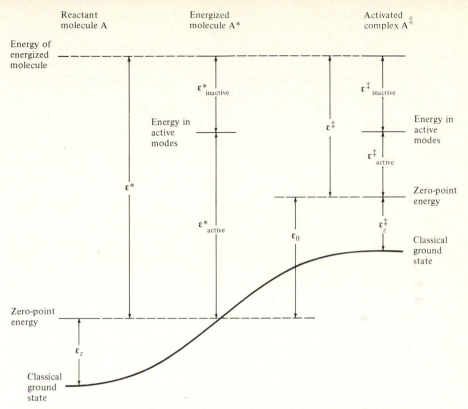

Figure 5.10 Energy scheme for the RRKM mechanism, in which (1) $f(\varepsilon^*)d\varepsilon^*$ is calculated using quantum-statistical mechanics, and (2) $k_2(\varepsilon^*)$ is expressed using CTST. The energy in A* and A‡ is either inactive (zero-point, overall translational, overall rotational) or active (all vibrations and free rotations).

A particularly satisfactory feature of RRKM theory is that it leads to the same expression for the limiting high-pressure first-order rate constant that is given by conventional transition-state theory:

$$k_\infty^1 = \frac{\mathbf{k}T}{h}\frac{q_\ddagger}{q_i}\, e^{-\varepsilon_0^*/\mathbf{k}T} \tag{5.57}$$

where $q_\ddagger$ and q_i are the partition functions for activated and initial states. Thus, the theory can explain the abnormally high pre-exponential factors that are sometimes obtained.

To make detailed calculations using RRKM theory it is necessary to decide on models for the energized and activated molecules. Vibrational frequencies for the various normal modes must be estimated and decisions made as to which energies are active and which inactive. Numerical methods are used to calculate rate coefficients k^1 at various reactant concentrations.

On the whole, the theory has been very successful in interpreting the experimental results. For a variety of reactions it has proved possible to formulate models for A* and $A^{\ddagger}$ that lead to good agreement with experiment over a range of pressures. Detailed calculations, with very satisfactory results, have been carried out for the following reactions: the cyclopropane isomerization,[55] the dissociation of cyclobutane into two ethylene molecules,[55,56] the isomerization of cyclobutane,[57] the dissociation of ethane,[58] and the decomposition of the ethyl radical into ethylene and a hydrogen atom.[58]

The isomerization of cyclopropane into propylene has been studied very extensively from both experimental and theoretical points of view. As previously noted, it was the first true unimolecular reaction to be studied. In the 1920s it was used to test the radiation hypothesis, always with negative results.[59] A careful study of the reaction over a considerable pressure range was made by Pritchard and coworkers,[60] whose results are shown plotted in Fig. 5.11. Note that the Hinshelwood treatment accounts for the falling off of k^1 but it does this only very approximately. The RRK theory accounts for the falling off much better, but s has to be taken as 12 instead of 21, which is the number of degrees of vibrational freedom in the molecule. This was shown by Slater,[61] who also applied his own treatment to the reaction. As seen from the figure, Slater was able to obtain equally good agreement by using all 21 modes. However, as previously discussed, Slater's theory is no longer considered to be satisfactory. A better interpretation of the results has been provided by the RRKM treatment.[62]

There has been some discussion of whether the cyclopropane isomerization proceeds by a concerted mechanism, in which a C—C bond is broken and a hydrogen

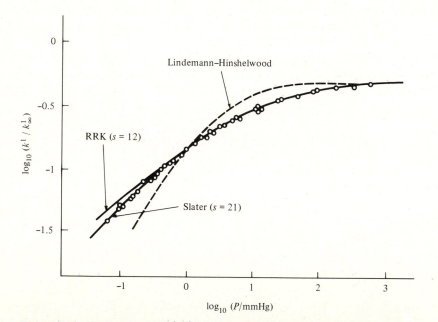

Figure 5.11 Plots of $\log_{10} (k^1/k^1_\infty)$ versus $\log_{10} (P/\text{mmHg})$ for the thermal isomerization of cyclopropane. The points are experimental, while the curves are those calculated from the theories, as indicated. The experimental values are displaced 0.3 units to the right to lie in the same range as the theoretical curves.

atom is transferred at the same time, or by a diradical mechanism. In the latter, a diradical is formed initially,

$$CH_2$$
$$H_2C-CH_2 \rightarrow -H_2C-CH_2-CH_2-$$

followed by a hydrogen atom transfer in a separate stage. The latter mechanism, however, can be excluded on the basis of symmetry arguments (Section 4.5) and with reference to a schematic potential-energy surface.[63]

Work on the isomerization of methyl isocyanide, CH_3NC, into methyl cyanide, CH_3CN, has raised some interesting questions. The experimental results[64] on the fall-off at low pressures of the first-order rate coefficient could be interpreted satisfactorily by RRKM theory, and the same is true of other isocyanides, such as CD_3NC and C_2H_5NC.[65] However, dynamical calculations[66] have suggested that the results cannot be interpreted in terms of the hypothesis that energized molecules are converted randomly into activated complexes. When this is the case, one speaks of non-RRKM kinetics; the theoretical treatment of it is rather complicated.[67]

On the whole, the RRKM theory has proved very satisfactory. A difficulty with applying it is that the vibrational frequencies for the activated complexes usually cannot be estimated very reliably. The fall-off behavior is not very sensitive to these frequencies, and this leads to the drawback that agreement with experiment can be obtained with various choices of activated complex frequencies, so that there is no way of knowing which choice is correct. The fall-off curves are more sensitive to the high-pressure Arrhenius parameters, which are sometimes not known accurately. The most stringent tests of RRKM theory are provided by chemical activation experiments, to be referred to later, and these provide good support for the theory in many cases. As will be seen, however, the chemical activation experiments do sometimes provide evidence for "non-RRKM behavior."

5.3.6 Influence of Foreign Gases

The rate of a unimolecular gas reaction can be influenced by the addition of a non-reacting or "foreign," gas, which can contribute to energization. If the reaction is occurring in the low-pressure region where there is a fall-off in the first-order rate coefficient, the addition of a foreign gas may raise the rate coefficient to its high-pressure limiting value.

If a gas A is undergoing reaction and a foreign gas M is added, the reactions are

$$A + A \underset{k_{-1}}{\overset{k_1}{\rightleftharpoons}} A^* + A$$

$$A + M \underset{k_1'}{\overset{k_1'}{\rightleftharpoons}} A^* + M$$

$$A^* \overset{k_2}{\rightarrow} \text{products}$$

With the simple Lindemann–Christiansen–Hinshelwood assumptions, the steady-state treatment leads to

$$v = \frac{k_2(k_1[A]^2 + k'_1[A][M])}{k_{-1}[A] + k'_{-1}[M] + k_2} \tag{5.58}$$

Corresponding but more complicated expressions are found for the RRK and RRKM mechanisms.

By studying rates at various concentrations of A and M, it is possible to obtain k'_1/k_1 and k'_{-1}/k_{-1}. These ratios represent the relative effectiveness with which M and A transfer energy to A. This method is therefore a useful one for obtaining information about intermolecular energy transfer.

5.3.7 Intermolecular Energy Transfer

Much evidence has accumulated about the efficiencies of energy transfers between molecules, and ony a very brief summary is given here. The noble gases and other simple molecules tend to be ineffective in transferring energy; complex molecules are more effective. Molecules that are able to react chemically transfer more energy than nonreacting molecules. This result can be interpreted in terms of some distortion of potential-energy surfaces when there is chemical interaction. This is illustrated schematically in Fig. 5.12, which shows how relative translational energy may be converted into the vibrational energy of a molecule A—B when it collides with a molecule C with which it has some chemical affinity.

A rate constant k^{sc}_{-1} can be defined for a process

$$M + A^* \rightarrow M + A$$

on the assumption that deactivation occurs on every collision, that is, if the collisions are strong collisions. In the general case, a collision efficiency B_c can be defined by

$$k_{-1} \equiv B_c k^{sc}_{-1} \tag{5.59}$$

where k_{-1} is the rate constant for a particular substance. For large molecules B_c is approximately unity, while for small molecules it is much less than unity.

Another quantity of interest, which can be estimated from the experimental results, is $\langle \Delta E_d \rangle$, the average energy transferred per collision. There is a rough correlation between B_c and $\langle \Delta E_d \rangle$. Molecules having B_c values close to unity tend to have $\langle \Delta E_d \rangle$ values of 40–60 kJ mol^{-1}. Small molecules, having low B_c values, tend to transfer only about 2 kJ mol^{-1} on each collision.

5.3.8 Intramolecular Energy Transfer

RRKM theory, now used almost exclusively for unimolecular gas reactions, is based on the assumption that when a molecule is energized the vibrational energy is exchanged rapidly between the normal modes. Specifically, it assumes that the rate of energy redistribution within a molecule A* is much greater than the rate with which A* is converted into products.

Two lines of evidence indicate that this assumption is usually valid. One is the success of conventional RRKM theory, which makes the assumption of rapid intramolecular energy transfer. As already discussed, the Slater theory, which assumed no

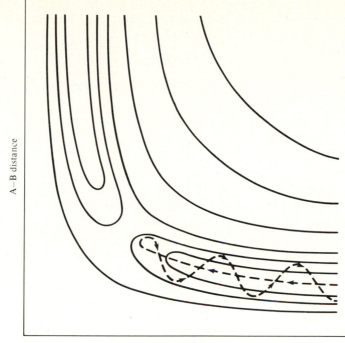

A–B distance

B–C distance

Figure 5.12 Conversion of translational energy into vibrational energy when a molecule A—B collides with C. Because there is the possibility of chemical reaction, there is a distortion of the energy valleys, and upon collision the A—B distance is altered, so that the molecule gains vibrational energy.

intramolecular energy distribution between collisions, was not satisfactory for a number of reactions.

The second line of evidence involves producing energized molecules in different ways and comparing their behavior. For example, Butler and Kistiakowsky[68] produced methylcyclopropane in two different ways and with different amounts of excess energy. They formed it by the addition of methylene to cyclopropane

$$CH_2 + H_2C\overset{\displaystyle}{\underset{\diagdown\diagup}{-}}CH_2 \rightarrow CH_3-CH\overset{\displaystyle}{\underset{\diagdown\diagup}{-}}CH_2^* \\ \qquad\quad CH_2 \qquad\qquad\qquad CH_2$$

and also by the process

$$CH_2 + CH_3-CH{=}CH_2 \rightarrow CH_3-CH\overset{\displaystyle}{\underset{\diagdown\diagup}{-}}CH_2^* \\ \qquad\qquad\qquad\qquad\qquad\qquad CH_2$$

The resulting methylcyclopropanes, although produced in different ways and having a range of energies, underwent decomposition in the same way, indicating that in spite of the different initial condition the energy is redistributed rapidly among the normal modes.

Similar investigations have been made on the unimolecular reactions of hot radicals produced by the addition of atoms to olefins. Thus, Rabinovitch and Diesen[69] caused hydrogen atoms to add on to *cis*-2-butene,

$$H + CH_3CH{=}CHCH_3 \rightarrow CH_3CH_2CHCH_3^{\ddagger}$$

and they studied the subsequent decomposition and de-energization of the excited *sec*-butyl radicals: for example,

$$CH_3CH_2CHCH_3^{\ddagger} \rightarrow CH_3 + CH_3CH{=}CH_2$$

$$CH_3CH_2CHCH_3^{\ddagger} + M \rightarrow CH_3CH_2CHCH_3 + M$$

Similar work has been done by the addition of deuterium atoms to *cis*-2-butene.[70] This produces radicals having slightly more energy than those from hydrogen atoms, owing to differences in zero-point energies, and the heavy radicals were found to decompose somewhat more rapidly. Hydrogen and deuterium atoms also have been caused to add on to 1-butene, with the formation of light and heavy *sec*-butyl radicals.[71] The radicals formed from 1-butene and 2-butene originally must have had significantly different distributions of vibrational energy, but the decomposition could be interpreted in terms of a treatment that considers only the total excess energy and not its distribution. The conclusion is that energy flows among the various vibrational modes before reaction occurs.

A good deal of work of this kind has now been carried out, and the general conclusion is that intramolecular energy transfer is rapid compared to chemical reaction. Relaxation constants for the transfer are estimated to be about 10^{12} s^{-1}; that is, redistribution occurs within about 10^{-12} s, which is a much shorter time than is required for the energized molecule to reach the activated state. However, some cases of slow transfer have been obtained, and one then speaks of non-RRKM behavior.[72]

5.3.9 Laser-Induced Unimolecular Reactions

A number of investigations have been carried out on the unimolecular reactions of substances energized by the use of lasers. Originally, such work was done with the object of seeing if energy could be introduced into particular modes of vibration, so as to bring about specific reaction patterns. It was thought that different products might be obtained by introducing energy into different modes. Such investigations are useful in leading to information about the speed of intramolecular energy transfer.

For example, cyclopropane has been irradiated[73] using lasers of widely different frequencies. Two reaction paths are possible, one involving a breakdown into ethylene and methylene,

$$\begin{array}{c} CH_2{-}CH_2^{\ddagger} \\ \diagdown \; \diagup \\ CH_2 \end{array} \rightarrow H_2C{=}CH_2 + CH_2$$

and the other, requiring less energy, an isomerization into propylene,

$$\begin{array}{c} CH_2{-}CH_2^{\ddagger} \\ \diagdown \; \diagup \\ CH_2 \end{array} \rightarrow CH_3CHCH_2$$

Irradiation at 3.22 μm, which caused excitation of the C—H asymmetric stretching vibrations in cyclopropane, gave rise to the second of these reactions. Irradiation at 9.50 μm caused excitation of the CH_2 wagging vibrations and led to the first reaction in addition to the second. Unfortunately, the interpretation of these results is not straightforward[74] and it is impossible to draw firm conclusions as to the extent of reaction selectivity.

During recent years there have been a number of investigations of reactions brought about by absorption of several photons of infrared radiation; this is referred to as *infrared multiphoton absorption* (IRMPA), or *multiphoton excitation* (see also Section 9.2.2). Infrared radiation of frequency 2×10^{13} s^{-1} has an energy of 12 kJ mol^{-1}, much less than usually required to induce reaction: an activation energy of 240 kJ mol^{-1} requires the absorption of 20 photons. Under conventional photo-chemical conditions (see Chapter 9), this is highly improbable, but it is much more probable with high-intensity lasers. The mechanisms of unimolecular reactions induced in this way have been treated by Ashfold and Hancock[75] and by Danen and Jang.[76]

5.3.10 Decomposition of Ions

With the development of the various techniques of mass spectrometry during recent years, much work has been done on the kinetics of the decompositions of ions, particularly of singly charged positive ions. The interpretation of the results is a matter of some complexity, and only a very brief account can be given here. For further details the reader is referred to the reviews listed in the bibliography at the end of this chapter.

Various experimental techniques, such as field ionization, are employed to produce positive ions in the mass spectrometer, and these may react with molecules to form other ions. The ions so formed have a range of energies, and in some experiments methods are used to produce ions having energies within a narrow range. In other investigations, ions have been characterized in terms of their lifetimes.

An early theoretical treatment[77] of the unimolecular decompositions of ions was based on the RRK theory and made use of Eq. (5.46) for the variation of the rate constant k_2 with the energy ε^* of the decomposing ion. More recent investigations have used the RRKM treatment.

5.4 COMBINATION AND DISPROPORTIONATION REACTIONS

Reactions in which atoms and free radicals combine together to give an addition product play a very important role in chemical kinetics. Such reactions, for example, have a significant bearing on the course of many composite reactions. Examples of combination reactions are

$$H + H \rightarrow H_2$$

$$CH_3 + CH_3 \rightarrow C_2H_6$$

$$H + C_2H_5 \rightarrow C_2H_6$$

Such reactions have zero activation energy and occur on virtually every collision. A related type of reaction comprises those in which an atom or radical adds on to a molecule containing a double bond, with the formation of a molecule or radical. Examples are

$$H + C_2H_4 \rightarrow C_2H_5$$

$$O + C_2H_4 \rightarrow CH_2—CH_2$$
$$\underset{O}{\diagdown\,\diagup}$$

$$CH_3 + C_2H_4 \rightarrow CH_3CH_2CH_2$$

Another type of process is disproportionation. Two ethyl radicals, for example, may produce a molecule of ethylene and one of ethane:

$$C_2H_5 + C_2H_5 \rightarrow C_2H_4 + C_2H_6$$

At first sight, such processes might seem to be straightforward examples of bimolecular reactions, showing second-order kinetics. The combination reactions, however, are the reverse of unimolecular dissociations and therefore must show special features. That this is so may be seen from the fact that the equilibrium expression must be the same under all conditions of pressure. Since at equilibrium the equilibrium constant is the ratio of the rate constants in the forward and reverse directions, it follows that if one of these "constants" shows some variation with total pressure, the other must show a parallel variation. Consider, for example, the combination of methyl radicals

$$CH_3 + CH_3 \rightarrow C_2H_6$$

and the reverse dissociation of ethane

$$C_2H_6 \rightarrow CH_3 + CH_3$$

Under all conditions of pressure, the equilibrium constant for this reaction is given by

$$K_c = \frac{[C_2H_6]}{[CH_3]^2} \tag{5.60}$$

At sufficiently high pressures, the dissociation is first-order, and the rate of dissociation is

$$v_{-1} = k_{-1}[C_2H_6] \tag{5.61}$$

This expression can be reconciled with Eq. (5.60) only if the rate of the combination reaction is given by

$$v_1 = k_1[CH_3]^2 \tag{5.62}$$

with k_1 constant. At equilibrium, v_1 is equal to v_{-1}, so that

$$k_1[CH_3]^2 = k_{-1}[C_2H_6] \tag{5.63}$$

and therefore

$$\frac{k_1}{k_{-1}} = \frac{[C_2H_6]}{[CH_3]^2} \tag{5.64}$$

For agreement with Eq. (5.60) it follows that if k_{-1} is constant, k_1 is also constant. Whenever the dissociation reaction is first order, the combination reaction is second order.

In a similar way it can be shown that if the dissociation reaction is in the second-order (low-pressure) region, the combination reaction must be third order. Thus, if the rate of the dissociation reaction is given by

$$v_1 = k_1'[C_2H_6]^2 \tag{5.65}$$

that of the combination reaction, for agreement with the equilibrium expression, must be given by

$$v_{-1} = k_{-1}'[CH_3]^2[C_2H_6] \tag{5.66}$$

Thus, the orders of the forward and reverse reactions are related to each other, that for the combination reaction always being one greater than that for the dissociation.

It follows that if there is information as to the pressure range over which there is a change in order for one of these reactions, there must be a corresponding change in order for the reverse reaction over the same pressure range.

5.4.1 Mechanisms of Atom and Radical Combinations

In Section 5.3 unimolecular reactions have been discussed on the basis of theories that are extensions of the original Lindemann-Christiansen mechanism. Thus, for the dissociation of a molecule R_2 into two radicals R, the initial step is taken to be the energization process,

$$R_2 + M \rightarrow R_2^* + M$$

and the second step is taken to be the dissociation of the energized molecule,

$$R_2^* \rightarrow 2R$$

In this scheme the energization has been brought about by a molecule M, which may be R_2 itself or may be a molecule of some added substance (a *third body*). If this is the mechanism of the dissociation, that of the reverse combination under the same conditions is

$$(1) \qquad 2R \rightleftharpoons R_2^*$$

$$(2) \quad R_2^* + M \rightarrow R_2 + M$$

For combination reactions the third body M is commonly known as a *chaperon*.§ The essential feature of this mechanism is that in the second step the energy of R_2^* is transferred to the chaperon, and the mechanism is referred to as an *energy-transfer mechanism*.

---

§ Some doubt has been cast on the propriety of using the word "chaperon" for a substance that facilitates union. N. K. Adam's definition (*Physical Chemistry*, p. 460, Clarendon Press, Oxford, 1956) provides the answer: "A chaperon is a third person who accompanies young girls on social occasions, to prevent too ardent affinity for a person of opposite sex resulting in a hasty and unstable association: in places where both sexes have acquired sufficient degrees of freedom chaperons have become obsolete."

Application of the steady-state treatment to this mechanism leads to

$$\frac{d[R^*]}{dt} = k_1[R]^2 - k_{-1}[R_2^*] - k_2[R_2^*][M] = 0 \tag{5.67}$$

Thus,

$$[R_2^*] = \frac{k_1[R]^2}{k_{-1} + k_2[M]} \tag{5.68}$$

and the rate of removal of R is

$$-\frac{d[R]}{dt} = 2k_2[R_2^*][M] = \frac{2k_1k_2[R]^2[M]}{k_{-1} + k_2[M]} \tag{5.69}$$

When [M] is sufficiently large ($k_2[M] \gg k_{-1}$) the rate is controlled by the rate of the combination of the radicals to form R_2^*, which is very rapidly de-energized:

$$v = 2k_1[R]^2 \tag{5.70}$$

At sufficiently low pressures, on the other hand,

$$v = \frac{2k_1k_2}{k_{-1}}[R]^2[M] = 2K_1k_2[R]^2[M] \tag{5.71}$$

where K_1 is the equilibrium constant for the formation of R_2^* from 2R.

It seems likely that the combinations of free radicals, as opposed to atoms, always occur largely by this type of mechanism. The energy of the species R_2^* is distributed among a number of normal modes of vibration, and there is a reasonable possibility that R_2^* will be struck by a molecule M before it has time to dissociate. The more complicated the radicals R, the longer is the lifetime of the complex R_2^*, since more vibrations will occur before dissociation takes place. The combinations of complex radicals R therefore remain second order down to lower pressures than do the combinations of simpler radicals or atoms.

The combinations of atoms represent an extreme situation. If the energy-transfer mechanism applies, two atoms will come together and unless a chaperon molecule arrives they will separate within the period of the first vibration (see Fig. 5.13). Since the period of a vibration is of the order of 10^{-13} s, to be effective the chaperon molecule must collide within that period of time. It is only at gas pressures of 10^4 or 10^5 atm that collision frequencies are sufficiently large for such stabilization to be likely. At ordinary pressures the collisions are so infrequent that most of the diatomic complexes formed will redissociate. The combination rates therefore are extremely low, and the reactions show third-order kinetics—second-order in the atoms and first-order in the chaperon.

Experimental measurements of atom combination rates have indicated that the behavior is sometimes consistent with these predictions based on the energy-transfer mechanism, but that sometimes the observed rates are much too high. For example, for the combination of iodine atoms, Porter[78] has found that at 27°C with argon as chaperon the third-order rate constant is 3.0×10^9 dm^6 mol^{-2} s^{-1}. Calculations[79] show that this is of the right order of magnitude to be explained by the energy-transfer mechanism. With benzene as chaperon, however, the rate constant is about 30 times higher, and with molecular iodine it is about 530 times higher. It seems impossible to explain these much higher rates in terms of an increased efficiency of energy transfer: such large differences have not been observed for unimolecular reactions, for example.

It therefore appears that certain more efficient chaperons can bring about atom combination by an alternative mechanism. In this mechanism, known as the *atom–molecule complex mechanism*, a complex is first formed between an atom and a chaperon molecule:

$$(1) \qquad R + M \rightleftharpoons RM*$$

The resulting complex RM* may lose its energy by collision with another molecule M:

$$(2) \quad RM* + M \rightleftharpoons RM + M$$

Finally, RM reacts with R to give the diatomic molecule R_2:

$$(3) \qquad RM + R \rightarrow R_2 + M$$

Application of the steady-state treatment to the mechanism leads to the conclusion that when, as is usual, the concentration of R is small, the rate is

$$v = 2k_3 K_1 K_2 [R]^2 [M] \qquad (5.72)$$

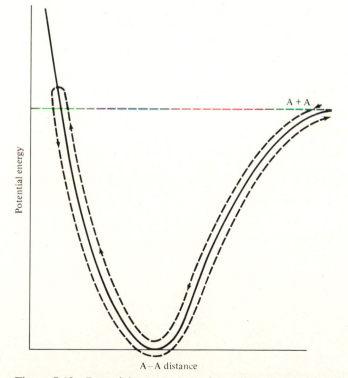

Figure 5.13 Potential-energy curve for a diatomic molecule A_2, showing that when the two A atoms come together, they separate in the first vibration.

where K_1 and K_2 are the equilibrium constants. The rate now does not become independent of [M] at high pressures, as it does in the energy-transfer mechanism; the kinetics remain third order at all pressures.

This mechanism is favored if K_1 is large, that is, if M forms a strong complex with the atom R. In an analysis of the data on the iodine atom recombinations, Eusuf and Laidler[80] concluded that with argon and hydrogen as chaperons, the reaction occurs mainly by the energy-transfer mechanism; these chaperons do not easily form complexes with iodine atoms. Benzene and molecular iodine, on the other hand, can become bound more readily to iodine atoms, by dispersion and charge-transfer forces. The theoretical calculations indicated that the binding is sufficiently strong to allow the recombination of iodine atoms to occur by the atom–molecule complex mechanism, and led to satisfactory agreement with the experimental rate constants. Support for this interpretation is provided by the fact that the rates of these reactions have negative temperature coefficients; these can be explained in terms of the increased dissociation of the RM* complexes at higher temperatures.

PROBLEMS

5.1. Ozone, whose molecule is nonlinear, undergoes unimolecular decomposition to give $O_2 + O$. Suppose that 40 quanta of energy are present in the molecule; calculate the probability that (a) 10, (b) 20, and (c) 30 quanta are present in a particular degree of freedom.

5.2. Carry out the same calculation as in Problem 5.1 for 40 quanta present in an ethane molecule.

5.3. Suppose that the Lindemann equation [Eq. (5.25)] applies, and that k^1 has reached 80% of the limiting value (k^1_∞) at a concentration of 10^{-4} mol dm^{-3}. Calculate k_2/k_{-1}.

5.4. The following values of the first-order rate coefficient k^1 were obtained for the isomerization of cyclopropane by H. O. Pritchard, R. G. Sowden, and A. F. Trotman-Dickenson [*Proc. R. Soc. London,* **A217**, 563 (1953)]:

Pressure/mmHg	84.1	11.0	2.89	0.569	0.120	0.067
$k^1/10^{-4}$ s^{-1}	2.98	2.23	1.54	0.857	0.392	0.303

Use these data to make a test of the simple Lindemann equation.

5.5. On the basis of CTST, what temperature dependence of the pre-exponential factor would be expected for the following reactions?

(a) $2 ClO \rightarrow Cl_2 + O_2$
(b) $NO + O_3 \rightarrow NO_2 + O_2$
(c) $NO_2 + F_2 \rightarrow NO_2F + F$
(d) $2 NOCl \rightarrow 2 NO + Cl_2$
(e) $2 NO + Br_2 \rightarrow 2 NOBr$

5.6. The following results were obtained by E. C. Freiling, H. C. Johnston, and R. A. Ogg [*J. Chem. Phys.,* **20**, 327 (1952)] for the reaction $NO + ClNO_2 \rightarrow NO_2 + ClNO$:

T/K	300.0	311.0	323.0	334.0	344.0
$k/10^3$ dm^3 mol^{-1} s^{-1}	7.9	12.5	16.4	25.6	34.0

The results can be interpreted in terms of the following alternative equations:

(a) $k = Ae^{-E/RT}$

(b) $k = A'T^{1/2}e^{-E'/RT}$

(c) $k = A''Te^{-E''/RT}$

(d) $k = A'''T^{-1.5}e^{-E'''/RT}$

Make appropriate plots to obtain the parameters E, E', E'', and E'''. Confirm that the values are consistent with Eq. (2.85).

REFERENCES

1. D. R. Herschbach, H. S. Johnston, K. S. Pitzer, and R. E. Powell, *J. Chem. Phys.*, **25**, 736 (1956).

2. K. H. Geib and P. Harteck, *Z. Phys. Chem. Bodenstein-Festband*, 849 (1931).

3. R. W. Wood, *Proc. R. Soc. London A*, **97**, 455 (1920); **102**, 1 (1922).

4. K. F. Bonhoeffer, *Z. Phys. Chem.*, **113**, 199, (1924); **116**, 391 (1926).

5. P. Harteck, *Z. Phys. Chem. A*, **139**, 98 (1928).

6. E. Wrede, *Z. Phys.*, **54**, 53 (1929); H. M. Chadwell and T. Titani, *J. Am. Chem. Soc.*, **55**, 1363 (1933).

7. W. R. Schulz and D. J. Le Roy, *J. Chem. Phys.*, **44**, 3344 (1966); D. J. Le Roy, B. A. Ridley, and K. A. Quickert, *Discuss. Faraday Soc.*, **44**, 92 (1967); K. A. Quickert and D. J. Le Roy, *J. Chem. Phys.*, **53**, 1325 (1970); D. N. Mitchell and D. J. Le Roy, *J. Chem. Phys.*, **58**, 3449 (1973).

8. A. A. Westenberg and N. de Haas, *J. Chem. Phys.*, **47**, 1393 (1967).

9. B. C. Garrett and D. G. Truhlar, *J. Chem. Phys.*, **72**, 3460 (1980). For interesting earlier calculations based on CTST see G. W. Koeppl, *J. Chem. Phys.*, **59**, 2425 (1973); R. E. Weston, *Science*, **158**, 332 (1967); I. Shavitt, *J. Chem. Phys.*, **49**, 4084 (1968).

10. P. Siegbahn and B. Liu, *J. Chem. Phys.*, **68**, 2457 (1978).

11. For references, see the bibliography at the end of this chapter.

12. J. A. Kerr and M. J. Parsonage, *Evaluated Kinetic Data on Gas-Phase Hydrogen-Transfer Reactions of Methyl Radicals*, Butterworth, London, 1976.

13. S. W. Benson, *Thermochemical Kinetics*, Wiley, New York, 1968 (1st ed.), 1976 (2nd ed.).

14. P. J. Boddy and E. W. R. Steacie, *Can. J. Chem.*, **39**, 13 (1961).

15. J. R. Cao and M. H. Back, *Can. J. Chem.*, **60**, 3039 (1982).

16. See, for example, W. C. Gardiner, *Acc. Chem. Res.*, **10**, 326 (1977).

17. P. D. Pacey, *J. Chem. Educ.*, **58**, 612 (1981).

18. J. J. Solomon, M. Meot-Ner, and F. H. Field, *J. Am. Chem. Soc.*, **95**, 3727 (1974).

19. M. McFarland, D. T. Albritton, F. C. Fehsenfeld, E. E. Ferguson, and A. C. Schmeltekopf, *J. Chem. Phys.*, **59**, 6620 (1973).

20. M. Meot-Ner, J. J. Solomon, F. H. Field, and H. Gershinowitz, *J. Phys. Chem.*, **78**, 1773 (1974).

21. E. E. Ferguson, in J. L. Franklin (Ed.), *Ion–Molecule Reactions*, chap. 8, Plenum, New York, 1972. For an alternative treatment of ion-molecule reactions see A. H. Viggiano, *J. Chem. Phys.*, **84**, 244 (1986).

22. M. Trautz, *Z. Anorg. Chem.*, **88**, 285 (1914).

23. W. Krauss and M. Saracini, *Z. Phys. Chem. A* **178**, 245 (1937).

24. M. Bodenstein. *Z. Elektrochem.*, **24**, 183 (1918); *Z. Phys. Chem.*, **100**, 68 (1922). See also E. Briner, W. Pfeiffer, and G. Malet, *J. Chim. Phys.*, **21**, 25 (1924); G. Kornfeld and E. Klinger, *Z. Phys. Chem.*, **4**, 37 (1929); M. Trautz and V. P. Dalal, *Z. Anorg. Chem.*, **102**, 149 (1918).

25. M. Bodenstein, *Z. Phys. Chem.,* **100,** 118 (1922). See also K. F. Herzfeld, *Z. Phys.,* **8,** 132 (1921).

26. H. Gershinowitz and H. Eyring, *J. Am. Chem. Soc.,* **57,** 985 (1935).

27. M. Meot-Ner, J. J. Solomon, F. H. Field, and G. Gershinowitz, *J. Phys. Chem.,* **78,** 1773 (1974).

28. J. H. Van't Hoff, *Etudes de dynamique chimique,* Muller, Amsterdam, 1884.

29. F. Daniels and E. H. Johnston, *J. Am. Chem. Soc.,* **43,** 72 (1921).

30. M. Trautz and K. Winkler, *J. Prakt. Chem.,* **104,** 53 (1922).

31. M. Bodenstein and H. Lutkemeyer, *Z. Phys. Chem.,* **114,** 208 (1924).

32. D. F. Smith, *J. Am. Chem. Soc.,* **47,** 1862 (1925).

33. D. F. Smith, *J. Am. Chem. Soc.,* **47,** 43 (1927).

34. J. Perrin, *Ann. Phys.* **11,** 5 (1919).

35. For a detailed account of the radiation hypothesis, with special reference to unimolecular reactions, see M. C. King and K. J. Laidler, *Arch. Hist. Exact Sci.,* **30,** 45 (1984).

36. For example, R. C. Tolman, W. C. McC. Lewis, F. Daniels (until 1926) and G. N. Lewis (until 1925).

37. For example, A. Einstein, J. J. Thomson, F. A. Lindemann, I. Langmuir, C. N. Hinshelwood, H. C. Urey, and H. S. Taylor.

38. F. A. Lindemann, *Trans. Faraday Soc.,* **17,** 598 (1922). See M. H. Back and K. J. Laidler (Eds.), *Selected Readings in Chemical Kinetics,* p. 930, Pergamon, New York, 1967. Lindemann's oral presentation was on Sept. 28, 1921.

39. J. A. Christiansen, Ph.D. thesis, University of Copenhagen, Oct. 1921.

40. C. N. Hinshelwood, *The Kinetics of Chemical Change in Gaseous Systems,* Clarendon, Oxford, 1926 (1st ed.), 1929 (2nd ed.), 1933 (3rd ed.).

41. C. N. Hinshelwood, *Proc. R. Soc. London A,* **113,** 230 (1927).

42. O. K. Rice and H. C. Ramsperger, *J. Am. Chem. Soc.,* **49,** 1616 (1927); **50,** 617 (1928); most of this paper is reproduced in M. H. Back and K. J. Laidler (Eds.), *Selected Readings in Chemical Kinetics,* Pergamon, Oxford, 1967.

43. L. S. Kassel, *J. Phys. Chem.,* **32,** 225 (1928); *Kinetics of Homogeneous Gas Reactions,* chap. 5, Reinhold, New York, 1932.

44. Tabulated values have been given by E. M. Willbanks, Los Alamos Lab. Rept., LA-2178 (1958), available from the Office of Technical Services, U.S. Department of Commerce, Washington, DC 20234.

45. C. Steel and K. J. Laidler, *J. Chem. Phys.,* **34,** 1827 (1961).

46. N. B. Slater, *Proc. Camb. Philos. Soc.,* **35,** 56 (1939).

47. N. B. Slater, *Theory of Unimolecular Reactions,* Cornell University Press., Ithaca, NY, 1959.

48. For example, N. B. Slater, in *The Transition State,* Chemical Society Special Publication No. 16, p. 29 (1962).

49. E. K. Gill and K. J. Laidler, *Proc. R. Soc. London A* **250,** 121 (1959); **251,** 66 (1959); *Can. J. Chem.,* **36,** 1570 (1958); *Trans. Faraday Soc.,* **55,** 753 (1959). A number of other reactions are treated in these papers.

50. D. Retzloff and J. Coull, *J. Chem. Phys.,* **47,** 3827 (1967).

51. E. K. Gill and K. J. Laidler, *Proc. R. Soc. London A* **250,** 121 (1959); M. Solc, *Mol. Phys.,* **11,** 519 (1966); **12,** 101 (1967); *Z. Phys. Chem. (Leipzig),* **224,** 185 (1967); *Chem. Phys. Lett.,* **1,** 160 (1967); N. B. Slater, *Mol. Phys.,* **12,** 107 (1967).

52. N. B. Slater, *Proc. R. Soc. Edinb.,* **64,** 161 (1955).

53. R. A. Marcus, *J. Chem. Phys.,* **20,** 359 (1952); G. M. Wieder and R. A. Marcus, *J. Chem. Phys.,* **37,** 1835 (1962); R. A. Marcus, *J. Chem. Phys.,* **43,** 2658 (1965). See also R. A. Marcus and O. K. Rice, *J. Phys. Colloid Chem.,* **55,** 894 (1951).

54. See the bibliography at the end of this chapter.

55. G. M. Wieder and R. A. Marcus, *J. Chem. Phys.,* **37,** 1835 (1962); M. C. Lin and K. J. Laidler, *Trans. Faraday Soc.,* **64,** 927 (1968).

56. M. C. Lin and K. J. Laidler, *Trans. Faraday Soc.,* **64,** 94 (1968).

57. C. S. Elliot and H. M. Frey, *Trans. Faraday Soc.,* **62,** 895 (1966); M. C. Lin and K. J. Laidler, *Trans. Faraday Soc.,* **64,** 94 (1968).

58. M. C. Lin and K. J. Laidler, *Trans. Faraday Soc.,* **64,** 79 (1968).

59. For example, F. Daniels and E. H. Johnston, *J. Am. Chem. Soc.,* **43,** 72 (1921); F. Daniels, *J. Am. Chem. Soc.,* **48,** 607 (1926); H. A. Taylor, *J. Am. Chem. Soc.,* **48,** 577 (1926); F. O. Rice, H. C. Urey, and R. N. Washburne, *J. Am. Chem. Soc.,* **50,** 2402 (1928).

60. H. O. Pritchard, R. G. Sowden, and A. F. Trotman-Dickenson, *Proc. R. Soc. London A* **217,** 563 (1953).

61. N. B. Slater, *Proc. R. Soc. London A* **218,** 224 (1953).

62. G. M. Wieder and R. A. Marcus, *J. Chem. Phys.,* **37,** 1835 (1962); M. C. Lin and K. J. Laidler, *Trans. Faraday Soc.,* **64,** 927 (1968).

63. K. J. Laidler, in J. E. Dubois (Ed.), *Reaction Transition States,* p. 23, Gordon & Breach, New York, 1972.

64. F. W. Schneider and B. S. Rabinovitch, *J. Am. Chem. Soc.,* **84,** 4215 (1962).

65. F. W. Schneider and B. S. Rabinovitch, *J. Am. Chem. Soc.,* **85,** 2365 (1963); B. S. Rabinovitch, P. W. Gilderson, and F. W. Schneider, *J. Am. Chem. Soc.,* **87,** 158 (1965); K. M. Maloney and B. S. Rabinovitch, *J. Phys. Chem.,* **73,** 1652 (1969); K. M. Maloney, S. P. Pavlov, and B. S. Rabinovitch, *J. Phys. Chem.,* **73,** 2756 (1969).

66. D. L. Bunker and W. L. Hase, *J. Chem. Phys.,* **59,** 4621 (1973); H. L. Harris and D. L. Bunker, *Chem. Phys. Lett.,* **11,** 433 (1971).

67. See, for example, P. J. Robinson and K. A. Holbrook, *Unimolecular Reactions,* chap. 10, Wiley-Interscience, London, 1972.

68. J. N. Butler and G. B. Kistiakowsky, *J. Am. Chem. Soc.,* **82,** 759 (1960).

69. B. S. Rabinovitch and R. W. Diesen, *J. Chem. Phys.,* **30,** 735 (1959).

70. R. E. Harrington, B. S. Rabinovitch, and R. W. Diesen, *J. Chem. Phys.,* **32,** 1245 (1960).

71. R. E. Harrington, B. S. Rabinovitch, and H. M. Frey, *J. Chem. Phys.,* **33,** 1271 (1960). Compare H. M. Frey, *Trans. Faraday Soc.,* **56,** 51 (1960); B. S. Rabinovitch, D. H. Dills, B. H. Mahan, and J. H. Current, *J. Chem. Phys.,* **32,** 493 (1960).

72. For further details see the reviews by Holbrook and by Oref and Rabinovitch, listed in the bibliography at the end of this chapter.

73. R. B. Hall and A. Kaldor, *J. Chem. Phys.,* **70,** 4072 (1979).

74. E. Thiele, M. F. Goodman, and J. Stone, *Chem. Phys. Lett.,* **69,** 18 (1980); *Opt. Eng.,* **19,** 10 (1980); M. N. R. Ashfold and G. Hancock, in P. G. Ashmore and R. J. Donovan (Eds.), *Gas Kinetics and Energy Transfer,* vol. 4, p. 73, Royal Society of Chemistry, London, 1980.

75. M. N. R. Ashfold and G. Hancock, in P. G. Ashmore and R. J. Donovan (Eds.), *Gas Kinetics and Energy Transfer* vol. 4, p. 73, Royal Society of Chemistry, London, 1980.

76. W. C. Danen and J. Jang, in J. I. Steinfeld (Ed.), *Laser-Induced Chemical Processes,* p. 45, Plenum, New York, 1981.

77. H. M. Rosenstock, M. B. Wallenstein, A. L. Wahthaftig, and H. Eyring, *Proc. Nat. Acad. Sci. USA,* **38,** 667 (1952).

78. G. Porter, *Discuss. Faraday Soc.,* **33,** 198 (1962).

79. M. Eusuf and K. J. Laidler, *Trans. Faraday Soc.,* **59,** 2750 (1963).

80. M. Eusuf and K. J. Laidler, *Trans. Faraday Soc.,* **59,** 2750 (1963). A similar theoretical treatment of this problem, with essentially the same conclusions, was given by H. S. Johnston, *Gas Phase Reaction Rate Theory,* chap. 14, pp. 253–262, Ronald, New York, 1966.

BIBLIOGRAPHY

The following older books on gas kinetics contain information that is still useful:

C. N. Hinshelwood, *Kinetics of Chemical Change,* Clarendon, Oxford, 1940.

E. W. R. Steacie, *Atomic and Free Radical Reactions,* Reinhold, New York, 1954.

A. F. Trotman-Dickenson, *Gas Kinetics,* Butterworths, London, 1955.

For more recent reviews of gas kinetics see the following publications:

C. H. Bamford and C. F. H. Tipper (Eds.), *Comprehensive Chemical Kinetics,* Various volumes, especially vol. 18 (1974), Elsevier, New York.

H. S. Johnston, *Gas Phase Reaction Rate Theory,* Ronald, New York, 1966.

V. N. Kondratiev and E. E. Nikitín, *Gas-Phase Reactions,* Springer-Valley, Berlin, 1981.

I. W. M. Smith, *Kinetics and Dynamics of Elementary Gas Reactions,* Butterworths, London, 1980.

For accounts of experimental methods in gas kinetics see:

L. Batt, Experimental Methods for the Study of Slow Reactions, in C. H. Bamford and C. F. H. Tipper (Eds.), *Comprehensive Chemical Kinetics,* vol. 1, pp. 1–111, Elsevier, New York, 1969.

D. N. Hague, Experimental Methods for the Study of Fast Reactions, in C. H. Bamford and C. F. H. Tipper (Eds.) *Comprehensive Chemical Kinetics,* vol. 1, pp. 112–179, Elsevier, New York, 1969.

H. W. Melville and B. G. Gowenlock, *Experimental Methods in Gas Reactions,* Macmillan, London, 1963.

Kinetic data are compiled, and in some cases critically evaluated, in the following publications:

D. L. Alldra and R. Shaw, A Compilation of Kinetic Parameters for the Thermal Degradation of *n*-Alkane Molecules, *J. Phys. Chem. Ref. Data,* **9,** 523 (1980).

D. L. Baulch, R. A. Cox, R. F. Hampson, J. A. Kerr, J. Troe, and R. T. Watson, Evaluated Kinetic and Photochemical Data for Atmospheric Chemistry, *J. Phys. Chem. Ref. Data,* **9,** 295 (1980); **11,** 327 (1982); **13,** 1259 (1984).

D. L. Baulch and J. Duxbury, Ethane Decomposition and the Reference Rate Constant for Methyl Radical Recombination, *Combust. Flame,* **37,** 313 (1980).

D. L. Baulch et al., *Evaluated Kinetic Data for High Temperature Reactions,* vols. 1–3, Butterworths, London, 1972–1976.

R. F. Hampson and D. Gavin, *Reaction Rate and Photochemical Data for Atmospheric Chemistry,* NBS Technical Note 513, National Bureau of Standards, 1978.

J. T. Herron and R. E. Huie, Rate Constants for the Reactions of Atomic Oxygen (O^3P) with Organic Compounds in the Gas Phase, *J. Phys. Chem. Ref. Data,* **2,** 467 (1974).

J. A. Kerr, Metathetical Reactions of Atoms and Radicals, in C. H. Bamford and C. F. H. Tipper (Eds.), *Comprehensive Chemical Kinetics,* vol. 18, pp. 39–109, Elsevier, New York, 1974.

J. A. Kerr and M. J. Parsonage, *Evaluated Kinetic Data on Gas-Phase Addition Reactions,* Butterworth, London, 1972.

J. A. Kerr and E. Ratajczak, *Second Supplementary Tables of Bimolecular Gas Reactions,* University of Birmingham, Birmingham, U.K., 1972.

V. N. Kondratiev, in R. M. Tristan (Ed.), *Rate Constants of Gas-Phase Reactions,* Report No. COM-72-10014, Office of Standard Reference Data, U.S. National Bureau of Standards, 1972.

E. Ratajczak and A. F. Trotman-Dickenson, *Supplementary Tables of Bimolecular Gas Reactions,* University of Wales Institute of Science and Technology, Cardiff, 1970.

A. F. Trotman-Dickenson and G. S. Milne, *Tables of Bimolecular Gas Reactions,* NSRDS, vol. 9, National Bureau of Standards, 1967.

W. E. Wilson, A Critical Review of the Gas-Phase Reaction Kinetics of the Hydroxyl Radical, *J. Phys. Chem. Ref. Data,* **1,** 535 (1972).

Ion–molecule reactions in the gas phase are treated in the following publications:

J. H. Beynon and M. L. McGlashan (Eds.), *Current Topics in Mass Spectrometry and Chemical Kinetics,* Heyden, London, 1982.

M. T. Bowers (Ed.), *Gas-Phase Ion Chemistry,* Academic, New York, 1979.

J. L. Franklin (Ed.), *Ion–Molecule Reactions,* Plenum, New York, 1972.

For reviews of unimolecular reactions see:

F. F. Crim, Selective Excitation Studies of Unimolecular Reaction Dynamics, *Ann. Rev. Phys. Chem.,* **35,** 657 (1984).

W. Forst, *Theory of Unimolecular Reactions,* Academic, New York, 1973.

K. A. Holbrook, Current Aspects of Unimolecular Reactions, *Chem. Soc. Rev.,* **12,** 163 (1983).

H. O. Pritchard, *The Quantum Theory of Unimolecular Reactions,* Cambridge University Press, 1984.

P. J. Robinson and K. A. Holbrook, *Unimolecular Reactions,* Wiley-Interscience, London, 1972.

N. B. Slater, *Theory of Unimolecular Reactions,* Cornell University Press, Ithaca, NY, 1959.

Unimolecular decompositions of ions are reviewed in the following publications:

A. G. Brenton, R. P. Morgan, and J. H. Beynon, Unimolecular Ion Decomposition, *Ann. Rev. Phys. Chem.,* **30,** 51 (1974).

J. L. Franklin, in M. R. Bowers (Ed.), *Gas-Phase Ion Chemistry,* vol. 1, p. 273, Academic, New York 1979.

Problems of energy transfer in gases are reviewed in:

P. G. Ashmore and R. J. Donovan (Eds.), *Gas Kinetics and Energy Transfer,* vols. 1–4, Royal Society of Chemistry, London, 1976–1980.

J. D. Lambert, *Vibrational and Rotational Relaxation in Gases,* Clarendon, Oxford, 1977.

I. Oref and B. S. Rabinovitch, Do Highly Excited Reactive Polyatomic Molecules Behave Ergodically?, *Acc. Chem. Res.,* **12,** 166 (1979).

D. C. Tardy and B. S. Rabinovitch, Intermolecular Vibrational Energy Transfer in Thermal Unimolecular Systems, *Chem. Rev.,* **77,** 369 (1977).

Laser-induced unimolecular reactions are reviewed in the following publications:

R. G. Harrison and S. R. Butcher, Multiple Photon Infrared Processes in Polyatomic Molecules, *Contemp. Phys.,* **21,** 19 (1980).

K. L. Kompa and S. D. Smith (Eds.), *Laser-Induced Processes in Molecules,* Springer-Verlag, Berlin, 1979.

J. I. Steinfeld (Ed.), *Laser-Induced Chemical Processes,* Plenum, New York, 1981.

Chapter 6

Elementary Reactions in Solution

When a reaction takes place in solution, the solvent usually is present in much larger amounts than the reactants. The solvent concentration therefore remains essentially unchanged during the course of a reaction, and if the method of integration is used to analyze the results, the effect of the solvent will not be revealed. The same is true if the differential method is used to analyze initial rates at different reactant concentrations.

For many reactions the stoichiometric equation does not involve the solvent. For others the solvent does enter into the equation, or there may be reason to believe that the solvent takes part in the chemical change but is regenerated at the end of the process. In these cases the solvent exerts some chemical effect as well as a purely environmental effect.

Theories of liquids are intrinsically more complicated than those of gases. Therefore, the understanding of reactions in solution is less straightforward than that of gas reactions, and sometimes it is preferable to apply different theoretical procedures. So far the treatments involving partition functions have rarely been applied to solution reactions; thermodynamic approaches have proved more useful.

6.1 SOLVENT EFFECTS ON REACTION RATES

Valuable evidence concerning the influence of the solvent on reactions in solution has come from investigations of two kinds. A few measurements have been made on reactions that occur both in solution and in the gas phase. However, the number of such studies is not large, since many reactions in solution (e.g., those involving ions) will not easily occur in the gas phase. Therefore, much of the information on solvent

TABLE 6.1   COMPARISON OF KINETIC PARAMETERS FOR THE DECOMPOSITION
OF NITROGEN PENTOXIDE

Medium	$k/10^{-5}$ dm^3 mol^{-1} s^{-1} at 25°C	$\log_{10} (A/\text{dm}^3 \text{ mol}^{-1} \text{ s}^{-1})$	E_a/kJ mol^{-1}
Gas phase[a]	3.38	13.6	103.3
Carbon tetrachloride[b]	4.09	13.8	106.7
Carbon tetrachloride[c]	4.69	13.6	101.3
Chloroform[b]	3.72	13.6	102.5
Chloroform[c]	5.54	13.7	102.9
Ethylene dichloride[c]	4.79	13.6	102.1
Ethylidene dichloride[c]	6.56	14.2	104.2
Pentachloroethane[c]	4.30	14.0	104.6
Nitromethane[c]	3.13	13.5	102.5
Bromine[c]	4.27	13.3	100.4
Nitrogen tetroxide[c]	7.05	14.2	104.6
Nitric acid[c]	0.147	14.8	118.4
Propylene dichloride[c]	0.510	14.6	113.0

[a] F. Daniels and E. H. Johnston, *J. Am. Chem. Soc.,* **43,** 53 (1921).
[b] R. H. Lueck, *J. Am. Chem. Soc.,* **44,** 757 (1922).
[c] H. Eyring and F. Daniels, *J. Am. Chem. Soc.,* **52,** 1473 (1930).

influences has come from the second type of investigation, in which the kinetics of reactions are compared in different solvents.

6.1.1 Comparison between Gas-Phase and Solution Reactions

Reactions that occur in the gas phase as well as in solution often show similar behavior in solution as in the gas phase and are affected little by a change of solvent. An example of such a reaction is the thermal decomposition of nitrogen pentoxide,§ some data for which are shown in Table 6.1. The rate constants, pre-exponential factors, and activation energies are much the same in most of the solvents as they are in the gas phase. An exception is with nitric acid, which plays a more active role in the reaction than the other solvents. Similar results are found with the dimerization of cyclopentadiene and the reverse dissociation of the dimer.[1] As seen in Table 6.2, solvents have relatively little effect.

It appears that when a reaction occurs in the gas phase as well as in solution, the solvent usually plays a relatively subsidiary role; it seems to act merely as a space filler and has only a minor influence on the kinetics.

6.1.2 Comparisons between Different Solvents

It would be unwise, however, to conclude from this evidence that solvent influences are not important in general. Reactions that occur in the gas phase as well as in

§ This reaction is actually composite (Section 8.4.6). The fact that its rate is affected very little by the solvent suggests that this is also true of the elementary reactions involved in the mechanism.

TABLE 6.2 COMPARISON OF KINETIC PARAMETERS FOR THE DIMERIZATION
OF CYCLOPENTADIENE AND THE REVERSE DISSOCIATION

Medium	Dimerization		Dissociation	
	$\log_{10}(A/\text{dm}^3\text{ mol}^{-1}\text{ s}^{-1})$	$E/\text{kJ mol}^{-1}$	$\log_{10}(A/\text{s}^{-1})$	$E/\text{kJ mol}^{-1}$
Gas phase	6.1	69.9	13.1	146.0
Pure liquid	5.7	66.9	13.0	144.3
Carbon tetrachloride	5.9	67.8	—	—
Carbon disulfide	5.7	74.1	—	—
Benzene	7.1	29.7	—	—
Paraffin	8.1	33.9	13.0	143.1

solution are a special type of reaction. When reactions that do not occur in the gas phase are studied, rates usually vary much more widely from solvent to solvent.

A good example is provided by the reaction between triethylamine and ethyl iodide, studied many years ago by Menschutkin[2] in 22 different solvents. He measured no activation energies, but this was done later by Grimm, Ruf and Wolff,[3] some of whose results are shown in Table 6.3. There is a considerable variation in k, A, and E_a from solvent to solvent. A similar wide variation was found in the reaction between pyridine and methyl iodide.[4]

An interesting feature of such reactions is that the pre-exponential factors are often markedly different from those predicted by simple collision theory, namely, 10^{10}–10^{11} dm^3 mol^{-1} s^{-1}; similar values are predicted by transition-state theory for reactions between simple molecules (Section 4.6.2). The values in Table 6.3, however, are much lower. In many cases the low A factors appear to be an intrinsic property of the reaction, arising from complexity of the reactants, rather than a specific solvent effect. This is certainly true of the dimerization of cyclopentadiene (Table 6.2), where low A factors occur for the reaction in the gas phase as well as in various solvents.

6.2 FACTORS DETERMINING REACTION RATES IN SOLUTION

A reaction between two molecules in solution can be thought of as involving three steps: (1) diffusion of the reactant molecules to each other, (2) the actual chemical

TABLE 6.3 COMPARISON OF KINETIC PARAMETERS FOR THE REACTION
BETWEEN TRIETHYLAMINE AND ETHYL IODIDE
IN VARIOUS SOLVENTS

Solvent	$k/10^{-5}$ dm^3 mol^{-1} s^{-1} at 100°C	$\log_{10}(A/\text{dm}^3\text{ mol}^{-1}\text{ s}^{-1})$	$E_a/\text{kJ mol}^{-1}$
Hexane	0.5	4.0	66.9
Toluene	25.3	4.0	54.4
Benzene	39.8	3.3	47.7
Bromobenzene	166.0	4.6	52.3
Acetone	265.0	4.4	49.8
Benzonitrile	1125.0	5.0	49.8
Nitrobenzene	1383.0	4.9	48.5

transformation, and (3) the diffusion of the products away from each other. For most processes with which chemists are concerned, the chemical transformation, step 2, is much slower than the diffusion processes: the overall rate therefore is controlled by the chemical change. If this were not the case, the viscosity of the solvent would exert an important influence on the rate, since diffusion and viscosity are related closely: this, however, is found rarely. Another line of evidence relates to activation energies, which for a diffusion process are not usually greater than about 20 kJ mol^{-1}. Most chemical reactions have values much higher than this and therefore are not controlled by diffusion.

There are some rapid reactions, however, that are controlled partly or almost completely by the rate at which the reactants come together: examples are ion combinations (such as $H^+ + OH^- \rightarrow H_2O$) and certain fluorescence quenching processes in solution. The theory of these diffusion-controlled reactions is dealt with in Section 6.7. In the meantime, a discussion is given of some of the theoretical treatments of solution reactions for which the rate is controlled by the chemical interaction that takes place when the reactants come together.

6.2.1 Collisions in Solution

In 1924 Christiansen[5] suggested that collision frequencies in solution should be similar to those in the gas phase. During the next few years Hinshelwood, Moelwyn-Hughes, and others investigated a number of reactions in a variety of solvents, and whenever possible they studied the same reactions in the gas phase. The general conclusion that they reached is that certain solvents, such as carbon tetrachloride and benzene, do appear to have a relatively minor effect on rates, but that others have a much stronger effect. They realized that simple collision theory, in which the reactants are regarded as spheres, often greatly exaggerates collision frequencies: therefore, abnormally low A factors may be due to this cause rather than to specific solvent effects. For example, for certain reactions between tertiary amines and alkyl iodides, Hinshelwood and Moelwyn-Hughes concluded[6] that in solvents such as carbon tetrachloride the A factors are approximately those that would be found if the reactions occurred in the gas phase. Larger A factors found for other solvents were attributed to specific effects, such as solvation.

Another important collision-theory approach to reactions in solution was made by Rabinowitch,[7] who based a treatment on a distribution function for mercury that had been given by Debye and Menke.[8] He made a theoretical calculation of the frequency of collisions between a given pair of mercury atoms and compared this frequency with the frequency in the gas phase. He found that in the liquid the frequency is two to three times greater than that in the gas phase. It must be noted that mercury has less structure than solvents such as water and alcohols, so that this conclusion may not apply to them.

The important question of both the frequency and the *distribution* of collisions in solution was investigated by Rabinowitch and Wood[9] by means of a model experiment involving a tray on which balls were allowed to roll. Agitation of the tray caused the balls to move around, and an electrical device was employed to record collisions between a given pair of balls. When very few balls were on the tray, the collisions between a given pair occurred individually, with relatively long intervals between successive collisions (see Fig. 6.1, record I). This represents the behavior in the gas phase. The effect of solvents was simulated by adding additional balls until there was fairly close packing. Then it was found that the frequency of collisions was not changed much when the additional spheres were added, but that the distribution of collisions between a given pair of spheres was very different when many spheres were present. When the gas-phase behavior was simulated, individual collisions usually were separated by fairly long periods of time. When the additional spheres were added, collisions instead tended to occur in sets, as shown schematically in Fig. 6.1, record II. Such a set of collisions, occurring in rapid succession, is referred to as an *encounter*. For example, there might be an average of four collisions in each encounter, but the interval between successive encounters would tend to be about four times the interval between collisions in the gas phase. The overall collision frequency therefore is much the same in solution as in the gas phase. The reason for the successive collisions is that in the liquid phase the surrounding solvent molecules form a "cage," which holds the colliding spheres together and causes them to collide a number of times before

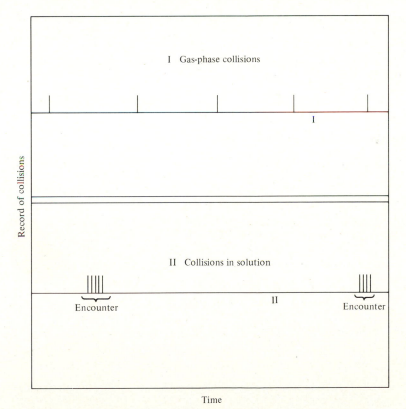

Figure 6.1 Distribution of collisions between solute molecules, as shown by the demonstration of Rabinowitch and Wood. Record I shows schematically the gas-phase collisions, while record II shows the distribution of collisions when a solvent is present.

they finally separate. This tendency for collisions to occur in sets has no effect on ordinary reactions, which involve an activation energy, since reaction may occur at any collision within the set. With reactions that do not involve an activation energy, such as free-radical combinations, this tendency of collisions to occur in sets makes a difference to the pre-exponential factors since reaction occurs at the first collision in any set, with the result that the remaining collisions do not contribute to the rate. Therefore, the pre-exponential factor is related to the reciprocal of the average time elapsing between successive sets of collisions, that is, between successive *encounters.*

This *cage effect,* also known as the *Franck–Rabinowitch effect,*[10] has other important consequences. In photochemical reactions in solution, for example, a pair of free radicals produced initially may recombine before they can separate from each other, owing to their being caged in by the surrounding solvent molecules. This phenomenon is known as *primary recombination,* as opposed to *secondary recombination* which occurs after the free radicals have separated.

6.2.2 Transition-State Theory

Transition-state theory[11] was applied to reactions in solution first by Wynne-Jones and Eyring[12] and later by Evans and Polanyi[13] and by Bell.[14] Since it is difficult to formulate partition functions for substances in the liquid state, it has proved more fruitful to use the thermodynamic formulation of CTST [Eq. (4.97)] and to make use of activity coefficients.

The basis of CTST is that rates are proportional to concentrations of activated complexes [Eq. (4.45)]:

$$v \propto [X^{\ddagger}] \tag{6.1}$$

For a reaction between two species A and B, the activated complexes are in quasi-equilibrium with A and B:

$$\frac{[X^{\ddagger}]}{[A][B]} \frac{\alpha_{\ddagger}}{\alpha_A \alpha_B} = K^{\ddagger} \tag{6.2}$$

where $K^{\ddagger}$ is the equilibrium constant and the α's are the activity coefficients. It is satisfactory to neglect these for reactions in the gas phase, but they must now be included since for a solution they may differ markedly from unity. Equations (6.1) and (6.2) give

$$v \propto [A][B]K^{\ddagger} \frac{\alpha_A \alpha_B}{\alpha_{\ddagger}} \tag{6.3}$$

and the rate constant is thus

$$k \propto K^{\ddagger} \frac{\alpha_A \alpha_B}{\alpha_{\ddagger}} \tag{6.4}$$

For an ideal gaseous system, the activity coefficients are unity, so that the rate constant reduces to

$$k_g \propto K^{\ddagger} \tag{6.5}$$

Therefore, the rate constant in solution can be related to that in the gas phase by the equation

$$k = k_g \frac{\alpha_A \alpha_B}{\alpha_\ddagger} \qquad (6.6)$$

This equation can be used for relating the rate constant in solution not only to that in the gas phase but also to that in very dilute solution, where the substances behave ideally; this will be considered later.

If the rates in solution and in the gas phase are to be equal, the activity-coefficient factor $\alpha_A \alpha_B / \alpha_\ddagger$ must be equal to unity. This may arise easily for a unimolecular reaction, for which the appropriate factor is $\alpha_A / \alpha_\ddagger$. If the reactant and the activated complex have similar structures, as is often the case, α_A and $\alpha_\ddagger$ will not differ greatly, and the rate in solution will be similar to that in the gas phase. This similarity in rates applies to the decomposition of nitrogen pentoxide in a number of solvents (Table 6.1). The abnormally slow rate of decomposition in nitric acid and propylene dichloride and the correspondingly high energies of activation may be due to the formation of a complex between the reactant and the solvent. Such compound formation decreases α_A and consequently the rate of reaction.

An alternative and very useful approach is to express the rate constant in terms of entropies and energies of activation [Eq. (4.97)]:

$$k = e \frac{kT}{h} e^{\Delta \ddagger S/R} e^{-E_a/RT} \qquad (6.7)$$

Empirical values for entropies of nonpolar molecules in solution were used by Bell[15] to make estimates of entropies of activation for reactions between such molecules. He deduced in this way that the pre-exponential factor for a reaction in solution should be approximately three times that in the gas phase. This is in agreement with the conclusion that Rabinowitch drew on the basis of collision theory.

6.2.3 Influence of Internal Pressure

Application of transition-state theory also leads to some useful conclusions about the effect on the rate of the internal pressure of the solvent. The activity coefficient α which appears in Eq. (6.6) is proportional to the activity coefficient y which relates the behavior of a solute in any solution to its behavior in an ideal solution. The rate therefore can be expressed as

$$k = k_0 \frac{y_A y_B}{y_\ddagger} \qquad (6.8)$$

where k_0 is now the rate constant in the ideal solution [rather than that in the gas phase as in Eq. (6.6)]. A *regular* solution is one in which the molecular distribution is random in spite of nonideality, and for such a solution the activity coefficient y_2 of a solute can be expressed as[16]

$$RT \ln y_2 = V_{m,2} \left(\frac{x_1 V_{m,1}}{x_1 V_{m,1} + x_2 V_{m,2}} \right)^2 \left[\left(\frac{E_{v,1}}{V_{m,1}} \right)^{1/2} - \left(\frac{E_{v,2}}{V_{m,2}} \right)^{1/2} \right]^2 \qquad (6.9)$$

where x is the mole fraction, V_m is the molar volume, and E_v is the energy of vaporization, and the subscripts 1 and 2 refer to solvent and solute, respectively. For a sufficiently dilute solution, $x_2 V_{2,m}$ can be neglected in comparison with $x_1 V_{1,m}$, and the equation reduces to

$$RT \ln y_2 = V_{m,2}\left[\left(\frac{E_{v,1}}{V_{m,1}}\right)^{1/2} - \left(\frac{E_{v,2}}{V_{m,2}}\right)^{1/2}\right]^2 \qquad (6.10)$$

To a crude approximation the quantities E_v/V_m may be set equal to the *internal pressures* P_i. Thus,

$$RT \ln y_2 \approx V_{m,2}(P_{i,1}^{1/2} - P_{i,2}^{1/2})^2 \qquad (6.11)$$

This is written conveniently as

$$RT \ln y_2 \approx V_{m,2}\Delta \qquad (6.12)$$

where Δ, always a positive quantity, is equal to $(P_{i,1}^{1/2} - P_{i,2}^{1/2})^2$. Taking logarithms of Eq. (6.8) and inserting the expressions for $\ln y$ using Eq. (6.12), we obtain

$$RT \ln k = RT \ln k_0 + V_A \Delta_A + V_B \Delta_B - V_\ddagger \Delta_\ddagger \qquad (6.13)$$

This equation gives a useful interpretation of rates in solution in terms of internal pressures.

The molar volumes do not vary greatly, so that the internal pressure factors Δ are the most important. If the internal pressures of the solvent, of A and B, and of the activated complexes are all similar, the Δ terms in Eq. (6.10) are all small, and the rate is the same in a particular solvent as it is in a solvent in which there is ideal behavior. If the internal pressure of the solvent is similar to the values for the reactants (i.e., if Δ_A and Δ_B are small) but is different from that of the activated complex ($\Delta_\ddagger$ is large), k is less in the solvent than in an ideal solvent. Conversely, if the solvent is such that its internal pressure is similar to that of the activated complex but different from those of the reactants, the rate will be high. In general, the internal pressure of the activated complexes is intermediate between that of the reactants and of the products. It therefore follows as a useful rule that if the reaction is one in which the products are of higher internal pressure than the reactants, it will be accelerated by solvents of high internal pressure: if the products are of lower internal pressure than the reactants, the reaction will be accelerated by solvents of low internal pressure. This rule was first found empirically by Richardson and Soper,[17] and it was arrived at theoretically by Glasstone,[18] whose argument was similar to that just given.

The rate constants listed in Table 6.3 provide a useful example of the application of this rule. The product, the quaternary ammonium salt, is ionized and has a much higher internal pressure than the reactants. The activated complex therefore is expected to have a higher internal pressure than the reactants. The rule predicts that the reaction is accelerated by solvents of high internal pressure, and this is generally the case.

6.2.4 Influence of Solvation

Another useful approach to solvent effects is in terms of the extent of solvation of the reactants and activated complex. The treatment of reactions from this point of view is due largely to Hughes and Ingold.[19] Consider, for example, the reaction between a tertiary amine such as pyridine and an alkyl iodide such as methyl iodide:

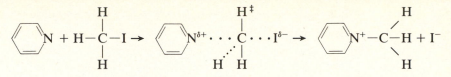

The products are two separated ions, and in the activated complex there is partial ionization, as represented above. In a polar solvent such as nitrobenzene there is more solvation of the activated complex than of the reactants. The effect of solvation is to lower the activity coefficient: thus, $\alpha_{\ddagger}$ in Eq. (6.6) is expected to be low compared to α_A and α_B, and as a result the rate is high, in agreement with experiment (compare Table 6.3 for an analogous reaction).

On the other hand, in a reaction of the type

$$Y^- + RX^+ \rightarrow (Y^{\delta-} \cdots R \cdots X^{\delta+})^{\ddagger} \rightarrow YR + X$$

there is a decrease in polarity as the activated complex is formed. A polar solvent therefore solvates the activated complex less than the reactants; now the activity coefficients are smaller for the reactants than for the solvent, and such a solvent reduces the rate.

6.3 REACTIONS BETWEEN IONS

The electrostatic forces between ions are much stronger than nonelectrostatic forces, and they are better understood. The theoretical treatment to be outlined in this section is applicable particularly to reactions between ions in which covalent chemical bonds are broken and formed. As a result, they have a significant activation energy and are not too rapid. Simple ion-combination reactions are much more rapid and are diffusion controlled; they are treated in Section 6.7.3.

The pre-exponential factors for ionic reactions depend in a rather simple way on the ionic charges. If the ions are of opposite signs, the pre-exponential factors are abnormally high, whereas if the charges are the same they are abnormally low. Obviously, the electrostatic forces are playing an important role. A simple explanation is provided by the kinetic theory of collisions; if the ions are of opposite signs, the frequency of collisions is increased by the attractive forces, while if they are of the same sign the frequency of collisions is reduced. Theories of reactions in solution based on collision theory were given by Scatchard[20] and by Moelwyn-Hughes.[21] An alternative treatment of ionic reactions is based on transition-state theory. The influence of solvent on rates and pre-exponential factors is discussed now from this point of view; later, with the aid of Debye–Hückel theory, the influence of the ionic strength is treated.

6.3.1 Influence of Solvent Dielectric Constant

A simple but useful treatment of the influence of the solvent on the rates of reactions between ions is based on electrostatic theory, the solvent being treated as a continuum

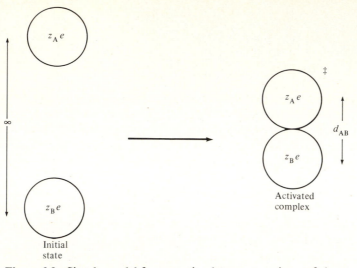

Figure 6.2 Simple model for a reaction between two ions, of charges $z_A e$ and $z_A e$, in a medium of dielectric constant ϵ. This is known as the "double-sphere" model.

having a dielectric constant ϵ. This treatment represents a gross oversimplification, but it has proved surprisingly useful because it leads to conclusions that are semi-quantitatively correct.

The model is represented in Fig. 6.2. The charges on the ions are $z_A e$ and $z_B e$, where z_A and z_B (which may be positive or negative) are the charge numbers and e is the elementary charge (1.602×10^{-19} C). Initially, the ions are an infinite distance apart. In this particular model, known as the *double-sphere* model, the ions are considered to remain intact as they approach one another and form an activated complex in which the centers of the ions are separated by a distance d_{AB}.

When the ions are separated by a distance x, the force acting between them is, according to Coulomb's law,

$$f = \frac{z_A z_B e^2}{4 \pi \epsilon_0 \epsilon x^2} \tag{6.14}$$

where ϵ is the dielectric constant and ϵ_0 is the permittivity of a vacuum (8.854×10^{-12} C^2 N^{-1} m^{-2}).§ The work done on the system in moving them together a distance dx is

$$dw = -\frac{z_A z_B e^2}{4 \pi \epsilon_0 \epsilon x^2} dx \tag{6.15}$$

(The negative sign appears because x decreases by dx.) The work done on the system in moving the ions from $x = \infty$ to $x = d_{AB}$ is therefore

§ In SI, with e having the unit of coulombs and x that of meters, the factor $4\pi\epsilon_0$ must be included. The unit of force is then the newton ($N \equiv$ kg m s^{-2}).

$$w = -\int_{\infty}^{d_{AB}} \frac{z_A z_B e^2}{4\pi\epsilon_0\epsilon x^2}\, dx \tag{6.16}$$

$$= \frac{z_A z_B e^2}{4\pi\epsilon_0\epsilon d_{AB}} \tag{6.17}$$

This work is positive if the ionic charges are of the same sign; if they are different, it is negative.

This work w is the electrostatic contribution to the Gibbs energy of activation when two ions form an activated complex. Multiplication by the Avogadro constant L gives the molar quantity,

$$\Delta^\ddagger G_{es}^\circ = \frac{L z_A z_B e^2}{4\pi\epsilon_0\epsilon d_{AB}} \tag{6.18}$$

There is also a nonelectrostatic contribution $\Delta^\ddagger G_{nes}^\circ$, and the total molar Gibbs energy of activation is thus

$$\Delta^\ddagger G^\circ = \Delta^\ddagger G_{nes}^\circ + \frac{L z_A z_B e^2}{4\pi\epsilon_0\epsilon d_{AB}} \tag{6.19}$$

Introduction of this into Eq. (4.85), which relates k to $\Delta^\ddagger G$, then gives

$$k = \frac{kT}{h} e^{-\Delta^\ddagger G^\circ/RT} = \frac{kT}{h} e^{-\Delta^\ddagger G_{nes}^\circ/RT} \exp\left(-\frac{z_A z_B e^2}{4\pi\epsilon_0\epsilon d_{AB}kT}\right) \tag{6.20}$$

since $R/L = \mathbf{k}$. Taking natural logarithms, we obtain

$$\ln k = \ln \frac{kT}{h} - \frac{\Delta^\ddagger G_{nes}^\circ}{RT} - \frac{z_A z_B e^2}{4\pi\epsilon_0\epsilon d_{AB}kT} \tag{6.21}$$

This may be written as

$$\ln k = \ln k_0 - \frac{z_A z_B e^2}{4\pi\epsilon_0\epsilon d_{AB}kT} \tag{6.22}$$

where k_0 is the value of k in a medium of infinite dielectric constant, in which the electrostatic forces have become zero. An equation of this form was first obtained by Scatchard,[22] whose method of derivation was quite different.

According to this equation, the logarithm of the rate constant of a reaction between ions should vary linearly with the reciprocal of the dielectric constant. This relationship has been tested many times, particularly by causing a reaction to occur in a series of mixed solvents of varying dielectric constant. On the whole, the relationship is obeyed to a reasonable approximation, although there are usually deviations at low dielectric constants. An example of a test of Eq. (6.22) is shown in Fig. 6.3.

The slope of the line obtained by plotting $\ln k$ versus $1/\epsilon$ is given by Eq. (6.22) as $z_A z_B e^2/4\pi\epsilon_0 d_{AB}\mathbf{k}T$. Since everything in this expression is known except d_{AB}, it is possible to calculate d_{AB} from the experimental slope. This has been done in a number of cases, and the values obtained, of the order of a few hundred picometers (1 pm = 10^{-12} m), are reasonable. For the data shown in Fig. 6.3, the value of d_{AB} is 510 pm.

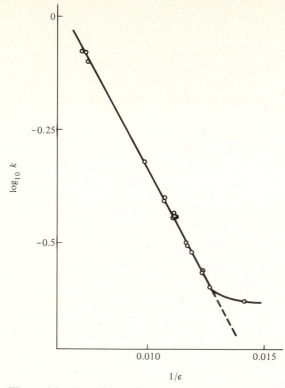

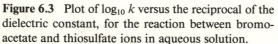

Figure 6.3   Plot of $\log_{10} k$ versus the reciprocal of the dielectric constant, for the reaction between bromoacetate and thiosulfate ions in aqueous solution.

6.3.2 Pre-exponential Factors

This treatment can be extended to give an interpretation of the pre-exponential factors of ionic reactions. The thermodynamic relationship between the entropy and the Gibbs energy is

$$S = -\left(\frac{\delta G}{\delta T}\right)_P \tag{6.23}$$

The electrostatic contribution to the Gibbs energy is given by Eq. (6.18), and in that expression only ϵ is temperature dependent. The electrostatic contribution to the entropy of activation is therefore

$$\Delta^{\ddagger}S_{es}^{\circ} = -\frac{Lz_Az_Be^2}{4\pi\epsilon_0 d_{AB}}\left(\frac{\delta(1/\epsilon)}{\delta T}\right)_P \tag{6.24}$$

$$= \frac{Lz_Az_Be^2}{4\pi\epsilon_0 d_{AB}\epsilon^2}\left(\frac{\delta\epsilon}{\delta T}\right)_P \tag{6.25}$$

$$= \frac{Lz_Az_Be^2}{4\pi\epsilon_0 d_{AB}\epsilon}\left(\frac{\delta\ln\epsilon}{\delta T}\right)_P \tag{6.26}$$

In aqueous solution ϵ is about 78.5, and $(\delta \ln \epsilon/\delta T)_P$ is constant at -0.0046 K^{-1} over a considerable temperature range. With d_{AB} taken as 200 pm, it follows from Eq. (6.26) that the entropy of activation arising from electrostatic effects is

$$\Delta^{\ddagger}S^{\circ}_{es} \approx -41 z_A z_B \text{ J K}^{-1} \text{ mol}^{-1} \qquad (6.27)$$

Thus, the entropy of activation in aqueous solution should decrease by about 41 J K^{-1} mol^{-1} for each unit of $z_A z_B$. Moreover, since the pre-exponential factor is proportional to $e^{\Delta^{\ddagger}S^{\circ}/R}$, which equals $10^{\Delta^{\ddagger}S^{\circ}/2.303R}$ or $10^{\Delta^{\ddagger}S^{\circ}/19.12}$, it follows that the factor should decrease by a factor of $10^{41/19.12}$, that is, about one-hundredfold, for each unit of $z_A z_B$. Table 6.4 shows that these relationships are obeyed in a very approximate manner. The treatment is too crude to allow detailed predictions to be made, but it is evidently along the right lines.

The physical model that lies behind these relationships is represented schematically in Fig. 6.4. In (a) the ions bear single positive charges, and the activated complex therefore bears a double positive charge. The solvent molecules in the neighborhood of an ion are acted on by strong electrostatic forces, which restrict their freedom of motion. This effect, known as *solvent binding* or *electrostriction,* leads to a reduction in entropy, the reduction being greater the larger the charge. In the case under consideration, there is some loss of entropy in the initial state, resulting from the fact that the reactant species are ions. In the activated state, however, there is a much greater loss of entropy since the charge is greater. Therefore, a loss of entropy occurs when the activated complex is formed.

In case (b) the reacting ions are of opposite signs, and therefore less charge exists on the activated complex than on the reactant molecules. There is a consequent decrease in electrostriction and an increase in entropy when the activated complex is formed.

6.3.3 Single-Sphere Activated Complex

Slightly different relationships have been obtained using a different model, known as the single-sphere model,[23] for the activated complex. In the treatment just described,

TABLE 6.4 SOME OBSERVED AND PREDICTED PRE-EXPONENTIAL FACTORS AND ENTROPIES OF ACTIVATION[a]

Reactants	Experimental		Estimated	
	A/dm^3 mol^{-1} s^{-1}	$\Delta^{\ddagger}S^{\circ}$/J K^{-1} mol^{-1}	A/dm^3 mol^{-1} s^{-1}	$\Delta^{\ddagger}S^{\circ}$/J K^{-1} mol^{-1}
$[Cr(H_2O)_6]^{3+} + CNS^-$	$\sim 10^{19}$	~ 126	10^{19}	126
$Co(NH_3)_5Br^{2+} + OH^-$	5×10^{17}	92	10^{17}	84
$CH_2BrCOOCH_3 + S_2O_3^{2-}$	1×10^{14}	25	10^{13}	0
$CH_2ClCOO^- + OH^-$	6×10^{10}	-50	10^{11}	-42
$ClO^- + ClO_2^-$	9×10^{8}	-84	10^{11}	-42
$CH_2BrCOO^- + S_2O_3^{2-}$	1×10^{9}	-71	10^{9}	-84
$Co(NH_3)_5Br^{2+} + Hg^{2+}$	1×10^{8}	-100	10^{5}	-167
$S_2O_4^{2-} + S_2O_4^{2-}$	2×10^{4}	-167	10^{5}	-167
$S_2O_3^{2-} + SO_3^{2-}$	2×10^{6}	-126	10^{5}	-167

[a] For references to the original literature see C. T. Burris and K. J. Laidler, *Trans. Faraday Soc.,* **15**, 1497 (1955); D. T. Y. Chen and K. J. Laidler, *Can. J. Chem.,* **37**, 599 (1959).

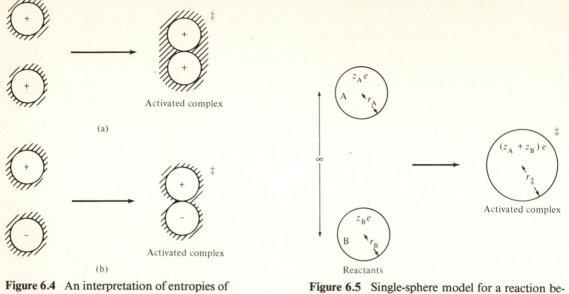

Figure 6.4 An interpretation of entropies of activation in terms of the electrostriction of solvent molecules. In (a) the ions are of the same sign and there is more electrostriction in the activated complex; thus, there is a decrease in entropy. In (b) there is less electrostriction in the activated complex.

Figure 6.5 Single-sphere model for a reaction between two ions.

the activated complex was regarded as having the form of a double sphere. In the single-sphere model, on the other hand, the reactant ions are regarded as becoming merged into one single sphere which has a charge equal to the algebraic sum of the charges on the ions. The results to which these alternative treatments lead are quite similar to each other, and it is likely that the truth usually lies somewhere between the two.

The single-sphere activated complex is represented schematically in Fig. 6.5, which is to be contrasted with Fig. 6.2. The derivation of the rate equation for this case is based on an expression obtained by Born[24] for the energy of charging an ion in solution. Born's expression is derived as follows. Consider the process of charging a conducting sphere of radius r from an initial charge of zero to a final charge equal to ze. This process is carried out by transporting, from an infinite distance, small increments of charge equal to $e\, d\lambda$, where λ is a parameter that varies from zero to z. At any time the charge on the sphere may be written as λe, and if at a given instant the increment of charge is at a distance x from the ion, the force acting on it is

$$df = \frac{\lambda e^2 \, d\lambda}{4\pi\epsilon_0\epsilon x^2} \tag{6.28}$$

The work of moving the increment from x to $x - dx$ is

$$dw = \frac{\lambda e^2 \, d\lambda \, dx}{4\pi\epsilon_0\epsilon x^2} \tag{6.29}$$

The total work of charging is obtained by carrying out the double integration, x being allowed to vary from infinity to r, and λ from zero to z. The total work is thus

$$w = \frac{e^2}{4\pi\epsilon_0\epsilon} \int_0^z \int_\infty^r \frac{\lambda \, d\lambda \, dx}{x^2} \tag{6.30}$$

$$= \frac{z^2 e^2}{8\pi\epsilon_0\epsilon r} \tag{6.31}$$

This work is the electrostatic contribution to the Gibbs energy of the ion,

$$G_{es} = \frac{z^2 e^2}{8\pi\epsilon_0\epsilon r} \tag{6.32}$$

For the process represented in Fig. 6.5, the electrostatic contributions to the Gibbs energies of the reactant ions and the activated complex are

$$G_{es}^\circ(A) = \frac{z_A^2 e^2}{8\pi\epsilon_0\epsilon r_A} \tag{6.33}$$

$$G_{es}^\circ(B) = \frac{z_B^2 e^2}{8\pi\epsilon_0\epsilon r_B} \tag{6.34}$$

$$G_{es}^\circ(\ddagger) = \frac{(z_A + z_B)^2 e^2}{8\pi\epsilon_0\epsilon r_\ddagger} \tag{6.35}$$

The change in the electrostatic contribution, $\Delta^\ddagger G_{es}^\circ$, is thus

$$\Delta^\ddagger G_{es}^\circ = \frac{e^2}{8\pi\epsilon_0\epsilon} \left(\frac{(z_A + z_B)^2}{r_\ddagger} - \frac{z_A^2}{r_A} - \frac{z_B^2}{r_B} \right) \tag{6.36}$$

Use of the same procedure that gave Eq. (6.22) then leads to

$$\ln k = \ln k_0 - \frac{e^2}{8\pi\epsilon_0\epsilon \mathbf{k}T} \left(\frac{(z_A + z_B)^2}{r_\ddagger} - \frac{z_A^2}{r_A} - \frac{z_B^2}{r_B} \right) \tag{6.37}$$

This equation is to be compared with Eq. (6.22), which is based on the double-sphere model. The two equations are identical if in Eq. (6.37) the radii r_A, r_B, and $r_\ddagger$ are all equal. The predictions of the two models are thus very similar, since in fact radii do not vary widely. The experimental results can be fitted equally well by the two equations.

6.3.4 Influence of Ionic Strength

It is found experimentally that the rates of second-order reactions between charged species are affected strongly by the *ionic strength* of the solution, this quantity being defined as

$$I = \tfrac{1}{2} \sum c_j z_j^2 \tag{6.38}$$

where c_j is the concentration of each ion present in the solution and z_j is its charge number (e.g., $+1$ for Na^+, -2 for SO_4^{2-}). When the reacting species are of the same sign, an increase in ionic strength increases the rate; when ions of opposite signs react, there is a decrease in rate with increasing ionic strength. The elucidation of this effect is important since it shows how activity coefficients are to be introduced into rate equations.

Brønsted[25] suggested that the rate of a reaction between species A and B should be expressed as

$$v = k_0[A][B] \frac{y_A y_B}{y_X} \tag{6.39}$$

where k_0 is a constant at a given temperature, y_A and y_B are the activity coefficients of the reactants, and y_X is that of a collision complex formed when A and B come together. Brønsted referred to the ratio $y_A y_B/y_X$ as the *kinetic activity factor,* and he showed how the use of this factor gives satisfactory agreement with experiment. He did this by making use of the fact that, according to Debye–Hückel theory,[26] the activity coefficient y_i of an ion is related to the ionic strength by the equation[27]

$$\log_{10} y_i = -Bz_i^2 I^{1/2} \tag{6.40}$$

where z_i is the charge number of the ion and B is a quantity whose value can be calculated from the Debye–Hückel equations; its value is approximately 0.51 dm$^{-3/2}$ mol$^{-1/2}$ for aqueous solutions at 25°C.

The application of this equation to a reaction

$$A^{z_A} + B^{z_B} \longrightarrow X^{z_A+z_B} \longrightarrow \text{products}$$

proceeds as follows. From Eq. (6.39) the second-order rate constant k ($=v/[A][B]$) is

$$k = k_0 \frac{y_A y_B}{y_X} \tag{6.41}$$

which in logarithmic form is

$$\log_{10} k = \log_{10} k_0 + \log_{10} y_A + \log_{10} y_B - \log_{10} y_X \tag{6.42}$$

Introduction of Eq. (6.40) then gives

$$\log_{10} k = \log_{10} k_0 - B[z_A^2 + z_B^2 - (z_A + z_B)^2]I^{1/2} \tag{6.43}$$

$$= \log_{10} k_0 + 2Bz_A z_B I^{1/2} \tag{6.44}$$

For an aqueous solution at 25°C this equation becomes

$$\log_{10} \frac{k}{k_0} = 1.02 z_A z_B (I/\text{mol dm}^{-3})^{1/2} \tag{6.45}$$

In these equations k_0 is the rate constant extrapolated to zero ionic strength.

Equation (6.45) has been tested a considerable number of times. Usually, the procedure has been to measure the rates of ionic reactions in media of varying ionic strength. According to Eq. (6.45), a plot of $\log_{10} k$ versus $(I/\text{mol dm}^{-3})^{1/2}$ will give a straight line of slope $1.02z_A z_B$. Figure 6.6 shows a plot of results for reactions of various types; the lines drawn are those with theoretical slopes, and the points lie close to them. If one of the reactants is a neutral molecule, $z_A z_B$ is zero, and the rate constant is expected to be independent of the ionic strength. This is approximately true, for example, for the base-catalyzed hydrolysis of ethyl acetate, shown in the figure. Small effects, however, are found for ion–dipole reactions, as discussed in Section 6.4.2.

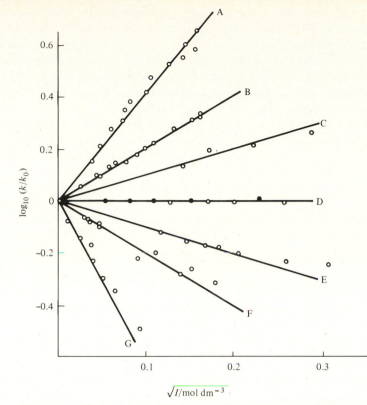

Figure 6.6 Plots of $\log_{10}(k/k_0)$ versus the square root of the ionic strength for ionic reactions of various types. The lines are drawn with slopes equal to $z_A z_B$. The reactants are:

A	$Co(NH_3)_5Br^{2+} + Hg^{2+}$	$(z_A z_B = 4)$
B	$S_2O_8^{2-} + I^-$	$(z_A z_B = 2)$
C	$CO(OC_2H_5)N{:}NO_2^- + OH^-$	$(z_A z_B = 1)$
D	$[Cr(urea)_6]^{3+} + H_2O$ (open circles)	$(z_A z_B = 0)$
	$CH_3COOC_2H_5 + OH^-$ (closed circles)	$(z_A z_B = 0)$
E	$H^+ + Br^- + H_2O_2$	$(z_A z_B = -1)$
F	$Co(NH_3)_5Br^{2+} + OH^-$	$(z_A z_B = -2)$
G	$Fe^{2+} + Co(C_2O_4)^{3-}$	$(z_A z_B = -6)$

Thus, Brønsted's equation (6.39) is satisfactory, but the justification for introducing the kinetic activity factor $y_A y_B/y_X$ was not at first clear. Bjerrum[28] presented a derivation of Eq. (6.39): he considered the equilibrium between A, B, and the collision complex X, the equilibrium constant being

$$K = \frac{a_X}{a_A a_B} = \frac{[X]}{[A][B]} \frac{y_X}{y_A y_B} \tag{6.46}$$

He then pointed out that if the rate is proportional to the *concentration* [X] of the complexes, Eq. (6.39) follows. Objections were raised to this derivation on the grounds that the rate should in some way involve the *activity* of X and not its concentration.[29] The later development of transition-state theory removed this dilemma provided the assumption is made that X is an *activated complex*.[30] Since this is only in a state of quasiequilibrium and cannot revert to reactants, the rate of formation of products depends on its concentration and not on its activity.

A derivation of Eq. (6.39) that differed in principle from that of Bjerrum was given by Christiansen,[31] who focused attention on the frequency of collisions between reacting species. He made use of the Debye–Hückel treatment of the electric potential between two ions, taking account of their ionic atmospheres. Thus, this derivation is concerned with the influence of ionic strength on the frequency of collisions between ions.

Equation (6.44) is extended readily to reactions of other orders. For a third-order reaction

$$A^{z_A} + B^{z_B} + C^{z_C} \longrightarrow (ABC^{\ddagger})^{z_A + z_B + z_C} \longrightarrow products$$

one obtains, instead of Eq. (6.44),

$$\log_{10} k = \log_{10} k_0 - B[z_A^2 + z_B^2 + z_C^2 - (z_A + z_B + z_C)^2]I^{1/2} \qquad (6.47)$$

$$= \log_{10} k_0 + 2B(z_A z_B + z_B z_C + z_A z_C)I^{1/2} \qquad (6.48)$$

It is important to note that the equation representing the ionic-strength effect always corresponds to the kinetic equation and is independent of the mechanism. The experimental study of an ionic-strength effect therefore cannot lead to any conclusions about the reaction mechanism. However, it can provide information about the charges that come together during reaction: this is often useful in enzyme kinetics, where the charge at the active center of the enzyme may be difficult to establish in other ways (see also Problem 6.8).

The Brønsted equation also has been used to determine the charge on muonium, Mu. At one time this species appeared to react in aqueous solution in a similar way to the hydrated electron (Section 9.4.3), and it was thought that it might be charged.[32] However, in its reactions with Cu^{2+} and other ionic species, the ionic strength effect was negligible,[33] showing that muonium bears no charge (see Problem 6.13). It is a positive muon plus an electron, that is, a light isotope of the hydrogen atom (see Section 11.2.5). Other applications of the procedure, in connection with the hydrated electron, are referred to in Section 9.4.3.

A careful study by Davies[34] of the applicability of Eq. (6.45) to bimolecular reactions between ions led to the conclusion that the equation is very satisfactory. The Debye–Hückel equation in its simple form [Eq. (6.40)] holds only at very low concentrations, and deviations from Eq. (6.40) inevitably occur at higher concentrations. Improvement can be obtained by adding terms to the Debye–Hückel equation, and deviations usually can be explained in a qualitative way on the basis of ion-pair formation.

Sometimes ion-pair formation is purely electrostatic, and the term *Bjerrum ion pairs* is employed. In other cases, the association product has a definite chemical structure. Ion association can affect rates in a number of ways, of which the following are the most important:

1. There is a reduction in the true ionic strength of the solution.
2. The ion pairing may involve one or both of the reactant ions, in which case there is a change in the electrostatic interactions between the ions that react together. In a reaction between ions of like sign, for example, association with an oppositely charged ion leads to acceleration by reducing the electrostatic repulsion.

6.3.5 More Advanced Treatments

In the procedures that have been outlined, the solvent was regarded as having no structure, that is, as continuous, and the dielectric constant was considered to be the same throughout. The assumption of a uniform dielectric constant gives rise to very considerable simplification, and it leads to a treatment that is valuable as a first approximation. In reality, a liquid is an assembly of molecules, and the dielectric constant is a gross property that is determined by making measurements over a large assembly of molecules; it therefore corresponds to an average behavior. Treatments of ionic reactions that avoid any of the simplifications of the dielectric continuum theory are much more complicated and can be applied only with considerably more difficulty to the experimental results. Two main approaches have been made and they are described now very briefly.

The first type of improvement to the treatment of ions in solution still regards the solvent as continuous but takes account of the change in dielectric constant in the neighborhood of an ion. The dielectric behavior of a liquid is due to the orientation of molecules in an electric field, and at very high field strengths the system approaches saturation, so that the dielectric constant approaches a low limiting value. Solvent molecules that are very close to an ion are subjected to a very high field and become aligned in the field. Since they cannot be oriented further, the dielectric constant close to an ion has the very low value of about 2 which is characteristic of hydrocarbons and other nonpolar substances. This effect is known as *dielectric saturation.*

A quantitative treatment of the effect has been given,[35] based on the known variation of dielectric constant with field strength. Figure 6.7 shows the results of such calculations for simple ions of various valence values in aqueous solution. The bulk dielectric constant of water at 25°C is 78 5, but close to an ion it is 1.78, there being a steep rise at short distances from the ion. As a result of this behavior, the electrostatic forces between ions that are close together in aqueous solution are very different from what they would be if the dielectric constant had a uniform value of 78.5. For example, if two ions of the same sign, such as Fe^{2+} and Fe^{3+} ions, are brought together in aqueous solution, the repulsive forces at short distances are much greater than if the dielectric constant were 78.5: there is therefore an enhanced Gibbs energy increase for the process.[36] The entropy change is also modified by the dielectric saturation.

The alternative procedure abandons the idea of treating the solvent as continuous and instead regards it as having a molecular structure. It considers in detail the interactions between the ions and the individual solvent molecules. Using this viewpoint,

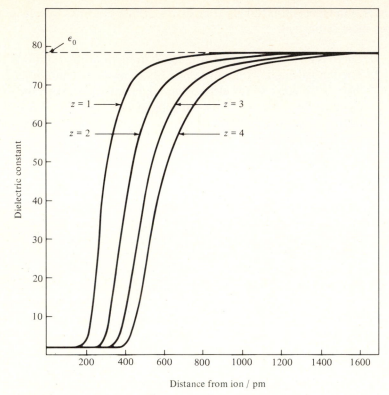

Figure 6.7 Dependence of the dielectric constant of water on the distance from ions having charge numbers of 1, 2, 3, and 4.

Frank and his coworkers[37] have considered the thermodynamic and other properties of ions in solution. They have shown that next to an ion in aqueous solution there is generally a layer of water molecules that are held fairly rigidly and that can be regarded as forming a *hydration shell.* These molecules have low partial molar entropies and volumes. Outside this layer there is a *disorder zone,* where the water molecules are arranged in a less orderly manner than in the remainder of the solution. This zone arises because of the competition between the electrical forces produced by the ion and the hydrogen-bonding forces resulting from the surrounding water molecules. The molecules in this disorder zone have a higher entropy than those in the bulk of the solution. This line of attack on the problem has been followed in quantitative treatments of the thermodynamics of simple ions in aqueous solution.[38] Levine and Bell[39] have calculated interaction energies between two ions in aqueous solution, allowing the hydration shell to have a molecular structure but treating the rest of the solvent as continuous. Unfortunately, the application of these ideas to kinetic problems is a matter of great difficulty, and not much progress has been made.

    The two types of approach lead to quite similar conclusions since the layer of frozen water molecules corresponds roughly to the region of very low dielectric constant. From some points of view, the first type of improvement, in which the change in dielectric constant is considered, is easier to apply in complicated cases and may perhaps be more fruitful in dealing with the kinetics of reactions between ions.

6.4 ION–DIPOLE AND DIPOLE–DIPOLE REACTIONS

The preceding section was concerned with simple ions, which were assumed to have zero dipole moments. In reality, many ions also have a distribution of charges, and this makes a difference to the electrostatic interactions. Also, many reactions of interest are between ions and dipolar molecules, or between two dipolar molecules. The present section gives a more general treatment,[40] which takes into account all these possibilities and again considers only the electrostatic interactions.

The treatments are based on an expression obtained by Kirkwood[41] for the Gibbs energy of charging a sphere that has charges imbedded in it in a number of specified positions. To a good first approximation, the problem can be reduced to one of a sphere of radius r, net charge ze, and dipole moment μ. The Gibbs energy increase when such a sphere is charged (i.e., the work required to charge it) is

$$\Delta G_{es}^{\circ} = \frac{z^2 e^2}{8\pi\epsilon_0 \epsilon r} + \frac{3\mu^2}{16\pi\epsilon_0 \epsilon r^3} \tag{6.49}$$

The first term corresponds to the Born expression [Eq. (6.32)]; the second is the additional contribution corresponding to the dipole moment μ.

For the bimolecular reaction

$$A + B \rightarrow X^{\ddagger} \rightarrow \text{products}$$

the electrostatic contributions to the Gibbs energies of charging A, B, and $X^{\ddagger}$ are

$$G_{es}^{\circ}(A) = \frac{z_A^2 e^2}{8\pi\epsilon_0 \epsilon r_A} + \frac{3\mu_A^2}{16\pi\epsilon_0 \epsilon r_A^3} \tag{6.50}$$

$$G_{es}^{\circ}(B) = \frac{z_B^2 e^2}{8\pi\epsilon_0 \epsilon r_B} + \frac{3\mu_B^2}{16\pi\epsilon_0 \epsilon r_B^3} \tag{6.51}$$

$$G_{es}^{\circ}(\ddagger) = \frac{(z_A + z_B)^2 e^2}{8\pi\epsilon_0 \epsilon r_{\ddagger}} + \frac{3\mu_{\ddagger}}{16\pi\epsilon_0 \epsilon r_{\ddagger}^3} \tag{6.52}$$

The electrostatic contribution to the Gibbs energy of activation is thus

$$\Delta^{\ddagger}G_{es}^{\circ} = \frac{e^2}{8\pi\epsilon_0 \epsilon}\left(\frac{(z_A + z_B)^2}{r_{\ddagger}} - \frac{z_A^2}{r_A} - \frac{z_B^2}{r_B}\right) + \frac{3}{16\pi\epsilon_0 \epsilon}\left(\frac{\mu_{\ddagger}^2}{r_{\ddagger}^3} - \frac{\mu_A^2}{r_A^3} - \frac{\mu_B^2}{r_B^3}\right) \tag{6.53}$$

Thus, the rate constant, proportional to $e^{-\Delta^{\ddagger}G^{\circ}/kT}$, is given by

$$\ln k = \ln k_0 - \frac{e^2}{8\pi\epsilon_0 kT\epsilon}\left(\frac{(z_A + z_B)^2}{r_{\ddagger}} - \frac{z_A^2}{r_A} - \frac{z_B^2}{r_B}\right) - \frac{3}{16\pi\epsilon_0 kT\epsilon}\left(\frac{\mu_{\ddagger}^2}{r_{\ddagger}^3} - \frac{\mu_A^2}{r_A^3} - \frac{\mu_B^2}{r_B^3}\right) \tag{6.54}$$

Here k_0 is the hypothetical rate constant arising from the nonelectrostatic contribution $\Delta^{\ddagger}G_{nes}^{\circ}$, that is, corresponding to $\epsilon = \infty$. The first and second terms correspond to Eq. (6.37).

Since the radii do not vary as much as the charges and dipole moments, it is a useful approximation to treat them as all the same and equal to d_{AB}. Equation (6.54) then reduces to

$$\ln k = \ln k_0 - \frac{z_A z_B e^2}{4\pi\epsilon_0\epsilon d_{AB}\mathbf{k}T} - \frac{\mu_\ddagger^2 - \mu_A^2 - \mu_B^2}{16\pi\epsilon_0\epsilon d_{AB}^3\mathbf{k}T} \tag{6.55}$$

Now the first two terms correspond to Eq. (6.22), for the double-sphere model.

If both species A and B are charged, the second term in Eq. (6.54) or Eq. (6.55) is usually much greater than the third. If either reactant has no charge, however, the final term becomes predominant. Equations (6.54) and (6.55) are useful in leading to predictions of the effect of changing the dielectric constant of the solvent. For example, they predict that if a reaction occurs between uncharged species with the formation of an activated complex that is more polar than the reactants (i.e., $\mu_\ddagger$ is large compared to μ_A and μ_B), the rate constant increases with increasing dielectric constant. The physical significance of this is that a high dielectric constant favors the formation of a highly dipolar species, just as it favors ionization. This point of view is consistent with that of Hughes and Ingold (Section 6.2.4), whose interpretation was in terms of solvation: the dielectric constant and solvating power vary in much the same way.

6.4.1 Pre-exponential Factors

An extension of the argument used in Section 6.3.2 for the pre-exponential factors of ion–ion reactions leads to the result that, in aqueous solution at 25°C,

$$\Delta^\ddagger S_{es}^\circ / \text{J K}^{-1} \text{ mol}^{-1} = -41 z_A z_B - 7.7\,\Delta\mu_0^2 \tag{6.56}$$

The quantity $\Delta\mu_0^2$, which corresponds to the product $z_A z_B$ of charge numbers, is defined by

$$\Delta\mu_0^2 \equiv \frac{\mu_\ddagger^2 - \mu_A^2 - \mu_B^2}{(1.602 \times 10^{-29} \text{ C m})^2} \tag{6.57}$$

where the quantity 1.602×10^{-29} C m is the dipole moment of two elementary charges (1.602×10^{-19} C) separated by a distance of 10^{-10} m; it is equal to 4.8 D (debye units). Equation (6.56) must be modified for solutions other than aqueous solutions.

Ester hydrolyses are examples of ion–dipole reactions, and in aqueous solution their entropies of activation are often -40 J K^{-1} mol^{-1} or so. This is too negative to be accounted for satisfactorily by Eq. (6.56): $z_A z_B$ is zero, and an unreasonably large value of μ_0 would be required to predict such a value. It would appear therefore that for these reactions there are important entropy losses arising from nonelectrostatic effects. In acid catalysis, for example, the structure of the complex may be something like

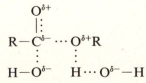

and there is much less freedom of motion than in the reactants. In agreement with Eq. (6.56), with $\Delta\mu_0^2$ positive, the hydrolyses are accelerated by an increase in the dielectric constant of the solvent.

The conclusions are summarized in Table 6.5, which also includes ionic strength effects (Sections 6.3.4 and 6.4.2) and pressure effects (Section 6.5).

6.4.2 Influence of Ionic Strength

Equation (6.45) predicts that a reaction between an ion and a neutral molecule, or between two neutral molecules, should not be affected by the ionic strength of the solution. However, this equation was obtained using approximations that hold only in very dilute solutions, and an extension of the treatment is necessary. The activity coefficient of an ion is given to a good approximation if, in addition to the Debye–Hückel term, the term bI introduced by Hückel[42] is added. The equation for the ion A therefore should be written as

$$\log_{10} y_A = -Bz_A^2 I^{1/2} + b_A I \tag{6.58}$$

If B has no net charge, its activity coefficient may be expressed by an approximate equation due to Debye and McAulay,[43]

$$\log_{10} y_B = b_B I \tag{6.59}$$

The activated complex must have the same net charge as A, and its activity coefficient is therefore given by an equation of the same form as Eq. (6.58). The introduction of these equations into Eq. (6.42) gives rise to

$$\log_{10} k = \log_{10} k_0 + (b_A + b_B - b_{\ddagger})I \tag{6.60}$$

The Debye–Hückel term involving the square root of the ionic strength has disappeared, since it occurs in the expressions for both A and the activated complex; as a result, it is important to include the terms that are linear in the ionic strength, as was done above. Equation (6.60) predicts that the logarithm of the rate constant of a reaction between an ion and a neutral molecule should vary with the first power of the ionic strength instead of with the square root: the effect, however, is much smaller than for

TABLE 6.5 SUMMARY OF EFFECTS FOR DIFFERENT TYPES OF REACTIONS

Reaction	Effect of increasing		$\Delta^{\ddagger}S^{\circ}$	A	$\Delta^{\ddagger}V^0$	Examples
	dielectric constant	ionic strength				
Ionic reactions						
Ions of same sign	↑	↑	<0	Small	<0	$CH_2BrCOO^- + S_2O_3^{2-}$, OH$^-$ + bromphenol blue
Ions of opposite signs	↓	↓	>0	Large	>0	$OH^- + Me_3S^+$, reverse Menschutkin reactions
Ion–dipole or dipole–dipole reactions						
$\Delta\mu_0^2 > 0$	↑	↑	<0	Small	<0	Menschutkin reaction, unimolecular solvolyses, hydrolysis of ester
$\Delta\mu_0^2 < 0$	↓	↓	>0	Large	>0	

reactions between ions. This conclusion has received experimental support from the work of Brønsted and Wynne-Jones[44] on the hydrolysis of acetals by hydroxide ions.

Equation (6.60) may be written as

$$k = k_0 e^{b'I} \tag{6.61}$$

If $b'I$ is much smaller than unity, the exponential may be expanded and only the first term accepted, the result being

$$k = k_0(1 + b'I) \tag{6.62}$$

Therefore, the rate constant should vary linearly with I under these conditions, and this has been found to be the case for several reactions.[45]

There is still no completely satisfactory treatment that permits the calculation of the magnitude of the coefficients b, in terms of the structures of the molecules concerned. The term $b_A I$ that occurs in the expression for the activity coefficient of an ion was introduced by Hückel on the basis of an argument that takes into consideration the effects of dielectric saturation, and a quantitative treatment has been given by Stokes and Robinson.[46] The b coefficient that appears in the expression for a neutral molecule, as in Eq. (6.59), has been discussed by many authors,[47] but reliable estimates of its magnitude can not be made. There is much scope for additional work in this important field.

6.5 INFLUENCE OF HYDROSTATIC PRESSURE

6.5.1 Van't Hoff's Equation

Valuable information about reactions in solution is provided by studies of the influence of hydrostatic pressure on their rates. The theory of pressure effects on rates was first formulated by van't Hoff,[48] who started with the equation for the effect of pressure on the equilibrium constant K_c:

$$\left(\frac{\delta \ln K_c}{\delta P}\right)_T = -\frac{\Delta V^\circ}{RT} \tag{6.63}$$

where ΔV° is the overall standard molar volume change. The argument is similar to the one van't Hoff used for temperature dependence [Eqs. (2.68)–(2.70)] and can be reformulated in terms of the concept of the activated complex as follows. The quantity ΔV° is the difference between the volumes of the products and reactants:

$$\Delta V^\circ = V_p - V_r \tag{6.64}$$

If $V^\ddagger$ is the partial molar volume of the activated complexes,

$$\Delta V^\circ = (V^\ddagger - V_r) - (V^\ddagger - V_p) \tag{6.65}$$

$$= \Delta^\ddagger V_1^\circ - \Delta^\ddagger V_{-1}^\circ \tag{6.66}$$

where $\Delta^\ddagger V_1^\circ$ is the standard volume change in going to the activated state in the forward direction, and $\Delta^\ddagger V_{-1}^\circ$ is that in the reverse direction. These quantities are known as *volumes of activation*. Since K_c is k_1/k_{-1}, Eq. (6.63) becomes

$$\left(\frac{\delta \ln k_1}{\delta P}\right)_T - \left(\frac{\delta \ln k_{-1}}{\delta P}\right)_T = -\frac{\Delta^\ddagger V_1^\circ}{RT} + \frac{\Delta^\ddagger V_{-1}^\circ}{RT} \qquad (6.67)$$

The argument is that the behavior in the forward direction can depend only on $\Delta^\ddagger V_1^\circ$, and that in the reverse direction only on $\Delta^\ddagger V_{-1}^\circ$. Thus, Eq. (6.67) can be split into the two equations

$$\left(\frac{\delta \ln k_1}{\delta P}\right)_T = -\frac{\Delta^\ddagger V_1^\circ}{RT} \qquad (6.68)$$

and

$$\left(\frac{\delta \ln k_{-1}}{\delta P}\right)_T = -\frac{\Delta^\ddagger V_{-1}^\circ}{RT} \qquad (6.69)$$

An alternative derivation starts with Eq. (4.84):

$$k = \frac{\mathbf{k}T}{h} K^\ddagger \qquad (6.70)$$

The equilibrium constant $K^\ddagger$ between initial and activated states varies with pressure according to

$$\left(\frac{\delta \ln K^\ddagger}{\delta P}\right)_T = -\frac{\Delta^\ddagger V^\circ}{RT} \qquad (6.71)$$

and therefore

$$\left(\frac{\delta \ln k}{\delta P}\right)_T = -\frac{\Delta^\ddagger V^\circ}{RT} \qquad (6.72)$$

It follows that a rate constant increases with increasing pressure if $\Delta^\ddagger V^\circ$ is negative, that is, if the activated complex has a smaller volume than the reactants. Conversely, pressure has an adverse effect on rates if there is a volume increase when the activated complex is formed. By use of Eq. (6.72), values of $\Delta^\ddagger V^\circ$ can be determined from experimental studies of rates at different pressures. In practice, it is usually necessary to use fairly high pressures for this purpose (at least 100 atm, or 10^4 kPa) since otherwise the changes of rate are too small for accurate $\Delta^\ddagger V^\circ$ values to be obtained.

According to Eq. (6.72), if $\ln k$ is plotted against pressure, the slope at any pressure is $-\Delta^\ddagger V^\circ/RT$. In some cases the plots are linear, which means that $\Delta^\ddagger V^\circ$ is independent of pressure. If this is so, Eq. (6.72) can be integrated to give

$$\ln k = \ln k_0 - \frac{\Delta^\ddagger V^\circ}{RT} P \qquad (6.73)$$

where k_0 is the rate constant at zero pressure (this is always very close to the value at atmospheric pressure). Examples of such linear plots are shown in Fig. 6.8. Table 6.6 shows some values of $\Delta^\ddagger V^\circ$ obtained from pressure studies and also gives the entropies of activation $\Delta^\ddagger S^\circ$.

6.5.2 Volumes of Activation

It was pointed out by Evans and Polanyi[49] that two distinct effects must be considered in connection with the interpretation of volumes of activation. First, there may be a

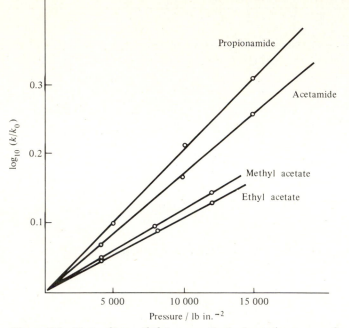

Figure 6.8 Plots of $\log_{10}(k/k_0)$ versus the hydrostatic pressure for the alkaline hydrolyses of some esters and amides. [*Source:* K. J. Laidler and D. Chen, *Trans. Faraday Soc.,* **54**, 1020 (1958).]

change, due to structural factors, in the volume of the reactant molecules themselves as they pass into the activated state. For a bimolecular process this always leads to a volume decrease, while for a unimolecular process there is a volume increase. Second, there may be a volume change resulting from reorganization of the solvent molecules. Studies of a variety of reactions[50] have led to the conclusion that, for reactions in which ions or fairly strong dipoles are concerned, the solvent effects are generally more important than the structural ones.

These effects of solvent on volumes are explained in a manner similar to the effects of solvent on entropies of activation (Section 6.3.2). Thus, if a reaction occurs with the approach of ions of the same sign, or a separation of ions of the opposite signs, there is an intensification of the electric field and therefore an increase in electrostriction and a resulting decrease in volume. There is also a decrease in entropy owing to the loss of freedom of the solvent molecules. Conversely, if the electric field is weakened when the activated complex is formed (as when two ions of opposite signs come together), there is some release of bound solvent molecules, and the volumes and entropies of activation are positive. All the $\Delta^{\ddagger}V^{\circ}$ values shown in Table 6.6 can be interpreted from this point of view.

For reactions between ions or in which there are substantial changes in polarity during the reaction, the volumes of activation and the entropies of activation depend on very similar factors, involving the orientation of solvent molecules. Therefore, some correlation between $\Delta^{\ddagger}V^{\circ}$ and $\Delta^{\ddagger}S^{\circ}$ values is expected. Figure 6.9 shows that there is indeed a rough correlation.

TABLE 6.6 VOLUMES AND ENTROPIES OF ACTIVATION[a]

Reaction	Solvent	$\Delta^{\ddagger}V°/cm^3$ mol^{-1}	$\Delta^{\ddagger}S°/J\ K^{-1}$ mol^{-1}
$Co(NH_3)_5Br^{2+} + OH^- \rightarrow Co(NH_3)_5OH^{2+} + Br^-$	H_2O	8.5	92
$(CH_3)(C_2H_5)(C_6H_5)(C_6H_5CH_2)N^+Br^- \rightarrow (CH_3)(C_6H_5)(C_6H_5CH_2)N + C_2H_5Br$	H_2O	3.3	63
$CH_2BrCOOCH_3 + S_2O_3^{2-} \rightarrow CH_2(S_2O_3^-)COOCH_3 + Br^-$	H_2O	3.2	25
Sucrose $+ H_2O \xrightarrow{H^+}$ glucose + fructose	H_2O	2.5	33
$C_2H_5O^- + C_2H_5I \rightarrow C_2H_5OC_2H_5 + I^-$	C_2H_5OH	-4.1	-42
$CH_2ClCOO^- + OH^- \rightarrow CH_2OHCOO^- + Cl^-$	H_2O	-6.1	-56
$CH_2BrCOO^- + S_2O_3^{2-} \rightarrow CH_2(S_2O_3^-)COO^- + Br^-$	H_2O	-4.8	-71
$CH_3COOCH_3 + H_2O \xrightarrow{H^+} CH_3COOH + CH_3OH$	H_2O	-8.7	-42
$CH_3CONH_2 + H_2O \xrightarrow{OH^-} CH_3COOH + NH_3$	H_2O	-14.2	-142
$C_5H_5N + C_2H_5I \rightarrow C_5H_5(C_2H_5)N^+I^-$	CH_3COCH_3	-16.8	-146
$C_6H_5CCl_3 \rightarrow C_6H_5CCl_2^+ + Cl^-$	80% C_2H_5OH	-14.5	-146

[a] For references to the original literature see C. T. Burris and K. J. Laidler, *Trans. Faraday Soc.,* **15,** 1497 (1955); D. T. Y. Chen and K. J. Laidler, *Can. J. Chem.,* **37,** 599 (1959).

Whalley[51] made use of the results of pressure studies to arrive at conclusions about reaction mechanisms. He showed that volumes of activation are more reliable than entropies of activation in suggesting the type of reaction that is taking place. This is because entropies of activation are quite sensitive to factors such as the loosening or strengthening of chemical bonds, whereas volumes depend to a much greater extent on electrostriction effects than on any other effects. Therefore, volume changes are much more constant for reactions of a given type than are entropies, and lead to more clear-cut conclusions about the processes taking place.

6.6 SUBSTITUENT AND CORRELATION EFFECTS

It is outside the scope of this book to cover the very extensive and important work that has been done on the effects of substituents on reaction rates.[52] Brief mention, however, is made of correlations that have been found between kinetic parameters for different types of reaction. The most fundamental of these are certain *linear Gibbs energy relationships.*

6.6.1 Hammett Equation

One of the first and most useful of these is an equation proposed by Hammett.[53] This relates equilibrium and rate constants for the reactions of meta- and para-substituted benzene derivatives. For example, one might compare the rate constants for two types of reaction involving a given set of substituents in the meta position. According to Hammett's relationship, a rate or equilibrium constant for the reaction of one compound is related to that for the unsubstituted ("parent") compound in terms of two parameters ρ and σ. For rate constants the relationship is

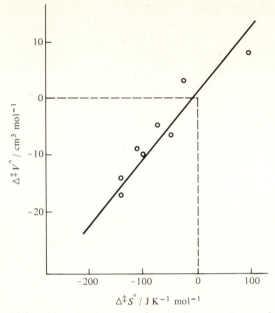

Figure 6.9 Correlation between volumes and entropies of activation. The sources of data are to be found in K. J. Laidler and D. T. Y. Chen, *Can. J. Chem.,* **37,** 599 (1959).

$$\log_{10} k = \log_{10} k_0 + \sigma\rho \tag{6.74}$$

where k_0 is the rate constant for the parent compound. For equilibrium constants

$$\log_{10} K = \log_{10} K_0 + \sigma\rho \tag{6.75}$$

where K_0 is for the parent compound. Of the two parameters, σ depends only on the *substituent* while ρ is a *reaction* constant, varying with the nature of the reaction and the external conditions such as the solvent. A value of unity is chosen arbitrarily for ρ for the ionization equilibrium constant of benzoic acid in aqueous solution and for the substituted benzoic acids. It follows that σ is the logarithm of the ratio of the dissociation constant of a substituted benzoic acid to that of benzoic acid itself.

Now it will be shown that the Hammett equations imply linear relationships between the Gibbs energies, of reaction or of activation, for different series of reactions. According to Eq. (4.85), the rate constant is related to the Gibbs energy of activation by

$$k = \frac{\mathbf{k}T}{h} e^{-\Delta^{\ddagger}G^{\circ}/RT} \tag{6.76}$$

and therefore

$$\log_{10} k = \log_{10} \frac{\mathbf{k}T}{h} - \frac{\Delta^{\ddagger}G^{\circ}}{2.303RT} \tag{6.77}$$

Equation (6.74) therefore may be written as

$$\Delta^{\ddagger}G^{\circ} = \Delta^{\ddagger}G_0^{\circ} - 2.303RT\rho\sigma \tag{6.78}$$

This equation, with a particular value of ρ, applies to any reaction involving a reactant having a series of substituents. For a second series of homologous reactions, with the reaction constant ρ',

$$\Delta^{\ddagger}G^{o\prime} = \Delta^{\ddagger}G_0^{o\prime} - 2.303RT\rho'\sigma \tag{6.79}$$

Equations (6.78) and (6.79) may be written as

$$\frac{\Delta^{\ddagger}G^{\circ}}{\rho} = \frac{\Delta^{\ddagger}G_0^{\circ}}{\rho} - 2.302RT\sigma \tag{6.80}$$

and

$$\frac{\Delta^{\ddagger}G^{o\prime}}{\rho'} = \frac{\Delta^{\ddagger}G_0^{o\prime}}{\rho'} - 2.303RT\sigma \tag{6.81}$$

Subtraction gives

$$\frac{\Delta^{\ddagger}G^{\circ}}{\rho} - \frac{\Delta^{\ddagger}G^{o\prime}}{\rho'} = \frac{\Delta^{\ddagger}G_0^{\circ}}{\rho} - \frac{\Delta^{\ddagger}G_0^{o\prime}}{\rho'} \tag{6.82}$$

or

$$\Delta^{\ddagger}G^{\circ} - \frac{\rho}{\rho'}\Delta^{\ddagger}G^{o\prime} = \text{const} \tag{6.83}$$

Thus, there is a linear relationship between the Gibbs energies of activation for one homologous series of reactions and those for another. An equivalent relationship can be derived for overall Gibbs energy changes.

The Hammett relationship does not apply well to the reactions of aliphatic compounds, partly because there is usually some steric interference between the substituent and the reactant site. An alternative equation was put forward by Taft[54] for aliphatic compounds. For details of this and more recent correlations the reader is referred to books and reviews listed in the bibliography at the end of this chapter.

6.6.2 Compensation Effects

It was assumed originally that linear Gibbs energy relationships such as those of Hammett and Taft are associated with the existence of linear relationships between energies of activation and enthalpies of reaction, the entropies remaining constant within a homologous series. It has now become apparent, on the contrary, that Gibbs energies are much simpler functions than energies, which are more sensitive to external factors, such as those brought about by the solvent. Examples are known in which Gibbs energies show linear relationships and exhibit additivity, but in which the corresponding energy and enthalpy changes show no such relationship. This is possible because there is a general tendency in processes in solution for enthalpies and entropies to compensate each other, so that the changes in Gibbs energy are much smaller.

In a considerable number of instances, plots of $T\,\Delta^{\ddagger}S^{\circ}$ versus $\Delta^{\ddagger}H^{\circ}$ have been found to be straight lines of approximately unit slope. This is often the case for a given reaction investigated in a series of solvents and also for homologous reactions in which substituents are introduced into a reactant.

The Gibbs energy of activation $\Delta^{\ddagger}G^{\circ}$ is equal to $\Delta^{\ddagger}H^{\circ} - T\,\Delta^{\ddagger}S^{\circ}$, and it follows that if there is an exact linear relationship between $\Delta^{\ddagger}H^{\circ}$ and $T\,\Delta^{\ddagger}S^{\circ}$, with unit slope, there will be no variation of $\Delta^{\ddagger}G^{\circ}$. If the relationship is only an approximate one, it means that the dependence of $\Delta^{\ddagger}G°$ on solvent or substituent is much smaller than that of $\Delta^{\ddagger}H°$ or $T\,\Delta^{\ddagger}S°$.

A similar compensation effect frequently has been found between $\Delta H°$ and $T\,\Delta S°$ for *overall* processes in solution; therefore, one cannot explain the compensation between $\Delta^{\ddagger}H°$ and $T\,\Delta^{\ddagger}S°$ in terms of purely kinetic effects. The true explanation must lie in terms of solvent–solute interactions.[55] Any effect, for example, that leads to a stronger binding between a solute molecule and the solvent molecules will lower the enthalpy; it also will lower the entropy by restricting the freedom of vibration and rotation of the solvent molecules. Application of more exact theories to these effects leads to the result that they generally will give rise to a fairly exact compensation between $\Delta H°$ and $T\,\Delta S°$ and therefore to a very small effect on $\Delta G°$.

Although changes of substituent and solvent often exert their influence on $\Delta^{\ddagger}H°$ in a rather complex manner, the partial compensation between $\Delta^{\ddagger}H°$ and $T\,\Delta^{\ddagger}S°$ is of such a nature that their influence on $\Delta^{\ddagger}G°$ is much simpler. It is for this reason that relatively simple concepts have been very successful in explaining the effects of solvents and substituents on rates, that is, on $\Delta^{\ddagger}G°$. Much more complicated explanations, involving detailed considerations of solvent–solute interactions, must be invoked to explain the effects on enthalpies and entropies of activation.

6.7 DIFFUSION-CONTROLLED REACTIONS

If a rapid bimolecular reaction in solution is initiated by mixing solutions of the reactants, the observed rate may depend on the rate with which the solutions mix. This effect is known as *mixing control* or *macroscopic diffusion control.*

The effect of mixing can be eliminated by the use of special methods, such as the relaxation techniques (Section 2.8.2). However, even when this is done the rate of reaction may be influenced by the rate with which the reactant molecules diffuse towards each other. This effect is known as *microscopic diffusion control* or *encounter control.* If the measured rate is almost exactly equal to the rate of diffusion, we speak of *full* microscopic diffusion control or of *full* encounter control. An example is provided by the combination of H^+ and OH^- ions in solution, which has a rate constant of 1.4×10^{11} dm^3 mol^{-1} s^{-1} at 25°C. For some reactions the rates of chemical reaction and of diffusion are similar to each other, and we then speak of *partial* microscopic diffusion control or *partial* encounter control.

6.7.1 Full Microscopic Diffusion Control

The rate equation for a reaction in which electrostatic forces are unimportant, and for which there is full microscopic diffusion control,§ can be derived as follows. According to Fick's first law of diffusion, if dn moles of a substance diffuse through an area A in time dt, the rate dn/dt, known as the *diffusive flux* and given the symbol J, is given by

§ The theory was first formulated by M. V. Smoluchowski [*Z. Phys. Chem.*, **92**, 129 (1917)] who applied it in particular to the rate of coagulation of colloids.

$$J \equiv \frac{dn}{dt} = -DA \frac{dc}{dx} \qquad (6.84)$$

where D is the diffusion coefficient and dc/dx is the concentration gradient; the negative sign appears because the diffusion is in the direction of decreasing concentration.

Suppose that in a solution there are two uncharged molecules A and B, having radii r_A and r_B. We first regard one molecule, A, as stationary and consider the rate of diffusion of B molecules towards it (Fig. 6.10). A sphere of radius r can be constructed around A; its surface area is $4\pi r^2$. The rate of diffusion of B molecules through this surface is

$$J_B = \frac{dn_B}{dt} = 4\pi r^2 D_B \frac{dc_B}{dr} \qquad (6.85)$$

where c_B is the local concentration of B molecules. Here the concentration gradient dc_B/dr is taken with reference to increasing radius, so that the negative sign is dropped. The concentration gradient is

$$\frac{dc_B}{dr} = \frac{J_B}{4\pi D_B r^2} \qquad (6.86)$$

and integration gives

$$c_B = -\frac{J_B}{4\pi D_B r} + I \qquad (6.87)$$

The constant of integration I is obtained by considering the boundary conditions. When $r = \infty$, c_B is equal to the bulk concentration of B, which we write as [B]. Thus, $I =$ [B] and therefore

$$c_B = -\frac{J_B}{4\pi D_B r} + [B] \qquad (6.88)$$

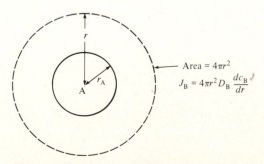

Figure 6.10 Diffusion of molecule B toward molecule A, which is considered to be stationary.

This result relates to a single molecule of A and is based on the assumption that this molecule is stationary and that only the B molecules are moving. In reality, there is diffusion of both molecules, and Chandrasekhar[56] has shown that this is taken care of by replacing D_B in Eq. (6.88) by $D_A + D_B$:

$$c_B = -\frac{J_B}{4\pi(D_A + D_B)r} + [B] \tag{6.89}$$

The variation of c_B with r is shown in Fig. 6.11a.

When there is full microscopic diffusion control, a molecule of B *immediately* reacts when it reaches a certain critical distance d_{AB} from the A molecule. Thus, when $r = d_{AB}$, $c_B = 0$, and from Eq. (6.89),

$$0 = -\frac{J_B}{4\pi(D_A + D_B)d_{AB}} + [B] \tag{6.90}$$

This particular situation is shown in Fig. 6.11b. From Eq. (6.90) it follows that

$$J_B = 4\pi(D_A + D_B)d_{AB}[B] \tag{6.91}$$

This is the diffusive flux of B molecules up to a given A molecule, and since there is reaction at every encounter this is the rate of reaction of B molecules with a given A molecule. The rate of reaction with all the A molecules is thus

$$v = 4\pi(D_A + D_B)d_{AB}[A][B] \tag{6.92}$$

and the rate constant for the fully diffusion-controlled reaction is

$$k_D = 4\pi(D_A + D_B)d_{AB} \tag{6.93}$$

The SI unit of $D_A + D_B$ is $m^2\ s^{-1}$ and that of k_D is therefore $m^3\ s^{-1}$. Conversion to the more usual units of $dm^3\ mol^{-1}\ s^{-1}$ involves multiplication by $L \times 10^3\ dm^3\ m^{-3}$, that is, by $6.022 \times 10^{26}\ dm^3\ m^{-3}\ mol^{-1}$.

Equation (6.93) is the basic equation for full microscopic diffusional control, but some approximate relationships based on Stokes's law are also useful. According to this law, which applies approximately to solute molecules that are much larger than the solvent molecules, and less satisfactorily under other circumstances, a diffusion coefficient D is related to the radius r of the molecule by

$$D = \frac{kT}{6\pi\eta r} \tag{6.94}$$

where η is the viscosity of the solvent. The sum of the diffusion coefficients is therefore given approximately by

$$D_A + D_B = \frac{kT}{6\pi\eta}\left(\frac{1}{r_A} + \frac{1}{r_B}\right) \tag{6.95}$$

With d_{AB} taken as equal to $r_A + r_B$, introduction of Eq. (6.95) into Eq. (6.93) leads to

$$k_D = \frac{2kT}{3\eta}\frac{(r_A + r_B)^2}{r_A r_B} \tag{6.96}$$

If the molecular radii are equal, Eq. (6.96) reduces to the simple and useful form

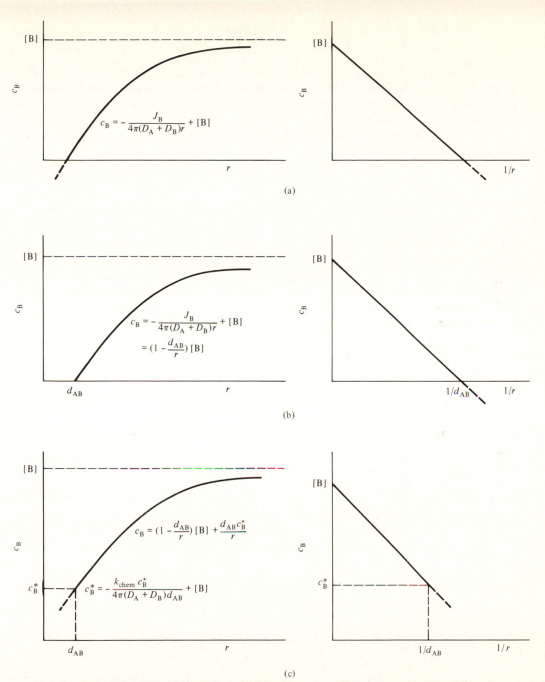

Figure 6.11 Relationships between the concentration c_B of B molecules at a distance r from a given A molecule: (a) general relationship [Eq. (6.89)]; (b) relationship for full microscopic diffusion control [Eq. (6.91)]; and (c) relationship for partial microscopic diffusion control [Eq. (6.98)].

$$k_D = \frac{8\mathbf{k}T}{3\eta} \tag{6.97}$$

For water at 25°C, this expression leads to a predicted rate constant of 7.0×10^9 dm^3 mol^{-1} s^{-1}. This value should therefore be an approximate upper limit for reactions between neutral molecules; the situation is different for reactions between ions, as will be considered in Section 6.7.3.

Equations (6.96) and (6.97) predict that, at a given temperature, k_D varies inversely with the viscosity η of the solvent, and this is roughly true for a number of reactions. The size of the molecules A and B does not have a strong effect on k_D, because $(r_A + r_B)^2/r_A r_B$, equal to

$$2 + \frac{r_A}{r_B} + \frac{r_B}{r_A}$$

does not vary greatly with r_A/r_B. Thus, if one radius is twice the other, this function is 4.5, as compared with 4.0 for equal radii. The reason for this insensitivity of rate to molecular size is that larger molecules move more slowly than smaller ones but present a larger target for encounter with other molecules: the two effects roughly compensate for one another.

6.7.2 Partial Microscopic Diffusion Control

Full diffusion control is found only if the rate of the chemical interaction is very much greater than the rate of diffusion. If this is not the case the concentration of B molecules at a distance d_{AB} is not equal to zero (as was shown in Fig. 6.11b), because the chemical process is no longer able to remove the B molecules completely as they reach the critical distance d_{AB}. The variation of c_B with r is now as shown in Fig. 6.11c, where c_B^* represents the concentration of B molecules at a distance d_{AB} from a given A molecule. If k_{chem} is the rate constant for the chemical interaction between A and B molecules, the rate of removal of B molecules by the A molecule is $k_{chem}c_B^*$, and this is equal to the diffusive flux J_B. Equation (6.89) then becomes

$$c_B^* = -\frac{k_{chem}c_B^*}{4\pi(D_A + D_B)d_{AB}} + [B] \tag{6.98}$$

and therefore

$$c_B^* = \frac{[B]}{1 + k_{chem}/4\pi(D_A + D_B)d_{AB}} \tag{6.99}$$

This applies to a given A molecule. For all the A molecules, the concentration of B molecules at a distance d_{AB} (i.e., the concentration c_{AB} of A—B pairs) is

$$c_{AB} = \frac{[A][B]}{1 + k_{chem}/4\pi(D_A + D_B)d_{AB}} \tag{6.100}$$

The rate is therefore

$$v = k_{chem}c_{AB} = \frac{k_{chem}[A][B]}{1 + k_{chem}/4\pi(D_A + D_B)d_{AB}} \tag{6.101}$$

and the rate constant is

$$k = \frac{k_{\text{chem}}}{1 + k_{\text{chem}}/4\pi(D_A + D_B)d_{AB}} \tag{6.102}$$

This general equation reduces to Eq. (6.93) when $k_{\text{chem}} \gg 4\pi(D_A + D_B)d_{AB}$. When, on the other hand, $4\pi(D_A + D_B)d_{AB} \gg k_{\text{chem}}$,

$$k = k_{\text{chem}} \tag{6.103}$$

and there is no diffusion control.

The way that k varies with k_{chem}, with k_D taken as 7.0×10^9 dm^3 mol^{-1} s^{-1}, is shown in Fig. 6.12. When k_{chem} has this value, there is 50% diffusion control and k is then 3.5×10^9 dm^3 mol^{-1} s^{-1}. For the diffusion control to be essentially complete, k_{chem} must be greater than 10^{11} dm^3 mol^{-1} s^{-1}. If it is less than 10^9 dm^3 mol^{-1} s^{-1}, k is very close to k_{chem}.

6.7.3 Ionic Reactions

The treatments just given require modification for reactions involving two ions A and B, in order to take into account the electrostatic interactions. An approximate treatment

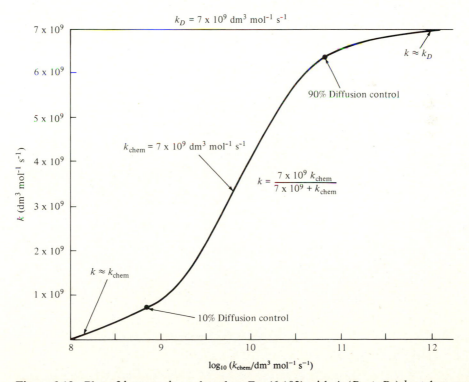

Figure 6.12 Plot of k versus k_{chem}, based on Eq. (6.102) with $4\pi(D_B + D_B)d_{AB}$ taken as 7.0×10^9 dm^3 mol^{-1} s^{-1}.

of such processes was given as long ago as 1903 by Langevin.[57] A more general treatment was given in 1942 by Debye,[58] whose treatment is outlined here.

As in the preceding treatment (Fig. 6.10), A is considered to be initially at rest, and B is diffusing towards it, eventually passing through the surface of a sphere of radius r. Three forces are acting on the moving B ion as follows:

1. The force due to the electrostatic interaction between the ions A and B. If the electric potential is U_{es}, the force is given by

$$F_{es} = -\frac{dU_{es}}{dr} \tag{6.104}$$

2. The diffusive force, arising from the concentration gradient. If c_B is the concentration at a given distance r, this diffusive force is[59]

$$F_d = -\frac{\mathbf{k}T}{c_B}\frac{dc_B}{dr} \tag{6.105}$$

3. The force F_r of frictional resistance on the moving ion. This force is proportional to the speed of the ion, dr/dt, the proportionality factor being the frictional coefficient f_B:

$$F_r = -f_B\frac{dr}{dt} \tag{6.106}$$

The negative signs in Eqs. (6.104)–(6.106) arise from the fact that the motion is in the direction of decreasing r.

After a steady state is established, the net force $F_{es} + F_d + F_r$ on B is zero, and therefore

$$-\frac{dU_{es}}{dr} - \frac{\mathbf{k}T}{c_B}\frac{dc_B}{dr} - f_B\frac{dr}{dt} = 0 \tag{6.107}$$

The velocity of B is

$$\frac{dr}{dt} = \frac{\mathbf{k}T}{f_B}\left(\frac{1}{c_B}\frac{dc_B}{dt} + \frac{1}{\mathbf{k}T}\frac{dU_{es}}{dr}\right) \tag{6.108}$$

$$= -D_B\left(\frac{1}{c_B}\frac{dc_B}{dt} + \frac{1}{\mathbf{k}T}\frac{dU_{es}}{dr}\right) \tag{6.109}$$

since $\mathbf{k}T/f_B = D_B$ by Einstein's equation.[60]

The diffusive flux J_B through the surface of the sphere of radius r (Fig. 6.10), that is, the number of ions passing into the sphere in unit time, is dr/dt multiplied by $4\pi r^2 c_B$, with the sign changed since the flux relates to decreasing r:

$$J_B = -\frac{dr}{dt}4\pi r^2 c_B \tag{6.110}$$

$$= 4\pi r^2 D_B\left(\frac{dc_B}{dr} + \frac{c_B}{\mathbf{k}T}\frac{dU_{es}}{dr}\right) \tag{6.111}$$

The first term corresponds to that in Eq. (6.85); the second term arises from the electrostatic interactions.

This equation leads to an explicit expression for J_B, by use of the following procedure. Equation (6.111) is mathematically equivalent to

$$J_B = 4\pi r^2 D_B e^{-U_{es}/kT} \frac{d}{dr}(c_B e^{U_{es}/kT}) \qquad (6.112)$$

which rearranges to

$$\frac{J_B e^{U_{es}/kT}}{r^2} = 4\pi D_B \frac{d}{dr}(c_B e^{U_{es}/kT}) \qquad (6.113)$$

Integration of both sides between the limits d_{AB} (the distance of closest approach) and infinity gives

$$J_B \int_{d_{AB}}^{\infty} \frac{e^{U_{es}/kT}}{r^2}\, dr = 4\pi D_B \int_{d_{AB}}^{\infty} \frac{d}{dr}(c_B e^{U_{es}/kT})dr \qquad (6.114)$$

$$= 4\pi D_B c_B e^{U_{es}/kT} \qquad (6.115)$$

The concentration c_B obeys the Boltzmann distribution equation§

$$c_B = [B]e^{-U_{es}/kT} \qquad (6.116)$$

the energy U_{es} being zero in the bulk solution where $c_B = [B]$. Equation (6.115) thus becomes

$$J_B \int_{d_{AB}}^{\infty} \frac{e^{U_{es}/kT}}{r^2}\, dr = 4\pi D_B[B] \qquad (6.117)$$

or

$$J_B = \frac{4\pi D_B[B]}{\displaystyle\int_{d_{AB}}^{\infty} \frac{e^{U_{es}/kT}}{r^2}\, dr} \qquad (6.118)$$

This expression relates to a single A ion. For all the A ions we multiply by [A] and replace D_B by $D_A + D_B$:

$$_A J_B = \frac{4\pi(D_A + D_B)[A][B]}{\displaystyle\int_{d_{AB}}^{\infty} \frac{e^{U_{es}/kT}}{r^2}\, dr} \qquad (6.119)$$

The rate constant (in molecular units: SI unit = $m^3\ s^{-1}$) is thus

$$k_D = \frac{4\pi(D_A + D_B)}{\displaystyle\int_{d_{AB}}^{\infty} \frac{e^{U_{es}/kT}}{r^2}\, dr} \qquad (6.120)$$

In the absence of electrostatic forces, the denominator of this expression becomes $1/d_{AB}$, and Eq. (6.93) is obtained. Otherwise, the integral must be evaluated by using the expression for U_{es}, which is

---

§ An approximation is involved here. In the absence of electrostatic effects, c_B varies with distance from A as a result of the diffusive flux (Fig. 6.11). It is assumed that the effects of diffusion are unimportant compared to the electrostatic effect.

**TABLE 6.7   VALUES OF THE ELECTROSTATIC
INTERACTION FUNCTION
[EQ. (6.124)] WITH d_{AB} = 200 pm**

$z_A z_B$	$\dfrac{z_A z_B e^2/4\pi\epsilon_0\epsilon d_{AB}kT}{e^{z_A z_B e^2/4\pi\epsilon_0\epsilon d_{AB}kT} - 1}$
0	1
−1	3.65
−2	7.08
1	0.106
2	0.006

$$U_{es} = \frac{z_A z_B e^2}{4\pi\epsilon_0\epsilon r} \tag{6.121}$$

The integration of the denominator of Eq. (6.120) proceeds as follows:§

$$\int_{d_{AB}}^{\infty} \frac{e^{z_A z_B e^2/4\pi\epsilon_0\epsilon rkT}}{r^2}\, dr = -\left[\frac{e^{z_A z_B e^2/4\pi\epsilon_0\epsilon rkT}}{z_A z_B e^2/4\pi\epsilon_0 kT}\right]_{d_{AB}}^{\infty} \tag{6.122}$$

$$= \frac{4\pi\epsilon_0\epsilon kT(e^{z_A z_B e/4\pi\epsilon_0\epsilon d_{AB}kT} - 1)}{z_A z_B e^2} \tag{6.123}$$

The expression for the rate constant is therefore

$$k_D = 4\pi(D_A + D_B)d_{AB}\left(\frac{z_A z_B e^2/4\pi\epsilon_0\epsilon d_{AB}kT}{e^{z_A z_B e^2/4\pi\epsilon_0\epsilon d_{AB}kT} - 1}\right) \tag{6.124}$$

The expression outside the brackets corresponds to Eq. (6.93) which applies when $z_A z_B = 0$. Values for the expression in the brackets, for various values of $z_A z_B$, are given in Table 6.7, for $d_{AB} = 2 \times 10^{-10}$ m.

The experimental results on ion combinations can be interpreted satisfactorily in terms of Eq. (6.124). Some data for the reactions between hydrogen ions and various anions are shown in Table 6.8. As was noted earlier, in the absence of electrostatic factors and with $d_{AB} = 200$ pm, a typical value of k_D is about 7×10^9 dm^3 mol^{-1} s^{-1}. However, the hydrogen ion has an abnormally high diffusion coefficient—9.3×10^{-9} m^2 s^{-1} as compared to 1–2×10^{-9} m^2 s^{-1} for many other substances in water. This leads to an estimate of ~1–2×10^{10} dm^3 mol^{-1} s^{-1} for the reactions in Table 6.1 if the electrostatic attraction were not involved. The electrostatic factor of ~3.7 therefore allows the values to be understood except that for H$^+$ + OH$^-$.

To explain the very large value of 1.4×10^{11} dm^3 mol^{-1} s^{-1} for the H$^+$ + OH$^-$ reaction, it is necessary to assume[61] that d_{AB} has the abnormally high value of about 850 pm (see Problem 6.11). The mechanism suggested for the process is that the H$^+$ and OH$^-$ ions diffuse together to a distance of about 850 pm, after which neutralization occurs by proton jumps from one neighboring water molecule to another, as represented in Fig. 6.13. This process of proton transfer corresponds to the mechanism that has been proposed to explain the very high conductivity of the H$^+$ ion in solution. Since

§ The integration is performed easily by setting $x = 1/r$; $dx = -(1/r^2)dr$.

TABLE 6.8 RATE CONSTANTS FOR THE
COMBINATIONS OF HYDROGEN
IONS WITH VARIOUS ANIONS
IN WATER AT 25°C[a]

Reacting ions	$k/10^{10}$ dm^3 mol^{-1} s^{-1}
$H^+ + OH^-$	14
$H^+ + CH_3COO^-$	5.1
$H^+ + C_6H_5COO^-$	3.7
$H^+ + o\text{-}NH_2C_6H_4COO^-$	5.8

[a] These results were obtained by M. Eigen and E. M. Eyring,
J. Am. Chem. Soc., **84**, 3254 (1962).

this mechanism bears some resemblance to that suggested in 1805 by Grotthuss[62] as a general explanation of conductance, it is known as a *Grotthuss mechanism*.

A parallel treatment of the rates of dissociation of molecules into ions, with the rate controlled by the diffusion of the ions away from each other, has been given by Eigen.[63] If two ions A and B are reacting together the process can be represented as

$$A + B \underset{k_{-1}}{\overset{k_1}{\rightleftharpoons}} A \cdots B \xrightarrow{k_{chem}} \text{products}$$

where $A \cdots B$ is an encounter complex, formed at the diffusion-controlled rate. The extent of diffusion control in the overall process then depends on the relative magnitudes of k_{-1}, which is related to the rate with which A and B diffuse away from one another, and k_{chem}, which relates to the rate of the chemical process.

The activation energies for diffusion-controlled reactions are always low compared to those of chemical processes; in water the activation energies are usually 12–15 kJ mol^{-1} at ordinary temperatures. According to the theory, the activation energy of a diffusion-controlled reaction is that of the diffusion processes, which means that it corresponds to the activation energy for the viscous flow of the solvent. For

(a)

(b)

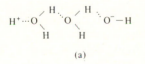

(c)

Figure 6.13 Mechanism of the $H^+ + OH^-$ reactions. (a) The H^+ and OH^- ions separated by a distance of 850 pm. (b) An intermediate state, with partial proton transfer. (c) The final state.

water the activation energy for viscous flow varies considerably with temperature, because of the changes in the hydrogen-bonded structure; the value is 21.1 kJ mol^{-1} at 0°C, 18.0 kJ mol^{-1} at 20°C, and 12.1 kJ mol^{-1} at 100°C. Corresponding values are found for reactions in aqueous solution that are diffusion controlled.

PROBLEMS

6.1. The rate constant k for the reaction between persulfate ions and iodide ions varies with ionic strength I as follows:

$I \times 10^3$/mol dm^{-3}	2.45	3.65	4.45	6.45	8.45	12.45
k/dm^3 mol^{-1} s^{-1}	1.05	1.12	1.16	1.18	1.26	1.39

Estimate the value of $z_A z_B$.

6.2. The following rate constants were obtained[64] for the reaction

$$[CoBr(NH_3)_5]^{2+} + OH^- \rightarrow [Co(NH_3)_5OH]^{2+} + Br^-$$

at 25°C and at the following reactant and salt concentrations:

Concentration/mol dm^{-3}			
$[CoBr(NH_3)_5]^{2+}$ 2Br$^-$	NaOH	NaCl	k/dm^3 mol^{-1} s^{-1}
5.0×10^{-4}	7.95×10^{-4}	0	1.52
5.96×10^{-4}	1.004×10^{-3}	0	1.45
6.00×10^{-4}	0.696×10^{-3}	0.005	1.23
6.00×10^{-4}	0.696×10^{-3}	0.020	0.97
6.00×10^{-4}	0.691×10^{-3}	0.030	0.91

Make an estimate of the rate constant of the reaction at zero ionic strength. Are the results consistent with $z_A z_B = -2$?

6.3. Suppose that the rates of ionic reactions in solution were proportional to the *activity* rather than the concentration of activated complexes. Derive an equation relating the logarithm of the rate constant to the ionic strength and the charge numbers of the ions, and contrast it with Eq. (6.45). Can the results in Fig. 6.6 be reconciled with the equation you have derived?

6.4. The rate of a reaction at 300 K is doubled when the pressure is increased from 1 to 2000 atm. Calculate $\Delta^{\ddagger}V$, assuming it to be independent of pressure.

6.5. The following results were obtained[65] for the solvolysis of benzyl chloride in an acetone–water solution at 25.0°C:

Pressure $\times 10^{-2}$/kPa	1.00	345	689	1033
$k \times 10^6$/s^{-1}	7.18	9.58	12.2	15.8

Make an appropriate plot and estimate $\Delta^{\ddagger}V°$.

6.6. The fading of bromphenol blue in alkaline solution is a second-order reaction between hydroxide ions and the quinoid form of the dye:

$$\text{Quinoid form (blue)}^{2-} + OH^- \rightarrow \text{carbinol form (colorless)}^{2-}$$

The following results[66] show the variation of the second-order rate constant k with the hydrostatic pressure P, at 25°C:

P/kPa	101.3	2.76×10^4	5.51×10^4	8.27×10^4	11.02×10^4
$k/10^{-4}$ dm^3 mol^{-1} s^{-1}	9.30	11.13	13.1	15.3	17.9

Estimate $\Delta^{\ddagger}V°$. Is its sign consistent with the reaction type?

6.7. The fading of bromphenol blue in alkaline solution (see Problem 6.6) has been studied at 25°C in solvents of different dielectric constants, with the following results:

ϵ	60	65	70	75	78.5
$k/10^{-4}$ dm^3 mol^{-1} s^{-1}	0.85	2.80	4.86	7.40	9.30

Check that these results are consistent with Eq. (6.22) and estimate d_{AB}.

6.8. The rate of the acid-catalyzed reaction between vanadium (V) and iodide ions follows the rate equation

$$v = k[V][I^-][H^+]^2$$

and k is found[67] to be independent of the ionic strength. Deduce from this information the charge on the predominant vanadium species that is involved in the reaction.

6.9. The following reactions occur in aqueous solution:
(1) $Co(NH_3)_5Br^{2+} + NO_2^- \rightarrow Co(NH_3)_5NO_2^{2+} + Br^-$
(2) $CH_2ClCOO^- + OH^- \rightarrow CH_2OHCOO^- + Cl^-$
(3) $CH_2BrCOOC_2H_5 + S_2O_3^{2-} \rightarrow CH_3(S_2O_3)COOC_2H_5 + Br^-$
(4) $CH_3Br + OH^- \rightarrow CH_3OH + Br^-$
(5) $Co(NH_3)_5Br^{2+} + OH^- \rightarrow Co(NH_3)_5OH^{2+} + Br^-$
(6) $Cr(H_2O)_6^{3+} + CNS^- \rightarrow Cr(H_2O)_5CNS^{2+} + H_2O$
(7) $CH_3COOCH_3 + OH^- + H_2O \rightarrow CH_3COOH + CH_3OH + OH^-$
In each case, deduce qualitatively the following:
(a) The effect on the rate of decreasing the dielectric constant.
(b) The effect on the rate of increasing the ionic strength.
(c) The effect on the rate of increasing the hydrostatic pressure.
(d) The sign of the entropy of activation.

6.10. The following reactions show essentially full diffusion control:
(a) The combination of iodine atoms in water.
(b) The combination of methyl radicals in toluene.
If the viscosities of water and toluene at 20°C are 1.002×10^{-3} kg m^{-1} s^{-1} and 5.90×10^{-4} kg m^{-1} s^{-1}, respectively, estimate the ratio of the rate constants of the two reactions at that temperature.

6.11. The diffusion coefficient D of an ion is related to the molar ionic conductivity λ by the Nernst–Einstein equation

$$D = \frac{\mathbf{k}T}{|z|eF}\lambda$$

where $\mathbf{k}$ is the Boltzmann constant, F is the Faraday constant, e is the elementary charge, and $|z|$ is the absolute value of the charge number. The molar ionic conductivities of the H^+ and OH^- ions are, at 25°C,

$$\lambda_{H^+} = 349.8 \ \Omega^{-1} \ cm^2 \ mol^{-1}$$

$$\lambda_{OH^-} = 198.6 \ \Omega^{-1} \ cm^2 \ mol^{-1}$$

Use these values, with a value of 850 pm for d_{AB}, to estimate the rate constant for the reaction

$$H^+ + OH^- \rightarrow H_2O$$

at 25°C. Compare the result with the experimental value of $1.4 \times 10^{11} \ dm^3 \ mol^{-1} \ s^{-1}$.

6.12. The following rate constants have been obtained[68] for the exchange reaction

$$Ir(NH_3)_5 \ {}^{18}OH_2^{3+} + H_2O \rightleftarrows Ir(NH_3)_5OH_2^{3+} + H_2 \ {}^{18}O \quad \text{at } 60.5°C$$

$P/10^5$ Pa	1	59	1027	2020	3310
$k/10^{-5} \ s^{-1}$	2.82	4.01	4.56	4.95	5.80

Estimate the volume of activation $\Delta^{\ddagger}V°$.

6.13. Confirm that the following ionic-strength results[69] at 25.0°C indicate that muonium is electrically neutral:

	$k/10^8 \ dm^3 \ mol^{-1} \ s^{-1}$	
Reaction	$I < 0.02 \ mol \ dm^{-3}$	$I = 0.9 \ mol \ dm^{-3}$
Mu + Cu^{2+}	65	63.5
Mu + CNS^-	0.62	0.74

REFERENCES

1. B. S. Khambata and A. Wassermann, *Nature,* **137,** 496 (1936); **138,** 368 (1936); A. Wassermann, *J. Chem. Soc.,* 1028 (1936); G. A. Benford, B. S. Khambata, and A. Wassermann, *Nature,* **139,** 669 (1937); A. Wassermann, *Trans. Faraday Soc.,* **34,** 128 (1938).
2. N. Menschutkin, *Z. Phys. Chem.,* **6,** 41 (1890).
3. H. G. Grimm, H. Ruf, and H. Wolff, *Z. Phys. Chem. B* **13,** 301 (1931).
4. N. J. T. Pickles and C. N. Hinshelwood, *J. Chem. Soc.,* 1353 (1936); R. A. Fairclough and C. N. Hinshelwood, *J. Chem. Soc.,* **538,** 1573 (1937).
5. J. A. Christiansen, *Z. Phys. Chem.,* **113,** 35 (1924).
6. C. N. Hinshelwood and E. A. Moelwyn-Hughes, *J. Chem. Soc.,* 230 (1932).
7. E. Rabinowitch, *Trans. Faraday Soc.,* **33,** 1225 (1937).
8. P. Debye and H. Menke, *Phys. Z.,* **31,** 797 (1930); **33,** 593 (1932).
9. E. Rabinowitch and W. C. Wood, *Trans. Faraday Soc.,* **32,** 1381 (1936).
10. J. Franck and E. Rabinowitch, *Trans. Faraday Soc.,* **30,** 120 (1934).
11. For a review of early applications of CTST to reactions in solution see S. Glasstone, K. J. Laidler, and H. Eyring, *The Theory of Rate Processes,* chap. 8, McGraw-Hill, New York, 1941.
12. W. F. K. Wynne-Jones and H. Eyring, *J. Chem. Phys.,* **3,** 493 (1935).
13. M. G. Evans and M. Polanyi, *Trans. Faraday Soc.,* **32,** 1333 (1936).
14. R. P. Bell, *Trans. Faraday Soc.,* **33,** 496 (1937); J. A. V. Butler, *Trans. Faraday Soc.,* **33,** 171, 229 (1937); I. M. Barklay, *Trans. Faraday Soc.,* **34,** 1445 (1938).

15. R. P. Bell, *Trans. Faraday Soc.,* **33,** 496 (1937).

16. G. Scatchard, *Chem. Rev.,* **8,** 321 (1931); J. H. Hildebrand and S. E. Wood, *J. Chem. Phys.,* **1,** 817 (1933); J. H. Hildebrand and R. L. Scott, *Regular Solutions,* Prentice-Hall, Englewood Cliffs, NJ, 1962.

17. M. Richardson and F. G. Soper, *J. Chem. Soc.,* 1873 (1929).

18. S. Glasstone, *J. Chem. Soc.,* 723 (1936).

19. E. D. Hughes and C. K. Ingold, *J. Chem. Soc.,* 244 (1935).

20. G. Scatchard, *Chem. Rev.,* **10,** 229 (1932).

21. E. A. Moelwyn-Hughes, *Proc. R. Soc. London A,* **155,** 308 (1936); *Kinetics of Reactions in Solution,* chap. 4, Clarendon, Oxford, 1947.

22. G. Scatchard, *Chem. Rev.* **10,** 229 (1932).

23. K. J. Laidler and H. Eyring, *Ann. N.Y. Acad. Sci.,* **39,** 303 (1940).

24. M. Born, *Z. Phys.,* **1,** 45 (1920).

25. J. N. Brønsted, *Z. Phys. Chem.,* **102,** 169 (1922). According to an earlier treatment, suggested by A. Lapworth, *J. Chem. Soc.,* **93,** 2187 (1908), the rate should be multiplied simply by $y_A y_B$; however, this *activity rate theory* is inconsistent with the results (see Problem 6.3).

26. P. J. W. Debye and E. Hückel, *Phys. Z.,* **24,** 185, 305 (1923).

27. J. N. Brønsted and V. K. La Mer, *J. Am. Chem. Soc.,* **46,** 555 (1924).

28. N. Bjerrum, *Z. Phys. Chem.,* **108,** 82 (1924).

29. J. N. Brønsted, *Z. Phys. Chem.,* **115,** 337 (1925); V. K. La Mer, *Chem. Rev.,* **10,** 179 (1932).

30. This point has been discussed by R. P. Bell, *Acid–Base Catalysis,* pp. 28–31, Clarendon, Oxford, 1944.

31. J. A. Christiansen, *Z. Phys. Chem.,* **113,** 35 (1924).

32. Y. C. Jean, J. H. Brewer, D. G. Fleming, D. M. Garner, R. J. Mikula, L. C. Vaz, and D. C. Walker, *Chem. Phys. Lett.,* **57,** 293 (1978).

33. Y. C. Jean, J. H. Brewer, D. G. Fleming, D. M. Garner, and D. C. Walker, *Hyperfine Interactions,* **6,** 409 (1979); Y. C. Jean, D. G. Fleming, B. W. Ng, and D. C. Walker, *Chem. Phys. Lett.,* **66,** 187 (1979).

34. C. W. Davies, in G. Porter (Ed.), *Progress in Reaction Kinetics,* vol. 1, p. 161, Pergamon, Oxford, 1961.

35. K. J. Laidler, *Can. J. Chem.,* **37,** 138 (1959); K. J. Laidler and C. Pegis, *Proc. R. Soc. London A,* **241,** 80 (1957); J. S. Muirhead-Gould and K. J. Laidler, *Trans. Faraday Soc.,* **63,** 953 (1967).

36. K. J. Laidler, *Can. J. Chem.,* **37,** 138 (1959); E. Sacher and K. J. Laidler, *Trans. Faraday Soc.,* **59,** 396 (1963).

37. H. S. Frank and M. W. Evans, *J. Chem. Phys.,* **13,** 507 (1945); H. S. Frank and W. V. Wen, *Discuss. Faraday Soc.,* **24,** 133 (1957).

38. J. S. Muirhead-Gould and K. J. Laidler, in B. E. Conway and R. G. Barradas (Eds.), *Chemical Physics of Ionic Solutions,* p. 75, Wiley, New York, 1967; J. S. Muirhead-Gould and K. J. Laidler, *Trans. Faraday Soc.,* **63,** 944 (1967).

39. S. Levine and G. M. Bell, in B. Pesce (Ed.), *Electrolytes,* p. 77, Pergamon, Oxford, 1962. See also S. Levine and H. E. Wrigley, *Discuss. Faraday Soc.,* **24,** 43 (1957).

40. K. J. Laidler and H. Eyring, *Ann. N.Y. Acad. Sci.,* **39,** 303 (1940); K. J. Laidler and P. A. Landskroener, *Trans. Faraday Soc.,* **52,** 200 (1956); K. J. Laidler, *Suomen Kemistilehti,* **A33,** 44 (1960); K. Hiromi, *Bull. Chem. Soc. Jpn.,* **33,** 1251, 1264 (1960).

41. J. G. Kirkwood, *J. Chem. Phys.,* **2,** 351 (1934). Spheroidal dipolar molecules have been treated by J. S. Muirhead-Gould and K. J. Laidler, *Trans. Faraday Soc.,* **63,** 958 (1966).

42. E. Hückel, *Phys. Z.,* **26,** 93 (1925).

43. P. Debye and J. McAulay, *Phys. Z.,* **26,** 22 (1925).

44. J. N. Brønsted and W. F. K. Wynne-Jones, *Trans. Faraday Soc.,* **25,** 59 (1929).

45. J. N. Brønsted and C. Grove, *J. Am. Chem. Soc.,* **52,** 1394 (1930).

46. R. H. Stokes and R. A. Robinson, *J. Am. Chem. Soc.,* **10,** 1870 (1948); R. A. Robinson and R. H. Stokes, *Electrolyte Solutions,* p. 246, Butterworth, London, 1959.

47. For a review see F. A. Long and W. F. McDewit, *Chem. Rev.,* **51,** 119 (1952).

48. J. H. van't Hoff, *Vorlesungen uber theoretische und physikalische Chemie,* Braunschweig, 1898; English translation by R. A. Lehrfeld, *Lectures on Theoretical and Physical Chemistry,* Edward Arnold, London, 1899. Equation (6.63) was first given by Max Planck, *Wied. Ann.,* **32,** 495 (1893).

49. M. G. Evans and M. Polanyi, *Trans. Faraday Soc.,* **34,** 144 (1938).

50. J. Buchanan and S. D. Hamann, *Trans. Faraday Soc.,* **49,** 1425 (1953); C. T. Burris and K. J. Laidler, *Trans. Faraday Soc.,* **51,** 1497 (1955).

51. E. Whalley, *Trans. Faraday Soc.,* **55,** 798 (1959).

52. For reviews see the bibliography at the end of this chapter.

53. L. P. Hammett, *Physical Organic Chemistry,* pp. 184–199, McGraw-Hill, New York, 1940.

54. R. W. Taft, *J. Am. Chem. Soc.,* **74,** 2729, 3120 (1952); **75,** 4231 (1953).

55. This is further discussed in K. J. Laidler, *Trans. Faraday Soc.,* **55,** 1725 (1959).

56. S. Chandrasekhar, *Rev. Mod. Phys.,* **15,** 1 (1943).

57. P. Langevin, *Ann. Chim. Phys.,* **28,** 28 (1903); *Comptes Rendus,* **146,** 1011 (1908).

58. P. Debye, *Trans. Electrochem. Soc.,* **82,** 265 (1942). See also J. Q. Umberger and V. K. La Mer, *J. Am. Chem. Soc.,* **67,** 1099 (1945).

59. This relationship is derived in physical chemistry texts; see, for example, K. J. Laidler and J. H. Meiser, *Physical Chemistry,* p. 830, Benjamin-Cummings, Menlo Park, CA, 1982.

60. A. Einstein, *Ann. Phys.,* **17,** 549 (1905); **19,** 371 (1906).

61. M. Eigen, *Z. Elektrochem.,* **64,** 115 (1960); M. Eigen and L. DeMaeyer, *Proc. R. Soc. London A,* **247,** 505 (1958).

62. C. J. D. Grotthuss, *Memoire sur la décomposition de l'eau et des corps, quelle tient en dissolution, á l'aide de l'electricité galvanique,* Rome, 1805; *Ann. Chim.,* **58,** 54 (1806); *Philos. Mag.,* **25,** 330 (1806).

63. M. Eigen, *Z. Phys. Chem.,* **1,** 176 (1954).

64. J. N. Brønsted and R. S. Livingston, *J. Am. Chem. Soc.,* **49,** 435 (1927).

65. K. J. Laidler and R. Martin, *Int. J. Chem. Kinet.,* **1,** 113 (1969).

66. D. T. Y. Chen and K. J. Laidler, *Can. J. Chem.,* **37,** 599 (1959).

67. D. R. Rosseinsky, *J. Inorg. Nucl. Chem.,* **33,** 3976 (1971).

68. S. B. Tong and T. W. Swaddle, *Inorg. Chem.,* **13,** 1538 (1974).

69. Y. C. Jean, J. H. Brewer, D. G. Fleming, D. M. Garner, and D. C. Walker, *Hyperfine Interactions,* **6,** 409 (1979).

BIBLIOGRAPHY

The following older books contain discussions of solution kinetics that are still useful:

E. S. Amis, *Solvent Effects on Reaction Rates and Mechanisms,* Academic, New York, 1966.

S. Glasstone, K. J. Laidler, and H. Eyring, *The Theory of Rate Processes,* chap. 8, McGraw-Hill, New York, 1941.

E. A. Moelwyn-Hughes, *The Kinetics of Reactions in Solution,* Clarendon, Oxford, 1947.

E. A. Moelwyn-Hughes, *Chemical Statics and Kinetics in Solutions,* Academic, London, 1971.

K. B. Wiberg, *Physical Organic Chemistry,* Wiley, New York, 1964.

For more recent accounts of solvent effects on reaction rates see:

M. R. J. Dack, The Influence of Solvent on Chemical Reactivity: An Alternative Approach, *J. Chem. Educ.,* **51,** 231 (1974).

J. T. Hynes, Chemical Reaction Dynamics in Solution, *Ann. Rev. Phys. Chem.,* **36,** 573 (1985).

C. G. Swain, M. S. Swain, A. L. Powell, and S. Alunni, Solvent Effects in Chemical Reactivity. Evaluation of Anion and Cation Solvation Components, *J. Am. Chem. Soc.,* **105,** 502 (1983).

D. G. Truhlar, W. L. Hase, and J. T. Hynes, The Current Status of Transition-State Theory, *J. Phys. Chem.,* **87,** 2664 (1983).

Pressure effects on reaction rates are treated in:

S. D. Hamann, *Physico-Chemical Effects of Pressure,* Butterworth, London, 1957.

For accounts of substituent effects and of the more organic aspects of solution kinetics see the following publications:

N. B. Chapman and J. Shorter, *Correlation Analysis in Chemistry: Recent Advances,* Plenum, New York, 1978.

R. Gallo, Treatment of Steric Effects, *Prog. Phys. Org. Chem.,* **14,** 85 (1983).

H. Maskill, *The Physical Basis of Organic Chemistry,* Clarendon, Oxford. 1985.

J. Shorter, *Correlation Analysis of Organic Reactivity,* Research Studies Press, Chichester, 1982.

G. C. Swain, S. H. Unger, N. R. Rosenquist, and M. S. Swain, Substituent Effects on Chemical Reactivity. Improved Evaluation of Field and Resonance Components, *J. Am. Chem. Soc.,* **105,** 492 (1983).

R. D. Tomson, The Contribution of Theoretical Chemistry to an Understanding of Electronic Substituent Effects, *Acc. Chem. Res.,* **16,** 292 (1983).

For treatments of rapid reactions and of diffusional effects in solution kinetics see the following publications:

I. Amdur and G. G. Hammes, *Chemical Kinetics: Principles and Selected Topics,* chap. 6, McGraw-Hill, New York, 1966.

E. F. Caldin, *Fast Reactions in Solution,* Blackwell, Oxford, 1964.

D. F. Calef and J. M. Deutch, Diffusion-Controlled Reactions, *Ann. Rev. Phys. Chem.,* **34,** 493 (1983).

R. M. Clegg, Derivation of Diffusion-Controlled Chemical Rate Constants with the Help of Einstein's Original Derivation of the Diffusion Constant, *J. Chem. Educ.,* **63,** 571 (1986).

E. A. Moelwyn-Hughes, *Chemical Statics and Kinetics in Solution,* chap. 5, pp. 99–123, Academic, London, 1971.

A. M. North, Diffusion-Controlled Reactions, *Q. Rev.,* **20,** 421 (1966).

R. M. Noyes, Effects of Diffusion Rates on Chemical Kinetics, in G. Porter (Ed.), *Progress in Reaction Kinetics,* vol. 1, p. 129, Pergamon, Oxford, 1961.

For general treatments of the theory of liquids and related topics see:

A. Ben-Naim, *Hydrophobic Interactions,* Plenum, New York, 1980.

C. A. Croxton, *Liquid State Physics: A Statistical Mechanical Introduction,* Cambridge University Press, Cambridge, U.K., 1974.

D. Eisenberg and W. Kauzmann, *The Structure and Properties of Water,* Clarendon, Oxford, 1969.

J. S. Rowlinson, *Liquids and Liquid Mixtures,* Butterworth, London, 1959.

For treatments of ionic solutions see:

B. E. Conway, *Ionic Hydration in Chemistry and Biophysics,* Elsevier, Amsterdam, 1981.

B. E. Conway and R. G. Barradas (Eds.), *Chemical Physics of Ionic Solutions,* Wiley, New York, 1966.

R. R. Dogonadze, E. Kálmán, A. A. Kornyshev, and J. Ulstrup (Eds.), *The Chemical Physics of Solvation, Part A, Theory of Solvation,* Elsevier, Amsterdam, 1985.

M. D. Newton, Electron Transfer Reactions in Condensed Phases, *Ann. Rev. Phys. Chem.,* **35,** 437 (1984).

Chapter 7

Reactions on Surfaces

It was known early in the 19th century that solids introduced into mixtures of certain gases had the power of inducing chemical reaction. In 1834, for example, Faraday[1] described investigations on the reaction between hydrogen and oxygen brought about by platinum, and he had previously studied the adsorption of gases on solid surfaces. In 1836, Berzelius[2] proposed the term "catalyst"§ to describe a substance that can bring about chemical reaction without itself apparently undergoing any change; he explained the effect in terms of a "catalytic force," a concept that did not prove useful. Later, Ostwald[3] made a detailed study of the mechanisms of catalyzed reactions and proposed a classification of them. The present chapter is concerned with catalysis brought about by surfaces, that is, heterogeneous catalysis; other types of catalysis are considered in Chapter 10.

During more recent years, a great deal of work has been devoted to the study of the kinetics and mechanisms of reactions catalyzed by solid surfaces. This is a subject of very great technical importance, since many reactions occur at an inconveniently low rate in the absence of a catalyst but can be accelerated greatly by a solid surface.

Aside from studies in which a solid catalyst has been introduced deliberately into a reaction system, incidental catalysis is often brought about by the walls of the reaction vessel. This was first realized by van't Hoff,[4] who studied the decompositions of phosphine and arsine in glass vessels and noted that reaction occurs to some extent on the walls. He pointed out that by varying the ratio of the surface area to the volume of the vessel (the "S/V ratio") it is possible to separate the reaction occurring on the surface from the reaction occurring homogeneously in the gas phase. Later, this procedure was applied successfully by Chapman and Jones[5] in an investigation of the decomposition of ozone. The S/V ratio can be varied by changing the size of the vessel, but it is more effective to pack the vessel with beads or rods.

§ The word "catalysis" comes from the Greek *kata,* meaning wholly, and *lyein,* meaning to loosen.

7.1 ADSORPTION

The most important result that has emerged from investigations of surface catalysis is that there must be specific adsorption of reactant molecules. This was first suggested by Faraday,[6] who carried out studies of the adsorption of gases at surfaces and of some surface-catalyzed reactions. Originally, it was thought that the main effect of the catalyst was to cause the reactants to be present at much higher concentrations than in the gas phase. However, this cannot be generally true, since in some cases different surfaces cause a substance to react in different ways; for example, on an alumina catalyst ethanol decomposes mainly into ethylene and water, while on copper it decomposes mainly into acetaldehyde and hydrogen. These and many other results indicate that specific forces are involved at surfaces.

Two main types of adsorption may be distinguished. In the first type the forces are of a physical nature and the adsorption is relatively weak. The forces in this type of adsorption correspond to those considered by van der Waals[7] in connection with his equation of state for gases and are known as van der Waals forces or as dispersion forces. This type of adsorption is called *van der Waals adsorption, physical adsorption,* or *physisorption.* The heat evolved in van der Waals adsorption is usually small, less than 20 kJ mol^{-1}. Van der Waals adsorption plays a role in catalysis only for some special types of reaction, such as in some atom or radical combinations.

In the second type of adsorption, first considered in 1916 by Langmuir,[8] the adsorbed molecules are held to the surface by covalent forces of the same general type as those occurring between bound atoms in molecules. The heat evolved for this type of adsorption, known as *chemisorption,* is usually comparable to that evolved in chemical bonding, namely, 300–500 kJ mol^{-1}.

An important consequence of this concept of chemisorption is that after a surface has become covered with a single layer of adsorbed molecules, it is saturated: additional adsorption can occur only on the layer already present, and this is generally physisorption. Langmuir thus emphasized that chemisorption involves the formation of a single layer.

It was suggested later by H. S. Taylor[9] that chemisorption is usually associated with an appreciable activation energy and therefore may be a relatively slow process. For this reason chemisorption often is referred to as *activated adsorption.* By contrast, van der Waals adsorption requires no activation energy and therefore occurs more rapidly than chemisorption.

Solid surfaces are never completely smooth at the atomic level (Section 7.3), and it was pointed out by Taylor[10] and by Constable[11] that adsorbed molecules will be attached more strongly to some surface sites than to others. Also, chemical reaction occurs predominantly on certain surface sites, which Taylor referred to as *active centers.*

7.2 ADSORPTION ISOTHERMS

Work on chemisorption has been concerned with equilibria and with rates. The amount of gas adsorbed after equilibrium is established depends on various factors, including the nature of the surface and the substance adsorbed, the temperature, and the pressure.

If for a given system one keeps the temperature constant and studies the amount of adsorption as a function of pressure, the resulting relationship is known as an *adsorption isotherm*. A number of such isotherms have been suggested, some being empirical and others theoretical. Of the theoretical equations the simplest is that of Langmuir,[12] whose isotherm has the special significance of being the one that applies to the ideal case of chemisorption on a perfectly smooth surface with no interactions between adsorbed molecules. The Langmuir isotherm therefore has an importance in adsorption theory similar to that of the ideal-gas equation $PV = nRT$, and it is convenient to speak of adsorption that obeys Langmuir's isotherm as *ideal adsorption*. The equations for ideal adsorption play an important role in surface kinetics, and they are derived now from different points of view and for several different conditions.

7.2.1 Simple Langmuir Isotherm

The simplest situation is when the gas atoms or molecules occupy single sites on the surface and are not dissociated; the process can be represented as

$$A + \ -\overset{\displaystyle |}{S}- \ \rightleftharpoons \ -\overset{\displaystyle \overset{\textstyle A}{|}}{\underset{|}{S}}-$$

Langmuir's kinetic derivation of the isotherm is essentially as follows. Let θ be the fraction of surface that is covered; $1 - \theta$ is the fraction bare. The rate of adsorption is then $k_a[A](1 - \theta)$, where $[A]$ is the gas concentration and k_a is a constant. The rate of desorption is $k_{-a}\theta$. At equilibrium the rates are equal so that§

$$k_a[A](1 - \theta) = k_{-a}\theta \tag{7.1}$$

or

$$\frac{\theta}{1 - \theta} = \frac{k_a}{k_{-a}} [A] \tag{7.2}$$

$$= K[A] \tag{7.3}$$

where K, equal to k_a/k_{-a}, is an equilibrium constant for the adsorption process. This equation can be written as

$$\theta = \frac{K[A]}{1 + K[A]} \tag{7.4}$$

The variation of θ with $[A]$ is shown in Fig. 7.1a. At sufficiently low concentrations $K[A]$ is small in comparison with unity, and then θ is proportional to $[A]$. Equation (7.4) can be written as

$$1 - \theta = \frac{1}{1 + K[A]} \tag{7.5}$$

so that at sufficiently high concentrations the bare fraction is given by

$$1 - \theta \approx \frac{1}{K[A]} \tag{7.6}$$

§ All the equations can be expressed alternatively in terms of the gas pressure, with an appropriate change in the significance of the constants.

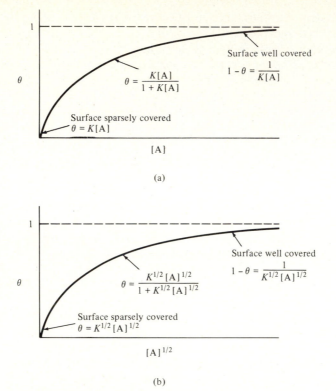

Figure 7.1   Plots of θ (fraction of surface covered) versus concentration [A] for a system obeying the Langmuir equations: (a) adsorption without dissociation [Eq. (7.4)] and (b) adsorption with dissociation [Eq. (7.11)].

A distinctive feature of this isotherm is that the surface becomes saturated ($\theta \approx 1$) with adsorbed molecules at high concentrations of A.

7.2.2 Adsorption with Dissociation

In certain cases there is evidence that the process of adsorption is accompanied by the dissociation of the molecule on the surface. It is found, for example, that hydrogen is adsorbed on the surfaces of many metals in the form of atoms, each of which occupies a surface site. Similarly, methane adsorbed on metals usually is dissociated into CH_3, CH_2, and hydrogen atoms.

The dissociation of a molecule into two species (e.g., H_2 into 2H) may be represented as

$$A_2 + \overset{\mid}{-}S\overset{\mid}{-}S- \rightleftharpoons \overset{A}{\underset{\mid}{}}\overset{A}{\underset{\mid}{}} \\ -S-S-$$

The process of adsorption is then a reaction between the gas molecule and two surface sites, and the rate of adsorption therefore may be written as

$$v_a = k_a[\text{A}](1 - \theta)^2 \tag{7.7}$$

The desorption process involves reaction between two adsorbed atoms, and the rate therefore is proportional to the square of the fraction of covered surface,

$$v_{-a} = k_{-a}\theta^2 \tag{7.8}$$

At equilibrium the rates are equal and therefore

$$\frac{\theta}{1 - \theta} = \left(\frac{k_a}{k_{-a}}[\text{A}]\right)^{1/2} \tag{7.9}$$

$$= K^{1/2}[\text{A}]^{1/2} \tag{7.10}$$

where K is equal to k_a/k_{-a}. This equation can be written as

$$\theta = \frac{K^{1/2}[\text{A}]^{1/2}}{1 + K^{1/2}[\text{A}]^{1/2}} \tag{7.11}$$

A plot of θ versus $[\text{A}]^{1/2}$ is shown in Fig. 7.1b. When the concentration is very small, $K^{1/2}[\text{A}]^{1/2}$ is much smaller than unity, and θ is then proportional to $[\text{A}]^{1/2}$. Equation (7.11) may be written as

$$1 - \theta = \frac{1}{1 + K^{1/2}[\text{A}]^{1/2}} \tag{7.12}$$

so that at high concentrations, when $K^{1/2}[\text{A}]^{1/2} \gg 1$,

$$1 - \theta \approx \frac{1}{K^{1/2}[\text{A}]^{1/2}} \tag{7.13}$$

The fraction of the surface that is bare at high concentrations is therefore inversely proportional to the square root of the concentration or pressure.

7.2.3 Competitive Adsorption

The isotherm for two substances adsorbed on the same surface is of importance in connection with inhibition and with the kinetics of surface reactions involving two reactants. Suppose that the fraction of surface covered by molecules of type A is θ_A and that the fraction covered by B is θ_B. The bare fraction is $1 - \theta_\text{A} - \theta_\text{B}$. If both substances are adsorbed without dissociation, the rates of adsorption of A and B are

$$v_a^\text{A} = k_a^\text{A}[\text{A}](1 - \theta_\text{A} - \theta_\text{B}) \tag{7.14}$$

and

$$v_a^\text{B} = k_a^\text{B}[\text{B}](1 - \theta_\text{A} - \theta_\text{B}) \tag{7.15}$$

The rates of desorption are

$$v_{-a}^\text{A} = k_{-a}^\text{A}\theta_\text{A} \tag{7.16}$$

$$v_{-a}^\text{B} = k_{-a}^\text{B}\theta_\text{B} \tag{7.17}$$

Equating Eqs. (7.14) and (7.16) leads to

$$\frac{\theta_\text{A}}{1 - \theta_\text{A} - \theta_\text{B}} = K_\text{A}[\text{A}] \tag{7.18}$$

where $K_A = k_a^A/k_{-a}^A$. From Eqs. (7.15) and (7.17) it follows that

$$\frac{\theta_B}{1 - \theta_A - \theta_B} = K_B[B] \tag{7.19}$$

where K_B is k_a^B/k_{-a}^B. Equations (7.18) and (7.19) are two simultaneous equations that can be solved to give, for the fractions covered by A and B, respectively,

$$\theta_A = \frac{K_A[A]}{1 + K_A[A] + K_B[B]} \tag{7.20}$$

$$\theta_B = \frac{K_B[B]}{1 + K_A[A] + K_B[B]} \tag{7.21}$$

Equation (7.20) reduces to Eq. (7.4) if [B] = 0 or if $K_B = 0$, which means that substance B is not adsorbed. It follows from Eqs. (7.20) and (7.21) that the fraction of the surface covered by one substance is reduced if the amount of the other substance is increased. This is because the molecules of A and B are competing with one another for a limited number of surface sites, and we speak of *competitive adsorption*. There is evidence that sometimes two substances are adsorbed on two different sets of surface sites, in which case there is no competition between them.

7.2.4 Nonideal Adsorption

Deviations from the Langmuir equations are often observed. This may be because the surface is not uniform, and also there may be interactions between adsorbed molecules. A molecule attached to a surface may make it more difficult, or less difficult, for another molecule to become attached to a neighboring site, and this will lead to a deviation from the ideal adsorption equation.

Another reason for deviations from the Langmuir isotherm is that there may be adsorption in more than one layer, that is, *multilayer adsorption*. When such deviations occur, the results can sometimes be fitted to an empirical adsorption isotherm due to Freundlich.[13] According to this equation, the amount of a substance adsorbed, *x,* is related to the concentration *c* by the equation

$$x = kc^n \tag{7.22}$$

where k and n are empirical constants. This equation does not give saturation of the surface: the amount adsorbed keeps increasing as c increases. If Eq. (7.22) applies, a plot of $\log_{10} x$ versus $\log_{10} c$ gives a straight line of slope n.

Another useful adsorption isotherm is that of Slygin and Frumkin:[14]

$$\theta = \frac{1}{f} \ln ac \tag{7.23}$$

where f and a are constants. Both the Freundlich and the Slygin–Frumkin isotherms have been obtained theoretically on the basis of distributions of surface sites of different energies.[15]

7.2.5 Thermodynamics and Statistical Mechanics of Adsorption

Many measurements have been made of heats evolved when gases are adsorbed at solid surfaces, and from the results the enthalpies and entropies of adsorption can be obtained. Heat is always liberated when a gas is adsorbed at a surface, which means that the enthalpy change ΔH is negative. The reason is that the adsorption of a gas inevitably involves an entropy decrease, because a molecule in the gas phase (or indeed in solution) has more freedom than one attached to a surface. For a process to occur to an appreciable extent, the Gibbs energy change ΔG must be negative. Thus, in view of the relationship

$$\Delta G = \Delta H - T \Delta S \tag{7.24}$$

ΔH must be negative if ΔS is negative.

By the methods of statistical mechanics, it is possible to derive isotherms of the Langmuir form and to express the constant K in terms of partition functions. This was first done by Fowler.[16] Consider the case of the adsorption of a gas without dissociation, to which Eq. (7.4) applies. Suppose that the volume of the gas is V and that the area of the surface is S. The total number of molecules in the gas phase may be written as N_g, the number of adsorbed molecules as N_a, and the number of bare sites at equilibrium as N_s. The concentrations of these species are as follows:

$$\text{Concentration in gas phase} \qquad c_g = \frac{N_g}{V} \tag{7.25}$$

$$\text{Concentration of adsorbed molecules} \quad c_a = \frac{N_a}{S} \tag{7.26}$$

$$\text{Concentration of bare sites} \qquad c_s = \frac{N_s}{S} \tag{7.27}$$

The equilibrium constant for the adsorption process is

$$K_c = \frac{c_a}{c_s c_g} = \frac{N_a/S}{(N_s/V)(N_g/S)} = \frac{N_a}{(N_s/V)N_g} \tag{7.28}$$

An equilibrium constant is equal to the ratio of the partition functions multiplied by an exponential term involving the difference between the zero-point levels of reactants and products (Section 4.5.2):

$$K_c = \frac{q_a}{q_g q_s} e^{-\Delta E_0/RT} \tag{7.29}$$

The partition functions q_a and q_s to be used in this expression are those for unit surface area; q_g is for unit volume. Then

$$\frac{c_a}{c_s} = c_g \frac{q_a}{q_g q_s} e^{-\Delta E_0/RT} \tag{7.30}$$

If θ is the fraction of the surface that is covered,

$$\frac{c_a}{c_s} = \frac{\theta}{1 - \theta} \tag{7.31}$$

and therefore

$$\frac{\theta}{1 - \theta} = c_g \frac{q_a}{q_g q_s} e^{-E_0/RT} \tag{7.32}$$

The partition function q_s per unit volume may be written as

$$q_g = \frac{(2\pi m\mathbf{k}T)^{3/2}}{h^3} b_g \tag{7.33}$$

where b_g represents the rotational and vibrational factors in the partition function. The adsorption sites have very little freedom of motion and their partition function q_s may be taken as unity. The partition function q_a for the adsorbed molecules only involves internal factors, which may be written as b_a. The adsorption isotherm thus becomes

$$\frac{\theta}{1 - \theta} = c_g \frac{h^3}{(2\pi m\mathbf{k}T)^{3/2}} \frac{b_a}{b_g} e^{-\Delta E_0/RT} \tag{7.34}$$

This equation has the same form as Eq. (7.3) but now the constant K is given in explicit form.

This equation is applicable to the situation where the adsorbed molecules are localized on the surface. This is usually the case with chemisorption, because of the strength of the binding of the adsorbed molecules to the surface. In some systems where there is weak binding and sparse surface coverage, there is evidence that the adsorbed molecules can move fairly freely on the surface; that is, they have two degrees of translational freedom. When this is the case, the equation that applies is

$$\theta = c_g \frac{h}{L_s(2\pi m\mathbf{k}T)^{1/2}} \frac{b_a}{b_g} e^{-\Delta E_0/\mathbf{R}T} \tag{7.35}$$

where L_s is the total number of molecules absorbed on unit area of surface when it is fully covered. This equation applies only to sparse coverage, since when there is a greater degree of coverage there can no longer be a free movement of the molecules on the surface.

For adsorption with dissociation [Eq. (7.10)], the equation is

$$\frac{\theta}{1 - \theta} = c_g^{1/2} \frac{h^{3/2}}{(2\pi m\mathbf{k}T)^{3/4}} \frac{b_a^{1/2}}{b_g^{1/2}} e^{-\Delta E_0/2RT} \tag{7.36}$$

where ΔE_0 is the change in energy when 1 mol of gas is adsorbed. For the adsorption of two gases on the same surface, Eqs. (7.20) and (7.21) apply, with K_A given by

$$K_A = \frac{h^3 b_a}{(2\pi m_A\mathbf{k}T)^{3/2} b_g} e^{-\Delta E_A/RT} \tag{7.37}$$

A corresponding expression applies to K_B.

7.3 STRUCTURES OF SOLID SURFACES AND ADSORBED LAYERS

The equations derived in the previous section have involved the assumption that surfaces are smooth, all surface atoms behaving in the same way. However, it has been known for many years that surfaces are usually quite irregular on the atomic scale, a fact that was first pointed out by Constable[17] and was further emphasized by H. S. Taylor[18] in his discussion of active centers.

Various lines of evidence suggest that surfaces show variability. For example, if a metal is heated for a period of time, its capacity for adsorption and catalysis often is reduced, even though the apparent surface area may remain the same. This effect is attributed to sintering of the surface, resulting in a decrease in the number of atoms that constitute the most active centers.

There is also much kinetic evidence for the variability of surfaces. For example, the decomposition of ammonia on molybdenum is retarded by nitrogen, but as the surface becomes saturated by nitrogen, the rate of the decomposition does not fall to zero.[19] This suggests that the reaction can occur on certain surface sites on which the nitrogen cannot be adsorbed.

Also, the heat evolved on chemisorption usually falls as the surface is covered progressively. This may be due to the fact that the gas initially becomes adsorbed on the most active sites: when these are covered the gas becomes adsorbed on less active sites, with less evolution of heat. However, this effect also is produced by repulsive interactions between adsorbed molecules, a matter that is considered later (Section 7.3.2.).

Until about 1970 there was little direct evidence as to the detailed nature of surfaces. Long before that, however, it had been realized that adsorption and catalysis depend in an important way on the surface interatomic distances. In 1926, for example, Burk[20] suggested that the adsorption of a molecule may occur on more than one surface site, an effect referred to as "multiple adsorption." This idea was developed by Balandin[21] in his "multiplet hypothesis." According to this hypothesis, a molecule of ethanol, for example, may become adsorbed as

$$
\begin{array}{ccc}
H_2C & -*- & CH_2 \\
| & & | \\
H & * & OH
\end{array}
$$

the asterisks denoting attachment to the surface. This particular type of attachment makes it easy for a water molecule to be split off. Balandin's picture of what occurs on a surface is undoubtedly too simple, but it proved valuable in leading to useful predictions.

These ideas naturally require that chemisorption and catalytic activity depend on the interatomic spacings on a surface and therefore that different crystal faces may behave quite differently. This conclusion was confirmed by some early quantum-mechanical calculations by Sherman and Eyring[22] on the activation energies for the adsorption of hydrogen on a carbon surface. It was found that the activation energy depended to a considerable extent on the $C-C$ distance in the surface layer, a distance of 360 pm being most favorable. This distance is considerably larger than the normal

H—H distance in molecular hydrogen. Similar results were obtained from calculations of the activation energy for the adsorption of hydrogen on nickel.[23] It was concluded that the hydrogen would be adsorbed primarily on the (100) and (110) planes of nickel and much less on the (111) plane where the internuclear separations are less suitable. The conclusions from these early calculations, that interatomic spacing is important, received experimental support from work on catalysis on different crystal faces, particularly from the work of Beeck, Gwathmey, and their coworkers.[24,25]

7.3.1 Detailed Structural Studies

Since about 1970, there have been many important advances in this field, particularly as a result of the development of new techniques for studying the detailed nature of solid surfaces and of adsorbed molecules. Details cannot be given here, and the reader is referred to the books and review articles listed in the bibliography at the end of this chapter. One technique that has revealed particularly valuable information about solid surfaces is field-ion microscopy (FIM),[26] in which a stream of helium gas enters an evacuated chamber and impinges on the surface of a crystal to which a high electric field is applied. This produces He^+ ions which are repelled from the surface, and analysis of the image produced on a screen by these ions permits an identification of the positions of the atoms on the surface.

Another widely used technique is low-energy electron diffraction (LEED),[27] which involves the backscattering of low-energy electrons from a surface. In order for there to be effective diffraction, the de Broglie wavelength of the radiation must be similar to the interatomic distances. It is also important that the electrons should not penetrate much below the surface, which requires the electron beam to be of low energy. The backscattered electrons are caused to impinge on a fluorescent screen, and analysis of the pattern provides detailed information about the atomic configuration of the surface.

The experimental studies of solid surfaces and adsorbed layers have revealed a number of very important results, which can be mentioned only briefly. Some of the main features of a solid surface are shown schematically in Fig. 7.2. There are seen

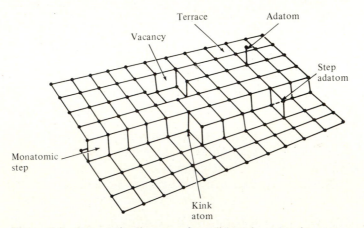

Figure 7.2 Schematic diagram of a solid surface showing some commonly observed features.

to be various kinds of surface site: there are atoms in terraces, atoms at steps, atoms at kinks, and adatoms, which project out of the surface. Atoms in terraces have several neighboring surface atoms, whereas adatoms have none. On most surfaces there are few adatoms and many more step, kink, and terrace atoms (10^{14}–10^{15} atoms cm^{-2}). The various types of surface atom differ very markedly in their ability to chemisorb and to catalyze chemical reactions.

Different crystal faces show a different distribution of the different types of surface atom. For example, Somerjai[28] has described platinum surfaces, as investigated by LEED, as follows (see Fig. 7.3):

1. The ($\bar{1}11$) face is relatively smooth. Most of the Pt atoms (i.e., about 10^{15} cm^{-2}) are on terraces or at steps, and there are fewer than 10^{12} adatoms per square centimeter.
2. The ($\bar{5}57$) surface is less smooth. There are steps every six atoms or so, and there are about 2.5×10^{14} step atoms per square centimeter.

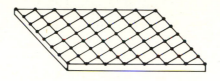

(a)

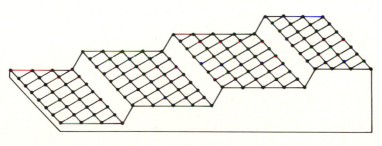

(b)

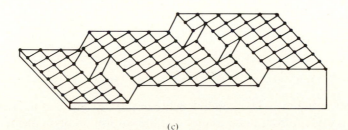

(c)

Figure 7.3 Schematic diagrams of platinum surfaces, as revealed by low-energy electron diffraction (LEED) experiments. (a) ($\bar{1}11$) face; (b) ($\bar{5}57$) face; (c) ($\bar{6}79$) face.

3. The ($\bar{6}$79) surface is less smooth again, there being more kinks. In each square centimeter there are about 2.3×10^{14} step atoms and about 7×10^{13} kink atoms.

Because of these different features, the three surfaces behave quite differently in chemisorption and in catalytic activity (see Section 7.4.1).

An experimental technique that recently has revealed valuable information about the detailed nature of surfaces and of adsorbed films is infrared spectroscopy. As an example of its use, mention may be made of studies of the nature of adsorbed ethylene molecules. The important result that has emerged from this work is that the manner in which the ethylene is adsorbed depends in an important way on the availability of surface hydrogen. When ethylene is adsorbed on bare palladium supported on silica, weak infrared bands are observed[29] and it appears that the ethylene has retained its double bonding. It is concluded that the adsorption is of the "dissociative" type, hydrogen atoms having split off:

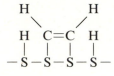

Admission of hydrogen, however, results in the disappearance of bands due to C=C, and to the appearance of bands characteristic of adsorbed ethyl groups.[30] One then speaks of "associative" chemisorption:

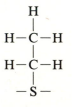

Evidence also has been obtained[31] for the presence of structures such as

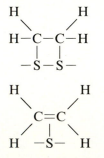

and

In the latter case there is π bonding between the ethylene and the surface.

It is evident from these studies that the adsorption of an olefin on a surface is a process of some complexity and that the nature of the adsorbed species varies widely with the type of surface and with the availability of adsorbed hydrogen atoms.

7.3.2 Induced Heterogeneity

Another complication with chemisorption and catalysis is that, in general, there are interactions between atoms or molecules adsorbed side by side on a surface. The consequences of this were first considered by Roberts,[32] who thought that he had obtained evidence for repulsive interactions between hydrogen atoms on a tungsten surface. He took precautions to obtain a smooth surface and measured the heats liberated when hydrogen was adsorbed. He found the heat liberated to vary from about 200 kJ mol^{-1} for a sparsely covered tungsten surface to about 60 kJ mol^{-1} for a surface that he believed to be covered fully. Later work,[33] however, indicated that the surfaces believed by Roberts to be covered fully were only 70% covered and that the heats liberated approached zero as the surface became covered completely.

The conclusion drawn from these results was that the decrease in heat liberated was due to repulsive interactions: when the surface is covered sparsely the molecules are sufficiently far apart as not to interact with each other, but the repulsive interactions become more important as the coverage increases. However, in view of modern work on the nature of surfaces at the atomic level, it is clear that the surfaces used in the earlier experiments were by no means smooth. The effects observed were therefore due at least in part to the fact that the hydrogen was adsorbed initially on more active sites, and later on less active sites.

Nevertheless, the effects of interactions between adsorbed molecules must be taken into account. Because the effects are similar to those resulting from the inherent heterogeneity of a surface, the expression *induced heterogeneity* is used to refer to repulsive interactions. Reliable evidence for repulsions has been provided by experimental studies of rates of adsorption.[34]

7.4 MECHANISMS OF SURFACE REACTIONS

A surface reaction may be regarded as occurring in five consecutive steps as follows:

1. Diffusion of the reactant molecules to the surface.
2. Adsorption of the gases on the surface.
3. Reaction on the surface.
4. Desorption of the products.
5. Diffusion of the desorbed products into the main body of the gas.

At one time it was believed that one of the diffusion processes, 1 or 5, was the slowest process and therefore determined the overall rate. More detailed investigations of surface reactions showed that this could not be the case, except perhaps in certain technical processes involving porous catalysts. This is evident from the fact that heterogeneous processes nearly always involve appreciable activation energies, whereas diffusion in the gaseous state involves no activation energy; the diffusion process therefore is much more rapid than the overall process and cannot constitute its slow step. Langmuir[35] presented the same type of argument in a somewhat different form when he showed that, in order for diffusion to be the slow step, it would be necessary to postulate such a thick diffusion layer on the surface that it would be visible. Since diffusion is slower in solution than in the gas phase, it sometimes may be the rate-determining step in solid–liquid reactions.

The processes of adsorption or desorption are much more likely to be the slow steps in heterogeneous reactions, since both may involve appreciable energies of activation. The activation energies for desorption are generally high, and it appears that in most reactions the desorption of the products is the rate-determining step. In practice, however, it is not always convenient to separate steps 3 and 4 because one usually does not know the rate of desorption of the products. It is usual, therefore, to regard the reaction on the surface, forming the gaseous products, as a single step. This concept is the basis of the treatments of surface reactions due to Langmuir[36] and Hinshelwood.[37] These treatments involve first obtaining an expression for the concentrations of reactant molecules on the surface, and then expressing the rate of formation of gaseous products in terms of these surface concentrations. The rate is then expressible in terms of the concentrations of the gaseous reactants. If there is a single reactant, the surface process is a simple unimolecular change; if there are two reactants, A and B, reaction may take place between two molecules adsorbed on neighboring surface sites, and the probability of this happening is proportional to the individual concentrations of adsorbed A and adsorbed B. The *Langmuir–Hinshelwood mechanism* for a reaction between A and B may be formulated as follows:

$$A + B + \begin{array}{c}|\quad| \\ -S-S- \end{array} \rightleftharpoons \begin{array}{c} A\quad B \\ |\quad | \\ -S-S- \end{array} \rightleftharpoons \begin{array}{c} A\cdot\cdot B^{\ddagger} \\ |\quad | \\ -S-S- \end{array} \rightleftharpoons \begin{array}{c}|\quad| \\ -S-S- \end{array} + Y + Z$$

Another type of mechanism for surface reactions was also considered by Langmuir. According to this mechanism, the reaction occurs between a gas molecule and an adsorbed molecule, so that only one of the reactants has to be adsorbed. This mechanism may be represented as

$$A + \begin{array}{c} B \\ | \\ -S- \end{array} \rightarrow \begin{array}{c} A^{\ddagger} \\ \vdots \\ B \\ \vdots \\ -S- \end{array} \rightarrow \begin{array}{c} | \\ -S- \end{array} + Y + Z$$

It is not necessary that A is not adsorbed at all; it is simply postulated, in this mechanism, that an adsorbed A molecule does not react. Interest in this mechanism was revived by Rideal,[38] and, as will be seen, it probably does apply to certain atom and radical combinations. It also may play a role in other reaction systems, but on the whole the *Langmuir–Rideal mechanism* does not appear to be as common as the Langmuir–Hinshelwood one, in which reaction occurs between two adsorbed molecules.

In connection with the mechanisms of surface reactions, an important concept is the *molecularity,* which is the number of molecules that come together during the course of reaction; by convention, the surface sites are not included. The molecularity of a surface reaction is deduced from the kinetics on the basis of the experimental results and of theoretical considerations. The relationships between molecularity and order will be considered in the following sections, in which reactions are classified according to their molecularity. Here one or two examples are mentioned briefly. Reactions involving a single reacting substance are usually, but not invariably, unimolecular. The surface-catalyzed ammonia decomposition (Section 7.5), for example, is usually unimolecular. On the other hand, the kinetics of the decomposition of acetaldehyde on various surfaces[39] can be interpreted only on the hypothesis that two acetaldehyde molecules, adsorbed on neighboring surface sites, undergo a bimolecular reaction. Reactions involving two reacting substances, such as the reaction between nitric oxide and oxygen on glass,[40] are usually bimolecular. When reactant molecules are dissociated on the surface, the reaction may involve interaction between an atom or radical and a molecule: for example, under some conditions the exchange reaction between ammonia and deuterium on iron[41] must be regarded as a bimolecular interaction between a deuterium atom and an ammonia molecule (Section 7.6.8).

7.4.1 Kinetic Effects of Surface Heterogeneity

In the treatment of the kinetics of surface reactions in Sections 7.5–7.7, use is made of the various forms of the Langmuir isotherm. These equations apply only to a smooth surface, but most kinetic studies have been made on surfaces that are by no means smooth. It is therefore necessary to consider the validity of interpreting the kinetics of surface reactions on the basis of the equations for ideal adsorption that were derived in Section 7.2.

When a reaction takes place on a surface on which there is a variation of activity, the overall rate is the sum of the rates on the various types of site. Suppose first for simplicity that there are only two kinds of site: a group of sites having a constant high activity and a second group of low activity. Suppose that the concentrations of the two kinds of sites are c_1 and c_2 and that the activation energies for reaction on the two types are E_1 and E_2. The rate of reaction is then

$$v = k(c_1 e^{-E_1/RT} + c_2 e^{-E_2/RT}) \qquad (7.38)$$

where k is a factor that in general includes reactant concentrations.

For the sites of type 1 to be more active, the activation energy E_1 will be less than E_2. If c_1 and c_2 are of comparable magnitude, and E_1 and E_2 are sufficiently different, the first term in Eq. (7.38) is greater than the second. The reaction therefore occurs predominantly on sites of type 1, and the measured activation energy is close to the lower value E_1. For the more general case in which there is a range of surface activity, Eq. (7.38) is replaced by

$$v = k \sum c_i e^{-E_i/RT} \qquad (7.39)$$

Again, unless there is a very unusual distribution of surface sites, the reaction occurs predominantly on the more active sites, and the observed activation energy is similar to the values for these more active sites.

This type of argument was first given by Constable,[42] who considered probable distributions of the activities of surface sites. His general conclusion, that it is likely that on a nonuniform surface the kinetics will correspond to reaction on the most active sites, may require modification under certain special circumstances. For example, it is quite possible for a reaction to occur mainly by a Langmuir–Hinshelwood mechanism on some of the surface sites and a Langmuir–Rideal mechanism on others. Also, some types of site may be covered fully, while others may be covered sparsely, under the experimental conditions. Such effects would introduce complexities into the kinetic behavior. Some special cases of this type of situation have been considered by Halsey,[43] who has shown how the kinetic equations are affected by such complexities. Because of the factors just discussed, there is not necessarily a simple correlation between the experimental adsorption isotherms and the kinetic behavior that is observed. The results of adsorption studies may indicate a wide range of surface activity, while the kinetic equations may be much simpler as a result of the fact that reaction is occurring predominantly on a fairly homogeneous group of surface sites of high activity.

7.4.2 Kinetic Effects of Interactions

A second matter to consider is to what extent the kinetic behavior is affected by interactions between adsorbed molecules, that is, by induced heterogeneity. A theoretical treatment of this problem[44] has led to the conclusion that the effect should not be important. Consider, as an example, a reaction involving the adsorption of hydrogen molecules on neighboring pairs of surface sites:

$$H_2 + \ -\underset{|}{S}-\underset{|}{S}- \ \rightarrow \ \overset{\overset{\displaystyle H \quad H}{\displaystyle | \quad |}}{-S-S-}$$

The effect of repulsions is to spread the adsorbed hydrogen atoms over the surface and thus to decrease the concentration of dual sites on which additional molecules may become adsorbed. This effect alone would reduce the rate of reaction. However, the partition function for the dual sites also is reduced by the repulsions, and the theory suggests that the two effects approximately compensate for one another.

On the whole, it appears that neither the inherent nor the induced heterogeneity of a surface has an important effect on the kinetic behavior, except in special circumstances.

7.5 UNIMOLECULAR SURFACE REACTIONS[45]

Surface reactions involving single adsorbed molecules, and therefore classified as unimolecular, are treated in terms of the Langmuir adsorption isotherm [Eq. (7.4)] as follows. The rate is proportional to the fraction θ of surface that is covered and is thus

$$v = k\theta = \frac{kK[A]}{1 + K[A]} \tag{7.40}$$

The dependence of rate on [A], shown in Fig. 7.4a, is exactly the same as that given by the Langmuir isotherm [Eq. (7.4) and Fig. 7.1a]. At sufficiently high concentrations

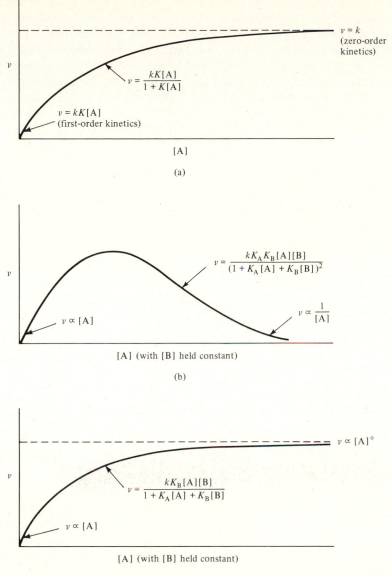

Figure 7.4 Variation of rate with concentration for various types of surface reaction: (a) simple unimolecular process: (b) bimolecular process occurring by a Langmuir–Hinshelwood mechanism; and (c) bimolecular process occurring by a Langmuir–Rideal mechanism.

of A the rate is independent of the concentration, which means that the kinetics are zero order. At low concentrations, when $K[A] \ll 1$, the kinetics are first order.

A good example of a unimolecular surface reaction is the decomposition of ammonia on tungsten. This was first investigated in 1925 by Hinshelwood and Burk,[46] who used a commercial tungsten-filament lamp as their reaction vessel. Under the conditions of their experiments the reaction was zero order, indicating that the surface was saturated with ammonia. Another example is the decomposition of phosphine on glass; this reaction was first investigated by van't Hoff and Kooij,[47] who found zero-order kinetics. Later, Barrer[48] studied this reaction over a range of pressures and found the expected change to first-order kinetics at low pressures. A considerable number of other unimolecular surface reactions, showing the behavior predicted by Eq. (7.40), are now known.

7.5.1 Inhibition

Sometimes a substance other than the reactant is adsorbed on the surface, with the result that the effective surface area and therefore the rate are reduced. Suppose that a substance A is undergoing a unimolecular reaction on a surface and that a nonreacting substance I, known as an *inhibitor* or a *poison,* is also adsorbed. According to Eq. (7.20), the fraction of the surface covered by A is

$$\theta = \frac{K_A[A]}{1 + K_A[A] + K_I[I]} \tag{7.41}$$

where K_A and K_I are the adsorption constants for A and I. The rate of reaction, equal to $k\theta$, is thus

$$v = \frac{kK_A[A]}{1 + K_A[A] + K_I[I]} \tag{7.42}$$

In the absence of inhibitor this equation reduces to Eq. (7.40).

A case of special interest is when the surface is covered only sparsely by the reactant but is covered fairly fully by the inhibitor. In other words,

$$K_I[I] \gg 1 + K_A[A] \tag{7.43}$$

and the rate is then

$$v = \frac{kK_A[A]}{K_I[I]} \tag{7.44}$$

This behavior arises from the fact that since the surface is covered sparsely by A, the coverage is proportional to [A]; since it is covered almost fully by I, the bare fraction is inversely proportional to [I] [Eq. (7.6)]. An example of a reaction obeying Eq. (7.44) is the decomposition of ammonia on platinum, inhibited by hydrogen.[49] The rate equation is

$$v = \frac{k[NH_3]}{[H_2]} \tag{7.45}$$

and since hydrogen is a product of reaction there is progressive inhibition as reaction proceeds.

7.5.2 Activation Energies

The rate constant k appearing in Eq. (7.40) is expected to obey the Arrhenius equation to a good approximation

$$\frac{d \ln k}{dT} = \frac{E_2}{RT^2} \tag{7.46}$$

where E_2 is the activation energy corresponding to the reaction of the absorbed species, that is, to the reaction

$$A-S \rightarrow S + \text{products}$$

The temperature dependence of the equilibrium constant K, if it is expressed as a concentration equilibrium constant, will follow the analogous van't Hoff relationship

$$\frac{d \ln K}{dT} = \frac{\Delta U^\circ}{RT^2} \tag{7.47}$$

where ΔU° is the standard molar internal energy change in the adsorption process.

If the reactant pressure is low, the rate expression [Eq. (7.40)] reduces to

$$v = kK[A] \tag{7.48}$$

and the first-order rate coefficient k^1 is given by

$$k^1 = kK \tag{7.49}$$

From Eq. (7.46) and (7.47) it follows that

$$\frac{d \ln v}{dT} = \frac{d \ln k^1}{dT} = \frac{d \ln k}{dT} + \frac{d \ln K}{dT} \tag{7.50}$$

$$= \frac{E_2 + \Delta U^\circ}{RT^2} \tag{7.51}$$

The apparent activation energy E_a is thus $E_2 + \Delta U^\circ$.

If the pressure is high the rate equation is simply

$$v = k[A] \tag{7.52}$$

The observed activation energy is then E_2.

These relationships are illustrated by the potential-energy diagram shown in Fig. 7.5. Reaction first involves the passage of the system over an initial energy barrier to give the adsorbed state, the energy of which is always lower than that of the initial state; that is, ΔU° is always negative. The system then passes over a second barrier of height E_2. If the pressure is low, most of the reactant molecules are not adsorbed, and to pass to the second activated state they have to acquire the energy $E_2 + \Delta U^\circ$, which is less than E_2. At high pressures, however, the equilibrium favors the adsorbed state, and the activation energy is E_2.

When a reaction is inhibited, the activation energy is modified by the energy of adsorption of the inhibitor. There are several possibilities, a simple one corresponding to Eq. (7.44), for which the reactant is adsorbed weakly and the inhibitor is adsorbed strongly. The inhibition constant K_I varies with the temperature according to

$$\frac{d \ln K_I}{dT} = \frac{\Delta U_I^\circ}{RT^2} \tag{7.53}$$

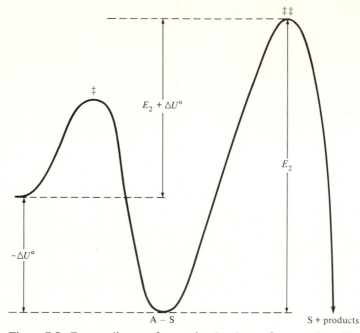

Figure 7.5 Energy diagram for a unimolecular surface reaction. The difference ΔU° between the energy of A—S and that of A + S is always negative (adsorption is exothermic).

where ΔU_I° (always negative) is the standard molar change in internal energy when the inhibitor is adsorbed. It follows that

$$\frac{d \ln v}{dT} = \frac{d \ln k}{dT} + \frac{d \ln K_A}{dT} - \frac{d \ln K_I}{dT} \qquad (7.54)$$

$$= \frac{E_2 + \Delta U_A^\circ - \Delta U_I^\circ}{RT^2} \qquad (7.55)$$

The apparent activation energy is thus

$$E_a = E_2 + \Delta U_A^\circ - \Delta U_I^\circ \qquad (7.56)$$

which, since ΔU_I° is negative, is greater than $E_2 + \Delta U_I^\circ$ for the uninhibited process. The reason for the increase is that a molecule of the inhibitor must be desorbed (an endothermic process) when a reactant molecule is adsorbed and undergoes reaction.

7.6 BIMOLECULAR SURFACE REACTIONS

As already noted, there are two distinctly different mechanisms for a surface reaction between two reactants A and B. In the Langmuir–Hinshelwood mechanism, reaction occurs between A and B molecules when both are adsorbed on the surface. Alternatively, in the Langmuir–Rideal mechanism, the reaction occurs between an adsorbed molecule and a molecule in the gas phase. These mechanisms may be distinguished on the basis of the different kinetic equations to which they give rise.

7.6.1 Reaction between Two Adsorbed Molecules

In a Langmuir–Hinshelwood mechanism the rate is proportional to the fractions of the molecules A and B that are adsorbed. These fractions are given by Eqs. (7.20) and (7.21), and the rate is therefore

$$v = k\theta_A\theta_B \tag{7.57}$$

$$= \frac{kK_AK_B[A][B]}{(1 + K_A[A] + K_B[B])^2} \tag{7.58}$$

If [B] is held constant and [A] is varied, the rate varies in accordance with Fig. 7.4b. The rate first increases, then passes through a maximum, and finally decreases. The explanation of the fall off in the rate at high concentrations is that one reactant displaces the other as its concentration is increased. The maximum rate corresponds to the existence of the maximum number of neighboring A—B pairs on the surface.

Two special cases are of interest:

1. *Sparsely Covered Surface.* If [A] and [B] are both sufficiently low that $K_A[A]$ and $K_B[B]$ may be neglected in comparison with unity, the rate equation becomes

$$v = kK_AK_B[A][B] \tag{7.59}$$

The reaction is therefore second order, being first order in both A and B. This type of behavior is observed frequently.

2. *One Reactant Very Weakly Adsorbed.* If reactant A is very weakly adsorbed, $K_A[A]$ in the denominator of Eq. (7.58) may be neglected, and the rate equation becomes

$$v = \frac{kK_AK_B[A][B]}{(1 + K_B[B])^2} \tag{7.60}$$

The rate still passes through a maximum as [B] increases, but as long as the condition $K_A[A] \ll 1 + K_B[B]$ remains satisfied the rate is proportional to [A]. Such behavior has been observed in the reaction between hydrogen and carbon dioxide on platinum[50] and in various other reactions.

If reactant B is adsorbed sufficiently strongly that $K_B[B] \gg 1$, Eq. (7.58) becomes

$$v = \frac{kK_A[A]}{K_B[B]} \tag{7.61}$$

The rate is now inversely proportional to [B], and the order with respect to B is −1. This behavior is shown under various circumstances by the reaction between carbon monoxide and oxygen on quartz[51] and on platinum.[52] In both cases the rate is inversely proportional to the pressure of carbon monoxide, which must be strongly adsorbed.

7.6.2 Reaction between a Gas Molecule and an Adsorbed Molecule

Suppose that the mechanism is of the Langmuir–Rideal type, and that the reaction occurs between an adsorbed B molecule and a molecule of A in the gas phase. The fraction of surface covered by B is given by Eq. (7.21), and the rate is now proportional to this fraction and to the concentration of A:

$$v = k\theta_B[A] \tag{7.62}$$

$$= \frac{kK_B[A][B]}{1 + K_A[A] + K_B[B]} \tag{7.63}$$

In this mechanism it is not assumed that A is not at all adsorbed, and the term $K_A[A]$ in the denominator corresponds to the adsorption of A. Adsorbed A molecules do not enter directly into reaction, but they reduce the rate by occupying surface that might otherwise be occupied by B molecules.

Equation (7.63) is to be contrasted with Eq. (7.58) for the Langmuir–Hinshelwood mechanism. There is now no maximum as [A] or [B] increases; instead, the rate varies with the concentration of either reactant in the manner shown in Fig. 7.4c. A decision between the two mechanisms may be made by seeing if the rate decreases at higher concentrations of reactants.

There are not many clear-cut examples of reactions occurring by a Langmuir–Rideal mechanism. The reaction between ethylene and hydrogen appears to occur under certain circumstances by the Langmuir–Hinshelwood and Langmuir–Rideal mechanisms occurring side by side.[53] The decomposition of acetaldehyde into methane and carbon monoxide on various surfaces[54] may occur by a Langmuir–Rideal mechanism. At low pressures the kinetics are second order, so that the reaction involves two molecules. The rate equation for the Langmuir–Hinshelwood mechanism is therefore

$$v = \frac{kK_A^2[A]^2}{(1 + 2K_A[A])^2} \tag{7.64}$$

while that for the Langmuir–Rideal mechanism is

$$v = \frac{kK_A[A]^2}{1 + 2K_A[A]} \tag{7.65}$$

At higher concentrations the kinetics become first order and not zero order, so that Eq. (7.65) appears to apply.

7.6.3 Adsorption of Two Gases without Mutual Displacement

A third possibility, which seems to occur in some cases, is that there is reaction between two molecules that are adsorbed on two different types of surface site, so that they do not displace one another from the surface. The Langmuir isotherm for the adsorption of A molecules on sites of type 1 may be written as

$$\theta_A = \frac{K_A[A]}{1 + K_A[A]} \tag{7.66}$$

and for the adsorption of B molecules on sites of type 2 as

$$\theta_B = \frac{K_B[B]}{1 + K_B[B]} \tag{7.67}$$

The rate is proportional to $\theta_A \theta_B$ and is thus

$$v = \frac{kK_A K_B[A][B]}{(1 + K_A[A])(1 + K_B[B])} \tag{7.68}$$

This type of behavior is experimentally distinguishable from that corresponding to Eqs. (7.58) and (7.63).

Equation (7.68) appears to be applicable to the reaction between hydrogen and nitrous oxide on gold,[55] and to the reaction between hydrogen and carbon dioxide on tungsten.[56]

7.6.4 Inhibition

A number of inhibition equations for bimolecular surface reactions are possible and are derived easily; only one simple case need be considered. Consider a reaction between A and B occurring by a Langmuir–Hinshelwood mechanism and inhibited by a gaseous substance of concentration [I]. If K_I is the adsorption constant for I, the fractions of surface covered by A and B are

$$\theta_A = \frac{K_A[A]}{1 + K_A[A] + K_B[B] + K_I[I]} \tag{7.69}$$

and

$$\theta_B = \frac{K_B[B]}{1 + K_A[A] + K_B[B] + K_I[I]} \tag{7.70}$$

The rate is therefore

$$v = k\theta_A \theta_B$$

$$= \frac{kK_A K_B[A][B]}{(1 + K_A[I] + K_B[B] + K_I[I])^2} \tag{7.71}$$

If there is sparse coverage by the reactants and large coverage by the inhibitor, this reduces to

$$v = \frac{kK_A K_B[A][B]}{K_I^2[I]^2} \tag{7.72}$$

7.6.5 Activation Energies

As with unimolecular surface reactions, the observed activation energy for a bimolecular reaction varies with the conditions. Suppose that the activation energy corresponding to the rate constant k is E_2 and that the internal energy changes for the adsorption of

A, B, and I are ΔU_A°, ΔU_B°, and ΔU_I°, respectively. Then, if Eq. (7.58) or (7.63) applies and A and B are at low concentrations,

$$E_a = E_2 + \Delta U_A^\circ + \Delta U_B^\circ \tag{7.73}$$

If Eq. (7.61) applies,

$$E_a = E_2 + \Delta U_A^\circ - \Delta U_B^\circ \tag{7.74}$$

If Eq. (7.72) applies

$$E_a = E_2 + \Delta U_A^\circ + \Delta U_B^\circ - 2\Delta U_I^\circ \tag{7.75}$$

Other cases are worked out easily.

7.6.6 Parahydrogen Conversion

A few particular types of reaction are considered now very briefly; for further details the reader is referred to books and reviews listed in the bibliography at the end of this chapter. It is important to appreciate that a given reaction may occur by entirely different mechanisms on different surfaces and under different conditions of pressure and temperature. Moreover, there are a number of reactions which, even under a given set of conditions, occur by two or more simultaneous mechanisms.

Parahydrogen becomes converted into the equilibrium mixture of the ortho and para forms when brought into contact with metal surfaces, particularly those of the transition metals. The order of reaction is usually close to zero,[57] and this can be explained by a mechanism originally proposed by Bonhoeffer and Farkas:[58]

$$p\text{-H}_2 + \;-\overset{|}{\text{S}}-\overset{|}{\text{S}}- \;\rightleftharpoons\; \overset{\text{H}\quad\text{H}}{\underset{-\text{S}-\text{S}-}{\overset{|}{\text{}}\;\overset{|}{\text{}}}} \;\rightleftharpoons\; -\overset{|}{\text{S}}-\overset{|}{\text{S}}- \;+\; o\text{-H}_2$$

The hydrogen becomes adsorbed on the surface and is then desorbed; since adsorption involves dissociation, the desorption process leads to the equilibrium mixture. At higher pressures the surface is covered fully and therefore the rate is independent of concentration.[59]

7.6.7 Combination and Formation of Atoms at Surfaces

Atoms and free radicals combine on most surfaces, and kinetic studies have been made of a number of systems. In almost all the cases investigated the reactions are first-order processes[60] which become second order at higher temperatures.[61] One mechanism that is consistent with this behavior is that reaction occurs between a gaseous atom or radical and an adsorbed one. Thus, for the recombination of hydrogen atoms, this mechanism is

$$\text{H} + \overset{\text{H}}{\underset{-\text{S}-}{\overset{|}{\text{}}}} \;\rightarrow\; -\overset{|}{\text{S}}- \;+\; \text{H}_2$$

If the surface is covered fully by atoms, the rate is proportional to the pressure of the atoms; that is, the kinetics are first order. At higher temperatures the surface is covered sparsely, and the fraction covered is proportional to the atomic concentration; the kinetics are then second order.

The kinetic equations for this mechanism may be formulated as follows. The fraction covered depends on the atomic concentration according to the isotherm

$$\theta_H = \frac{K_H[H]}{1 + K_H[H]} \tag{7.76}$$

The rate is proportional to θ_H and to [H], and thus

$$v = \frac{kK_H[H]^2}{1 + K_H[H]} \tag{7.77}$$

Since the adsorption is exothermic, at sufficiently low temperatures there is full coverage of the surface ($K_A[H] \gg 1$) and the kinetics are first order. At higher temperatures $K_A[H] \ll 1$ and then the kinetics are second order. This mechanism is of the Langmuir–Rideal type.

It was first shown by Langmuir[62] that hydrogen atoms are produced from hydrogen molecules at a hot tungsten surface, and Bryce[63] showed that the rate of reaction is approximately proportional to the square root of the hydrogen pressure. Roberts and Bryce[64] proposed a mechanism in which a hydrogen molecule strikes a bare site, whereupon one atom is adsorbed and the other passes into the gas phase:

$$H-H + -S- \longrightarrow \begin{array}{c} H \\ | \\ -S- \end{array} + H$$

This mechanism is the reverse of that considered above for the combination reaction. It accounts for the pressure dependence of the rate, provided the surface is covered fully. The fraction of bare surface is then inversely proportional to the square root of the pressure, so that the rate, being proportional to both the pressure and the bare fraction, is proportional to the square root of the pressure. A more detailed analysis of the results has provided further support for this mechanism.[65]

Another investigation[66] of the reaction, under different experimental conditions, has suggested an alternative mechanism. It is proposed that the hydrogen first becomes adsorbed in the atomic form,

$$H_2 + \begin{array}{c} | \quad | \\ -S-S- \end{array} \rightleftharpoons \begin{array}{c} H \quad H \\ | \quad | \\ -S-S- \end{array}$$

The hydrogen atoms form a mobile layer in which they have full translational motion. The next step is the desorption of hydrogen atoms:

$$\begin{array}{c} H \\ | \\ -S- \end{array} \rightarrow \begin{array}{c} | \\ -S- \end{array} + H$$

Different mechanisms apply to the process under different experimental conditions.

7.6.8 Exchange Reactions

The exchange reaction between hydrogen and deuterium,

$$H_2 + D_2 \rightleftharpoons 2HD$$

occurs very rapidly on many metal surfaces, even at temperatures well below room temperature. The original mechanism for the process, proposed by Bonhoeffer and Farkas,[67] is that hydrogen and deuterium are adsorbed dissociatively,

$$
(1) \qquad H_2 + D_2 + \;-\!S\!-\!S\!-\!S\!-\!S\!- \;\rightleftharpoons\; \begin{matrix} H & H & D & D \\ | & | & | & | \\ -S & -\,S & -S & -\,S\,- \end{matrix}
$$

and that neighboring H and D atoms recombine

$$
(2) \qquad \begin{matrix} H & D \\ | & | \\ -S- & S- \end{matrix} \;\rightarrow\; \begin{matrix} | & | \\ -S- & S- \end{matrix} \;+\; HD
$$

However, since the process can occur rapidly at temperatures at which desorption is slow, Rideal[68] proposed that reaction occurs between an adsorbed atom and a molecule in the gas phase:

$$
(3) \qquad \begin{matrix} H \\ | \quad | \\ -S-S- \end{matrix} + D_2 \rightarrow \begin{matrix} D^{\ddagger} \\ H \quad D \\ | \quad | \\ -S-S- \end{matrix} \rightarrow \begin{matrix} D \\ | \quad | \\ -S-S- \end{matrix} + HD
$$

Again, it is likely that different mechanisms are predominant under different conditions.

The H_2—D_2 exchange reaction has been studied by Bernasek and Somerjai[69] on different surfaces of a platinum single crystal, which was illustrated in Fig. 7.3. The $(\bar{1}11)$ surface is quite smooth, most of the surface atoms being on terraces, and no exchange reaction was observed. On the $(\bar{5}57)$ surface, where there are many steps, the exchange reaction occurred readily. In some of the experiments a beam of D_2 molecules was allowed to impinge on a $(\bar{5}57)$ surface on which hydrogen atoms had been chemisorbed; at higher temperatures the rate of exchange was then found to be proportional to the first power of the D_2 concentration and to the square root of the hydrogen concentration. This result suggests that reaction (3) is the rate-controlling step. At lower temperatures, on the other hand, the rate-controlling process was found to be the diffusion of the incident D_2 molecules on the surface toward a step at which the exchange reaction can take place.

On these platinum surfaces H_2 and D_2 are adsorbed with no activation energy. On other surfaces an activation energy is required;[70] for example, on a copper surface the activation energy is about 40 kJ mol^{-1}.

Some kinetic studies have been made[71] of the exchange reaction between ethylene and deuterium on a nickel surface, and it appears that the rate is proportional to the first power of the ethylene pressure and to the square root of the deuterium pressure. The mechanism[72] may involve the formation of mixed ethyl radicals by reaction between adsorbed deuterium atoms and ethylene molecules:

$$
(1) \qquad C_2H_4 + \begin{matrix} D \\ | \\ -S- \end{matrix} \;\rightleftharpoons\; \begin{matrix} CH_2D \\ \diagup \\ CH_2 \\ | \\ -\,S\,- \end{matrix}
$$

or

$$
(2) \qquad
\begin{array}{ccc}
& & \text{CH}_2\text{D} \\
& & \diagup \\
\text{CH}_2{-}\text{CH}_2 \;\; \text{D} & \rightleftharpoons & \text{CH}_2 \\
\;|\qquad\;| \qquad | & & \;|\quad\,|\;\,| \\
-\text{S}{-\!-}\text{S}{-\!-}\text{S}- & & -\text{S}{-}\text{S}{-}\text{S}-
\end{array}
$$

Surface exchange may occur by interaction between adsorbed alkyl radicals and adsorbed ethylene molecules:

$$
(3) \qquad
\begin{array}{ccccc}
\text{CH}_2\text{D} & & & & \text{CH}_3 \\
| & & & & | \\
\text{CH}_2 & \text{CH}_2{-}\text{CH}_2 & \rightleftharpoons & \text{CH}_2{-}\text{CH}_2\text{D} & \text{CH}_2 \\
| & |\qquad\; | & & |\qquad\quad | & | \\
-\text{S}{-\!-}\text{S}{-\!-}\text{S}- & & & -\text{S}{-\!-}\text{S}{-\!-}\text{S}-
\end{array}
$$

Species such as C_2HD_3 and C_2D_4 are obtained in the initial stages of the reaction,[73] and this is explained if reactions (2) and (3) occur rapidly. The evidence suggests that there is a complete mixing of the H and D atoms on the surface. Since the concentration of adsorbed D atoms is proportional to the square root of the pressure, the exchange rate is proportional to the square root of the pressure.

The exchange reactions between methane and deuterium on evaporated nickel surfaces have been studied by Kemball.[74] He found that all the products, CH_3D, CH_2D_2, and CD_4, were produced in the initial stages of reaction; this is in contrast to the reaction between D_2 and ammonia, where the products are formed by consecutive reactions. The rates of formation of all the methanes are proportional to the first power of the methane pressure. The rate of formation of CH_3D is inversely proportional to the square root of the deuterium pressure, whereas the rates of formation of CH_2D_2, CHD_3, and CD_4 are inversely proportional to the first power of the deuterium pressure.

These results suggest that two different exchange mechanisms are involved, one leading to the production of CH_3D, and the other to the production of CH_2D_2, CHD_3, and CD_4. Kemball suggested that CH_3D is formed from adsorbed CH_3 radicals, and the other products from adsorbed CH_2 radicals which rapidly exchange with adsorbed deuterium atoms. The surface is covered by deuterium atoms, and the number of bare sites is inversely proportional to the square root of the deuterium pressure [Eq. (7.13)]. The methyl radicals are adsorbed on single sites, and therefore their concentration on the surface is proportional to $[CH_4]/[D_2]^{1/2}$. The formation of CH_3D is explained as due to the reaction

$$
\begin{array}{ccc}
\text{CH}_3 \;\;\; \text{D} & & |\;\; | \\
| \qquad\quad | & \rightleftharpoons & -\text{S}{-}\text{S}- + \text{CH}_3\text{D} \\
-\text{S}{-\!-}\text{S}-
\end{array}
$$

Since practically every adsorbed methyl radical has a deuterium atom as a near neighbor, the rate of production of CH_3D is proportional to $[CH_4]/[D_2]^{1/2}$.

The different rate law for the other products indicates that they are not formed from CH_3, which therefore cannot exchange readily with D atoms. When methane is adsorbed it is believed that some CH_2 radicals are formed on the surface:

$$
\begin{array}{ccc}
& & \text{H} \quad \text{CH}_2 \;\text{H} \\
\text{CH}_4 + \;\;-\text{S}{-}\text{S}{-}\text{S}{-}\text{S}- & \rightleftharpoons & |\quad\;\diagup\diagdown\; | \\
& & -\text{S}{-}\text{S}{-}\text{S}{-}\text{S}-
\end{array}
$$

Since these are adsorbed on a pair of sites, their concentration is proportional to the square of the number of bare sites, that is, to $1/[D_2]$. The adsorbed CH_2 radicals are assumed to exchange readily with neighboring D atoms, so that CH_2, CHD, and CD_2 radicals are present on the surface in amounts that depend on statistical factors. Their concentrations will be proportional to $[CH_4]/[D_2]$. Therefore, the rates of production of CH_2D_2, CHD_3, and CD_4, by reactions such as

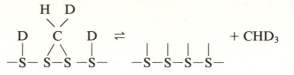

will be proportional to $[CH_4]/[CD_2]$.

The exchange reaction between ammonia and deuterium, with the formation of NH_2D, NHD_2, and ND_3, offers an interesting contrast in that, on a number of surfaces (including nickel and platinum), the products are formed in consecutive reactions.[75] The initial product is NH_2D, and NHD_2 is formed from it in a subsequent independent reaction; finally ND_3 is formed from NHD_2. These results are explained most simply on the assumption that ammonia upon adsorption is dissociated only into NH_2 and H and that the former can add on a deuterium atom:

$$\begin{array}{ccc} NH_2 & D \\ | & | \\ -S\!-\!\!-\!\!-\!S- \end{array} \rightleftharpoons \begin{array}{ccc} | & | \\ -S\!-\!S- \end{array} + NH_2D$$

The adsorbed radical NH_2 apparently cannot undergo exchange with an adsorbed deuterium atom.

A number of studies have been made of the dependence of the exchange rate on the concentrations of ammonia and deuterium. Farkas[76] used a pure iron surface and found that the rate was proportional to the square root of the deuterium pressure and independent of the ammonia pressure. Weber and Laidler[77] worked with an activated iron surface and also observed the square-root dependence on deuterium pressure. They found, however, that the rate passed through a maximum as the ammonia pressure was increased and could be fitted to the Langmuir–Hinshelwood equation,

$$v = \frac{k[D_2]^{1/2}[NH_3]}{(1 + K[NH_3])^2} \tag{7.78}$$

The conclusion is that the surface is covered rather heavily by NH_2 radicals and only sparsely by D atoms, and that reaction occurs between these two adsorbed species.

Somewhat different results were obtained by Singleton, Roberts, and Winter,[78] who used evaporated films of iron, tungsten, and nickel. Their results are consistent with the hypothesis that the Langmuir–Hinshelwood mechanism applies at lower pressures of ammonia. At higher pressures of ammonia the rate does not fall towards zero but only to a constant level, and the rate is then proportional to the first power of the deuterium pressure. It appears that under these conditions a Langmuir–Rideal mechanism predominates, reaction being between an adsorbed NH_2 radical and a gaseous deuterium molecule:

$$\begin{array}{c} NH_2 \\ | \\ -S- \end{array} + D_2 \rightleftharpoons \begin{array}{c} D \\ | \\ -S- \end{array} + NH_2D$$

Again, different kinetic behavior is found on different surfaces.

7.6.9 Addition of Hydrogen to Ethylene

Much work has been done on the hydrogenation of ethylene, on a variety of surfaces. Very different types of kinetic behavior have been observed under different conditions, and a number of different mechanisms can occur, sometimes simultaneously under a single set of conditions. Only a brief outline of the main features can be included here.[79]

An early investigation by Pease[80] of the reaction on a copper surface led to the result that the rate may be expressed approximately as

$$v = \frac{k[H_2][C_2H_4]}{(1 + K[C_2H_4])^2} \tag{7.79}$$

The rate passes through a maximum as the ethylene concentration increases, and the results are consistent with a Langmuir–Hinshelwood mechanism [Eq. (7.58)].

On a nickel surface, on the other hand, several workers[81] have obtained results consistent with the rate equation

$$v = \frac{k[H_2][C_2H_4]}{1 + K[C_2H_4]} \tag{7.80}$$

This type of behavior is consistent with a Langmuir–Rideal mechanism, and Beeck[82] and Jenkins and Rideal[83] suggested that reaction occurs between a pair of adsorbed hydrogen atoms and a gaseous ethylene molecule:

$$\begin{array}{cc} H & H \\ | & | \\ -S-S- \end{array} + C_2H_4 \rightarrow \begin{array}{cc} | & | \\ -S-S- \end{array} + C_2H_6$$

Toyama,[84] on the other hand, obtained definite evidence that the rate fell at high ethylene pressures, a fact that requires that the Langmuir–Hinshelwood mechanism plays a role. A similar result was obtained by an investigation, using an evaporated nickel surface, that was concerned primarily with distinguishing between the two mechanisms.[85] The kinetic behavior depended to a considerable extent on which gas was first admitted to the surface (or whether both were admitted together). It seems likely that two mechanisms are taking place simultaneously. One mechanism, favored by prior addition of ethylene or by the presence of excess of ethylene, is the Langmuir–Hinshelwood mechanism. The second, favored by prior addition of hydrogen or excess of hydrogen, is the Langmuir–Rideal mechanism between adsorbed hydrogen and gaseous ethylene.

On the whole, it appears that the following mechanism[86] applies. Ethylene may become adsorbed "associatively" on a pair of bare surface sites:

(1)
$$C_2H_4 + \underset{\displaystyle -S-S-}{\overset{\displaystyle |\ \ |}{}} \rightleftharpoons \overset{\displaystyle CH_2-CH_2}{\underset{\displaystyle -S\ -\ S-}{\overset{\displaystyle |\ \ \ \ \ |}{}}}$$

In addition, an ethylene molecule from the gas phase may add on to an adsorbed hydrogen atom to form an alkyl radical:

(2)
$$C_2H_4 + \overset{\displaystyle H}{\underset{\displaystyle -S-}{\overset{\displaystyle |}{}}} \rightleftharpoons \overset{\displaystyle CH_3}{\underset{\displaystyle -S-}{\overset{\displaystyle \diagup}{\underset{\displaystyle |}{CH_2}}}}$$

Adsorbed alkyl radicals also may be formed from adsorbed ethylene molecules and neighboring adsorbed hydrogen atoms. Ethane is produced in the gas phase either by interaction between two neighboring adsorbed alkyl radicals,

(3)
$$\overset{\displaystyle CH_3\ CH_3}{\underset{\displaystyle -S-S-}{\overset{\displaystyle |\ \ \ \ |}{\underset{\displaystyle |\ \ \ \ |}{CH_2\ CH_2}}}} \rightleftharpoons \overset{\displaystyle |\ \ |}{-S-S-} + CH_3CH_3 + C_2H_4$$

or by interaction between an adsorbed alkyl radical and an adsorbed hydrogen atom,

(4)
$$\overset{\displaystyle CH_3}{\underset{\displaystyle -S-\ \ \ \ \ S-}{\overset{\displaystyle |}{\underset{\displaystyle |\ \ \ \ \ \ \ |}{CH_2\ \ \ H}}}} \rightleftharpoons \overset{\displaystyle |\ \ |}{-S-S-} + CH_3CH_3$$

This mechanism is of the Langmuir–Hinshelwood type.

In an investigation of the reaction on evaporated iridium films,[87] the rate equation was found to be

$$v = k[C_2H_4][H_2]^{1/2} \tag{7.81}$$

The rate is proportional to the square root of the hydrogen concentration, rather than to the first power as in the earlier work. The same type of mechanism may apply but a different step may be rate limiting.

7.7 TRANSITION-STATE THEORY OF SURFACE REACTIONS

The treatment of the rates of surface reactions by conventional transition-state theory (CTST) involves a knowledge of the activation energies and of the partition functions for reacting species and activated complexes. The calculation of activation energies by quantum-mechanical methods involves the usual difficulties outlined in Chapter 3; it also requires a knowledge of the detailed structure of surfaces.

The calculation of rates of surface reaction using CTST, with the use of experimental energies of activation, was first done in some detail by Laidler, Glasstone, and Eyring,[88] and other applications have been made subsequently. The general procedures have been reviewed elsewhere,[89] and only a brief discussion need be given here.

The theory can be applied profitably only to investigations in which the surface area is known and in which the surface is reasonably smooth. In any case, because of the inherent complications of the systems, only order-of-magnitude values of rates are to be expected.

7.7.1 Rates of Chemisorption

The process of adsorption is a bimolecular reaction between surface sites and gas molecules. It is first supposed that the activated complexes are localized on the surface. Let N_s, N_g, and $N_{\ddagger}$ be the numbers of bare sites, gas molecules, and activated complexes, respectively, and let the corresponding concentrations be $c_s\ (=N_s/S)$, $c_g\ (=N_g/V)$, and $c_{\ddagger}\ (=N_{\ddagger}/S)$. The usual assumption of equilibrium between the activated complexes and the reactants then gives rise to the equation

$$\frac{c_{\ddagger}}{c_g c_s} = \frac{q^{\ddagger}}{q_g q_s}\, e^{-\varepsilon_1/\mathbf{k}T} \tag{7.82}$$

As in Eq. (7.29), the partition functions $q^{\ddagger}$ and q_s relate to unit surface area, and q_g relates to unit volume. The energy ε_1 is the energy of the complexes with reference to the reactants, at the absolute zero, and is therefore the activation energy per molecule at that temperature.

The partition function q_g can be factorized into translational, rotational, and vibrational factors. If the activated complex is localized on the surface, it has no translational or rotational motion, but it undergoes various types of vibrational motion. As previously, one of the vibrational factors in the partition function corresponds to a very loose vibration and may be expressed as $\mathbf{k}T/h\nu$, where ν is the frequency of the vibration. Equation (7.82) therefore may be expressed as

$$\frac{c_{\ddagger}}{c_g c_s} = \frac{q_{\ddagger}(\mathbf{k}T/h\nu)}{q_g q_s}\, e^{-\varepsilon_1/\mathbf{k}T} \tag{7.83}$$

where the partition function $q_{\ddagger}$ lacks the contribution for the one degree of freedom. Equation (7.83) rearranges to

$$\nu c_{\ddagger} = c_g c_s\, \frac{\mathbf{k}T}{h}\, \frac{q_{\ddagger}}{q_g q_s}\, e^{-\varepsilon_1/\mathbf{k}T} \tag{7.84}$$

The frequency ν is the frequency of vibration of the activated complexes in the degree of freedom corresponding to their transformation into adsorbed molecules. The expression on the left-hand side of Eq. (7.84) is the product of the concentration $c_{\ddagger}$ of complexes and the frequency of their transformation; it is the rate of the adsorption process, which is given by the expression on the right-hand side of the equation, namely,

$$v_1 = c_g c_s\, \frac{\mathbf{k}T}{h}\, \frac{q_{\ddagger}}{q_g q_s}\, e^{-\varepsilon_1/\mathbf{k}T} \tag{7.85}$$

The partition functions may be evaluated in the same manner as in the equilibrium equations. There are no translational and rotational factors in $q_{\ddagger}$, and the remainder, due to vibration, are represented by $b_\ddagger$. The function q_s is taken as unity since the surface atoms can undergo only very restricted vibration, while q_g can be written $(2\pi m\mathbf{k}T)^{3/2}b_g/h^3$, where b_g represents the vibrational and rotational factors. Substitution of these values gives

$$v_1 = c_g c_s \frac{\mathbf{k}T}{h} \frac{b_\ddagger}{[(2\pi m\mathbf{k}T)^{3/2}/h^3]b_g} e^{-\varepsilon_1/\mathbf{k}T} \tag{7.86}$$

If the gas is diatomic, factor b_g contains the rotational factor $8\pi^2 I\mathbf{k}T/h^2$ and in addition contains a vibrational factor. At ordinary temperatures, the latter is close to unity and therefore can be omitted. Similarly, $b_\ddagger$ is very close to unity; the rate expression therefore can be written as

$$v_1 = c_g c_s \frac{\mathbf{k}T}{h} \frac{1}{[(2\pi m\mathbf{k}T)^{3/2}/h^3](8\pi^2 I\mathbf{k}T/h^2)} e^{-\varepsilon_1/\mathbf{k}T} \tag{7.87}$$

$$= c_g c_s \frac{h^4}{8\pi^2 I(2\pi m\mathbf{k}T)^{3/2}} e^{-\varepsilon_1/\mathbf{k}T} \tag{7.88}$$

In some cases a statistical factor (Section 4.5.4) must be included.

In the above, it has been assumed that in the formation of the activated complex the gas molecule becomes attached to only one adsorption site. In many cases two adjacent sites (i.e., a dual site) are necessary for adsorption. If this is so, Eq. (7.88) still gives the rate of adsorption provided that c_s is replaced by c_{s_2}, the concentration of such dual sites. If the surface is bare, the number of dual sites is related to the number of single sites as follows. Each single site has a certain number s of adjacent sites; s is known as the coordination number of the surface and depends on the particular type of surface lattice. If the number of dual sites were evaluated by counting s for each single site, the result would be sc_s, but in this procedure each pair is counted twice; the actual number of dual sites is $sc_s/2$. Thus, the rate of adsorption on dual sites is given by replacing c_s in the above expression by $sc_s/2$:

$$v_1 = \tfrac{1}{2}s c_g c_s \frac{h^4}{8\pi^2 I(2\pi m\mathbf{k}T)^{3/2}} e^{-\varepsilon_1/\mathbf{k}T} \tag{7.89}$$

This expression, however, is valid only for adsorption on an initially bare surface; in other words, it applies only to initial rates of adsorption on dual sites. If the fraction of surface already covered is θ, the average number of bare sites adjacent to any given site is, assuming random distribution, $s(1 - \theta)$. The total number of bare dual sites is therefore $c_s s(1 - \theta)/2$. If the concentration of covered single sites is c_a, θ is given by

$$\theta = \frac{c_a}{c_a + c_s} \tag{7.90}$$

so that the concentration of bare dual sites is

$$c_{s_2} = \frac{c_s^2 s}{2(c_a + c_s)} \tag{7.91}$$

$$= \frac{s c_s^2}{2L_s} \tag{7.92}$$

where $L_s = c_a + c_s$ is the total concentration of single sites when the surface is completely bare. Thus, the rate of adsorption on dual sites is given by replacing c_s in Eq. (7.88) by c_{s2}, using Eq. (7.92), and the expression obtained is

$$v_1 = \frac{sc_g c_s^2}{2L_s} \frac{h^4}{8\pi^2 I(2\pi mkT)^{3/2}} e^{-\varepsilon_1/kT} \tag{7.93}$$

This reduces to Eq. (7.89) if $L_s = c_s$, which is true if the surface is completely bare.

The rate equations are quite different if the molecules are not localized in the adsorbed state, and hence presumably in the activated state. The equilibrium between initial and activated states may be represented by

$$K_c = \frac{c_\ddagger}{c_g} = \frac{N_\ddagger/S}{N_g/V} = \frac{q^\ddagger}{q_g} \tag{7.94}$$

where q_g and $q^\ddagger$ are the partition functions for unit volume of gas and per unit area of activated complex, respectively. By exactly the same methods as used previously, it can be shown that the rate of adsorption is

$$v_1 = c_g \frac{kT}{h} \frac{q_\ddagger}{q_g} e^{-\varepsilon_1/kT} \tag{7.95}$$

where $q_\ddagger$ differs from $q^\ddagger$ in lacking the $kT/h\nu$ factor corresponding to reaction. The activated complexes now differ from the reactants by having translational freedom in only two dimensions, and the ratio $q_\ddagger/q_g$ is simply $h/(2\pi mkT)^{1/2}$. The rate of adsorption becomes

$$v_1 = c_g \frac{kT}{h} \frac{h}{(2\pi mkT)^{1/2}} e^{-\varepsilon_1/kT} \tag{7.96}$$

Replacement of $c_g kT$ by the pressure p, for the special case of zero activation energy, gives

$$v_1 = \frac{p}{(2\pi mkT)^{1/2}} \tag{7.97}$$

This equation is the classical Hertz–Knudsen equation for the number of gas molecules striking a unit area of surface in unit time. Equation (7.97) often is used for calculating rates of adsorption when there is no activation energy, but it is applicable only when the adsorbed molecules are not localized on the surface, and this is by no means always the case. For a localized layer, Eq. (7.89) or (7.93) should be used; these expressions correspond to much lower rates of adsorption than does Eq. (7.97).

Equations for rates of adsorption can be formulated for adsorption with dissociation; they are not derived here.

7.7.2 Rates of Desorption

We now consider the process of desorption of undissociated molecules held in a localized layer. Desorption from such a layer involves an activated state in which the molecule has acquired the necessary configuration and activation energy to escape from the surface. If N_a and $N_\ddagger$ are the numbers of adsorbed molecules and activated complexes, and c_a and $c_\ddagger$ are the corresponding concentrations, in molecules per unit area, the equilibrium between the initial and activated states may be written as

$$K_c = \frac{c_\ddagger}{c_a} = \frac{N_\ddagger/S}{N_a/S} = \frac{q^\ddagger}{q_a} e^{-\varepsilon_{-1}/kT} \tag{7.98}$$

where ε_{-1} is the molecular activation energy for desorption at the absolute zero. The concentration of activated complexes is given by

$$c_\ddagger = c_a \frac{q^\ddagger}{q_a} e^{-\varepsilon_{-1}/kT} \tag{7.99}$$

Application of the methods employed earlier gives the rate expression

$$v_{-1} = c_a \frac{kT}{h} \frac{q_\ddagger}{q_a} e^{-\varepsilon_{-1}/kT} \tag{7.100}$$

where $q_\ddagger$ differs from $q^\ddagger$ in excluding the partition function $(kT/h\nu)$ for passage across the potential-energy barrier.

7.7.3 Unimolecular Surface Reactions

The rates of ordinary chemical reactions can be formulated in a very similar way using CTST. Unimolecular reactions are considered first: these involve reaction between a gas molecule and a surface site, to give an activated complex at the surface.

Consider a reaction involving one molecule of the gaseous reactant A, which is undergoing reaction on the surface. If the process occurs on a single site S, it may be written as

$$A + S \rightleftharpoons AS^\ddagger \rightarrow \text{products}$$

and the equilibrium between initial and activated states is expressed by

$$\frac{c_\ddagger}{c_g c_s} = \frac{q^\ddagger}{q_g q_s} e^{-\varepsilon_0/kT} \tag{7.101}$$

where ε_0 is the molecular energy of activation at the absolute zero. The rate of reaction therefore is given by

$$v = c_g c_s \frac{kT}{h} \frac{q_\ddagger}{q_g q_s} e^{-\varepsilon_0/kT} \tag{7.102}$$

The expression is formally identical with that for the rate of adsorption [Eq. (7.85)], but the activated states are different in the two cases. However, in both cases the value of the partition function $q_\ddagger$ may be taken as unity, since the activated state consists of a molecule immobilized on the surface.

Two limiting cases of Eq. (7.102) are of interest, according to whether c_s is large or small. If the surface is covered only sparsely by adsorbed molecules, the concentration of bare surface sites c_s is approximately equal to L_s, the number of sites per unit area of completely bare surface. Under these conditions, c_s is approximately independent of c_g, so that the rate of reaction is directly proportional to c_g; therefore, the process is kinetically of the first order.

If a reaction of this type occurs on a dual surface site, the concentration c_s in Eq. (7.102) must be replaced by c_{s_2}. For a bare surface c_{s_2} is equal to $sc_s/2$, that is, $sL_s/2$, so that the rate is

$$v = \tfrac{1}{2} sc_g L_s \frac{kT}{h} \frac{q_\ddagger}{q_g q_s} e^{-\varepsilon_0/kT} \tag{7.103}$$

If the reaction involves a diatomic molecule, the rate equation becomes

$$v = c_g L_s \frac{\tfrac{1}{2} sh^4}{8\pi^2 I (2\pi m kT)^{3/2}} e^{-\varepsilon_0/kT} \tag{7.104}$$

where I and m are the moment of inertia and the mass of the reacting molecule. For a nonlinear polyatomic molecule the rate is

$$v = c_g L_s \frac{\tfrac{1}{2} sh^5}{8\pi^2 (8\pi^3 I_A I_B I_C)^{1/2} (2\pi m)^{3/2} (kT)^2} e^{-\varepsilon_0/kT} \tag{7.105}$$

where I_A, I_B, and I_C are the three moments of inertia of the reactant. This equation applies satisfactorily[90] to the decomposition of phosphine on glass.[91] The decompositions of hydrogen iodide on platinum and of nitrous oxide on gold, however, appear to occur at rates that are much greater than predicted by Eq. (7.105), and Robertson[92] has suggested that the activated complexes are formed in a mobile layer.

When the surface is covered almost completely by adsorbed molecules, the concentration c_g varies with the pressure of the gas, and the kinetic equation may be obtained by combining the rate equation (7.102) with the isotherm:

$$\frac{c_a}{c_s} = c_g \frac{q_a}{q_g q_s} e^{\varepsilon/kT} \tag{7.106}$$

The equation obtained is

$$v = c_a \frac{kT}{h} \frac{q_\ddagger}{q_a} \exp\left(-\frac{\varepsilon_0 + \varepsilon}{kT}\right) \tag{7.107}$$

When the surface is covered almost completely by adsorbed molecules, c_a may be taken as constant, so that the rate is independent of the pressure of the reactant; the kinetics are therefore zero order.

Since both $q_\ddagger$ and q_s may be taken as unity, the rate law may be written as

$$v = c_a \frac{kT}{h} e^{-E_0/RT} \tag{7.108}$$

where E_0 is the molar activation energy at the absolute zero. An equation of this general form, with a pre-exponential factor of 10^{12} s^{-1} in place of kT/h, was first proposed by Topley.[93]

In Table 7.1 are given some of the calculated and observed rates for reactions that exhibit zero-order kinetics; in all cases, c_s was taken to be 10^{15} cm^{-2}. The agreement is seen to be satisfactory except for the decomposition of hydrogen iodide on gold; the discrepancy here is due presumably to the fact that the reaction occurs on a small fraction of the surface.

TABLE 7.1 OBSERVED AND CALCULATED RATES FOR ZERO-ORDER SURFACE REACTIONS

Decomposition of	Surface	E/kJ mol^{-1}	Temperature/K	Rate/cm^{-2} s^{-1}		Reference
				Calculated	Observed	
NH_3	W	159	904	8.0×10^{18}	4×10^{17}	a
NH_3	W	174	1316	3.4×10^{21}	2×10^{19}	b
NH_3	Mo	223	1228	8.5×10^{18}	$5-20 \times 10^{18}$	c
HI	Au	105	978	5.2×10^{22}	1.6×10^{17}	d
$HCOOCH(CH_3)_2$	Glass	146	714	7.5×10^{17}	5.8×10^{16}	e

[a] C. N. Hinshelwood and R. E. Burk, *J. Chem. Soc.*, **127**, 1051 (1925).
[b] C. H. Kunsman, E. S. Lamar, and W. E. Deming, *Philos. Mag.*, **10**, 1015 (1930).
[c] R. E. Burk, *Proc. Natl. Acad. Sci. USA*, **13**, 67 (1927).
[d] C. N. Hinshelwood and C. R. Prichard, *J. Chem. Soc.*, **127**, 1552 (1925).
[e] G. M. Schwab, *J. Phys. Chem.*, **50**, 427 (1946).

7.7.4 Bimolecular Surface Reactions

A bimolecular reaction occurring by a Langmuir–Hinshelwood mechanism may be formulated as

$$A + B + S_2 \rightleftharpoons \begin{matrix} A & B \\ | & | \\ -S & -S- \end{matrix} \rightarrow products$$

where A and B are the reacting molecules, and S_2 is a dual site. The rate of reaction is then given by

$$v = c_g c'_g c_{s2} \frac{kT}{h} \frac{q_\ddagger}{q_g q'_g q_{s2}} e^{-\epsilon_0/kT} \tag{7.109}$$

where c_g and c'_g are the gas-phase concentrations of A and B, and q_g and q'_g are the corresponding partition functions per unit volume. The concentration of bare dual sites is related to the concentration of bare single sites by Eq. (7.92), so that the rate may be formulated as

$$v = \tfrac{1}{2}s \frac{c_g c'_g c_s^2}{L_s} \frac{kT}{h} \frac{q_\ddagger}{q_g q'_g q_s} e^{-\epsilon_0/kT} \tag{7.110}$$

This equation may be put into a more general form, using the isotherms

$$\frac{c_a}{c_g c_s} = K \tag{7.111}$$

and

$$\frac{c'_a}{c'_g c_s} = K' \tag{7.112}$$

where c_a and c'_a are the concentrations of adsorbed A and adsorbed B. Since, in addition,

$$c_a + c'_a + c_s = L_s \tag{7.113}$$

it is found that

$$c_s = \frac{L_s}{1 + Kc_g + K'c'_g} \tag{7.114}$$

Insertion of this expression into Eq. (7.110) gives

$$v = \frac{\frac{1}{2}sc_g c'_g L_s}{(1 + Kc_g + K'c'_g)^2} \frac{\mathbf{k}T}{h} \frac{q_{\ddagger}}{q_g q'_g q_s} e^{-\varepsilon_0/\mathbf{k}T} \tag{7.115}$$

as the general equation for a bimolecular reaction. The equation may be applied to the data in its various limiting forms as follows.

When the surface is covered sparsely, Kc_g and $K'c'_g$ can be neglected in comparison with unity; the rate equation is then

$$v = \frac{1}{2}sc_g c'_g L_s \frac{\mathbf{k}T}{h} \frac{q_{\ddagger}}{q_g q'_g q_s} e^{-\varepsilon_0/\mathbf{k}T} \tag{7.116}$$

so that the kinetics are second order.

This equation has been applied to the reaction between nitric oxide and oxygen on a glass surface. This reaction has been found experimentally to be of the second order; the rate at 85 K is given by[94]

$$v = (9.4 \times 10^{-27} \text{ cm}^4 \text{ s}^{-1}) \times c_{NO}c_{O_2}e^{-E_0/RT} \tag{7.117}$$

The rate was calculated with $q_{\ddagger}/q_s$ taken as unity, L_s as 10^{15} cm^{-2}, and s equal to 4. The resulting proportionality factor was 14.8×10^{-27}, in satisfactory agreement with the experimental value of 9.4×10^{-27}.

If B is adsorbed only weakly but A is not necessarily adsorbed weakly, $K'c'_g$ may be neglected but not Kc_g. Thus, the rate equation is

$$v = \frac{1}{2}s \frac{L_s c_g c'_g}{(1 + Kc_g)^2} \frac{\mathbf{k}T}{h} \frac{q_{\ddagger}}{q_g q'_g q_s} e^{-\varepsilon_0/\mathbf{k}T} \tag{7.118}$$

The rate therefore passes through a maximum as c_g is increased, the maximum occurring when $Kc_g = 1$.

If A is adsorbed strongly, Kc_g is large compared with unity, and the equation becomes

$$v = \frac{1}{2}s \frac{L_s}{K^2} \frac{c'_g}{c_g} \frac{\mathbf{k}T}{h} \frac{q_{\ddagger}}{q_g q'_g q_s} e^{-\varepsilon_0/\mathbf{k}T} \tag{7.119}$$

If K is expressed in terms of partition functions, Eq. (7.119) becomes

$$v = \frac{1}{2}sL_s \frac{c'_g}{c_g} \frac{\mathbf{k}T}{h} \frac{q_g q_{\ddagger} q_s}{q'_g q_a^2} \exp\left(-\frac{\varepsilon_0 + 2\varepsilon}{\mathbf{k}T}\right) \tag{7.120}$$

The activation energy is greater by 2ε than the value for a sparsely covered surface; the reason for this is that two molecules of A must be desorbed before the activated complex is formed.

Equation (7.120) has been applied quantitatively to the reaction between carbon monoxide and oxygen on a platinum surface. The observed rate at 572 K was found[95] to be

$$v/\text{cm}^{-2}\ \text{s}^{-1} = 7.1 \times 10^{-14} \frac{c_{O_2}}{c_{CO}} \tag{7.121}$$

with an activation energy of 139 kJ mol^{-1}. The calculated rate gave a factor of 4.33×10^{-14}, in satisfactory agreement.

When both reacting molecules are the same, the general rate equation (7.115) becomes

$$v = \tfrac{1}{2}s \frac{c_g^2 c_s^2}{L_s} \frac{kT}{h} \frac{q_{\ddagger}}{q_g^2 q_s} e^{-\varepsilon_0/kT} \tag{7.122}$$

If the surface is covered sparsely, Eq. (7.122) reduces to

$$v = \tfrac{1}{2}s c_g^2 L_s \frac{kT}{h} \frac{q_{\ddagger}}{q_g^2 q_s} e^{-\varepsilon_0/kT} \tag{7.123}$$

The general rate equation is

$$v = \tfrac{1}{2}s \frac{L_s c_g^2}{(1 + Kc_g)^2} \frac{kT}{h} \frac{q_{\ddagger}}{q_g^2 q_s} e^{-\varepsilon_0/kT} \tag{7.124}$$

When the surface is covered almost completely

$$v = \tfrac{1}{2}s \frac{L_s}{K^2} \frac{kT}{h} \frac{q_{\ddagger}}{q_g^2 q_s} e^{-\varepsilon_0/kT} \tag{7.125}$$

$$= \tfrac{1}{2}s L_s \frac{kT}{h} \frac{q_{\ddagger} q_s}{q_a^2} \exp\left(-\frac{\varepsilon_0 + 2\varepsilon}{kT}\right) \tag{7.126}$$

Rate equations also have been derived in a similar manner for the Langmuir–Rideal mechanism[96] and for more complex reactions.[97]

7.7.5 Comparison of Homogeneous and Heterogeneous Reaction Rates

Theoretical treatments of the kind just outlined can be used to make a comparison between the rates of heterogeneous and homogeneous reactions between the same reactants. The reaction rate per unit surface for a second-order heterogeneous reaction between A and B may be written as

$$v_{\text{het}} = c_A c_B c_s \frac{kT}{h} \frac{1}{q_A q_B} e^{-E_{\text{het}}/RT} \tag{7.127}$$

The partition functions of the activated complex and of the reaction centers have been taken as unity. For the corresponding homogeneous process the rate is

$$v_{\text{hom}} = c_A c_B \frac{kT}{h} \frac{q_{\ddagger}}{q_A q_B} e^{-E_{\text{hom}}/RT} \tag{7.128}$$

It follows therefore that

$$\frac{v_{\text{het}}}{v_{\text{hom}}} = \frac{c_s}{q_{\ddagger}} e^{\Delta E/RT} \tag{7.129}$$

where $\Delta E = E_{\text{hom}} - E_{\text{het}}$. The concentration c_s is about 10^{15} cm^{-2}, and $q_{\ddagger}$ is usually of the order of 10^{27} cm^{-3}. Hence,

$$\frac{v_{\text{het}}/\text{cm}^{-2}}{v_{\text{hom}}/\text{cm}^{-3}} \approx 10^{-12} \, e^{\Delta E/RT} \tag{7.130}$$

For the two rates to be comparable either a very large surface must be employed or the activation energy of the surface reaction must be much lower than that of the gas-phase reaction. At 300 K the quantities $v_{\text{het}}/\text{cm}^{-2}$ and $v_{\text{hom}}/\text{cm}^{-3}$ are comparable if the activation energy of the surface reaction is about 70 kJ mol^{-1} lower than that of the homogeneous reaction. At higher temperatures a larger difference is necessary, the difference required being proportional to the absolute temperature.

In view of this result, it is necessary for a reaction that proceeds at an appreciable rate on a surface of comparatively small size to have a much smaller activation energy

TABLE 7.2 ACTIVATION ENERGIES FOR SOME HOMOGENEOUS AND HETEROGENEOUS REACTIONS

Reaction	Surface	E_a/kJ mol^{-1}	Reference
Decomposition of HI	None	184	a
	Gold	105	b
	Platinum	140	c
Decomposition of N_2O	None	~250	d
	Gold	120	e
	Platinum	136	f
	Calcium oxide	146	g
	Aluminum oxide	123	g
Decomposition of NH_3	None	>340	h
	Tungsten	162	i
	Molybdenum	134–180	j
	Osmium	200	k
Para-hydrogen conversion	None	254	l
	Gold	20–85	m
	Copper	42–5	m
	Palladium	17	n

[a] M. Bodenstein, *Z. Phys. Chem.*, **13**, 56 (1894); **22**, 1 (1897); **29**, 295 (1899).

[b] C. N. Hinshelwood and C. R. Prichard, *J. Chem. Soc.*, **127**, 1552 (1925).

[c] C. N. Hinshelwood and R. E. Burk, *J. Chem. Soc.*, **127**, 2896 (1925).

[d] H. S. Johnston, *J. Chem. Phys.*, **19**, 663 (1951).

[e] C. N. Hinshelwood and C. R. Prichard, *Proc. R. Soc. London A*, **108**, 211 (1925).

[f] C. N. Hinshelwood and C. R. Prichard, *J. Chem. Soc.*, **127**, 327 (1925).

[g] G. M. Schwab, R. Stager, and H. H. von Baumbach, *Z. Phys. Chem. B*, **21**, 65 (1933).

[h] Apparently, no data are available; the lower limit of 340 kJ mol^{-1} is deduced from the known thermal stability of ammonia.

[i] C. N. Hinshelwood and R. E. Burk, *J. Chem. Soc.*, **127**, 1105 (1925).

[j] C. H. Kunsman, *J. Am. Chem. Soc.*, **50**, 2100 (1928).

[k] E. A. Arnold and R. E. Burk, *J. Am. Chem. Soc.* **54**, 23 (1932).

[l] R. E. Weston, *J. Chem. Phys.*, **31**, 892 (1959).

[m] D. D. Eley and D. R. Rossington, in W. E. Garner (Ed.), *Chemisorption*, p. 137, Butterworth, London, 1957.

[n] A. Couper and D. D. Eley, *Discuss. Faraday Soc.*, **8**, 172 (1950).

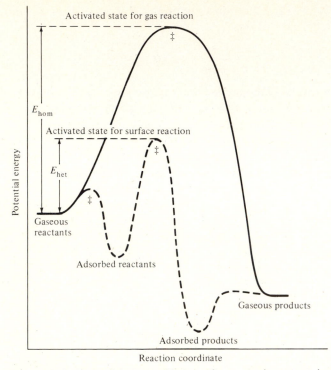

Figure 7.6 Potential-energy diagram for a reaction occurring homogeneously (solid curve) and heterogeneously (dashed curve).

on the surface than in the gas phase. An examination of the data reveals that in a number of instances this is the case; some values are given in Table 7.2.

The way in which the activation energy of the surface reaction may be less than that of the homogeneous reaction is illustrated with reference to the potential-energy diagram shown in Fig. 7.6. The solid curve in this figure represents the variation in the potential energy as the homogeneous reaction proceeds, and the dashed curve represents the corresponding change for the surface reaction. In general, the system passes over an energy barrier to reach the adsorbed state, and then a second barrier is surmounted to give products in the adsorbed state. The height of this second barrier with reference to the initial level is E_{het}. The difference between E_{hom} and E_{het} is the difference between the energy of the gaseous activated complex and that of the adsorbed activated complex. A good catalyst is therefore one on which the activated complex can be adsorbed strongly.

PROBLEMS

7.1. A first-order surface reaction is proceeding at a rate of 1.5×10^{-4} mol dm^{-3} s^{-1} and with a rate constant of 2.0×10^{-3} s^{-1}. What will be the rate and the rate constant if:

(a) The surface area is increased by a factor of 10?

(b) The amount of gas is increased tenfold at constant pressure and temperature?

If these values of v and k apply to a reaction occurring on the surface of a spherical vessel of radius 10 cm:

(c) What will be the rate and rate constant in a spherical vessel, of the same material, of radius 100 cm, at the same pressure and temperature?

(d) Define a new rate constant k' that is independent of the gas volume V and the area S of the catalyst surface.

(e) What would be its SI unit?

7.2. A zero-order reaction is proceeding at a rate of 2.5×10^{-3} mol dm^{-3} s^{-1} and a rate constant 2.5×10^{-3} mol dm^{-3} s^{-1}.

(a) How will the changes (a), (b), and (c) in Problem 7.1 affect the rate and the rate constant in this case?

(b) Again, define a rate constant that is independent of S and V.

(c) What would be its SI unit?

7.3. The decomposition of ammonia on platinum,

$$2NH_3 \rightarrow N_2 + 3H_2$$

is first order in NH_3 and the rate is inversely proportional to the hydrogen concentration [Eq. (7.45)]. Write the differential rate equation for the rate of formation of hydrogen, dx/dt, in terms of the initial concentration of ammonia, a_0, and the concentration x of hydrogen at time t.

7.4. A unimolecular surface reaction is inhibited by a poison I and obeys Eq. (7.44). If E_2 is the activation energy corresponding to the reaction of the adsorbed substrate molecule (i.e., corresponding to k) and ΔH_A and ΔH_I are the enthalpies of adsorption of A and I, what is the activation energy:

(a) At very low concentrations of A and I?

(b) At a very high concentration of A and a very low concentration of I?

(c) At a very low concentration of A and a very high concentration of I?

7.5. Suppose that a reaction,

$$A \rightarrow Y + Z$$

occurs initially as a homogeneous first-order reaction (rate constant k) but that the product Z is adsorbed on the surface and catalyzes the reaction according to a zero-order law (rate is $k_c z$). Obtain a differential equation for the rate of appearance of Z and integrate it to give z as a function of time.

7.6. Use the Langmuir isotherm for adsorption with dissociation to obtain an expression for the rate of the *para*-hydrogen conversion if it occurs by the Bonhoeffer–Farkas mechanism. What is the order of the reaction (a) at high surface coverage and (b) at low surface coverage?

7.7. Derive an equation for the rate of formation of H atoms by the following mechanism:

$$H_2 + \begin{array}{cc} | & | \\ -S & -S- \end{array} \rightleftharpoons \begin{array}{cc} H & H \\ | & | \\ -S & -S- \end{array}$$

$$\begin{array}{cc} H & H \\ | & | \\ -S & -S- \end{array} \rightarrow \begin{array}{cc} H & \\ | & | \\ -S & -S- \end{array} + H$$

Under what circumstances is the rate proportional to $[H_2]^{1/2}$?

7.8. Derive Eq. (7.36) for the case of adsorption with dissociation.

7.9. Derive Eq. (7.35) for a gas adsorbed in a mobile layer.

7.10. Derive the rate equation for a bimolecular reaction occurring by a Langmuir–Hinshelwood mechanism and inhibited by a diatomic molecule X_2 which is adsorbed with dissociation into the atoms. What equation applies if A and B are adsorbed very weakly and X_2 very strongly?

7.11. The rate of combination of hydrogen atoms on a surface is given, for one possible mechanism, by Eq. (7.77):

$$v = \frac{kK_H[H]^2}{1 + K_H[H]}$$

Derive, on the basis of conventional transition-state theory, the corresponding expression for the rate in terms of partition functions. Then write down the expression for the limiting case of high surface coverage.

7.12. The rate of desorption of CO from Pt may be expressed by Eq. (7.100). Langmuir[98] obtained for this process an activation energy of 134 kJ mol^{-1}. The concentration c_a may be taken as 10^{15} cm^{-2}, and $q_{\ddagger}$ and q_a as equal to each other. Calculate E_0, the activation energy at the absolute zero, and the rate of desorption per cm^2 at 600°C.

7.13. The rate of the reaction between carbon monoxide and oxygen on platinum is found under certain circumstances to be directly proportional to the oxygen pressure and inversely proportional to the carbon monoxide pressure.[98] Derive a rate equation, in terms of partition functions, that is consistent with this fact. If the activation energy is 140 kJ mol^{-1} at 600 K, calculate the value at the absolute zero. Calculate the rate of reaction at 600 K as a function of the concentrations of oxygen and carbon monoxide and corresponding to a volume of 1 dm^3 and a catalyst surface area of 1 cm^2. The interatomic distances in O_2 and CO are 121 and 113 pm, respectively; the vibrational partition functions may be taken as unity.

7.14. On the basis of CTST and for mechanisms discussed in Section 7.6.8, derive expressions for the rates of the following reactions in terms of the appropriate partition functions:
(a) The ethylene–deuterium exchange reaction.
(b) The methane–deuterium exchange reaction.
(c) The ammonia–deuterium exchange reaction.

7.15. A reaction between A and B occurs by a Langmuir–Hinshelwood mechanism and the rate equation is

$$s = \frac{kK_A K_B[A][B]}{(1 + K_A[A] + K_B[B])^2}$$

Show that if [A] is held constant and [B] is varied, the maximum rate is

$$v_{max} = \frac{kK_A[A]}{4(1 + K_A[A])}$$

REFERENCES

1. M. Faraday, *Philos. Trans.,* **124,** 55 (1834).
2. J. Berzelius, *Jahres-Bericht uber die Fortschritte der Physichen Wissenschaft,* p. 243, Tubingen, 1836.
3. W. Ostwald, *Uber Katalyze,* Leipzig, 1902; *Phys. Z.,* **3,** 313 (1902).

4. J. H. van't Hoff, *Etudes de dynamique chimique,* Muller, Amsterdam, 1884.

5. D. L. Chapman and H. E. Jones, *Trans. Chem. Soc.,* 2463 (1910).

6. M. Faraday, *Philos. Trans.,* **124,** 55 (1834).

7. J. H. van der Waals, Dissertation, Leipzig, 1873.

8. I. Langmuir, *J. Am. Chem. Soc.,* **38,** 2221 (1916).

9. H. S. Taylor, *J. Am. Chem. Soc.,* **53,** 578 (1931).

10. H. S. Taylor, *Proc. R. Soc. London A,* **108,** 105 (1925).

11. F. H. Constable, *Proc. R. Soc. London A,* **108,** 355 (1925).

12. I. Langmuir, *J. Am. Chem. Soc.,* **38,** 2221 (1916); **40,** 1361 (1918).

13. H. M. F. Freundlich, *Kappilarchemie,* Leipzig, 1909.

14. A. Slygin and P. Frumkin, *Acta Physicochim. URSS,* **3,** 791 (1935).

15. J. Zeldowitch, *Acta Physicochim. URSS,* **1,** 961 (1935); G. D. Halsey and H. S. Taylor, *J. Chem. Phys.,* **19,** 624 (1947); T. L. Hill, *J. Chem. Phys.,* **17,** 762 (1949); R. Sips, *J. Chem. Phys.,* **18,** 1024 (1950); F. C. Tompkins, *Trans. Faraday Soc.,* **46,** 569 (1950).

16. R. H. Fowler, *Proc. Cambr. Philos. Soc.,* **31,** 260 (1935); R. H. Fowler and E. A. Guggenheim, *Statistical Thermodynamics,* p. 426, Cambridge University Press, New York, 1939.

17. F. H. Constable, *Proc. R. Soc. London A,* **108,** 355 (1925).

18. H. S. Taylor, *J. Am. Chem. Soc.,* **53,** 578 (1931).

19. R. E. Burk, *Proc. Natl. Acad. Sci. USA,* **13,** 67 (1927).

20. R. E. Burk, *J. Phys. Chem.,* **30,** 1134 (1926).

21. A. A. Balandin, *Z. Phys. Chem.* **B,** *2,* 209 (1929); **3,** 167 (1929).

22. A. Sherman and H. Eyring, *J. Am. Chem. Soc.,* **54,** 2661 (1932).

23. G. Okamoto, J. Horiuti, and K. Hirota, *Sci. Papers Inst. Phys. Chem. Res. (Tokyo),* **29,** 223 (1936). Compare A. Sherman, C. E. Sun, and H. Eyring, *J. Chem. Phys.,* **3,** 49 (1934).

24. O. Beeck, A. Wheeler, and A. E. Smith, *Phys. Rev.,* **55,** 601 (1939); O. Beeck and A. Wheeler, *J. Chem. Phys.,* **7,** 631 (1939); A. E. Smith and O. Beeck, *Phys. Rev.,* **55,** 602 (1939); O. Beeck, A. E. Smith, and A. Wheeler, *Proc. R. Soc. London,* **177,** 64 (1940). See also G. H. Twigg and E. K. Rideal, *Trans. Faraday Soc.,* **36,** 533 (1940).

25. H. Leidheiser and A. T. Gwathmey, *J. Am. Chem. Soc.,* **70,** 1200, 1206 (1948).

26. E. W. Muler and T. T. Tsong, *Field Ion Microscopy,* Elsevier, New York, 1969.

27. G. A. Somerjai and H. H. Farrell, *Adv. Chem. Phys.,* **20,** 215 (1970).

28. G. A. Somerjai, *Acc. Chem. Res.,* **9,** 248 (1976).

29. L. H. Little, *Infrared Spectra of Adsorbed Species,* Academic, New York, 1966.

30. B. A. Morrow and N. Sheppard, *J. Phys. Chem.,* **70,** 2406 (1966); *Proc. R. Soc. London A,* **311,** 391 (1969); N. Sheppard, *Discuss. Faraday Soc.,* **41,** 171 (1966).

31. J. D. Prentice, A. Lesiuras, and N. Sheppard, *J. Chem. Soc. Chem. Commun.,* 76 (1976); R. C. Brady and R. Pettit, *J. Am. Chem. Soc.,* **102,** 6181 (1982).

32. J. K. Roberts, *Proc. R. Soc. London A,* **152,** 445 (1935).

33. E. K. Rideal and B. M. W. Trapnell, *J. Chim. Phys.,* **47,** 126 (1950); B. M. W. Trapnell, *Proc. R. Soc. London A,* **206,** 39 (1951).

34. P. H. Emmett and J. T. Kummer, *J. Chim. Phys.,* **47,** 67 (1950); J. T. Kummer and P. H. Emmett, *J. Am. Chem. Soc.,* **73,** 2886 (1951); J. Weber and K. J. Laidler, *J. Chem. Phys.,* **18,** 1418 (1950); **19,** 1089 (1951).

35. I. Langmuir, *J. Am. Chem. Soc.,* **38,** 1145 (1916).

36. I. Langmuir, *Trans. Faraday Soc.,* **17,** 621 (1921).

37. C. N. Hinshelwood, *Kinetics of Chemical Change in Gaseous Systems,* p. 145, Clarendon, Oxford, 1926.

38. E. K. Rideal, *Proc. Cambr. Philos. Soc.,* **35,** 130 (1939); *Chem. Ind.,* **62,** 335 (1943); D. D. Eley and E. K. Rideal, *Proc. R. Soc. London A,* **178,** 429 (1941). What is here called the Langmuir–Rideal mechanism is sometimes called the Rideal–Eley mechanism.

39. P. C. Allen and C. N. Hinshelwood, *Proc. R. Soc. London A,* **121,** 141 (1928).

40. M. Temkin and V. Pyzhev, *Acta Physicochim. URSS,* **2,** 473 (1935).
41. J. Weber and K. J. Laidler, *J. Chem. Phys.,* **19,** 1089 (1951).
42. F. H. Constable, *Proc. R. Soc. London A,* **108,** 355 (1925). See also E. Cremer and G. M. Schwab, *Z. Phys. Chem. A,* **144,** 243 (1929); G. M. Schwab, *Z. Phys. Chem. B,* **5,** 406 (1929).
43. G. D. Halsey, *J. Chem. Phys.,* **17,** 758 (1949).
44. K. J. Laidler, in P. H. Emmett (Ed.), *Catalysis,* vol. 1, pp. 239–241, Reinhold, New York, 1954.
45. The arguments in Sections 7.5 and 7.6 are presented in much greater detail by the author in P. H. Emmett (Ed.), *Catalysis,* vol. 1, pp. 119–194, Reinhold, New York, 1954.
46. C. N. Hinshelwood and R. E. Burk, *J. Chem. Soc.,* **127,** 1051, 1114 (1925).
47. J. H. van't Hoff and D. M. Kooij, *Z. Phys. Chem.,* **12,** 155 (1893).
48. R. M. Barrer, *Trans. Faraday Soc.,* **32,** 496 (1936).
49. C. N. Hinshelwood and R. E. Burk, *J. Chem. Soc.,* **127,** 1114 (1925).
50. C. R. Prichard and C. N. Hinshelwood, *J. Chem. Soc.,* **127,** 806 (1925).
51. M. Bodenstein and F. Ohlmer, *Z. Phys. Chem.,* **53,** 166 (1905).
52. I. Langmuir, *Trans. Faraday Soc.,* **17,** 621 (1922).
53. K. J. Laidler and R. E. Townshend, *Trans. Faraday Soc.,* **57,** 1590 (1961).
54. P. C. Allen and C. N. Hinshelwood, *Proc. R. Soc. London A,* **121,** 141 (1928).
55. W. H. Hutchison and C. N. Hinshelwood, *J. Chem. Soc.,* **129,** 1556 (1926).
56. C. R. Prichard and C. N. Hinshelwood, *J. Chem. Soc.,* **127,** 1546 (1925).
57. D. D. Eley and E. K. Rideal, *Proc. R. Soc. London A,* **178,** 429 (1941).
58. K. F. Bonhoeffer and A. Farkas, *Z. Phys. Chem. B,* **12,** 231 (1931); K. F. Bonhoeffer, A. Farkas, and K. W. Rummel, *Z. Phys. Chem. B,* **21,** 225 (1933).
59. For a detailed discussion of the mechanism, which involves some complex features, see K. J. Laidler, *J. Phys. Chem.,* **57,** 320 (1953); K. J. Laidler, in P. H. Emmett (Ed.), *Catalysis,* vol. 1, pp. 178–180, Reinhold, New York, 1954.
60. L. von Muffling, *Handbuch der Katalyse,* vol. VI, p. 94, Springer, Vienna, 1943; W. V. Smith, *J. Chem. Phys.,* **11,** 110 (1941).
61. W. Buben and A. Schechter, *Acta Physicochim. URSS,* **10,** 371 (1939); F. Paneth, W. Hofeditz, and A. Wunsch, *J. Chem. Soc.,* 372 (1935); F. Paneth and W. Lautsch, *Ber.,* **64B,** 2708 (1931).
62. I. Langmuir, *J. Am. Chem. Soc.,* **34,** 1310 (1912); **37,** 417 (1915).
63. G. Bryce, *Proc. Cambr. Philos. Soc.,* **32,** 648 (1936).
64. J. K. Roberts and G. Bryce, *Proc. Cambr. Philos. Soc.,* **32,** 653 (1936).
65. K. J. Laidler, *J. Phys. Colloid Chem.,* **55,** 1067 (1951).
66. D. Brennan and P. C. Fletcher, *Proc. R. Soc. London A,* **250,** 389 (1959).
67. K. F. Bonhoeffer and A. Farkas, *Z. Phys. Chem. B,* **12,** 231 (1931).
68. E. K. Rideal, *Proc. Cambr. Philos. Soc.,* **35,** 130 (1939).
69. S. L. Bernasek and G. A. Somerjai, *J. Chem. Phys.,* **62,** 3149 (1975).
70. R. L. Palmer, J. N. Smith, H. Saltzburg, and D. R. O'Keefe, *J. Chem. Phys.,* **53,** 1666 (1970); T. L. Bradley, A. E. Dabiri, and R. E. Stickney, *Surf. Sci.,* **29,** 590 (1912); M. Balosch and R. E. Stickney, *Surf. Sci.,* **44,** 310 (1974).
71. G. H. Twigg and E. K. Rideal, *Proc. R. Soc. London A,* **171,** 55 (1939); G. K. T. Conn and G. H. Twigg, *Proc. R. Soc. London A,* **171,** 70 (1939); C. Kemball, *J. Chem. Soc.,* 735 (1956).
72. K. J. Laidler, M. C. Wall, and M. C. Markham, *J. Chem. Phys.,* **21,** 949 (1953).
73. C. Kemball, *J. Chem. Soc.,* 735 (1956).
74. C. Kemball, *Proc. R. Soc. London A,* **207,** 539 (1951).
75. C. Kemball, *Proc. R. Soc. London A,* **217,** 376 (1953).

76. A. Farkas, *Trans. Faraday Soc., 32,* 416 (1936).
77. J. Weber and K. J. Laidler, *J. Chem. Phys., 19,* 1089 (1951).
78. J. W. Singleton, E. R. Roberts, and E. R. S. Winter, *Trans. Faraday Soc., 47,* 1318 (1951).
79. For further details see G. Webb, in C. H. Bamford and C. F. H. Tipper (Eds.), *Comprehensive Chemical Kinetics,* vol. 20, pp. 16–37, Elsevier, Amsterdam, 1978.
80. R. N. Pease, *J. Am. Chem. Soc., 45,* 1196 (1923).
81. E. K. Rideal, *J. Chem. Soc., 121,* 309 (1922); H. Zur Strassen, *Z. Phys. Chem. A, 169,* 81 (1934); A. Farkas, L. Farkas, and E. K. Rideal, *Proc. R. Soc. London A, 171,* 55 (1939); O. Beeck, *Rev. Mod. Phys., 17,* 61 (1945).
82. O. Beeck, *Rev. Mod. Phys., 20,* 127 (1948); *Discuss. Faraday Soc., 8,* 118, 126, 193 (1956).
83. G. I. Jenkins and E. K. Rideal, *J. Chem. Soc.,* 2490, 2496 (1955).
84. O. Toyama, *Rev. Phys. Chem. Jpn., 11,* 153 (1937).
85. K. J. Laidler and R. E. Townshend, *Trans. Faraday Soc., 57,* 1590 (1961).
86. M. C. Markham, M. C. Wall, and K. J. Laidler, *J. Chem. Phys., 20,* 1331 (1952); *21,* 949 (1953).
87. P. Mahaffy, P. B. Masterson, and R. S. Hanson, *J. Chem. Phys., 64,* 3911 (1976).
88. K. J. Laidler, S. Glasstone, and H. Eyring, *J. Chem. Phys., 8,* 667 (1940). Interesting earlier treatments were given by B. Topley, *Nature, 128,* 115 (1931) and by R. E. Burk, *J. Am. Chem. Soc., 57,* 1279 (1935).
89. K. J. Laidler, Absolute Rates of Surface Reactions, in P. H. Emmett (Ed.), *Catalysis,* chap. 5, Reinhold, New York, 1954.
90. K. J. Laidler, S. Glasstone, and H. Eyring, *J. Chem. Phys., 8,* 667 (1940); A. J. B. Robertson, *J. Colloid Sci., 11,* 308 (1956).
91. D. M. Kooij, *Z. Phys. Chem., 12,* 155 (1893).
92. A. J. B. Robertson, *J. Colloid Sci., 11,* 308 (1956).
93. B. Topley, *Nature, 128,* 115 (1931).
94. M. Temkin and V. Pyzhev, *Acta Physicochim. URSS, 3,* 473 (1935).
95. I. Langmuir, *Trans. Faraday Soc., 17,* 621 (1922).
96. K. J. Laidler, *Discuss, Faraday Soc., 8,* 47 (1950); *J. Phys. Colloid Chem., 55,* 1067 (1951).
97. M. C. Markham, M. C. Wall, and K. J. Laidler, *J. Phys. Colloid Chem., 57,* 321 (1953).
98. I. Langmuir, *Trans. Faraday Soc., 17,* 621 (1922).

BIBLIOGRAPHY

For the IUPAC recommendations on symbolism and terminology in heterogeneous catalysis and also a useful brief summary of the field, see:

Manual of Symbols and Terminology for Physicochemical Quantities and Units, Appendix II, Part 2 (*Heterogeneous Catalysis*), Pergamon, Oxford, 1976; also published in *Pure Appl. Chem., 46,* 73 (1976).

For reviews of the adsorption of gases on solid surfaces see the following publications:

S. Brunauer, *Physical Adsorption,* Princeton University Press, Princeton, NJ, 1943.

J. C. Dash, *Films on Solid Surfaces,* Academic, New York, 1975.

W. E. Garner (Ed.), *Chemisorption,* Butterworths, London, 1957.

D. O. Hayward and B. M. W. Trapnell, *Chemisorption,* Butterworths, London, 1964.

K. J. Laidler, Chemisorption, in P. H. Emmett (Ed.), *Catalysis,* vol. 1, pp. 75–118, Reinhold, New York, 1954.

A. R. Miller, *The Adsorption of Gases on Solids,* Cambridge University Press, New York, 1949.

H. Saltzburg, J. N. Smith, and M. Rogers (Eds.), *Fundamentals of Gas–Surface Interactions,* Academic, New York, 1967.

The structures of solid surfaces and adsorbed layers are reviewed in the following publications:

A. W. Adamson, *Physical Chemistry of Surfaces,* 4th ed., Wiley, New York, 1982.

A. Campion, Raman Spectroscopy of Molecules Adsorbed on Solid Surfaces, *Ann. Rev. Phys. Chem.,* **36,** 549 (1985).

R. K. Grasselli and J. F. Brazdil (Eds.), *Solid-State Chemistry in Catalysis, Am. Chem. Soc. Symp. Series* **279** (1985).

N. B. Hannay (Ed.), *Treatise on Solid State Chemistry,* vols. 6A and 6B, Plenum, New York, 1976.

M. Kerker, *Surface Chemistry and Colloids,* Butterworth, London, 1975.

L. H. Little, *Infrared Spectra of Adsorbed Species,* Academic, New York, 1966.

E. W. Muler and T. T. Tsong, *Field Ion Microscopy,* Elsevier, New York, 1969.

G. A. Somerjai, *Principles of Surface Chemistry,* Prentice-Hall, Englewood Cliffs, NJ, 1972.

G. A. Somerjai, *Chemistry in Two Dimensions: Surfaces,* Cornell University Press, Ithaca, NY, 1981.

G. A. Somerjai and H. H. Farrell, Low-Energy Electron Diffraction, *Adv. Chem. Phys.,* **20,** 215 (1970).

The Role of the Adsorbed State in Heterogeneous Catalysis, *Discuss. Faraday Soc.,* **41** (1966).

For general treatments of catalysis on surfaces see:

P. G. Ashmore, *Catalysis and Inhibition of Chemical Reactions,* Butterworth, London, 1963.

G. C. Bond, *Catalysis by Metals,* Academic, New York, 1962.

M. Boudart, Correlations in Heterogeneous Catalysis, in *Kinetics of Chemical Processes,* chap. 9, Prentice-Hall, Englewood Cliffs, NJ, 1968.

K. J. Laidler, Kinetic Laws in Surface Catalysis, in P. H. Emmett (Ed.), *Chemisorption,* vol. 1, chap. 4, pp. 119–194, Reinhold, New York, 1954.

C. L. Thomas, *Catalytic Processes and Proven Catalysts,* Academic, New York, 1970.

J. M. Thomas and W. J. Thomas, *Introduction to the Principles of Heterogeneous Catalysis,* Academic, New York, 1986 (2nd edition).

For reviews of the kinetics of individual types of reaction see:

L. Beránek and M. Kraus, Heterogeneous Eliminations, Additions, and Substitutions, in C. H. Bamford and C. F. H. Tipper (Eds.), *Comprehensive Chemical Kinetics,* vol. 20, chap. 3, pp. 263–398, Elsevier, New York, 1978.

K. van der Waals and P. J. van der Berg, Heterogeneous Oxidation Processes, in C. H. Bamford and C. F. H. Tipper (Eds.), *Comprehensive Chemical Kinetics,* vol. 20, chap. 2, pp. 123–262, Elsevier, New York, 1978.

G. Webb, Catalytic Hydrogenation, in C. H. Bamford and C. F. H. Tipper (Eds.), *Comprehensive Chemical Kinetics,* vol. 20, chap. 1, pp. 1–122, Elsevier, New York, 1978.

For treatments of surface-catalyzed reactions in terms of transition-state theory and of dynamical theories see the following publications:

K. J. Laidler, Absolute Rates of Surface Reactions, in P. H. Emmett (Ed.), *Catalysis,* vol. 1, chap. 5, pp. 195–244, Reinhold, New York, 1954.

J. C. Tully, Dynamics of Chemical Processes at Surfaces, *Acc. Chem. Res.,* **14,** 188 (1981).

Chapter 8

Composite Reactions

Upon detailed study, nearly all chemical processes are found not to be elementary; that is, they do not occur in a single step. Instead, they occur in more than one elementary step and are then called *composite, complex,* or *stepwise* reactions; the term *composite* is used in this book.

The present chapter is concerned with some of the general features of reactions occurring by composite mechanisms, and a number of examples of composite reactions are considered in some detail. Since so many overall reactions now are recognized to be composite, only a very small selection of reactions can be included here. Particular attention is given to a few reactions, mostly occurring in the gas phase, that have been investigated very thoroughly and whose mechanisms are understood reasonably well. It should be emphasized that even when a reaction has been studied very extensively, its mechanism is still not known with certainty; an alternative mechanism may explain the experimental results equally well. In particular, if evidence points towards a particular mechanism, one cannot exclude the possibility that additional reactions occur but do not affect the kinetic behavior.

Section 8.1 suggests a classification of reaction *mechanisms.* A useful classification of reactions themselves is in terms of whether they occur thermally, photochemically, or radiation-chemically. A thermal reaction occurs simply by virtue of the thermal energy present in the system; molecules undergo collisions and in this way acquire the energy they require for reaction. Photochemical and radiation-chemical reactions are brought about as a result of radiation absorbed by the reaction system. The distinction between the two types of reaction is discussed in Chapter 9; here it simply is mentioned that photochemical reactions occur with less energetic radiation, while radiation-chemical processes are brought about by more energetic radiation. Chapter 9 deals with some of the details of the interaction between radiation and reactant molecules; the present chapter includes some overall mechanisms induced by radiation, mainly to show the similarities and differences between thermal, photochemical, and radiation-chemical mechanisms.

The most obvious indication that a mechanism is composite is when the kinetic equation is inconsistent with the stoichiometric equation for the reaction. An example of this is the gas-phase reaction between nitric oxide and hydrogen, the process being represented by the equation

$$2NO + 2H_2 \rightarrow N_2 + 2H_2O$$

According to the stoichiometry, this reaction might be expected to be of the fourth order, with the rate proportional to the squares of the concentrations of both nitric oxide and hydrogen. Under usual conditions, however, the rate is proportional to the square of the nitric oxide concentration but only to the first power of the hydrogen concentration,[1]

$$v = k[NO]^2[H_2] \tag{8.1}$$

This result suggests that the initial step in the reaction is

$$2NO + H_2 \rightarrow N_2 + H_2O_2$$

and that this process is followed by

$$H_2O_2 + H_2 \rightarrow 2H_2O$$

If it is assumed that the second reaction is rapid compared with the first, the rate of the first reaction controls the rate of the second and therefore gives rise to the kinetic behavior observed.

Another example of a composite mechanism is the reaction between hydrogen and iodine monochloride,

$$2ICl + H_2 \rightarrow I_2 + 2HCl$$

Instead of being of the third order, which would be the case if it were elementary, this reaction obeys the rate equation[2]

$$v = k[ICl][H_2] \tag{8.2}$$

that is, it is of the second order. The mechanism suggested is that there is an initial slow process,

$$ICl + H_2 \rightarrow HI + HCl$$

and that this is followed by the rapid process

$$HI + ICl \rightarrow HCl + I_2$$

Because reactions can occur in steps, it is found very frequently that the order of a reaction is less than corresponds directly to the stoichiometric equation. It was seen in Section 5.2 that a reaction between three molecules is very much less probable than one between two because of the greater loss of entropy in the activated state. Collisions between four molecules are so improbable that it is unlikely that there are any elementary reactions having a molecularity of 4 or more. A reaction whose stoichiometric equation involves more than three molecules usually may proceed much more effectively by two or more elementary reactions of lower molecularity.

8.1 TYPES OF COMPOSITE MECHANISM

Composite reaction mechanisms can have a number of features. Reactions occurring in parallel, such as§

$$A \rightarrow Y$$

$$A \rightarrow Z$$

are called *simultaneous* reactions. When there are simultaneous reactions, there is sometimes *competition,* as in the scheme

$$A + B \rightarrow Y$$

$$A + C \rightarrow Z$$

where B and C compete with one another for A.

Reactions occurring in forward and reverse directions are called *opposing:*

$$A + B \rightleftharpoons Z$$

Reactions occurring in sequence, such as

$$A \rightarrow X \rightarrow Y \rightarrow Z$$

are known as *consecutive* reactions, and the overall process is said to occur by consecutive steps. Reactions are said to exhibit *feedback* if a substance formed in one step affects the rate of a previous step. For example, in the scheme

$$A \overset{1}{\rightarrow} X \overset{2}{\rightarrow} Y \overset{3}{\rightarrow} Z$$

the intermediate Y may catalyze reaction 1 (positive feedback) or inhibit reaction 1 (negative feedback). A final product as well as an intermediate may bring about feedback.

Chain reactions are a special type of composite reaction and are treated in Section 8.3.

8.2 RATE EQUATIONS FOR COMPOSITE MECHANISMS

In dealing with composite mechanisms it is generally inappropriate to define the rate of an individual step in terms of concentration changes that actually occur in the reaction system, since these changes may be brought about by more than one reaction. Instead it is necessary to consider the rate of an individual elementary reactions as if no other reaction were occurring.§ For example, for the composite mechanism

$$A \underset{-1}{\overset{1}{\rightleftharpoons}} X \underset{-2}{\overset{2}{\rightleftharpoons}} Z$$

§ Filled-in arrows are used to emphasize that reactions are elementary.

§ Such a rate has been referred to by V. Gold (*Nouveau Journal de Chemie, 3,* 69 (1979)) as the *chemical flux,* but this term has not proved popular; most workers have continued to use "rate".

there are four elementary reactions. The rate of reaction 1, denoted by v_1 or $v_{-X,A}$, is the rate of A $\rightarrow$ X if no other reaction were occurring; it is given by

$$v_1 = v_{-A,X} = k_1[A] \tag{8.3}$$

Similarly,

$$v_{-1} = v_{-X,A} = k_{-1}[X] \tag{8.4}$$

$$v_2 = v_{-X,Z} = k_2[X] \tag{8.5}$$

$$v_{-2} = v_{-Z,X} = k_{-2}[Z] \tag{8.6}$$

The sum of the rates of all reactions that produce a species X can be called the *rate into X* and given the symbol $\Sigma\, v_X$. For the scheme just considered it is given by

$$\Sigma\, v_X = v_1 + v_{-2} = k_1[A] + k_{-2}[Z] \tag{8.7}$$

The sum of the rates of all the reactions that remove X can be called the *rate out of X* and given the symbol $\Sigma\, v_{-X}$. For the scheme it is given by

$$\Sigma\, v_{-X} = v_{-1} + v_2 = (k_{-1} + k_2)[X] \tag{8.8}$$

For a system at equilibrium the rate into each species is equal to the rate out of it.

8.2.1 Simultaneous and Consecutive Reactions

A number of composite mechanisms have been considered by Capellos and Bielski,[3] who have given solutions for the various differential rate equations that apply to them. Only a brief treatment need be given here. Table 8.1 gives solutions for a few cases.

The simplest consecutive mechanism is

$$A \overset{1}{\rightarrow} X \overset{2}{\rightarrow} Z$$

Equations for the rates of change of the concentrations of A, X, and Z were first given by Harcourt and Esson.[4] If the initial concentration of A is $[A]_0$ and its concentration at any time t is $[A]$, the rate equation for A is

$$-\frac{d[A]}{dt} = k_1[A] \tag{8.9}$$

Integration of this equation, subject to the boundary condition that $[A] = [A]_0$ when $t = 0$, gives

$$[A] = [A]_0 e^{-k_1 t} \tag{8.10}$$

The net rate of formation of X is

$$\frac{d[X]}{dt} = k_1[A] - k_2[X] \tag{8.11}$$

which, with Eq. (8.10), is

$$\frac{d[X]}{dt} = k_1[A]_0 e^{-k_1 t} - k_2[X] \tag{8.12}$$

This contains only the variables $[X]$ and t, and integration gives

TABLE 8.1   RATE EQUATIONS FOR SOME SIMULTANEOUS AND
CONSECUTIVE REACTIONS

System	Differential equation	Integrated equation	Reference
$A \xrightarrow{k_1} Z$ (first order) $A \xrightarrow{k_2} Z$ (second order)	$\dfrac{d[Z]}{dt} = k_1[A] + k_2[A]^2$	$\ln \dfrac{k_1[A]_0 + k_2[A]_0[A]}{(k_1 + k_2[A]_0)[A]} = k_1 t$	a
$A \xrightarrow{k_1} Z_1$ $A \xrightarrow{k_2} Z_2$ (all first order) $A \xrightarrow{k_3} Z_3$	$\dfrac{d[Z_1]}{dt} = k_1[A]$ etc.	$[Z_1] = [Z_1]_0 + \dfrac{k_1[A]_0}{k_1 + k_2 + k_3}$ $\times \{1 - \exp[-(k_1 + k_2 + k_3)t]\}$ etc.	
$A \xrightarrow{k_1} Z$ $B \xrightarrow{k_2} Z$ (first order)	$\dfrac{d[Z]}{dt} = k_1[A] + k_2[B]$	$[A] = [A]_0 e^{-k_1 t}$ $[B] = [B]_0 e^{-k_2 t}$ $[Z] = [A]_0(1 - e^{-k_1 t})$	
$A \xrightarrow{k_1} X \xrightarrow{k_2} Z$	$\dfrac{-d[A]}{dt} = k_1[A]$ $\dfrac{d[X]}{dt} = k_1[A] - k_2[X]$ $\dfrac{d[Z]}{dt} = k_2[B]$	$[A] = [A]_0 e^{-k_1 t}$ $[X] = \dfrac{[A]_0 k_1}{k_2 - k_1}(e^{-k_1 t} - e^{-k_2 t})$ $[Z] = \dfrac{[A]_0}{k_2 - k_1}[k_2(1 - e^{-k_1 t}) - k_1(1 - e^{-k_2 t})]$	b

[a] R. Wegschneider, *Z. Phys. Chem.,* **41**, 52 (1902).
[b] A. V. Harcourt and W. Esson, *Proc. R. Soc. London A,* **14**, 470 (1865); *Philos. Trans.,* **156**, 193 (1866); **157**, 117 (1867).

$$[X] = [A]_0 \frac{k_1}{k_2 - k_1}(e^{-k_1 t} - e^{-k_2 t}) \tag{8.13}$$

The equation for the variation of [Z] can be obtained by noting that

$$[A] + [X] + [Z] = [A]_0 \tag{8.14}$$

so that

$$[Z] = [A]_0 - [A] - [X] \tag{8.15}$$

Insertion of the expressions for [A] and [X] into Eq. (8.15) leads to

$$[Z] = \frac{[A]_0}{k_2 - k_1}[k_2(1 - e^{-k_1 t}) - k_1(1 - e^{-k_2 t})] \tag{8.16}$$

Figure 8.1a shows the time variations in the concentrations of A, X, and Z as given by these equations. The concentration [A] falls exponentially, while [X] goes through a maximum. Since the rate of formation of Z is proportional to the concentration of X, this rate is initially zero and is a maximum when [X] reaches its maximum value. For an initial period of time it may be impossible to detect any of the product Z, and the reaction is said to have an *induction period*. Such induction periods are commonly observed for reactions occurring by composite mechanisms.

Kinetic equations like (8.13) and (8.16) are obeyed frequently by nuclides undergoing radioactive decay,[5] but there are not many examples of chemical reactions

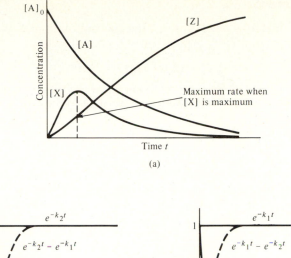

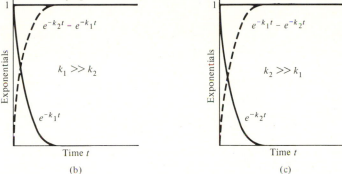

Figure 8.1 (a) Variations in the concentrations of A, X, and Z for a reaction occurring by the mechanism A → X → Z. (b) Variations with time of the exponentials when $k_1 \gg k_2$. (c) Variations of the exponentials when $k_2 \gg k_1$.

that show consecutive first-order behavior. Two good examples are the thermal isomerizations of 1,1-dicyclopropylene and 1-cyclopropylcyclopentene.[6]

Two limiting cases are of special interest. Suppose, first, that the rate constant k_1 is very large and that k_2 is very small. The reactant A is converted rapidly into the intermediate X, which slowly forms Z. Figure 8.1b shows plots of the exponentials $e^{-k_1 t}$ and $e^{-k_2 t}$ and of their difference. Since k_2 is small, the exponential $e^{-k_2 t}$ shows a very slow decay, while $e^{-k_1 t}$ shows a rapid fall. The difference

$$e^{-k_2 t} - e^{-k_1 t}$$

is shown by the dashed line in Fig. 8.1b. The concentration of X is, by Eq. (8.13), equal to this difference multiplied by $[A]_0$ (since $k_1 \gg k_2$), and $[X]$ therefore rises rapidly to the value $[A]_0$ and then slowly declines. The rise in $[Z]$ then follows approximately the simple first-order law.

The converse case, when $k_2 \gg k_1$, is a particularly interesting one, since it leads to the concept of the steady state, which we now discuss.

8.2.2 Steady-State Treatment

When k_1 is small and k_2 is large, $k_2 \gg k_1$, the concentration of X is given by [see Eq. (8.13)]

$$[X] = [A]_0 \frac{k_1}{k_2} (e^{-k_1 t} - e^{-k_2 t}) \tag{8.17}$$

At $t = 0$, $[X] = 0$, but after a very short time, relative to the duration of the reaction, the difference

$$e^{-k_1 t} - e^{-k_2 t}$$

has attained the value of unity, and the concentration of X is then $[A]_0 k_1 / k_2$, which is much less than $[A]_0$. After this short induction period the concentration of X remains practically constant, so that to a good approximation

$$\frac{d[X]}{dt} = 0 \tag{8.18}$$

This is the basis of the *steady-state treatment,* the use of which was first suggested by Chapman and Underhill.[7] It was subjected to some criticism, particularly by Skrabal,[8] but his objections were answered by Bodenstein.[9]

What has been shown for the very simple scheme of two consecutive first-order reactions is that if the conditions are such that the concentration of the intermediate X is always much smaller than the reactant concentration, the concentration of X rapidly reaches a value that remains practically constant during the course of the reaction. It is not possible to give a formal proof of this hypothesis, applicable to any reaction mechanism, because the rate equations for more complicated mechanisms are often impossible to solve. However, the derivation given for the two-stage system of first-order reactions leads to an important general conclusion. The rate of change of the concentration of an intermediate, to a good approximation, can be set equal to zero whenever the intermediate is formed slowly and disappears rapidly. In other words, whenever an intermediate X is such that it is always present in amounts much smaller than those of the reactants, the sum of the rates of all reactions that form X, $\sum v_X$, is approximately equal to the sum of the rates of all reactions that remove X, $\sum v_{-X}$:

$$\frac{d[X]}{dt} = \sum v_X - \sum v_{-X} = 0 \tag{8.19}$$

The steady-state treatment is of great importance in the analysis of composite mechanisms, since often there are mathematical difficulties that make it impossible to obtain an explicit solution of the rate equations. Consider, for example, the mechanism

$$(1) \quad A + B \rightleftharpoons X$$

$$(2) \quad\;\; X \rightarrow Z$$

The differential rate equations that apply to this set of reactions are

$$-\frac{d[A]}{dt} = -\frac{d[B]}{dt} = k_1[A][B] - k_{-1}[X] \tag{8.20}$$

$$\frac{d[X]}{dt} = k_1[A][B] - k_{-1}[X] - k_2[X] \tag{8.21}$$

$$\frac{d[Z]}{dt} = k_2[X] \tag{8.22}$$

To treat this problem exactly it would be necessary to eliminate $[X]$ and to solve the resulting differential equation to find $[Z]$ as a function of t. Unfortunately, however, in spite of the simplicity of the kinetic scheme, it is not possible to obtain an explicit solution; computers, however, provide numerical solutions for particular values of the rate constants.

The steady-state treatment, which is valid provided that the concentration of X is always small, involves using Eq. (8.18) so that, from Eq. (8.21),

$$k_1[A][B] - k_{-1}[X] - k_2[X] = 0 \tag{8.23}$$

The concentration of $[X]$ therefore is given by

$$[X] = \frac{k_1[A][B]}{k_{-1} + k_2} \tag{8.24}$$

and insertion of this into Eq. (8.22) gives

$$v = v_Z = \frac{d[Z]}{dt} = \frac{k_1 k_2[A][B]}{k_{-1} + k_2} \tag{8.25}$$

8.2.3 Rate-Determining (Rate-Controlling) Steps

Suppose that in the reaction scheme just considered the intermediate X is converted very rapidly into Y, much more rapidly than it can go back into A + B. In that case the rate of the reaction is the rate of formation of X from A + B; that is,

$$v = k_1[A][B] \tag{8.26}$$

since as soon as X is formed, it is transformed into Y. The initial step therefore is the *rate-determining step*. The exact condition, for this two-step mechanism, is

$$k_2 \gg k_{-1}$$

and the steady-state rate equation (8.25) becomes Eq. (8.26) if this inequality is satisfied.

Alternatively, suppose that the rate constant for the second reaction, X → Y, is very small compared to that for the reverse of the first reaction; that is,

$$k_2 \ll k_{-1}$$

The overall rate is

$$v = k_2[X] \tag{8.27}$$

and since reaction 2 is too slow to disturb the equilibrium A + B ⇌ X,

$$[X] = \frac{k_1}{k_{-1}}[A][B] \tag{8.28}$$

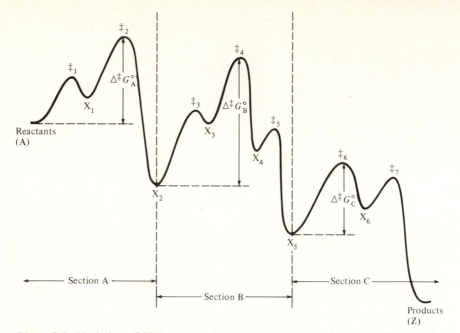

Figure 8.2   Variation of Gibbs energy throughout the course of a hypothetical reaction involving a number of intermediates, $X_1, X_2, \ldots, X_6$.

Insertion of Eq. (8.28) into (8.27) gives

$$v = \frac{k_1 k_2}{k_{-1}} [A][B] \tag{8.29}$$

Again, this is the expression to which Eq. (8.25) reduces if the inequality $k_{-1} \gg k_2$ is satisfied. Reaction 2 is now the rate-determining step or the rate-controlling step.§

It is important to work out each kinetic scheme separately because there are some pitfalls. Note in particular that when there is a chain reaction (see Section 8.3), there is usually no rate-determining step: the rate is not equal to or controlled by that of any particular step.

In more complicated situations it is by no means a straightforward matter to decide which is the rate-determining step. Consider, for example, the reaction scheme

$$A \rightleftharpoons X_1 \rightleftharpoons X_2 \rightleftharpoons X_3 \rightleftharpoons X_4 \rightleftharpoons X_5 \rightleftharpoons X_6 \rightleftharpoons Z$$

and suppose that the standard Gibbs energies for the reactants, products, and six intermediates are as represented in Fig. 8.2. It has been suggested[10] that the reaction rate would be controlled by the reaction leading to the highest activated complex; for

---

§ Physical-organic chemists sometimes make a distinction between rate-determining and rate-controlling steps. The former is used when the overall rate is equal to the rate of the first step in a consecutive mechanism, the latter when a later reaction is involved, the overall rate also involving equilibrium constants for pre-equilibria [as in Eq. (8.29)]. However, this distinction is not recognized universally; those who make the distinction should emphasize that they are doing so.

the system shown in Fig. 8.2 this is $\ddagger_2$, so that the rate-determining step would be $X_1 \to X_2$. This, however, is incorrect,[11] since the argument leading to the conclusion ignored the possibility of the buildup of concentrations of intermediates. The correct solution was given by Murdoch,[12] who analyzed the problem in detail and proposed an intuitive method of deciding on the rate-determining step.

Murdoch's procedure is to divide the reaction sequence into sections, as shown in Fig. 8.2. The first section, Section A, begins with the reactants and ends with the first intermediate that is more stable than the reactants; this is X_2. The second section, B, begins with X_2 and ends with the next intermediate, X_5, that is more stable than X_2. There is no intermediate more stable than X_5, and the third section, C, therefore begins at X_5 and ends at the products.

The processes within the sections can be lumped together, so that the scheme reduces to

$$A \rightleftharpoons X_2 \rightleftharpoons X_5 \rightleftharpoons Z$$

Now the Gibbs energy differences between the beginning of each section and the highest activated state are considered: these are $\Delta^{\ddagger}G_A^{\circ}$, $\Delta^{\ddagger}G_B^{\circ}$, and $\Delta^{\ddagger}G_C^{\circ}$. For the scheme shown in the figure the largest of these is $\Delta^{\ddagger}G_B^{\circ}$. This being so, there will be an accumulation of the intermediate X_2, since X_2 is converted into X_5 more slowly than it is formed from reactants. The intermediate X_5 will give products rapidly and therefore will be present in small amounts.

It follows from this analysis that the bottleneck in the reaction scheme lies in Section B. More specifically, it is the highest activated complex in this sequence that controls the overall rate; the rate-determining step for the entire process therefore is the reaction

$$X_3 \to X_4$$

The procedure may be summarized as follows. First divide the sequence into sections, which end at intermediates of steadily decreasing Gibbs energy. Then determine $\Delta^{\ddagger}G^{\circ}$ for each section; it is the difference between the Gibbs energy of the highest activated complex and that of the first intermediate in the section. The section with the largest $\Delta^{\ddagger}G^{\circ}$ contains the rate-determining step, which is that leading through the activated complex of highest Gibbs energy within that section.

This is the procedure for first-order reactions. The principles are essentially the same for processes of other orders, but concentration terms must be introduced into the Gibbs energy functions; for details, Murdoch's paper[12] should be consulted.

8.2.4 Microscopic Reversibility and Detailed Balance

We now consider certain consequences of the principles of microscopic reversibility and of detailed balance, which were introduced in Section 4.10.

For an *elementary* reaction the equilibrium constant must be the ratio of the rate constants in the forward and reverse directions. Thus, consider the process

$$A + B \rightleftharpoons Y + Z$$

in which the reactions in the forward and reverse directions, as indicated by the filled-in arrows, are elementary. The rates in the two directions are

$$v_1 = k_1[A][B] \tag{8.30}$$

and
$$v_{-1} = k_{-1}[Y][Z] \tag{8.31}$$

If the system is at equilibrium, these rates are equal; hence,

$$\frac{k_1}{k_{-1}} = \left(\frac{[Y][Z]}{[A][B]}\right)_{eq} = K_c \tag{8.32}$$

where K_c is the equilibrium constant.

This argument can be extended to a reaction that occurs in two or more stages. Consider, for example, the reaction

$$H_2 + 2ICl \rightleftharpoons I_2 + 2HCl$$

which occurs in two steps. At equilibrium the processes are occurring at equal rates in the forward and reverse directions:

$$(1) \quad H_2 + ICl \rightleftharpoons HI + HCl$$

$$(2) \quad HI + ICl \rightleftharpoons HCl + I_2$$

Thus, at equilibrium

$$k_1[H_2][ICl] = k_{-1}[HI][HCl] \tag{8.33}$$

$$k_2[HI][ICl] = k_{-2}[HCl][I_2] \tag{8.34}$$

The equilibrium constant for each reaction is thus

$$K_1 = \frac{k_1}{k_{-1}} = \left(\frac{[HI][HCl]}{[H_2][ICl]}\right)_{eq} \tag{8.35}$$

$$K_2 = \frac{k_2}{k_{-2}} = \left(\frac{[HCl][I_2]}{[HI][ICl]}\right)_{eq} \tag{8.36}$$

The product of these two equilibrium constants is

$$K_1 K_2 = \frac{k_1 k_2}{k_{-1} k_{-2}} = \left(\frac{[I_2][HCl]^2}{[H_2][ICl]^2}\right)_{eq} = K_c \tag{8.37}$$

where K_c is the equilibrium constant for the overall reaction. For any mechanism, involving any number of elementary and consecutive steps, the overall equilibrium constant is the product of the equilibrium constants for the individual steps and is the product of the rate constants for the reactions in the forward direction divided by the product of those for the reverse reactions:

$$K_c = K_1 K_2 K_3 \cdots = \frac{k_1 k_2 k_3 \cdots}{k_{-1} k_{-2} k_{-3} \cdots} \tag{8.38}$$

If a reaction occurs by a composite mechanism, and we measure a rate coefficient k_1 for the overall reaction from left to right and also measure a rate coefficient k_{-1} for the overall reaction from right to left, at the same temperature, the ratio k_1/k_{-1} is not necessarily the equilibrium constant for the overall reaction. The reason is that rate laws for composite reactions change with the experimental conditions, such as reactant concentrations, and the rate coefficients also change. The ratio of the rate coefficients k_1 and k_{-1} that apply when the system is at equilibrium is equal to the equilibrium constant, but rate coefficients determined away from equilibrium are not necessarily the same as those at equilibrium, and their ratio is not necessarily equal to K_c. Therefore, great caution should be used in deducing rate coefficients and rate laws for reactions from the equilibrium constant and the rate coefficient for the reverse reaction.

As an example of a situation where the ratio of rate constants is not the equilibrium constant, consider the reaction system

$$A \underset{-1}{\overset{1}{\rightleftharpoons}} X \underset{-2}{\overset{2}{\rightleftharpoons}} Z$$

If the system is at complete equilibrium,

$$\left(\frac{[X]}{[A]}\right)_{eq} = \frac{k_1}{k_{-1}} \quad \text{and} \quad \left(\frac{[Z]}{[X]}\right)_{eq} = \frac{k_2}{k_{-2}} \tag{8.39}$$

and

$$\left(\frac{[Z]}{[A]}\right)_{eq} = \frac{k_1 k_2}{k_{-1} k_{-2}} = K_c \tag{8.40}$$

If one makes measurements of the rate of consumption of A at the very beginning of the reaction, before any X and Z have accumulated,

$$-\frac{d[A]}{dt} = k_1[A] \tag{8.41}$$

and the first-order rate constant is k_1. Similarly, if one starts with pure Z and measures initial rates of consumption of Z, the rate constant obtained is k_{-2}. In general, the ratio k_1/k_{-2} is not equal to the equilibrium constant K_c.

The point is that the initial rate constants measured, k_1 and k_{-2}, are not the rate constants that apply when the system is at equilibrium. When the system is at equilibrium,

$$\frac{d[X]}{dt} = k_1[A] - (k_{-1} + k_2)[X] + k_{-2}[Z] = 0 \tag{8.42}$$

and therefore

$$[X] = \frac{k_1[A] + k_{-2}[Z]}{k_{-1} + k_2} \tag{8.43}$$

The net rate of consumption of A is

$$-\frac{d[A]}{dt} = k_1[A] - k_{-1}[X] \tag{8.44}$$

Introduction of the expression for [X] gives, with some rearrangement,

$$-\frac{d[A]}{dt} = \frac{k_1 k_2}{k_{-1} + k_2}[A] - \frac{k_{-1} k_{-2}}{k_{-1} + k_2}[Z] \tag{8.45}$$

The first term is the rate of the reaction from left to right, and the corresponding rate constant is $k_1 k_2/(k_{-1} + k_2)$. Similarly, the rate constant from right to left is

$k_{-1}k_{-2}/(k_{-1} + k_2)$. The ratio of these two rate constants, $k_1k_2/k_{-1}k_{-2}$, is equal to the equilibrium constant.

If the rate constants that are measured for forward and reverse directions are those that apply when the system is at equilibrium, their ratio is bound to be an equilibrium constant for the reaction. This follows from the fact that at equilibrium the rates must be the same in both directions. For a system not at equilibrium, however, the ratio of rate constants may be very different from the equilibrium constant.[13]

8.3 CHAIN REACTIONS

Atoms and free radicals play a special role as reaction intermediates. Because they have incomplete electron shells they are usually highly reactive, often being able to react with stable molecules at ordinary temperatures. They commonly bring about abstraction reactions, such as

$$Cl + H_2 \rightarrow HCl + H$$

and

$$CH_3 + H_2 \rightarrow CH_4 + H$$

In doing so they produce new atoms and radicals, which can bring about other reactions.

Because of their high reactivity, atoms and free radicals are present in reaction systems only at very low concentrations; it is usually permissible to apply the steady-state treatment to them.§ More important, they frequently participate in what are called *chain reactions*. For example, in the reaction between hydrogen and chlorine, for which the stoichiometric equation is

$$H_2 + Cl_2 = 2HCl$$

chlorine atoms bring about the following reaction sequence:

$$(1) \quad Cl + H_2 \rightarrow HCl + H$$

$$(2) \quad H + Cl_2 \rightarrow HCl + Cl$$

In reaction (1) a chlorine atom is lost, but in reaction (2) a chlorine atom is formed. Similarly, in reaction (2) a hydrogen atom is consumed, but one is formed in reaction (1). Therefore, reactions (1) and (2) together constitute a closed cycle or sequence of reactions, in which two molecules of the product HCl are formed. There is no net loss of H or Cl atoms when the two reactions occur, and if the cycle is repeated many times a considerable amount of product is formed from a single chlorine atom. Provided that such a cycle of reactions occurs more than once, the overall reaction is called a chain reaction. In the hydrogen–chlorine reaction, under some conditions, the cycle occurs a very large number of times (often of the order of 10^6).

This particular cycle of reactions for the hydrogen–chlorine system was first suggested by Nernst[14] in 1918. His proposal was put forward to explain the very surprising result, obtained by Bodenstein and Dux,[15] that the number of molecules of product formed per photon of light absorbed (known as the *quantum yield*) was

§ An exception is when an explosion occurs; see Section 8.6.

sometimes as high as 10^6. The significance of this is that one quantum of light produces two chlorine atoms (Section 9.1), and that two chlorine atoms can give rise to 10^6 molecules of the product HCl. This result is explained by the reaction cycle considered above, provided that the cycle, on average, is repeated many times.

Although Nernst was the first to propose the correct propagating steps for the hydrogen–chlorine reaction, he was not the first to suggest the idea of a chain reaction. This had been done in 1913 by Bodenstein[16] who, however, postulated ions such as Cl_2^+ as the chain carriers; later, he suggested electronically excited molecules.[17] Ions and excited molecules are involved only under rather exceptional circumstances; most chain reactions are composed of atomic and free-radical reactions.

The idea that organic free radicals are important in reaction mechanisms came a few years after Nernst proposed his mechanism for the hydrogen–chlorine reaction. In 1925, H. S. Taylor[18] described investigations of the reactions of hydrogen atoms with various substances, including ethylene. He proposed that the reaction

$$H + C_2H_4 \rightarrow C_2H_5$$

can occur and that the resulting ethyl radical plays an important role in hydrocarbon reactions. This idea was followed up by F. O. Rice and his coworkers,[19] who demonstrated the presence of organic free radicals in reaction systems involving organic compounds such as hydrocarbons. They did this by making use of the "mirror technique," introduced by Paneth and Hofeditz.[20] A thin metallic deposit on a glass surface, such as one of lead, is removed by free radicals which combine with the metal to form volatile compounds such as $Pb(C_2H_5)_4$.

Taylor[21] later reported that these early ideas on organic free radicals first "met with a chilly reception." The ideas became more palatable after Rice and Herzfeld[22] proposed specific reaction schemes involving the participation of free radicals, a matter considered in Section 8.5. Further strong support was provided by new experimental techniques, such as spectroscopy, electron-spin resonance spectroscopy, and mass spectrometry, which have demonstrated beyond question that free radicals are important in many reactions and that they can act as chain carriers.

A chain reaction must consist of at least four elementary processes, which will be illustrated for the reaction between hydrogen and bromine. First, there must be an *initiation* reaction, in which chain carriers are formed; for the thermal hydrogen–bromine reaction this is

$$Br_2 \rightarrow 2Br$$

This is followed by two *chain-propagating* reactions:

$$Br + H_2 \rightarrow HBr + H$$

$$H + Br_2 \rightarrow HBr + Br$$

Finally, there must be a *termination* reaction, in which chain carriers are removed from the system. There may be other reactions, but in a chain reaction there must always be an initiation reaction, a termination reaction, and a pair of chain-propagating reactions.

A useful quantity in connection with chain reactions is the *chain length,* often given the symbol γ. This is defined as the average number of times the closed cycle of reactions (comprising the chain-propagating steps) is repeated. Thus, it is equal to the rate of the overall reaction divided by the rate of the initiation step in which the chain carriers are formed.

8.3.1 Chain Initiation Processes

In a purely thermal reaction the initiation process must be a thermal process. For example, when a mixture of hydrogen and bromine gases is heated, some thermal dissociation of bromine molecules occurs:

$$Br_2 \rightleftharpoons 2Br$$

The initiation process is thus the thermal dissociation of bromine molecules:

$$Br_2 \rightarrow 2Br$$

Similarly, in the thermal decomposition of ethane (Section 8.5.3) the initiation process is

$$C_2H_6 \rightarrow 2CH_3$$

These unimolecular processes are first order at sufficiently high pressures and second order at low pressures. In the case of the dissociation of a diatomic molecule such as Br_2, first-order kinetics are obtained only at very high pressures; under ordinary conditions the kinetics are second order.

Photochemical and radiation-chemical processes are treated in more detail in Chapter 9, and only a few points are mentioned here. In the photochemical hydrogen–bromine reaction (Section 8.4.1) the initiation process is

$$Br_2 + h\nu \rightarrow 2Br$$

where $h\nu$ represents a photon of radiation. The process can be brought about only by a photon that is absorbed by the bromine molecule. If the energy $h\nu$ of the absorbed photon is sufficiently great the Br_2 molecule may be dissociated into atoms. According to Einstein's *law of photochemical equivalence,* there is a one-to-one relationship between photons absorbed and molecules decomposed. A measurement of the intensity of radiation absorbed, at wavelengths where dissociation is brought about, allows an estimate to be made of the number of active species (e.g., Br atoms) produced in a particular experiment. It is useful to express the intensity of radiation absorbed as mole-photons per unit volume per unit time. A mole-photon of radiation is often called an *einstein:* in the case of bromine, for example, 1 einstein of radiation absorbed produces 2 mol of Br atoms.

When the initiation is brought about by the absorption of high-energy electromagnetic or particle radiation (e.g., β or γ rays) the situation is a good deal more complicated. Ions are produced in addition to atoms and free radicals, and there is no longer a simple relationship between the amount of radiation absorbed and the number of active species formed in the initiation process. Also, the initiation process itself often occurs in a number of steps in a radiation-chemical reaction. This matter is considered in further detail in Section 9.3, which describes some of the elementary steps that occur in some relatively simple systems.

8.4 SOME INORGANIC MECHANISMS

Many reactions, in the gas phase and in solution, are composite, and only a few of them can be considered here. The present section deals with some inorganic reactions in the gas phase, while Section 8.5 deals with some organic decompositions.

8.4.1 Hydrogen–Bromine Reaction

It is convenient to start with the reaction between hydrogen and bromine in the gas phase, because its mechanism is fairly simple. The thermal reaction was first studied by Bodenstein and Lind[23] over the temperature range 205–302°C. They found empirically that the results could be fitted to the following expression for the rate of consumption of H_2 or Br_2:

$$v_t = \frac{k[H_2][Br_2]^{1/2}}{1 + [HBr]/m[Br_2]} \tag{8.46}$$

where k and m are constants; the value of m is about 10 and is practically independent of temperature. This equation shows that hydrogen bromide inhibits the reaction and that the extent of inhibition by HBr is reduced by the addition of Br_2. The photochemical H_2–Br_2 reaction also has been investigated and is considered later.

Bodenstein and Lind pointed out that the appearance of the bromine concentration to the one-half power suggested that bromine molecules were dissociated during the reaction. However, the reaction was not explained properly until 1919, the year after Nernst had suggested a mechanism for the hydrogen–chlorine reaction. The now generally accepted mechanism of the thermal hydrogen–bromine reaction was proposed simultaneously by Christiansen,[24] Herzfeld,[25] and Polanyi.[26] Their mechanism is

$$\begin{array}{lll} (1) & Br_2 \rightarrow 2Br & \text{Initiation} \\[4pt] (2) & Br + H_2 \rightarrow HBr + H & \left.\right\} \text{Chain propagation} \\ (3) & H + Br_2 \rightarrow HBr + Br & \\[4pt] (-2) & H + HBr \rightarrow H_2 + Br & \\[4pt] (-1) & 2Br \rightarrow Br_2 & \text{Termination} \end{array}$$

This scheme has the characteristic features of a chain reaction, namely, the initiation and termination steps and the pair of chain-propagation steps. It also has an additional reaction, reaction (-2), which accounts for the inhibition of the reaction by the product HBr. The fact that Br_2 and HBr compete with one another for H atoms explains why the inhibition by HBr is reduced by Br_2, that is, why $[HBr]/[Br_2]$ appears in the denominator of Eq. (8.46).

The steady-state hypothesis must be applied to the two intermediates Br and H, both of which are present at very low concentrations. The steady-state equation for H is

$$\frac{d[H]}{dt} = k_2[Br][H_2] - k_3[H][Br_2] - k_{-2}[H][HBr] = 0 \tag{8.47}$$

That for Br is

$$\frac{d[\text{Br}]}{dt} = k_1[\text{Br}_2] - k_2[\text{Br}][\text{H}_2] + k_3[\text{H}][\text{Br}_2] + k_{-2}[\text{H}][\text{HBr}] - k_{-1}[\text{Br}]^2 = 0 \quad (8.48)$$

A solution for [Br] is obtained by adding these two equations:§

$$k_1[\text{Br}_2] - k_{-1}[\text{Br}]^2 = 0 \quad (8.49)$$

and therefore

$$[\text{Br}] = \left(\frac{k_1}{k_{-1}}\right)^{1/2} [\text{Br}_2]^{1/2} \quad (8.50)$$

This is the equilibrium concentration of Br atoms. It is by no means always the case that atoms and free radicals in reaction systems are present at their equilibrium concentrations. In the present example, the H atoms are present at much higher concentrations than their equilibrium concentrations.

An expression for [H] can be obtained by inserting this expression for [Br] into either Eq. (8.47) or (8.48). Insertion into Eq. (8.47) leads to

$$k_2\left(\frac{k_1}{k_{-1}}\right)^{1/2} [\text{Br}_2]^{1/2}[\text{H}_2] - k_3[\text{H}][\text{Br}_2] - k_{-2}[\text{H}][\text{HBr}] = 0 \quad (8.51)$$

so that

$$[\text{H}] = \frac{k_2(k_1/k_{-1})^{1/2}[\text{H}_2][\text{Br}_2]^{1/2}}{k_3[\text{Br}_2] + k_{-2}[\text{HBr}]} \quad (8.52)$$

The rate of reaction is the rate of consumption of H_2, which is

$$v_t = k_2[\text{Br}][\text{H}_2] - k_{-2}[\text{H}][\text{HBr}] \quad (8.53)$$

Subtraction of Eq. (8.47) gives

$$v_t = k_3[\text{H}][\text{Br}_2] \quad (8.54)$$

and insertion of the expression for [H] gives

$$v_t = \frac{k_2 k_3 (k_1/k_{-1})^{1/2}[\text{H}_2][\text{Br}_2]^{3/2}}{k_3[\text{Br}_2] + k_{-2}[\text{HBr}]} \quad (8.55)$$

$$= \frac{k_2(k_1/k_{-1})^{1/2}[\text{H}_2][\text{Br}_2]^{1/2}}{1 + (k_{-2}/k_3)[\text{HBr}]/[\text{Br}_2]} \quad (8.56)$$

This is of the same form as the empirical equation (8.46), and we note that the empirical k is equal to $k_2(k_1/k_{-1})^{1/2}$ and that $m = k_3/k_{-2}$.

All the rate constants k_1, k_2, k_3, k_{-2}, and k_{-1} have been evaluated. The experimental quantity k, equal to $k_2(k_1/k_{-1})^{1/2}$, together with a value for the equilibrium constant k_1/k_{-1} leads to k_2; from its temperature dependence E_2 is obtained. Reaction (-2) is the reverse of reaction (2), so that from the equilibrium constant k_2/k_{-2} and from the thermochemistry for the reaction, k_{-2} and E_{-2} can be obtained. The values

---

§ In dealing wth composite reactions, labor often can be saved by adding or subtracting steady-state equations.

of k_3 and E_3 can be found from the value of m; since this is temperature independent, E_3 is equal to E_{-2}.

The values of k_1 and k_{-1} were obtained by comparing the thermal and photochemical hydrogen–bromine reactions. Bodenstein and Lutkemeyer[27] found that the photochemical reaction proceeds according to the equation

$$v_p = \frac{k'[H_2]I^{1/2}}{1 + [HBr]/m'[Br_2]} \tag{8.57}$$

where k' and m' are constants and I is the intensity of light absorbed. The similarity between this equation and Eq. (8.46) suggests that the mechanisms are similar, the only difference being that the initiation reaction in the photochemical reaction is the absorption of a photon by a bromine molecule:

$$Br_2 + h\nu \rightarrow 2Br$$

This process is followed by the same four reactions as in the thermal reaction. The steady-state equation for bromine atoms is now§

$$\frac{d[Br]}{dt} = 2I - k_2[Br][H_2] + k_3[H]Br_2 + k_{-2}[H][HBr] - k_{-1}[Br]^2 = 0 \tag{8.58}$$

instead of Eq. (8.48). Equation (8.47) is still the steady-state equation for H, and addition of Eqs. (8.47) and (8.58) gives

$$2I - k_{-1}[Br]^2 = 0 \tag{8.59}$$

or

$$[Br] = \left(\frac{2I}{k_{-1}}\right)^{1/2} \tag{8.60}$$

The rate equation is obtained from Eq. (8.56) by replacing $k_1[Br_2]$ by $2I$:

$$v_p = \frac{k_2(2/k_{-1})^{1/2}[H_2]I^{1/2}}{1 + (k_{-2}/k_3)[HBr]/[Br_2]} \tag{8.61}$$

in agreement with the empirical equation (8.57).

It follows from Eqs. (8.56) and (8.61) that the ratio of the photochemical and thermal rates is

$$\frac{v_p}{v_t} = \left(\frac{2}{k_1}\right)^{1/2} \frac{I^{1/2}}{[Br_2]^{1/2}} \tag{8.62}$$

at the same concentrations of H_2 and Br_2. Therefore, if v_p/v_t is measured at known values of I and $[Br_2]$, the rate constant k_1 can be obtained. If the experiments are repeated at different temperatures, the parameters A_1 and E_1 are obtained. For example, in one set of experiments carried out by Bodenstein and Lutkemeyer, the rate of the photochemical reaction was 300 times that of the thermal reaction. This means that [Br] in the photochemical reaction is 300 times that in the thermal reaction. The

---

§ If I is expressed in einsteins (mole-photons) per unit volume per unit time, $2I$ is the amount of Br produced under the same conditions.

concentration of Br atoms in the thermal reaction can be calculated from the equilibrium constant for the Br_2 dissociation; thus, the value of [Br] in the photochemical reaction is known. The rate of production of bromine atoms in the photochemical experiment is known from the intensity of light absorbed, and in the steady state this is the rate of removal of atoms. Therefore, the rate constant for the combination reaction, reaction (-1), may be calculated.

A point of some importance concerns the nature of the initiating and chain-ending steps. It is seen from Eq. (8.50) that the bromine atom concentration in the thermal reaction is the same as it would be in the absence of hydrogen, because k_1/k_{-1} is the equilibrium constant for the dissociation $Br_2 \rightleftharpoons 2Br$. This is by no means a necessary feature of chain reactions, in which atom and radical concentrations are frequently very different from their equilibrium values. In the ethane decomposition, for example, the concentration of ethyl radicals is considerably higher than if they were produced by the establishment of the equilibrium $C_2H_6 \rightleftharpoons H + C_2H_5$.

Since the bromine atoms are at equilibrium with the bromine molecules, the concentration of bromine atoms (and hence the concentration of hydrogen atoms and the rate of reaction) is unaffected by any catalysts that affect the rate of production of bromine atoms: such catalysts cannot affect the equilibrium concentration. Therefore, a change in the nature of the surface or in its area cannot affect the rate of the reaction, and this has been found to be the case. Furthermore, the presence of third bodies, which may accelerate the formation of atoms, cannot affect their concentration. For a molecule as simple as bromine, with only one degree of vibrational freedom, the dissociation is in the second-order region, and the reaction should have been represented as

$$(1') \quad Br_2 + M \rightarrow 2Br + M$$

where M is a third body that may be a bromine or hydrogen molecule. Similarly, reaction (-1) should be written as a third-order reaction,

$$(-1) \quad 2Br + M \rightarrow Br_2 + M$$

In deriving the steady-state equations, k_1 should have been written as $k'_1[M]$ and k_{-1} should have been written as $k'_{-1}[M]$. The constants k_1 and k_{-1} only appear in the rate equation as the ratio k_1/k_{-1}, which now becomes k'_1/k'_{-1}. No change is involved because this ratio is still the equilibrium constant. Therefore, the rate expression derived above is valid, and foreign gases can have no effect on the rate. This prediction is confirmed by experiment.

Table 8.2 shows activation energies for the various reactions that are involved in the hydrogen–bromine reaction. The table also gives a calculated value of 190 kJ mol^{-1} for the hypothetical molecular reaction. It is seen from Eq. (8.56) that initially, in the absence of the product HBr, the rate constant for the thermal reaction is $k_2(k_1/k_{-1})^{1/2}$; the activation energy is therefore $E_2 + \frac{1}{2}(E_1 - E_{-1})$. Use of the values in Table 8.2 leads to $96 + 72 = 168$ kJ mol^{-1}, which is considerably less than the value of 190 kJ mol^{-1} for the molecular reaction. This explains why the chain mechanism is much more important than the molecular one.

TABLE 8.2 EXPERIMENTAL AND ESTIMATED ACTIVATION
ENERGIES FOR HYDROGEN–HALOGEN REACTIONS

	Activation energy E/kJ mol^{-1}		
Reaction	X = Cl	X = Br	X = I
(1) $X_2 \rightarrow 2X$	234	192	142
(2) $X + H_2 \rightarrow HX + H$	25	72	140
(3) $H + X_2 \rightarrow HX + X$	~13	5	0
(−2) $H + HX \rightarrow H_2 + X$	21	5	6
(−1) $2X \rightarrow X_2$	0	0	0
(4) $H_2 + X_2 \rightarrow 2HX$	210^a	190^a	170
$E_2 + \frac{1}{2}(E_1 - E_{-1})$	142	168	211

^{a} Estimated by A. Wheeler, B. Topley, and H. Eyring, *J. Chem. Phys.*, **4**, 178 (1936).
The remaining values are experimental, assembled from various sources listed in the
bibliography at the end of Chapter 5.

8.4.2 Hydrogen–Chlorine Reaction

The thermal and photochemical reactions between hydrogen and chlorine show some
resemblance to the hydrogen–bromine reactions, but the mechanisms are less straightforward. The photochemical hydrogen–chlorine reaction is considered first because it
is a little easier to understand than the thermal reaction.

The reaction has a number of complicating experimental features, which at first
caused some confusion. For example, there often is a long induction period during
which no reaction occurs. This was shown by Burgess and Chapman[28] to be due to
the presence of nitrogenous impurities, and later Griffiths and Norrish[29] found that
the real inhibitor is nitrogen trichloride, formed from the impurities.

Oxygen, even in minute amounts, has a profound effect on the rate, and early
kinetic studies showed important discrepancies because different amounts of oxygen
were present in the reaction system as an impurity. To a good approximation the rate
of the reaction is inversely proportional to the concentration of oxygen,[30] and as the
gases are purified from oxygen the rate becomes extremely high.

A useful but slightly oversimplified expression for the rate of the photochemical
reaction was given by Bodenstein and Unger:[31]

$$\frac{d[\text{HCl}]}{dt} = \frac{kI[\text{H}_2][\text{Cl}_2]}{m[\text{Cl}_2] + [\text{O}_2]([\text{H}_2] + n[\text{Cl}_2])} \tag{8.63}$$

where I is the intensity of light absorbed and m and n are constants. In contrast to
the situation with the hydrogen–bromine reaction, I and $[\text{Cl}_2]$ appear to the first power
and not to the one-half power.

At oxygen pressures greater than about 0.04 mmHg, the first term in the denominator of Eq. (8.63) is negligible, and the rate is inversely proportional to the
oxygen pressure. In the complete absence of oxygen the rate is independent of the
chlorine pressure.

The initial step in the photochemical reaction is the dissociation of a chlorine
molecule into atoms. It is known from the spectral evidence that chlorine is dissociated

by wavelengths of less than 478.5 nm, and it is at such wavelengths that most of the absorption takes place. More direct evidence was obtained by Jost and Schweizer,[32] who exposed chlorine to radiation immediately before mixing it with hydrogen, the radiation being prevented from reaching the hydrogen; reaction was found to take place.

The Nernst scheme of reactions (mentioned in Section 8.3) is along the right lines, but it requires some modification to account for the effect of oxygen and of other impurities. The following scheme, similar to one proposed in 1921 by Gohring,[33] represents the behavior fairly satisfactorily:

$$(1) \quad Cl_2 + h\nu \rightarrow 2Cl$$

$$(2) \quad Cl + H_2 \rightarrow HCl + H$$

$$(3) \quad H + Cl_2 \rightarrow HCl + Cl$$

$$(4) \quad H + O_2 \rightarrow HO_2$$

$$(5) \quad Cl + O_2 \rightarrow ClO_2$$

$$(6) \quad Cl + X \rightarrow ClX$$

Here X is any substance that can remove chlorine atoms. The chain-ending step $Cl + Cl \rightarrow Cl_2$ is not included in this scheme, since under ordinary conditions Cl atoms are removed more effectively by reactions (5) and (6); the reason for this is considered later. The species HO_2 that is formed in reaction (4) is unstable and is converted eventually into water and oxygen.

The rate of formation of Cl atoms by reaction (1) is $2I$, and the steady-state equation for Cl atoms is

$$\frac{d[Cl]}{dt} = 2I - k_2[Cl][H_2] + k_3[H][Cl_2] - k_5[Cl][O_2] - k_6[Cl][X] = 0 \quad (8.64)$$

The steady-state equation for H atoms is

$$\frac{d[H]}{dt} = k_2[Cl][H_2] - k_3[H][Cl_2] - k_4[H][O_2] = 0 \quad (8.65)$$

from which it follows that

$$[Cl] = \frac{k_3[H][Cl_2] + k_4[H][O_2]}{k_2[H_2]} \quad (8.66)$$

Introduction of this expression into Eq. (8.64), with some rearrangement and the neglect of a small term involving $[O_2]^2$, gives

$$[H] = \frac{2Ik_2[H_2]}{k_3k_6[Cl_2][X] + [O_2](k_2k_4[H_2] + k_3k_5[Cl_2] + k_4k_6[X])} \quad (8.67)$$

The rate of formation of HCl is

$$v_{HCl} = \frac{d[HCl]}{dt} = k_2[Cl][H_2] + k_3[H][Cl_2] \quad (8.68)$$

and subtraction of Eq. (8.65) gives

$$v_{HCl} = 2k_3[H][Cl_2] + k_4[H][O_2] \tag{8.69}$$

At low concentrations of oxygen the second term can be neglected, and then, with the introduction of Eq. (8.67),

$$v_{HCl} = \frac{2k_2k_3[H_2][Cl_2]I}{k_3k_6[Cl_2][X] + [O_2](k_2k_4[H_2] + k_3k_5[Cl_2] + k_4k_6[X])} \tag{8.70}$$

$$= \frac{2k_3/k_4 I[H_2][Cl_2]}{(k_3k_6/k_2k_4)[Cl_2][X] + [O_2]\{[H_2] + (k_3k_5/k_2k_4)[Cl_2] + (k_6/k_2)[X]\}} \tag{8.71}$$

Apart from the final term in the denominator, this expression is of the form of the empirical equation (8.63), the constants being related by

$$k = \frac{2k_3}{k_4} \qquad m = \frac{[X]k_3k_6}{k_2k_4} \qquad n = \frac{k_3k_5}{k_2k_4} \tag{8.72}$$

The scheme of reactions therefore gives a kinetic law that is essentially in agreement with the experimental rate equation.

The thermal reaction between hydrogen and chlorine has been investigated extensively and has some complicating features.[34] To a useful approximation the rate may be expressed as

$$\frac{d[HCl]}{dt} = \frac{k'[H_2][Cl_2]^2}{m'[Cl_2] + [O_2]([H_2] + n'[Cl_2])} \tag{8.73}$$

This can be explained by the scheme considered earlier, with the photochemical initiating reaction (1) replaced by

$$(1') \quad Cl_2 + M \rightarrow 2Cl + M$$

where M is a third body.

The thermal reaction is complicated by the fact that the initial thermal decomposition of a molecule of chlorine may take place on the surface, so that different results are obtained with different sizes and shapes of vessels and with different types of surface. In the photochemical reaction the species X could be part of the surface, so that the theory allows for the surface removal of chlorine atoms; however, the photochemical formation of the atoms takes place entirely in the gas phase. We have seen that this surface complication does not arise in the thermal hydrogen–bromine reaction, since the rate of this reaction depends on the equilibrium concentration of bromine atoms, which are removed by the reverse of the reaction by which they are formed.

Pease[35] carried out experiments with the object of elucidating the wall effect in the thermal hydrogen–chlorine reaction; he compared the rates of reaction in packed and unpacked vessels. If the chains are started on the surface but end largely in the gas phase, increasing the surface increases the rate, but if they start in the gas phase and end on the surface, the rate is decreased by increasing the surface. On the other hand, if both formation and recombination of atoms take place in the gas phase, or if both take place on the surface, changing the amount or nature of the surface has little or no effect on the kinetics. In fact, changing the surface/volume ratio had little effect on the rate. This result alone cannot allow a decision as to whether the chains both start and stop in the gas phase or start and stop at the surface. However, with oxygen in excess more chain ending will take place in the gas phase, by the collision of hydrogen atoms with oxygen molecules, and Pease therefore investigated the effect of packing the vessels in the presence of 10% oxygen. Under these conditions packing gave a large increase in the rate, from which he concluded that chains start on the surface. With little oxygen present they therefore end primarily at the surface. This conclusion that chains are initiated on the surface is confirmed further by quantitative arguments. The energy of activation for the homogeneous dissociation of a chlorine molecule must be at least 240 kJ mol^{-1}, the energy of dissociation, and the rate of this process can be shown to be too low to account for the overall rate of the reaction. If the dissociation takes place on the surface, the activation energy may be as low as one-half the energy of dissociation (Section 7.6.7), and then the rate may be sufficient to account for the rate of the formation of hydrogen chloride.

Activation energies for the elementary processes occurring in the hydrogen–chlorine reactions, obtained in various ways, are shown in Table 8.2.

8.4.3 Hydrogen–Iodine Reaction

The reaction between hydrogen and iodine, the reverse decomposition of hydrogen iodide, and the equilibrium

$$H_2 + I_2 \rightleftharpoons 2HI$$

were investigated very extensively by Bodenstein.[36] The forward reaction obeys second-order kinetics,

$$v_1 = k_1[H_2][I_2] \tag{8.74}$$

as does the reverse reaction,

$$v_{-1} = k_{-1}[HI]^2 \tag{8.75}$$

It was natural to conclude that the reactions were elementary, and this was assumed for a long time. In his kinetic-theory treatment of reactions, W. C. McC. Lewis[37] obtained excellent agreement for the decomposition of hydrogen iodide (Section 4.1).

Certain results, however, have suggested that reactions involving the participation of hydrogen and iodine atoms play a small but significant role. Thus, Rosenbaum and Hogness[38] allowed the reaction $H_2 + I_2 \rightleftharpoons 2HI$ to proceed to equilibrium and studied the rate of conversion of *para*-hydrogen into *ortho*-hydrogen under these conditions. The object of their work was to see whether rates at equilibrium are the same as rates when the system is far from equilibrium, which was the case under the conditions employed by Bodenstein. Their conclusion was that the rate of conversion of *para*-hydrogen was about twice the rate that was predicted on the assumption that the conversion occurred by the process

$$p\text{-}H_2 + I_2 \rightarrow 2HI \rightarrow o\text{-}H_2 + I_2$$

An additional conversion mechanism therefore must exist, and they considered this to be the conversion by the paramagnetic iodine atoms. This was the first clear suggestion that these atoms play a role in the reaction.

Later, Benson and Srinivasan[39] noted that there is a significant temperature dependence of the activation energies for the reaction between hydrogen and iodine, and for the reverse decomposition of hydrogen iodide. They concluded that for the hydrogen–iodine reaction at temperatures above 600 K the following chain reaction plays a significant role:

$$I_2 \rightleftharpoons 2I$$

$$I + H_2 \rightarrow HI + H$$

$$H + I_2 \rightarrow HI + I$$

A similar conclusion was reached by Sullivan,[40] who concluded that this mechanism is dominant at temperatures above 750 K. The reverse decomposition of hydrogen iodide would occur by a chain reaction initiated by I atoms produced as soon as a little I_2 had been formed, and the chain-propagating steps would be the reverse of those for the H_2–I_2 reaction:

$$I + HI \rightarrow I_2 + H$$

$$H + HI \rightarrow H_2 + I$$

These chain mechanisms give rise to different rate equations from the simple second-order equations for the molecular mechanisms (see Problem 8.13).

Another mechanism for the hydrogen–iodine reaction, first suggested by Semenov,[41] is that the iodine atoms produced by the dissociation of I_2 undergo a trimolecular reaction with H_2:

$$I_2 \rightarrow 2I$$

$$I + I + H_2 \rightarrow 2HI$$

This mechanism gives rise to the simple second-order kinetic equation, and the overall kinetics therefore cannot distinguish it from the molecular mechanism. The same is true if the second step proceeds in two stages, via a loosely bound intermediate H_2I, a possibility first pointed out by Sullivan[42]:

$$I + H_2 \rightleftharpoons H_2I$$

$$H_2I + I \rightarrow 2HI$$

Sullivan carried out experiments with iodine atoms produced photochemically, causing them to react with H_2. He concluded that the dominant mechanism at temperatures below those in which the chain reaction is important is this nonchain process involving iodine atoms.

The matter is by no means as simple as had been supposed originally. All the above mechanisms, including the purely molecular mechanisms, play some role, to extents determined by the temperature and other conditions. Further clarification has been provided by molecular-dynamical calculations, particularly those of Anderson and coworkers.[43] They considered various configurations for activated complexes that may be important in the reactions. For example, a trapezoidal complex

$$\overset{H \cdots H^\ddagger}{\underset{I \cdots\cdots\cdots I}{\vdots \qquad \vdots}}$$

may be formed from $H_2 + I_2$ or from 2HI. Also, a linear complex

$$I \cdots H \cdots H \cdots I^{\ddagger}$$

may be formed from two HI molecules, and this may split into $I + H_2I$ or in one stage into $I + H_2 + I$. In the reverse reaction this complex may be formed from $I + H_2 + I$ in a ternary collision, or in two steps via $I + H_2I$. By considering the relevant potential-energy surfaces and making dynamical calculations, Anderson and coworkers were able to confirm that all the various suggested processes could play a significant role in the reactions.

8.4.4 Comparison of the Hydrogen–Halogen Reactions

Although chlorine, bromine, and iodine are homologues, there are significant differences in the way they react with hydrogen. The main differences are the following:

1. The H_2–I_2 reaction is largely a molecular reaction; the others are almost entirely chain reactions.
2. In the H_2–Br_2 reaction there is significant inhibition by the product HBr; in the H_2–Cl_2 reaction there is no noticeable inhibition by HCl.
3. The rate of the thermal H_2–Br_2 reaction is proportional to the square root of the Br_2 concentration, and that of the photochemical reaction to the square root of the intensity of light absorbed. In the H_2–Cl_2 system the dependence is on the first powers instead of the square roots.
4. The chain lengths and the quantum yields in the photochemical reactions are much higher for the hydrogen–chlorine reaction than for the hydrogen–bromine reaction.

These differences can be understood in terms of the activation energies for the elementary steps; these are listed in Table 8.2. If the H_2–I_2 reaction proceeded by a mechanism analogous to that for the H_2–Br_2 reaction, the overall activation energy would be $E_2 + \frac{1}{2}(E_1 - E_{-1}) = 140 + 71 = 211$ kJ mol^{-1}. The molecular process, however, has an activation energy of about 170 kJ mol^{-1} and therefore occurs more rapidly.

The inhibition by HBr of the H_2–Br_2 reaction is understandable in view of the fact that the activation energies for both $H + Br_2$ and $H + HBr$ are about 5 kJ mol^{-1}; the rates are therefore comparable and Br_2 and HBr compete with one another for H atoms. The activation energy for $H + HCl$, 21 kJ mol^{-1}, is significantly larger than that for $H + Cl_2$, about 13 kJ mol^{-1}. Hydrogen chloride molecules therefore do not compete effectively with Cl_2 for H atoms.

In the H_2–Br_2 reaction the reaction $H + Br_2$, with an activation energy of only 5 kJ mol^{-1}, is considerably faster than the reaction $H + O_2$. Oxygen therefore cannot compete with Br_2 for H atoms and does not inhibit the H_2–Br_2 reaction. Hydrogen atoms thus react largely with Br_2 with the formation of Br atoms, the concentration of which is fairly high in the system because of the high activation energy (72 kJ mol^{-1}) for their removal by reaction with H_2. Because of this high concentration of Br atoms in the steady state, the favored mechanism for their removal is by combination with one another.

In the H_2–Cl_2 system the situation is quite different. The activation energy for $H + Cl_2$ is much higher (~ 13 kJ mol^{-1}) than that of $H + Br_2$, and H atoms are more likely to be removed by oxygen molecules. Also, because of the much lower activation energy for $Cl + H_2$ (25 kJ mol^{-1}) as compared with 72 kJ mol^{-1} for $Br + H_2$, the Cl atoms are present in much lower concentrations in the steady state. As a result, the combination reaction $2Cl \rightarrow Cl_2$ is less important than the other chain-ending steps, such as $H + O_2$ and $Cl + O_2$.

It follows from this argument that in the complete absence of oxygen the square-root relationships might be expected in the H_2–Cl_2 reaction, since $2Cl \rightarrow Cl_2$ will provide the only chain-ending step. Experimental evidence for this has been obtained.[44]

The much higher chain lengths for the hydrogen–chlorine reaction than for the hydrogen–bromine reaction may be understood in terms of the much higher rate of the reaction

$$Cl + H_2 \rightarrow HCl + H$$

as compared with

$$Br + H_2 \rightarrow HBr + H$$

The activation energies are 25 kJ mol^{-1} and 72 kJ mol^{-1}, respectively. It is to be seen from Eq. (8.56) that in the absence of HBr the rate of the H_2–Br_2 reaction is proportional to k_2, so that the chain length is also proportional to k_2. The low value of k_2, which relates to $Br + H_2$, therefore makes the chain length small. For the hydrogen–chlorine reaction the rate and chain length depend on the rate constant for the corresponding reaction $Cl + H_2$; because this rate constant is large the chain length is large.

8.4.5 Formation and Decomposition of Phosgene

The rates of formation and decomposition of phosgene,

$$CO + Cl_2 \rightleftharpoons COCl_2$$

have been measured by Christiansen[45] and by Bodenstein and Plaut,[46] who obtained results in essential agreement.

The rate of formation of phosgene obeys the equation

$$\frac{d[COCl_2]}{dt} = k[Cl_2]^{3/2}[CO] \tag{8.76}$$

and the rate of its decomposition obeys the equation

$$-\frac{d[COCl_2]}{dt} = k'[Cl_2]^{1/2}[COCl_2] \tag{8.77}$$

Equating these rates gives the expression for the equilibrium constant,

$$K = \frac{k}{k'} = \frac{[COCl_2]}{[CO][Cl_2]} \tag{8.78}$$

the value of which was determined as a function of the temperature. The rate constants and activation energies of both reactions were also determined.

The expressions for the rates show that the mechanisms are composite, and in particular the presence of the concentration of chlorine molecules to the one-half and three-halves powers indicates that chlorine atoms play a part in the reactions. Bodenstein and Plaut pointed out that the observed results can be explained equally well in terms of two alternative mechanisms, one of which involves the intermediate formation of COCl and the other of Cl_3. The first scheme may be represented by

(1) $Cl_2 \rightleftharpoons 2Cl$

(2) $Cl + CO \rightleftharpoons COCl$

(3) $COCl + Cl_2 \rightleftharpoons COCl_2 + Cl$

Processes (1) and (2) are supposed to be rapid in both directions, and in addition process (2) involves a third body. The equilibria expressed in reactions (1) and (2) may be written as

$$\frac{[Cl]^2}{[Cl_2]} = K_1 \tag{8.79}$$

and
$$\frac{[COCl]}{[CO][Cl]} = K_2 \tag{8.80}$$

Solving for [COCl] and [Cl] gives

$$[COCl] = K_1^{1/2} K_2 [CO][Cl_2]^{1/2} \tag{8.81}$$

and
$$[Cl] = K_1^{1/2}[Cl_2]^{1/2} \tag{8.82}$$

Since the rate of formation of phosgene is controlled by reaction (3), the rate constant for which may be written as k_3, the overall rate is given by

$$\frac{d[COCl_2]}{dt} = k_3[COCl][Cl_2] \tag{8.83}$$

$$= k_3 K_1^{1/2} K_2 [CO][Cl_2]^{3/2} \tag{8.84}$$

which is of the same form as Eq. (8.76). The rate of decomposition of phosgene is given by

$$-\frac{d[COCl_2]}{dt} = k_{-3}[COCl_2][Cl] \tag{8.85}$$

$$= k_{-3} K_1^{1/2}[COCl_2][Cl_2]^{1/2} \tag{8.86}$$

which agrees with Eq. (8.77).

The alternative scheme for the reaction may be represented by

(1) $Cl_2 \rightleftharpoons 2Cl$

(2) $Cl + Cl_2 \rightleftharpoons Cl_3$

(3) $Cl_3 + CO \rightleftharpoons COCl_2 + Cl$

Again the rate of reaction is supposed to be controlled by reaction (3), reactions (1) and (2) being postulated to be rapid in both directions. The treatment is very similar to the one just given and gives rise to kinetic equations of the right form.

On the basis of the thermal reactions alone, a decision cannot be made between the two possible mechanisms. However, investigations of the photochemical formation of phosgene[47] suggest that $COCl$ is a more probable intermediate than Cl_3.

8.4.6 Decomposition of Nitrogen Pentoxide

The decomposition of nitrogen pentoxide,

$$2N_2O_5 \rightarrow 2N_2O_4 + O_2$$

is another reaction about which there originally was much confusion. The kinetics of the reaction was first investigated by Daniels and Johnston,[48] who believed that they had discovered the first true unimolecular gas reaction. They obtained first-order kinetics and assumed that the rate was controlled by the breakdown of N_2O_5 molecules in a single step. They demonstrated that the reaction is not affected by the nature or surface area of the reaction vessel and therefore occurs homogeneously.

At the time of their investigation, in the early 1920s, the radiation hypothesis (Section 4.1) was accepted widely, and Daniels and Johnston devoted much effort to this reaction as a test of the hypothesis. The activation energy was found to be 103 kJ mol^{-1}, which corresponds to a wavelength of 1160 nm, which is in the near infrared. However, radiation of this wavelength had no effect on the rate.

Later, Hunt and Daniels[49] investigated the N_2O_5 decomposition with the object of testing the Lindemann–Christiansen mechanism of unimolecular reactions, according to which the first-order rate coefficient will fall off at low pressures. However, they were unable to find any such fall off. An investigation down to lower pressures was carried out later by Ramsperger, Nordheim, and Tolman,[50] who found a significant fall off at 0.05 mmHg, the rate coefficient reaching half its high-pressure value at 0.005 mmHg. Similar results were obtained by others.[51] However, a difficulty remained, since the fall off occurred at much too low a pressure, even when Hinshelwood's modification [Eq. (5.30)] to the Lindemann–Christiansen hypothesis was taken into account.

That the N_2O_5 decomposition is composite is evident from the fact that there is no way in which a single N_2O_5 molecule can produce N_2O_4 and O_2. This was recognized by Busse and Daniels,[52] who suggested that the reaction occurs by the mechanism

$$N_2O_5 \rightarrow N_2O_3 + O_2 \quad \text{(slow)}$$

$$N_2O_3 \rightarrow NO + NO_2 \quad \text{(fast)}$$

$$NO + N_2O_5 \rightarrow 3NO_2 \quad \text{(fast)}$$

The rate of the reaction would then be that of the first step. However, this does not overcome the difficulty of the failure of the first-order rate coefficient to fall off over the predicted pressure range.

The mechanism now accepted for the nitrogen pentoxide decomposition was first proposed by Ogg[53] and is as follows:

$$(1) \quad N_2O_5 \rightarrow NO_2 + NO_3$$

$$(-1) \quad NO_2 + NO_3 \rightarrow N_2O_5$$

(2) $NO_2 + NO_3 \rightarrow NO_2 + O_2 + NO$

(3) $NO + N_2O_5 \rightarrow 3NO_2$

It is to be noted that NO_2 and NO_3 react together in two different ways, to give N_2O_5 [reaction (-1)] and also to give $NO_2 + O_2 + NO$ [reaction (2)].

The intermediates NO_3 and NO are present in smaller amounts than the N_2O_5, and the steady-state hypothesis therefore may be applied to them:

$$k_1[N_2O_5] - (k_{-1} + k_2)[NO_2][NO_3] = 0 \quad \text{(for } NO_3\text{)} \tag{8.87}$$

$$k_2[NO_2][NO_3] - k_3[NO][N_2O_5] = 0 \quad \text{(for } NO\text{)} \tag{8.88}$$

The rate of consumption of N_2O_5 is

$$-\frac{d[N_2O_5]}{dt} = k_1[N_2O_5] - k_{-1}[NO_2][NO_3] + k_3[NO][N_2O_5] \tag{8.89}$$

With Eq. (8.88) this becomes

$$-\frac{d[N_2O_5]}{dt} = k_1[N_2O_5] + (k_2 - k_{-1})[NO_2][NO_3] \tag{8.90}$$

Elimination of $[NO_2][NO_3]$ by means of Eq. (8.87) gives

$$-\frac{d[N_2O_5]}{dt} = k_1[N_2O_5] + \frac{k_1(k_2 - k_{-1})[N_2O_5]}{k_{-1} + k_2} \tag{8.91}$$

$$= \frac{2k_1k_2[N_2O_5]}{k_{-1} + k_2} \tag{8.92}$$

The mechanism thus explains the first-order kinetics. The rate constant k_2 is expected to be much smaller than k_{-1}, so that to a good approximation

$$-\frac{d[N_2O_5]}{dt} = \frac{2k_1k_2}{k_{-1}}[N_2O_5] \tag{8.93}$$

The rate constant $2k_1k_2/k_{-1}$ involves the product of an equilibrium constant k_1/k_{-1} and a second-order rate constant k_2, neither of which is expected to fall off at low pressures. The actual fall off at very low pressures has been explained[54] by the fact that k_{-1} falls at low pressures, ultimately becoming unimportant compared with k_2; the rate is then $2k_1[N_2O_5]$.

Confirmation of Ogg's mechanism has been provided in various ways. If it applies it should be possible to demonstrate isotopic exchange in a mixture of, for example, $N_2{}^{15}O_5$ and $N^{14}O_2$:

$$N_2{}^{15}O_5 + N^{14}O_2 \rightleftharpoons N_2{}^{14}O_5 + N^{15}O_2$$

Such an exchange was observed by Ogg[55] in the gas phase and in carbon tetrachloride solution. The exchange was first order in N_2O_5 at higher pressures and zero order in NO_2; the reaction therefore is controlled by reaction (1), the rate of which is given by the rate of the exchange. Ogg found that at 27°C the exchange rate fell by a factor of about 5 when the total pressure was reduced from 500 to 50 mmHg. This is the kind of behavior to be expected for a molecule of the size of nitrogen pentoxide.

Further evidence for the mechanism has been provided by studies[54] of the decomposition of nitrogen pentoxide in the presence of nitric oxide. The mechanism is

$$(1) \quad N_2O_5 \rightarrow NO_2 + NO_3$$

$$(-1) \quad NO_2 + NO_3 \rightarrow N_2O_5$$

$$(2) \quad NO + NO_3 \rightarrow 2NO_2$$

Application of the steady-state treatment leads to

$$-\frac{d[NO]}{dt} = -\frac{d[N_2O_5]}{dt} = \frac{k_1 k_2 [N_2O_5][NO]}{k_2[NO] + k_{-1}[NO_2]} \tag{8.94}$$

The initial rate (when $[NO_2]$ is zero) and any rate at a high concentration of nitric oxide are equal to $k_1[N_2O_5]$; an experimental study under these conditions therefore provides information about k_1 and its fall off at low pressures. In the investigation of Mills and Johnston[54] the reaction between N_2O_5 and NO was studied over a 100,000-fold range of total pressure and included studies with various inert gases. The results fully confirmed the mechanism and were consistent with the isotope-exchange studies.

8.4.7 Decomposition of Ozone

Ozone, O_3, decomposes thermally into oxygen, O_2, and the reaction has some confusing features. It usually occurs partly on the surface of the vessel and partly in the gas phase. In 1910 Chapman and Jones[56] were able to separate the homogeneous and heterogeneous reactions by varying the surface/volume ratio. This was the first time this had been done satisfactorily, although the idea was suggested much earlier by van't Hoff.[57] The homogeneous gas reaction is second order, while the surface reaction is first order.[58]

Chapman and his coworkers[56,59] studied the decomposition of ozone in the presence of an excess of oxygen and found the reaction to be second order in ozone. Jahn[60] found the rate to be inversely proportional to the oxygen concentration, and to explain this he proposed the mechanism

$$(1) \quad O_3 \rightleftharpoons O_2 + O$$

$$(2) \quad O + O_3 \rightarrow 2O_2$$

The decomposition of a molecule as simple as ozone, however, must be in its second-order region except at extremely high pressures, and Benson and Axworthy[61] therefore modified the Jahn mechanism as follows:

$$(1) \quad O_3 + M \rightleftharpoons O_2 + O + M$$

$$(2) \quad O + O_3 \rightarrow 2O_2$$

Application of the steady-state treatment to the concentration of oxygen atoms gives

$$k_1[O_3][M] - k_{-1}[O_2][O][M] - k_2[O][O_3] = 0 \tag{8.95}$$

and therefore

$$[O] = \frac{k_1[O_3][M]}{k_{-1}[O_2][M] + k_2[O_3]} \tag{8.96}$$

The rate of decomposition of ozone is

$$-\frac{d[O_3]}{dt} = k_1[O_3][M] - k_{-1}[O_2][O][M] + k_2[O][O_3] \qquad (8.97)$$

and insertion of Eq. (8.96) for [O] gives rise to

$$-\frac{d[O_3]}{dt} = \frac{2k_1k_2[O_3]^2[M]}{k_{-1}[O_2][M] + k_2[O_3]} \qquad (8.98)$$

Benson and Axworthy obtained additional kinetic results using both concentrated ozone and ozone diluted by oxygen and found that all the results were consistent with this kinetic equation. The equation accounts for the fact that the rate is inversely proportional to the oxygen concentration when sufficient oxygen is present and for the second-order behavior when oxygen is present in excess.

The Benson–Axworthy mechanism does not correspond to a chain reaction. There are no chain-propagating steps: the atoms instead lead directly to products by the reaction $O + O_3 \rightarrow 2O_2$. A chain reaction was suggested by Glissman and Schumacher[62] and discussed again by Schumacher.[63] It involves excited oxygen molecules which are supposed to be involved in the following chain-propagating steps:

$$O + O_3 \rightarrow O_2 + O_2^*$$

$$O_2^* + O_3 \rightarrow 2O_2 + O$$

The suggestion was that when the reaction $O + O_3$ occurs, one of the two oxygen molecules produced is in an excited (perhaps electronically excited) state; it may be deactivated by collisions but may also continue the propagation of the chains. It is difficult to determine the relative importance of such a mechanism, but on the basis of a critical examination of the problem Benson and Axworthy[61,64] conclude it to be unimportant. A similar conclusion has been reached on the basis of a consideration of the potential-energy surfaces for the reaction.[65]

The photochemical decomposition of ozone also has been studied in considerable detail, and there again has been some disagreement about the mechanism. The decomposition is brought about by red light and also by light of higher frequency, such as ultraviolet light. It appears[64–66] that the reaction brought about by red light does not involve energy chains, like the thermal decomposition, but that the reaction in ultraviolet light does involve energy chains. The important difference is that in the thermal decomposition and the decomposition in red light the atoms produced are in their ground ^3P states and do not have enough energy to give rise to excited oxygen molecules that are sufficiently energetic to propagate the chain. In ultraviolet light, on the other hand, $O^*(^1D)$ atoms are produced and these undergo the reaction

$$O^*(^1D) + O_3 \rightarrow O_2 + O_2^*(^3\Sigma_g^-)$$

The $O_2^*(^2\Sigma_g^-)$ molecules produced in this reaction then propagate the chain as follows:

$$O_2^*(^3\Sigma_g^-) + O_3 \rightarrow 2O_2 + O^*(^1D)$$

The overall mechanism for the photochemical decomposition in ultraviolet light therefore appears to be

$$(1) \quad O_3 + h\nu \rightarrow O_2^* + O^*(^1D)$$

$$(2) \quad O^*(^1D) + O_3 \rightarrow O_2 + O_2^*(^3\Sigma_g^-)$$

$$(3) \quad O_2^*(^3\Sigma_g^-) + O_3 \rightarrow 2O_2 + O^*(^1D)$$

$$(4) \quad O + O + M \rightarrow O_2 + M$$

$$(5) \quad O + O_2 + M \rightarrow O_3 + M$$

$$(6) \quad O_2^* + M \rightarrow O_2 + M$$

Reactions (2) and (3) constitute the chain-propagating steps, and (4), (5), and (6) are the chain-ending steps.

In visible (red) light the oxygen atoms are produced not in the ^1D state but in the less energetic ^3P state. These atoms have insufficient energy to produce oxygen molecules having enough energy to regenerate oxygen atoms. Energy chains therefore are not involved.

8.4.8 Thermal Para–Ortho Hydrogen Conversion

The conversion of *para*-hydrogen into *ortho*-hydrogen takes place homogeneously, the rate being convenient to measure between 700 and 800°C.[67] The order of the reaction is three-halves, suggesting the mechanism

$$(1) \quad H_2 \rightleftharpoons 2H$$

$$(2) \quad H + p\text{-}H_2 \rightarrow o\text{-}H_2 + H$$

The concentration of hydrogen atoms is in accordance with the equilibrium equation

$$[H]^2 = K[H_2] \tag{8.99}$$

since there is no net change in the number of hydrogen atoms when reaction (2) occurs. The rate of the conversion is therefore

$$\frac{d[o\text{-}H_2]}{dt} = k_2[H][p\text{-}H_2] \tag{8.100}$$

$$= k_2 K^{1/2}[p\text{-}H_2]^{3/2} \tag{8.101}$$

The activation energy of the process is given by

$$E = E_2 + \tfrac{1}{2}D \tag{8.102}$$

where D is the dissociation energy of hydrogen and has the value of approximately 435 kJ mol^{-1}. Since E_2 is about 38 kJ mol^{-1}, it follows that the overall activation energy should be about 256 kJ mol^{-1}. The experimental value is close to this.

8.5 ORGANIC DECOMPOSITIONS

Many organic decompositions obey simple kinetic laws and consequently were assumed for some time to be elementary processes. However, F. O. Rice and his coworkers showed that free radicals undoubtedly play an essential part in them. An important advance was made in 1934 when Rice and Herzfeld[68] pointed out that chain mechanisms sometimes lead to simple overall kinetics and proposed specific mechanisms for a number of reactions. There is no doubt that mechanisms of the type proposed by Rice and Herzfeld play an important role in the thermal decompositions, known as *pyrolyses,* of organic compounds. In many cases, however, it has been necessary to change some of the details of their proposed mechanisms.

The significant conclusion reached by Rice and Herzfeld was that the kinetic equations obeyed by many organic pyrolyses and the products of the reactions can be explained on the basis of free-radical mechanisms. An important fact that the mechanisms had to explain was that the kinetics were in many cases simple, the reactions frequently having orders of either unity or three-halves. Another feature of the reactions is that the production of radicals often involves the rupture of a carbon–carbon bond, which usually requires rather more than 340 kJ of energy per mole. The overall activation energy of most of the reactions, however, is considerably less than this. As will be seen, the Rice–Herzfeld mechanisms are consistent with both these points. In particular, the experimental activation energies can be explained in terms of the activation energies for the individual elementary processes believed to be playing a part.

The proposed mechanisms are here discussed in a general way, and three cases are considered, according to whether the order of the overall reaction is unity, one-half, or three-halves. It will be seen later that the order is influenced largely by the nature of the initiating and chain-ending steps.

A mechanism that gives rise to first-order overall kinetics is one in which there is a first-order initiation process and a termination process in which there is combination between the two different active species that are involved in the chain-propagating steps. This may be illustrated by a mechanism,§ no longer believed to be the correct one, for the ethane decomposition:

$$(1) \quad C_2H_6 \rightarrow 2CH_3$$

$$(2) \quad CH_3 + C_2H_6 \rightarrow CH_4 + C_2H_5$$

$$(3) \quad C_2H_5 \rightarrow C_2H_4 + H$$

$$(4) \quad H + C_2H_6 \rightarrow H_2 + C_2H_5$$

$$(5) \quad H + C_2H_5 \rightarrow C_2H_6$$

The chain-propagating steps are reactions (3) and (4), and the chain carriers are H and C_2H_5. These two species combine in reaction (5), the termination step. The CH_3 radicals produced in reaction (1) are not chain carriers; they are removed by reaction (2) and are not regenerated.

The steady-state equations for this reaction scheme are, for the methyl radicals,

$$2k_1[C_2H_5] - k_2[CH_3][C_2H_6] = 0 \qquad (8.103)$$

<hr/>

§ This was the original Rice–Herzfeld mechanism for the reaction [*J. Am. Chem. Soc.,* **56,** 284 (1934)]; the correct mechanism is considered later.

for the ethyl radicals,

$$k_2[CH_3][C_2H_6] - k_3[C_2H_5] + k_4[H][C_2H_6] - k_5[H][C_2H_5] = 0 \qquad (8.104)$$

and for hydrogen atoms,

$$k_3[C_2H_5] - k_4[H][C_2H_6] - k_5[H][C_2H_5] = 0 \qquad (8.105)$$

Addition of (8.103), (8.104), and (8.105) leads to

$$[H] = \frac{k_1}{k_5} \frac{[C_2H_6]}{[C_2H_5]} \qquad (8.106)$$

Insertion of this in (8.105) gives, after rearrangement,

$$k_3 k_5 [C_2H_5]^2 - k_1 k_5 [C_2H_6][C_2H_5] - k_1 k_4 [C_2H_6]^2 = 0 \qquad (8.107)$$

The general solution of this quadratic equation is

$$[C_2H_5] = \left\{ \frac{k_1}{2k_3} + \left[\left(\frac{k_1}{2k_3} \right)^2 + \left(\frac{k_1 k_4}{k_3 k_5} \right) \right]^{1/2} \right\} [C_2H_6] \qquad (8.108)$$

The constant k_1 is very small, since the initiating reaction has a very high activation energy. The terms involving $k_1/2k_3$ are therefore very small in comparison with $k_1 k_4 / k_3 k_5$ (which appears as the square root) and therefore

$$[C_2H_5] = \left(\frac{k_1 k_4}{k_3 k_5} \right)^{1/2} [C_2H_6] \qquad (8.109)$$

The rate of production of ethylene is

$$\frac{d[C_2H_4]}{dt} = k_3[C_2H_5] \qquad (8.110)$$

$$= \left(\frac{k_1 k_3 k_4}{k_5} \right)^{1/2} [C_2H_6] \qquad (8.111)$$

The reaction is thus of the first order. If the activation energies for reactions (1), (3), (4), and (5) are E_1, E_3, E_4, and E_5, and the corresponding pre-exponential factors are A_1, A_3, A_4, and A_5, Eq. (8.111) may be written as

$$\frac{d[C_2H_4]}{dt} = \left(\frac{A_1 e^{-E_1/RT} A_3 e^{-E_3/RT} A_4 e^{-E_4/RT}}{A_5 e^{-E_5/RT}} \right)^{1/2} [C_2H_6] \qquad (8.112)$$

$$= \left(\frac{A_1 A_3 A_4}{A_5} \right)^{1/2} \exp\left[\frac{-(E_1 + E_3 + E_4 - E_5)}{2RT} \right] [C_2H_6] \qquad (8.113)$$

The activation energy for the overall reaction is therefore

$$E = \tfrac{1}{2}(E_1 + E_3 + E_4 - E_5) \qquad (8.114)$$

Since E_1 is usually very much larger than E_3 and E_4, the overall activation energy is usually appreciably smaller than E_1. The chain length γ, defined as the rate of the overall reaction divided by the rate of the initiation reaction, is given by

$$\gamma = \frac{(k_1 k_3 k_4 / k_5)^{1/2}[C_2H_6]}{k_1[C_2H_6]} \tag{8.115}$$

$$= \frac{1}{2}\left(\frac{k_3 k_4}{k_1 k_5}\right)^{1/2} \tag{8.116}$$

Since k_1 is very small, this quantity is often very large. It represents the average amount of reactant transformed each time the initiating step occurs.

To obtain three-halves-order kinetics, there can still be first-order initiation, but now the termination step must be the combination of two identical chain carriers. Moreover, they must be those carriers that are involved in a second-order propagation reaction. A good example is the mechanism originally proposed by Rice and Herzfeld for the thermal decomposition of acetaldehyde:

(1) $CH_3CHO \rightarrow CH_3 + CHO$

(2) $CH_3 + CH_3CHO \rightarrow CH_4 + CH_3CO$

(3) $CH_3CO \rightarrow CH_3 + CO$

(4) $2CH_3 \rightarrow C_2H_6$

The radical CHO undergoes further reactions, but for simplicity they are ignored here. For this reaction the Rice–Herzfeld scheme is close to the truth; a somewhat improved mechanism is given later (Section 8.5.4).

The steady-state equations are now, for the methyl radicals,

$$k_1[CH_3CHO] - k_2[CH_3][CH_3CHO] + k_3[CH_3CO] - k_4[CH_3]^2 = 0 \tag{8.117}$$

and for the CH_3CO radicals,

$$k_2[CH_3][CH_3CHO] - k_3[CH_3CO] = 0 \tag{8.118}$$

Addition of these equations gives rise to

$$[CH_3] = \left(\frac{k_1}{k_4}\right)^{1/2}[CH_3CHO]^{1/2} \tag{8.119}$$

The rate of formation of methane is

$$\frac{d[CH_4]}{dt} = k_2[CH_3][CH_3CHO] \tag{8.120}$$

$$= k_2\left(\frac{k_1}{k_4}\right)^{1/2}[CH_3CHO]^{3/2} \tag{8.121}$$

Thus, the mechanism explains correctly the three-halves order. The overall activation energy is

$$E = E_2 + \tfrac{1}{2}(E_1 - E_4) \tag{8.122}$$

and this is again usually much less than E_1, since E_2 is very small compared with E_1. The chain length is

$$\gamma = k_2\left(\frac{1}{k_1 k_4}\right)^{1/2}[CH_3CHO]^{1/2} \tag{8.123}$$

and depends on the concentration of reactant.

When initiation is first-order, one-half-order kinetics are found when the terminating step is taken to be the second-order reaction between two radicals that undergo first-order propagation reactions. In the acetaldehyde decomposition, for example, these would be the CH_3CO radicals. Thus, the hypothetical scheme

(1) $CH_3CHO \rightarrow CH_3 + CHO$

(2) $CH_3 + CH_3CHO \rightarrow CH_4 + CH_3CO$

(3) $CH_3CO \rightarrow CH_3 + CO$

(4) $2CH_3CO \rightarrow CH_3COCOCH_3$

will give one-half-order kinetics, as will now be shown. The steady-state equations are, for CH_3,

$$k_1[CH_3CHO] - k_2[CH_3][CH_3CHO] + k_3[CH_3CO] = 0 \qquad (8.124)$$

and for CH_3CO,

$$k_2[CH_3][CH_3CHO] - k_3[CH_3CO] - k_4[CH_3CO]^2 = 0 \qquad (8.125)$$

Addition of these equations gives

$$[CH_3CO] = \left(\frac{k_1}{k_4}\right)^{1/2} [CH_3CHO]^{1/2} \qquad (8.126)$$

The rate of formation of carbon monoxide§ is therefore

$$\frac{d[CO]}{dt} = k_3[CH_3CO] \qquad (8.127)$$

$$= k_3\left(\frac{k_1}{k_4}\right)^{1/2} [CH_3CHO]^{1/2} \qquad (8.128)$$

The overall activation energy is now

$$E = E_3 + \tfrac{1}{2}(E_1 - E_4) \qquad (8.129)$$

and the chain length is $k_3(1/k_1k_4)^{1/2}[CH_3CHO]^{-1/2}$.

8.5.1 Goldfinger–Letort–Niclause Rules

The above examples show that the order of the overall reaction depends on the manner in which the chains are broken. This problem was treated systematically by Goldfinger, Letort, and Niclause,[69] who distinguished between two types of radicals:

1. Radicals that are involved in second-order propagation reactions; these are referred to as β radicals. In the above schemes hydrogen atoms and methyl radicals are acting as β radicals.
2. Radicals that undergo first-order reactions in the propagation steps. The radicals C_2H_5 and CH_3CO are of this type and are referred to as μ radicals.

§ In these schemes the rate of formation of carbon monoxide and methane are very close to each other but are not identical. In the present scheme it is simpler to consider the rate of formation of carbon monoxide, whereas in the three-halves-order scheme it was better to consider the rate of formation of methane.

Goldfinger and coworkers concluded that if the initiating reaction is first order and the termination steps do not require third bodies, the orders are as follows:

Termination	Overall order
$\beta\beta$	3/2
$\beta\mu$	1
$\mu\mu$	1/2

Their treatment also showed that if the chain-initiating reaction is second order rather than first order, the orders are higher by one-half.

Another possibility is that the chain-ending steps involve the participation of a third body and are therefore third-order reactions. The effect of this is to reduce the order by one-half. Thus, with first-order initiation and $\beta\beta$ termination requiring a third body (this may be represented by $\beta\beta M$), the order is unity.

The various possibilities are summarized in Table 8.3. This general type of treatment is useful in deciding on possible reaction mechanisms, since it avoids the necessity of working out steady-state equations in each case.

The decision as to the mechanism for a given reaction must be based on consideration of a number of factors. The order of the initiation reaction depends on the complexity of the molecule that is dissociating, the temperature of the experiment, and the pressure range. It follows from the treatment in Section 5.3 that a unimolecular reaction is more likely to be in its first-order region if:

1. The number of degrees of freedom is large.
2. The pressure is high.
3. The temperature is low.

With regard to the chain-ending step, two questions are important: the first is the nature of the radicals that recombine, and the second is whether a third body is required. What radicals are involved depends on the relative concentrations of the radicals. If β radicals are predominant, $\beta\beta$ recombination will be most important, but

TABLE 8.3 OVERALL ORDERS OF REACTION FOR DECOMPOSITIONS[a]

First-order initiation		Second-order initiation		
Simple termination	Third-body termination	Simple termination	Third-body termination	Overall order
		$\beta\beta$		2
$\beta\beta$		$\beta\mu$	$\beta\beta M$	3/2
$\beta\mu$	$\beta\beta M$	$\mu\mu$	$\beta\mu M$	1
$\mu\mu$	$\beta\mu M$		$\mu\mu M$	1/2
	$\mu\mu M$			0

[a] The arrows show the changes that may be expected if the pressure is lowered.

$\mu\mu$ recombination will predominate if the concentration of μ radicals is very large compared with that of β. In intermediate cases $\beta\mu$ recombination may be most important. An additional factor is the relative rate constants of the combination reactions. Radical combinations usually occur with no activation energy, and the rate constants are determined largely by the pre-exponential factors, which depend on the relative complexities of the molecules. On the whole, μ radicals are larger and more complex than β radicals, and thus the pre-exponential factors for $\mu\mu$ recombination are less than for $\beta\mu$, which in turn are less than for $\beta\beta$. In the ethane decomposition, for example, the pre-exponential factors are in the order

$$H + H > H + C_2H_5 > C_2H_5 + C_2H_5$$

The latter reaction is nevertheless preponderant, owing to the much higher concentration of ethyl radicals. The H–H recombination would in any case be unimportant since this reaction certainly occurs as a third-order reaction except at exceedingly high pressures.

The relative magnitudes of the radical concentrations depend on the relative rates of the chain-propagating steps. In the ethane decomposition the chain-propagating reactions are

$$(3) \quad C_2H_5 \rightarrow C_2H_4 + H$$

$$(4) \quad H + C_2H_6 \rightarrow H_2 + C_2H_5$$

The first reaction produces H and removes C_2H_5, while the second produces C_2H_5 and removes H. If k_3 is very large and k_4 is very small, the concentration of H is much greater than that of C_2H_5. Conversely, if k_4 is very large and k_3 is very small, the concentration of C_2H_5 predominates. Under usual conditions this second situation does apply in this case; the concentration of ethyl radicals is therefore very much greater than that of hydrogen atoms, and the predominant chain-breaking step is usually

$$C_2H_5 + C_2H_5 \rightarrow C_4H_{10}$$

What is believed now to be the correct mechanism for the ethane pyrolysis is considered later; it follows from that mechanism [Eq. (8.135)] that

$$\frac{[C_2H_5]}{[H]} = \frac{k_4}{k_3}[C_2H_6]^{1/2} \tag{8.130}$$

At sufficiently low pressures of ethane the relative concentration of ethyl radicals is therefore smaller, and the reaction

$$C_2H_5 + H \rightarrow C_2H_6$$

may then become the main terminating step. Evidence that this occurs has been obtained.

To summarize, the decision as to the most important terminating reaction must take account of the following factors:

1. The rate constants of the chain-propagating reactions, from which the relative radical concentrations can be deduced.

2. The pre-exponential factors of the possible termination steps; these depend on the relative complexities of the radicals.
3. The number of degrees of freedom in the terminating reaction; if this is large a third body is unnecessary, but otherwise it is required.

The following is the type of procedure that might be used for drawing preliminary conclusions as to the mechanism of a pyrolysis reaction. A study may be made first of the overall order of the reaction over a wide pressure and temperature range. This limits the number of possible mechanisms to the ones corresponding to the experimental order of the reaction, as shown in Table 8.3. The effect of added inert gas may be determined next. An increase in the overall rate due to increase in pressure indicates a mechanism with a second-order initiation and pressure-independent termination (column three in Table 8.3). A decrease of overall rate indicates a mechanism with pressure-dependent termination and first-order initiation (column two in Table 8.3). Both these effects, together with the overall order of the reaction, lead to a conclusion concerning the type of mechanism that applies. If no inert-gas effect is observed, the reaction may be of the type in column one or four, and it is usually necessary to decide, on the basis of further information, which of the two possibilities is correct. A useful clue may be provided by an observed change in the order of the overall reactions; the changes that are observed with decreasing pressure or increasing temperature follow the arrows in Table 8.3.

8.5.2 Molecular Processes

Since the 1930s it has been agreed that hydrocarbon decompositions proceed mainly by free-radical mechanisms. However, there has been some uncertainty as to whether purely molecular mechanisms play a significant role. For example, although it has been accepted for some time that the pyrolysis of ethane occurs mainly by the type of chain process given in Section 8.5.3, it was thought that ethane could also split off a hydrogen molecule in a single elementary step:

$$C_2H_6 \rightarrow C_2H_4 + H_2$$

Evidence that this was the case seemed to be provided by some experiments by Staveley and Hinshelwood,[70] who added small amounts of nitric oxide to decomposing organic compounds. They often found the type of behavior represented in Fig. 8.3; the rate falls to a value that is sometimes about one-tenth of the uninhibited rate. The addition of further amounts of nitric oxide did not lead to complete inhibition; indeed, there was sometimes an increase in rate at higher nitric oxide concentrations.

These results suggested that when there was maximal inhibition all free radicals had been removed from the system, by reactions such as

$$CH_3 + NO \rightarrow CH_3NO \rightarrow H_2O + HCN$$

and

$$C_2H_5 + NO \rightarrow C_2H_5NO \rightarrow H_2O + CH_3CN$$

According to this interpretation, the residual reaction was a molecular reaction. However, it has become clear that the hypothesis of a molecular reaction is not generally

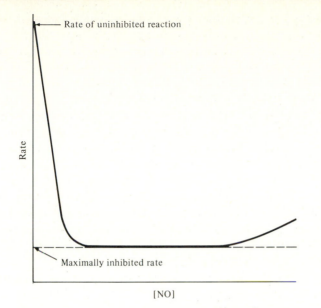

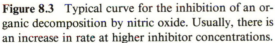

Figure 8.3 Typical curve for the inhibition of an organic decomposition by nitric oxide. Usually, there is an increase in rate at higher inhibitor concentrations.

acceptable, since for certain reactions the maximally inhibited reaction still involves the participation of free radicals. Evidence of this was first provided by isotopic-mixing experiments carried out by Poltorak and Voevodsky[71] on the propane pyrolysis. These workers decomposed C_3H_8 in the presence of D_2 and with different amounts of NO. The products C_2H_4 and C_2H_6 were separated from the reaction mixture, and their deuterium content was the same for the reaction inhibited by NO as in the uninhibited reaction. The possibility of a molecular exchange between C_3H_8 and D_2 was excluded, and the results therefore showed that the maximally inhibited pyrolysis is a free-radical process, any molecular process being unimportant.

More detailed evidence of the same kind was obtained for the ethane decomposition by Rice and Varnerin,[72] who decomposed C_2D_6 in the presence of CH_4 and investigated the rate of production of the mixed methanes. The mixed product CH_3D, for example, is formed in the following sequence of reactions:

$$C_2D_6 \rightarrow 2CD_3$$

$$CD_3 + C_2D_6 \rightarrow CD_4 + C_2D_5$$

$$C_2D_5 + CH_4 \rightarrow C_2D_5H + CH_3$$

$$CH_3 + C_2D_6 \rightarrow CH_3D + C_2D_5 \quad \text{and so on}$$

Rice and Varnerin found that the ratio of CH_3D to CH_4 varied linearly with the percentage of C_2D_6 that had decomposed, and that a plot of $[CH_3D]/[CH_4]$ against the amount of C_2D_6 decomposed was independent of the concentration of NO. This result shows that the addition of nitric oxide reduces the rates of the mixing processes in the same ratio as it reduces the rate of the overall reaction. The rates of mixing are proportional to the radical concentrations, which are not reduced to zero by the addition of NO: they are reduced in the same ratio as the overall rate is reduced. Thus, the reaction is primarily a free-radical process. Similar investigations on butane[73] and on dimethyl ether[74] have led to the same conclusion that there is no molecular mechanism.

The results of the nitric oxide experiments therefore were misleading at first and have to be interpreted in a different way; this is considered briefly in Section 8.5.5.

8.5.3 Decomposition of Ethane

Much work has been done on the pyrolysis of ethane; for reviews the reader is referred to the bibliography at the end of this chapter. The reaction is largely homogeneous and is of the first order, although a change of order towards 3/2 has been noted at higher pressures and lower temperatures. The main products are ethylene and hydrogen, but significant amounts of methane and butane are formed.

As previously discussed, the reaction occurs mainly by a free-radical chain mechanism. The weakest bond in the ethane molecule is the $C-C$ bond, so that there is no doubt that the initiation reaction is

$$(1) \quad C_2H_6 \rightarrow 2CH_3$$

This reaction is followed by the abstraction by a methyl radical of a hydrogen atom from ethane

$$(2) \quad CH_3 + C_2H_6 \rightarrow CH_4 + C_2H_5$$

Since there is no reason to believe that methane is formed in any other reaction, the rate of methane production is a measure of the rate of the initiation reaction (1). Lin and Back[75] made a careful study of the kinetics of methane formation over a range of pressure and temperature. At higher pressures the process is first order in ethane, but there is a fall off of the first-order rate coefficient at lower pressures. The first-order rate coefficient can be expressed as

$$k_1/s^{-1} = 1.0 \times 10^{16} \exp\left(-\frac{360 \text{ kJ mol}^{-1}}{RT}\right)$$

The fall off of the rate coefficient can be interpreted in terms of RRK theory with $s = 13$. RRKM calculations gave good agreement with experiment with all vibrations assumed to be active.[76]

The chain-propagating steps in the ethane pyrolysis are undoubtedly

$$(3) \quad C_2H_5 \rightarrow C_2H_4 + H$$

and

$$(4) \quad H + C_2H_6 \rightarrow H_2 + C_2H_5$$

It can be calculated on the basis of the known rates of these reactions that the ethyl radicals are present at much higher concentrations than hydrogen atoms under usual pyrolysis conditions, so that the predominant chain-ending step must be

$$(5) \quad 2C_2H_5 \rightarrow C_4H_{10} \quad \text{and} \quad C_2H_6 + C_2H_4$$

The combination/disproportionation ratio is known for this reaction, so that from the rate of butane production it is possible to calculate the total rate of this reaction between ethyl radicals.

In any chain mechanism the steady-state hypothesis requires that the rate of initiation is equal to the rate of termination. The rate of reaction (5), as given by the rate of butane formation, is therefore also a measure of the rate of the initiation reaction (1). Lin and Back also made a study of the kinetics of butane formation and deduced from the results information about the initiation process. The conclusions agree with those obtained from the methane measurements.

Lin and Back also obtained information about reaction (3), the decomposition of the ethyl radical, from measurements of the rates of formation of ethylene and butane. The scheme of reactions may be written as

$$(1) \quad C_2H_6 \rightarrow 2CH_3$$

$$(2) \quad CH_3 + C_2H_6 \rightarrow CH_4 + C_2H_5$$

$$(3) \quad C_2H_5 \rightarrow C_2H_4 + H$$

$$(4) \quad H + C_2H_6 \rightarrow H_2 + C_2H_5$$

$$(5a) \quad 2C_2H_5 \rightarrow C_4H_{10}$$

$$(5b) \quad 2C_2H_5 \rightarrow C_2H_6 + C_2H_4$$

The rate of formation of ethylene, v_E, is given by

$$v_E = k_3[C_2H_5] + k_{5b}[C_2H_5]^2 \tag{8.131}$$

and that of butane by

$$v_B = k_{5a}[C_2H_5]^2 \tag{8.132}$$

Since k_{5a}/k_{5b} is known, the values of k_3, k_{5a}, and k_{5b} can be obtained. By working over a wide pressure range, Lin and Back were able to show that reaction (3) is first order at higher pressures, and that there is a fall off of the first-order rate coefficient as the pressure is lowered, the reaction eventually becoming second order. The Lin–Back expressions for the rate constant of reaction (3) in the high- and low-pressure regions are

$$k_3^\infty /s^{-1} = 2.7 \times 10^{14} \exp\left(-\frac{160 \text{ kJ mol}^{-1}}{RT}\right)$$

$$k_3^o /\text{dm}^3 \text{ mol}^{-1} \text{ s}^{-1} = 6.8 \times 10^{14} \exp\left(-\frac{136 \text{ kJ mol}^{-1}}{RT}\right)$$

The scheme of reactions proposed for the ethane pyrolysis provides a satisfactory interpretation of the overall kinetic behavior. If it is assumed that reaction (1) is in its first-order region and that the rate of reaction (3) is $k_3[C_2H_5][C_2H_6]^{1/2}$, the steady-state equations are, for CH_3,

$$k_1[C_2H_6] - k_2[CH_3][C_2H_6] = 0 \tag{8.133}$$

and, for C_2H_5,

$$k_2[CH_3][C_2H_6] - k_3[C_2H_5][C_2H_6]^{1/2} + k_4[H][C_2H_6] - k_5[C_2H_5]^2 = 0 \tag{8.134}$$

where $k_5 = k_{5a} + k_{5b}$. The steady-state equation for H is

$$k_3[C_2H_5][C_2H_6]^{1/2} - k_4[H][C_2H_6] = 0 \qquad (8.135)$$

Addition of all three equations gives

$$k_1[C_2H_6] - k_5[C_2H_5]^2 = 0 \qquad (8.136)$$

and therefore

$$[C_2H_5] = \left(\frac{k_1}{k_5}\right)^{1/2}[C_2H_6]^{1/2} \qquad (8.137)$$

The rate of formation of ethylene, which is mainly by reaction (3), is

$$\frac{d[C_2H_4]}{dt} = k_3[C_2H_5][C_2H_6]^{1/2} \qquad (8.138)$$

$$= k_3\left(\frac{k_1}{k_5}\right)^{1/2}[C_2H_6] \qquad (8.139)$$

Thus, the overall first-order behavior is explained.

This interpretation of the kinetics, in terms of a three-halves order for reaction (3), was first suggested by Quinn[77] in 1963. Previously, there had been some confusion. Originally, Rice and Herzfeld[78] thought that the chain-ending step was

$$H + C_2H_5 \rightarrow C_2H_6$$

(see Eqs. (8.103)–(8.111)), but this cannot be justified in view of the fact that the concentration of C_2H_5 is very much greater than that of H atoms. Küchler and Theile[79] suggested that the initiation reaction $C_2H_6 \rightarrow 2CH_3$ is in its second-order region and that termination is $2C_2H_5 \rightarrow C_4H_{10}$. This also correctly explains the first-order overall kinetics but it was not supported by the later measurements of Lin and Back on the rates of formation of the minor products. What caused confusion is that the mechanism now known to be correct does not fit simply into the Goldfinger–Letort–Niclause scheme (Table 8.3). The ethyl radical is neither a β nor a μ radical but is about half-way between. First-order initiation with $\mu\mu$ termination leads to one-half-order kinetics and with $\beta\beta$ termination to three-halves-order kinetics; if the radicals are half-way between β and μ the result is first-order kinetics.

The photolysis and radiolysis of ethane have been investigated but not as extensively as the pyrolysis. Ethane does not absorb electromagnetic radiation in the visible range of the spectrum, but it does so in the ultraviolet at wavelengths of 147 and 123.6 nm. In both cases the initial step is

$$C_2H_6 + h\nu \rightarrow C_2H_6^*$$

and the excited ethane molecules can undergo various reactions such as[80]

$$C_2H_6^* \rightarrow CH_3CH_2^* + H$$

$$C_2H_6^* \rightarrow CH_4 + CH_2$$

$$C_2H_6^* \rightarrow 2H + C_2H_4$$

The species $CH_3CH_2^*$, H, and CH_2 then undergo further processes. No C_2D_6 is formed in the photolysis of CH_3CD_3, which means that processes such as

$$C_2H_6 + h\nu \rightarrow 2CH_3$$

followed by free-radical reactions including

$$2CH_3 \rightarrow C_2H_6$$

are relatively unimportant. The overall mechanism of the ethane photolysis is very different from that occurring in the pyrolysis and is much more complicated.

In the radiolysis of ethane a number of primary steps occur,[81] and these are indicated by wavy arrows:

$$C_2H_6 \rightsquigarrow C_2H_4^+ + H_2 + e^-$$

$$\rightsquigarrow C_2H_3^+ + H_2 + H + e^-$$

$$\rightsquigarrow C_2H_6^+ + e^-$$

$$\rightsquigarrow C_2H_5^+ + H + e^-$$

$$\rightsquigarrow C_2H_6^*$$

The products of these processes then undergo a variety of reactions; the mechanism is again very complicated.

8.5.4 Decomposition of Acetaldehyde

Acetaldehyde decomposes both thermally and photochemically, the overall reaction in each case being largely

$$CH_3CHO \rightarrow CH_4 + CO$$

Certain minor products are also found in significant amounts, as will be discussed later in connection with the reaction mechanism.

There was at first some controversy about the order of the pyrolysis reaction, which is largely homogeneous: some workers thought that the order was 3/2, others that it was 2. The problem was resolved by Letort,[82] who showed that the order of reaction with respect to concentration (the "true" order) is 3/2, while that with respect to time is 2 (Section 2.1). This difference is due to inhibition by products. The activation energy of the thermal reaction[82,83] is about 200 kJ mol^{-1}.

The presence of free radicals in the reaction system was first detected by Rice and coworkers,[84] although at higher temperatures than employed in the kinetic experiments ($\sim 500°C$). Later, Burton and coworkers[85] used radioactive lead mirrors to demonstrate the presence of radicals at 500°C.

In 1934, Rice and Herzfeld proposed for the pyrolysis of acetaldehyde the free-radical mechanism given at the beginning of Section 8.5. This successfully accounts for the three-halves order and for the fact that methane and carbon monoxide are the main products, these being formed in the chain-propagating steps. The radical CHO produced in the initiation step of the Rice–Herzfeld mechanism undergoes the following reactions:

$$CHO \rightarrow CO + H$$

$$H + CH_3CHO \rightarrow H_2 + CH_3CO$$

This accounts for the formation of hydrogen as a minor product. Another minor product is ethane, which is formed in the termination step of the Rice–Herzfeld mechanism.

In addition to these minor products, however, significant amounts of acetone, propionaldehyde, ethylene, and carbon monoxide are also formed.[86] The latter two substances are secondary products, formed in later stages of the reaction, but the acetone and propionaldehyde are certainly primary products.[87] Since they are not accounted for in the Rice–Herzfeld scheme, additional free-radical processes must be occurring. Also, some effects of added inert gases are not consistent with the original mechanism.

The matter is complicated, and for details the original publications must be consulted. To explain the main kinetic results the following extension of the original Rice–Herzfeld mechanism has been suggested:[87,88]

(1) $CH_3CHO \rightarrow CH_3 + CHO$

(2) $CHO \rightarrow CO + H$

(3) $H + CH_3CHO \rightarrow H_2 + CH_3CO$

(4) $CH_3 + CH_3CHO \rightarrow CH_4 + CH_3CO$

(5) $CH_3CO \rightarrow CH_3 + CO$

(6) $CH_3 + CH_3CHO \rightarrow CH_4 + CH_2CHO$

(7) $CH_2CHO \rightarrow CH_2CO + H$

(8) $CH_2CHO + CH_3CHO \rightarrow CH_3CHO + CH_3CO$

(9) $CH_3 + CH_3CHO \rightarrow CH_3COCH_3 + H$

(10) $CH_3 + CH_3 \rightarrow C_2H_6$

(11) $CH_3 + CH_2CHO \rightarrow CH_3CH_2CHO$

The additions to the original mechanism are reactions (6), (7), (8), (9), and (11). The last reaction is less important than reaction (10), but it accounts for the propionaldehyde formation. Reactions (6) and (7) produce ketene, CH_2CO, which, however, is unstable at the temperatures of the acetaldehyde pyrolysis and decomposes to produce ethylene and carbon dioxide as minor products. Reaction (9) accounts for the formation of acetone as a minor product.

Since reaction (10) is still the main chain-ending step, and CH_3 radicals are β radicals, the overall order is still approximately 3/2. The modified mechanism accounts satisfactorily for the inert-gas effects that were observed.

A number of studies have been made of the photolysis of acetaldehyde. An early investigation[89] led to the conclusion that the main primary process is usually

$$CH_3CHO + h\nu \rightarrow CH_3 + CHO$$

and this subsequently has been confirmed a number of times.[90] The products CH_3 and CHO are the same as those found in the initiation step in the pyrolysis, so that the subsequent reactions are the same. The mechanism of the photolysis is thus largely

(1) $CH_3CHO + h\nu \rightarrow CH_3 + CHO$

(2) $CHO \rightarrow CO + H$

(3) $CH_3 + CH_3CHO \rightarrow CH_4 + CH_3CO$

(4) $CH_3CO \rightarrow CH_3 + CO$

(5) $H + CH_3CHO \rightarrow H_2 + CH_3CO$

(6) $2CH_3 \rightarrow C_2H_6$

Application of the steady-state treatment to this mechanism leads to

$$\frac{d[CH_4]}{dt} = k_3\left(\frac{2I}{k_6}\right)^{1/2}[CH_3CHO] \tag{8.140}$$

where I is the intensity of light absorbed. Other processes, not included in this simplified mechanism, also occur.

It has been shown, in flash-photolysis studies,[91] that at high light intensities the initiation process

$$CH_3CHO + h\nu \rightarrow CH_3CO + H$$

occurs in addition.

8.5.5 Inhibition Mechanisms

Nitric oxide frequently inhibits organic decompositions, the rate being reduced not to zero but to a value that is a significant fraction of the uninhibited rate. Isotopic mixing experiments have shown that free-radical reactions still are occurring when there is maximal inhibition and that molecular reactions are unimportant.

This leaves the question of the way in which nitric oxide inhibits. If it did so by merely removing radicals and if there is no molecular reaction, it would be possible to reduce the rate to zero. To answer the question it has been suggested[92] that the effect of nitric oxide is to introduce free-radicals into the system as well as to remove them. The limiting rate is then explained as arising when the total rate of production of radicals is equal to the total rate of their removal.

For example, the following simplified mechanism was proposed for the decomposition of ethane in the presence of sufficient nitric oxide to give maximal inhibition:

(1) $C_2H_6 + NO \rightarrow C_2H_5 + HNO$

(2) $C_2H_5 \rightarrow H + C_2H_4$

(3) $H + C_2H_6 \rightarrow C_2H_5 + H_2$

(4) $H + NO \rightarrow HNO$

$$(-4) \quad HNO \rightarrow H + NO$$

$$(-1) \quad C_2H_5 + HNO \rightarrow C_2H_6 + NO$$

It is supposed that under these conditions the first reaction predominates over the reaction $C_2H_6 \rightarrow 2CH_3$ which is the initiation step for the uninhibited reaction. Similarly, the termination step $2C_2H_5 \rightarrow C_4H_{10}$ is supposed to be much less important than $C_2H_5 + HNO$. The species HNO has been identified spectroscopically[93] and has a dissociation energy (into H + NO) of about 205 kJ mol^{-1}.

Application of the steady-state treatment to this reaction scheme gives

$$\frac{d[C_2H_4]}{dt} = \left(\frac{k_1 k_2 k_3 k_{-4}}{k_{-1} k_4} \right)^{1/2} [C_2H_6] \qquad (8.141)$$

In agreement with experiment, the rate of the fully inhibited reaction is first order in ethane, and there is no dependence on the nitric oxide concentration. When there is partial inhibition, the reaction occurs to some extent by the uninhibited mechanism and in part by the inhibited one, and there is then a decrease in rate with increasing nitric oxide concentration.

One interesting feature of this mechanism for the inhibited reaction arises from the fact that the ratio $k_1 k_{-4}/k_{-1} k_4$ corresponds to the equilibrium constant for the reaction

$$C_2H_6 \rightleftharpoons C_2H_5 + H$$

If this constant is written as K, the rate expression can be written as

$$v = (K k_2 k_3)^{1/2} [C_2H_6] \qquad (8.142)$$

The quantity $K k_2 k_3$ in no way depends on the nitric oxide, and the conclusion is therefore that any inhibitor that can give rise to the same type of mechanism will give rise to the same limiting rate. Propylene gives the same maximal inhibition as nitric oxide (although a larger amount is needed), and the suggestion is made that the allyl radical C_3H_5 plays the same role as NO: it initiates chains by hydrogen-atom abstraction,

$$C_2H_6 + C_3H_5 \rightarrow C_2H_5 + C_3H_6$$

and terminates them by the reverse reaction. Propylene undergoes some decomposition when it acts as an inhibitor, and allyl radicals are known to be present.[94]

8.6 GAS-PHASE COMBUSTION

The gas-phase reactions between oxygen and various substances such as hydrogen, carbon monoxide, phosphorus, and phosphine have a special characteristic: under some conditions of temperature and pressure they occur in a normal way, but a slight change in temperature or pressure may cause the mixture to explode. The reaction between phosphine and oxygen was studied in some detail in van't Hoff's laboratories,[95] and it was noted for the first time that sometimes a decrease in pressure, at constant temperature, caused a steady reaction to change into an explosion. Similar behavior has been observed with several other gas-phase reactions involving oxygen.

Under certain conditions an explosion in a gaseous mixture is accompanied by an explosion wave at which there is a sudden pressure change, and one then speaks of a *detonation.* The manner in which a detonation travels through a gas mixture was investigated experimentally in great detail by Dixon.[96] Later, Chapman[97] used Dixon's results to develop the theory of gaseous detonations in essentially the form in which it is applied today.

8.6.1 Hydrogen–Oxygen Reaction

A good example of the changes from steady reaction to explosion is provided by the reaction between hydrogen and oxygen. A stoichiometric mixture of the two gases explodes spontaneously at any pressure if the temperature is above about 600°C; below about 460°C there is no explosion at any pressure unless a spark is passed through the mixture or a flame is introduced into it. Between 460 and 600°C, however, a stoichiometric mixture explodes spontaneously at certain pressures but not at others. Figure 8.4a illustrates schematically the kind of behavior often observed with a stoichiometric mixture of H_2 and O_2 at 500°C. If the pressure is below about 260 Pa (~2 mmHg) there is steady reaction, but if the pressure is increased to that value an explosion occurs. If the pressure is initially maintained at about 30 kPa (~230 mmHg) there is again a steady reaction; if the pressure is reduced the rate decreases, often to an extremely low value, until a pressure of about 13 kPa (~100 mmHg) is reached, when the mixture explodes. Thus, at this temperature, there is a pressure region within which there is explosion but above and below which there is steady reaction. The two explosion limits are referred to as the *first* and *second limits.* Figure 8.4b shows how the limits change with temperature. The first limit is very sensitive to the nature of the reaction vessel surface and to the vessel dimensions. The second limit, on the other hand, does not change much with the nature of the vessel.

There is also a *third limit,* at still higher pressures. For some reactions the explosion that occurs beyond this limit is due to the fact that the reaction rate has become so high that the heat evolved causes a substantial rise in temperature. Such an explosion limit is referred to as a *thermal limit,* and an explosion occurring under such circumstances is known as a *thermal explosion.* By contrast, the explosions occurring between the first and second limits occur in systems in which there has been no temperature rise; these are known as *isothermal explosions.* It has been suggested, however, that in the hydrogen–oxygen system the explosion occurring beyond the third limit is not a thermal explosion but is due to an increase in the concentration of free radicals. It appears, however, that the limit is distorted by self-heating.[98]

To explain the kinetic behavior of such systems, the idea of *branching chains* was suggested almost simultaneously by Semenov[99] and by Hinshelwood and Thompson.[100] When a pair of ordinary chain-propagating steps occurs, there is no change in the number of chain carriers. When a branching chain process occurs, however, there is an increase in the number of carriers. An example is the pair of reactions

$$(1) \quad H + O_2 \rightarrow OH + O$$

$$(2) \quad O + H_2 \rightarrow OH + H$$

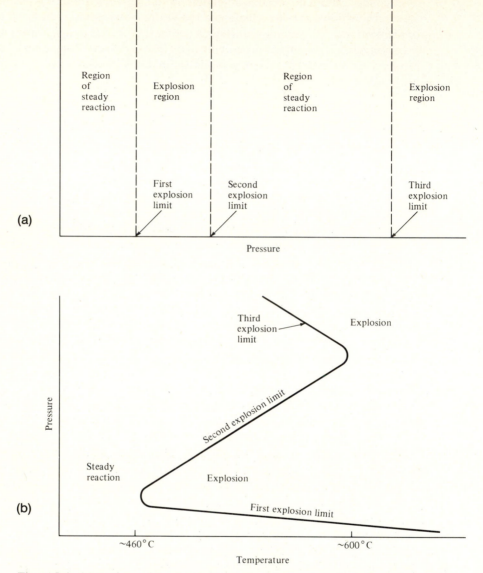

Figure 8.4 (a) Regions of steady reaction and explosion for a typical gas-phase combustion, such as the hydrogen–oxygen reaction. (b) Variation of the explosion limits with temperature. The temperatures given are approximately those for a stoichiometric $2H_2$–O_2 mixture. For such a mixture there is usually no explosion below ~460°C; above ~600°C explosion occurs at all pressures.

In each of these reactions two chain carriers have been formed from one. When the two reactions are added together the result is

$$H + O_2 + H_2 \rightarrow 2OH + H$$

so that the H atom is regenerated but has produced two OH radicals which can undergo further reactions. When such chain branching occurs, the number of atoms and free radicals in the system can increase extremely rapidly, and an explosion may result.

Suppose, for example, that each OH radical produced in the above pair of processes were to undergo the reaction

$$(3) \quad OH + H_2 \rightarrow H_2O + H$$

Each time the pair of reactions (1) and (2) occurs, reaction (3) occurs twice, so that the overall reaction is

$$H + 3H_2 + O_2 \rightarrow 2H_2O + 3H$$

Thus, in the first cycle of reactions, one H atom produces 3H atoms together with two molecules of water. The numbers produced in subsequent cycles are shown in Table 8.4. It is to be seen that after 50 cycles have occurred, roughly 1 mol of H_2O has been formed. Approximately that number of cycles is therefore sufficient to transform all the products under ordinary circumstances. Since the reactions are very fast, 50 cycles can occur in a minute fraction of a second.

In the above scheme the removal of radicals by termination processes has been ignored, the object being merely to explain the general idea behind chain-branching mechanisms. The role played by initiation and termination processes may be considered with reference to the following schematic and simplified mechanism:

(1) $? \rightarrow R$ Chain initiation

(2) $R \rightarrow \alpha R$ Chain branching

(3) $R \rightarrow Z + R$ Reaction to give final product Z and another radical

(4) $R \rightarrow ?$ Chain ending at surface

(5) $R \rightarrow ?$ Chain ending in gas phase

Here no distinction is made between different kinds of radical. The first reaction is the initiation reaction which produces a free radical R. Reaction (2) is a reaction in which the radical R undergoes an elementary process with the production of more

TABLE 8.4   HYDROGEN ATOMS AND H_2O MOLECULES PRODUCED IN A CHAIN-BRANCHING PROCESS

Cycle	H atoms entering	H atoms produced	H_2O molecules produced
1	1	3	2
2	3	9	6
3	9	27	18
4	27	81	54
5	81	243	162
10	19 683	59 049	39 366
20	1.16×10^9	3.49×10^9	2.32×10^9
30	6.86×10^{13}	2.05×10^{14}	1.37×10^{14}
40	4.05×10^{18}	1.22×10^{19}	8.11×10^{18}
50	2.39×10^{23}	7.18×10^{23}	4.79×10^{23}

than one radical. Reaction (3) is the reaction by which the product is formed; it may involve reaction between a radical and a molecule and usually produces a radical in addition to the final product. Examples of such processes are

$$OH + H_2 \rightarrow H_2O + H$$

and

$$HO_2 + H_2 \rightarrow H_2O + OH$$

Reactions (4) and (5) are chain-ending processes.

The rates of the five reactions may be written as:

$$(1) \quad v_1 \equiv \frac{d[R]}{dt} \tag{8.143}$$

$$(2) \quad v_b \equiv \frac{d[R]}{dt} = f_b(\alpha - 1)[R] \tag{8.144}$$

$$(3) \quad v_z \equiv \frac{d[Z]}{dt} = f_p[R] \tag{8.145}$$

$$(4) \quad v_s \equiv \frac{-d[R]}{dt} = f_s[R] \tag{8.146}$$

$$(5) \quad v_g \equiv \frac{-d[R]}{dt} = f_g[R] \tag{8.147}$$

The coefficients $f_b, f_z, f_s,$ and f_g may be functions of concentrations of reactants, products, and other substances present. The steady-state equation for R is

$$v_1 + f_b(\alpha - 1)[R] - (f_s + f_g)[R] = 0 \tag{8.148}$$

and therefore

$$[R] = \frac{v_1}{f_s + f_g - f_b(\alpha - 1)} \tag{8.149}$$

The overall rate of formation of the product Z is therefore

$$v_z = \frac{d[Z]}{dt} = f_b[R] \tag{8.150}$$

$$= \frac{f_b v_1}{f_s + f_g - f_b(\alpha - 1)} \tag{8.151}$$

In the branching reaction α is greater than unit (it is generally 2), so that the last term in the denominator is positive. Therefore, it is possible for the denominator to be zero, and under these circumstances the rate, in principle, becomes infinite. In reality, it becomes very large rather than infinite, steady-state conditions breaking down when the radical concentrations increase very rapidly. The condition that $f_b(\alpha - 1)$ is equal to $f_s + f_g$ nevertheless may be used as a condition for the explosion limit.

The argument may be presented in an alternative form, without making any use of the steady-state hypothesis. According to Eq. (8.143)–(8.147), the net rate of formation of radicals is

$$\frac{d[R]}{dt} = v_1 + f_b(\alpha - 1)[R] - (f_s + f_g)[R] \tag{8.152}$$

This is written conveniently as

$$\frac{dc}{dt} = \theta + \phi c \tag{8.153}$$

where c is the concentration of radicals; ϕ, equal to $f_b(\alpha - 1) - (f_s + f_g)$, is known as the *net branching factor*. Integration of this equation, subject to the boundary condition that $c = 0$ when $t = 0$, gives

$$c = \frac{\theta(e^{\phi t} - 1)}{\phi} \tag{8.154}$$

The variation of c with t is very different according to whether ϕ is positive or negative, as is shown in Fig. 8.5. If ϕ is negative, c rises exponentially to a value of $-\theta/\phi$ (a positive quantity); this corresponds to steady-state conditions and a positive value in the denominator of Eq. (8.151). If, on the other hand, ϕ is positive, c continues to increase and approaches infinity at large values of t. This is the situation when the branching processes dominate the termination processes.

The relative magnitudes of $f_s + f_g$ and of $f_b(\alpha - 1)$ depend on the concentrations of the various species. The denominator of Eq. (8.151) may sometimes be positive and may become zero at certain values of the concentrations. At any set of concentrations for which $f_s + f_g - f_b(\alpha - 1) = 0$, there is an explosion limit: as the concentration is changed, a steady reaction suddenly changes into an explosion.

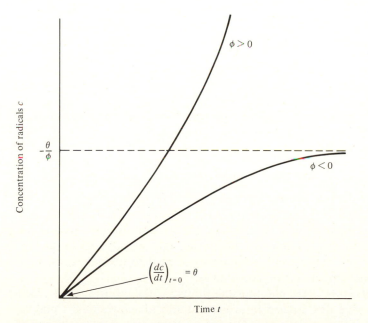

Figure 8.5 Variation of radical concentration c with the time t for a chain-branching reaction obeying the equation $dc/dt = \theta + \phi c$.

The reason for the first and second limits is as follows. The removal of radicals may occur primarily at the surface or primarily in the gas phase. At very low pressures the radicals have easy access to the surface as they do not meet many gas-phase molecules; surface recombination is therefore predominant. An increase in pressure, however, reduces the rate at which radicals can reach the walls and be removed. The term f_s therefore decreases and may do so to the extent that $f_s + f_g$ becomes as small as $f_b(\alpha - 1)$; explosion then occurs. At higher pressures, on the other hand, the removal of radicals takes place primarily in the gas phase, since there are sufficient molecules for gas-phase recombination to be important and to hinder radicals from diffusing to the walls. Under these circumstances a decrease in pressure leads to a decrease in the rate of removal of radicals; f_g thus decreases and $f_s + f_g$ may become equal to $f_b(\alpha - 1)$, with resulting explosion.

It follows from the above considerations that since the lower limit depends on surface recombination, it is sensitive to the nature of the surface and to its area. Coating the surface with material that inhibits recombination therefore increases the explodability and results in a lowering of the lower explosion limit. The same effect is obtained by increasing the size of the vessel; the radicals then have to diffuse further to the walls and recombine less rapidly. The phenomenon has been observed for several reaction systems, including the hydrogen–oxygen one. Roughly speaking, the product of the pressures of the two gases at the lower limit varies inversely with the square of the diameter of the vessel. Foreign gases also depress the lower limit by hindering diffusion to the surface.

As expected, the second limit is not sensitive to the shape and size of the vessel. Foreign gases contribute to gas-phase deactivation and inhibit explosion, which means that they lower the second limit.

8.6.2 Combustion of Hydrocarbons

Many investigations also have been made of the gas-phase combustion of hydrocarbons and of other organic substances. These reactions have some features in common with reactions such as that between hydrogen and oxygen, in that explosion occurs between certain limits of temperature and pressure. They also show the additional feature that under certain conditions a mild type of explosion occurs, accompanied by the emission of light. Since the temperatures are relatively low, the effect is referred to as a *cool flame.*

Like the other combustion reactions, the oxidations of organic compounds usually show an induction period, followed by rapid reaction. Traces of aldehydes are formed during this induction period, which is shortened by the initial addition of aldehydes. Organic peroxides also are formed as intermediates. As with the oxidation of hydrogen and other substances, organic combustions are sensitive to the size of the vessel and the nature of its surface. Electronically excited formaldehyde is mainly responsible for the light emission from cool flames.[101]

The temperature and pressure limits for the occurrence of cool flames and explosions were first studied by Townend and his coworkers[102] and have undergone extensive investigation since. The usual pattern of behavior, illustrated in Fig. 8.6, is quite different from that in the hydrogen–oxygen reaction, for which at a given tem-

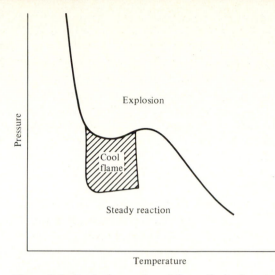

Figure 8.6 Explosion and cool-flame limits for a typical hydrocarbon–oxygen reaction.

perature explosion occurs when the *pressure* is within two limits (Fig. 8.4b). By contrast, for the organic oxidations there is an explosion peninsula between two *temperature* limits. The region of cool flames is represented by an island which adjoins this explosion peninsula.

To explain the main features of the combustion of hydrocarbons and other organic substances, Semenov[103] put forward the hypothesis of *degenerate branching*. His suggestion was that this type of branching is brought about by some intermediate X which has a relatively long lifetime compared with ordinary free radicals. This intermediate X can break down in two different ways, one of which is a branching process and the other not:

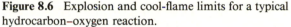

For example, it has been suggested[104] that formaldehyde, an intermediate in many organic oxidations, can undergo the following reactions:

$$\text{HCHO} \quad \begin{array}{c} +\text{OH} \\[4pt] \longrightarrow \text{H}_2\text{O} + \text{HCO} \quad \text{(nonbranching)} \\[4pt] \longrightarrow \text{HCO} + \text{HO}_2 \quad \text{(branching)} \\[4pt] +\text{O}_2 \end{array}$$

Since the intermediate X (HCHO in this example) has a long life compared with that of a radical, branching is delayed. There is a relatively slow increase in the number of radicals instead of the very rapid increase found in the hydrogen–oxygen reaction under conditions leading to explosion. In the hydrocarbon oxidations, the number of radicals increases more slowly, and the termination processes prevent their concentrations from becoming extremely large. Under cool flame conditions the reaction rate is high, but not as high as in an ordinary explosion, and one speaks of a *degenerate explosion.*

An enormous amount of work has been carried out on organic combustions, but there still is uncertainty as to the detailed mechanisms of particular reactions. It is agreed that cool flames result from a subtle interplay between chemical and thermal effects. Space does not permit a consideration of these problems; the reader is referred to books and articles listed in the bibliography at the end of this chapter.

8.7 POLYMERIZATION REACTIONS

The word *polymer* (Greek *poly,* meaning many, and *meros,* meaning part) was coined in 1830 by Berzelius to refer to a molecule of general formula M_n made up by the repetition of n identical units M, known as *monomers.* The word polymer now has a more extended meaning; it is no longer necessary for the polymer to consist of exactly an integral number of monomer units. When they do, they are called *addition polymers.* However, many polymers are formed from monomers with the splitting off of water molecules or other small molecules; these are known as *condensation polymers.* Furthermore, it is no longer necessary for the polymer to be formed from identical monomer molecules. Some polymers, called *copolymers,* formed from two or more different monomers, are distinguished from *homopolymers,* in which only one type of monomer is involved. Today, even the proteins, which are composed of more than 20 different amino acids, are called polymers.

Gas-phase polymerizations occur by composite mechanisms that may be of the molecular type or may involve free-radical processes. In solution or in the liquid phase, polymerizations frequently occur by ionic mechanisms.

8.7.1 Molecular Mechanisms

An example of a polymerization occurring by a molecular mechanism is the reaction between a glycol and a dicarboxylic acid. With ethylene glycol and succinic acid the mechanism is

(1) $\quad HOOC(CH_2)_2COOH + HO(CH_2)_2OH \rightarrow HOOC(CH_2)_2COO(CH_2)_2OH + H_2O$

(2) $\quad HOOC(CH_2)_2COO(CH_2)_2OH + HOOC(CH_2)_2COOH \rightarrow$

$$HOOC(CH_2)_2COO(CH_2)_2OOC(CH_2)_2COOH + H_2O$$

(3) $\quad HOOC(CH_2)_2COO(CH_2)_2OOC(CH_2)_2COOH + HO(CH_2)_2OH \rightarrow$

$$HOOC(CH_2)_2COO(CH_2)_2OOC(CH_2)_2COO(CH_2)_2OH + H_2O$$

and so on. If the rate of each reaction is simply proportional to the concentrations of glycol and acid (this will be the case in the presence of a constant amount of acid catalyst), the treatment is as follows. Let c be the concentration, at any time, of functional groups (either OH or COOH); then

$$-\frac{dc}{dt} = kc^2 \qquad (8.155)$$

For simplicity the rate constant k is assumed to be the same for each reaction. This equation, with the boundary condition $c = c_0$ when $t = 0$, integrates to

$$\frac{c_0 - c}{cc_0} = kt \qquad (8.156)$$

If f is the fraction of functional groups that are esterified at time t,

$$f = \frac{c_0 - c}{c_0} \qquad (8.157)$$

and elimination of c gives

$$\frac{1}{1-f} - 1 = c_0 kt \qquad (8.158)$$

A plot of $1/(1-f)$ versus t should therefore be linear, provided that the assumption of constant k is correct. Flory[105] has verified this relationship for a number of reactions.

A more general treatment of the kinetics of molecular polymerization was given by Dostal and Raff.[106] The individual reactions occurring may all be written as

$$M_m + M_n \rightarrow M_{m+n}$$

where M_m and M_n represent a chain containing m and n monomers, respectively; m and n can have any value from unity upward. The monomer M_1 may be removed by reaction with itself and with any other molecule present; its net rate of consumption is therefore

$$-\frac{d[M_1]}{dt} = k[M_1]^2 + k[M_1][M_2] + \cdots \qquad (8.159)$$

This may be written as

$$\frac{d[M_1]}{dt} = -k[M_1] \sum [M_n] \qquad (8.160)$$

A molecule containing two monomer molecules may be formed by reaction between two monomers and is consumed by reaction with a molecule of any length; the net rate of production of dimer is therefore

$$\frac{d[M_2]}{dt} = \tfrac{1}{2}k[M_1]^2 - k[M_2] \sum [M_n] \qquad (8.161)$$

The factor $\tfrac{1}{2}$ is required because the formation of one dimer requires the reaction of two monomers.

The net rate of formation of n-mers is given in general by

$$\frac{d[M_n]}{dt} = \tfrac{1}{2}k \sum [M_s][M_{n-s}] - k[M_n] \sum [M_n] \qquad (8.162)$$

Addition of all these equations gives

$$\frac{d \sum [M_n]}{dt} = -\tfrac{1}{2} k (\sum [M_n])^2 \tag{8.163}$$

If $[M_1]_0$ represents the total concentration of monomers at the beginning of the reaction, and $\sum [M_n]$ that of all the molecules at time t, the fraction of reaction that has occurred at time t is given by

$$f = \frac{[M_1]_0 - \sum [M_n]}{[M_1]_0} \tag{8.164}$$

Equations (8.163) and (8.164) give rise to

$$\frac{df}{dt} = \tfrac{1}{2} [M_1]_0 k (1 - f)^2 \tag{8.165}$$

which integrates to

$$f = \frac{[M_1]_0 k t}{2 + [M_1]_0 k t} \tag{8.166}$$

This is equivalent to Eq. (8.158). The concentration of n-mers is given by

$$[M_n] = [M_1]_0 f^{n-1} (1 - f)^2 \tag{8.167}$$

This equation is useful in giving the distribution of polymers corresponding to any extent of reaction.

8.7.2 Free-Radical Mechanisms

Olefinic substances, such as ethylene and styrene, usually polymerize by free-radical mechanisms. The reactions sometimes occur in the gas phase and sometimes in solution or in the liquid phase; the free-radical processes are similar in all cases. The nature of the initiation reaction varies with the conditions. Sometimes, as with ethylene, the initiation may be of the thermal type, the molecule forming the diradical $-CH_2CH_2-$. In other cases, as with vinyl acetate, light of certain frequencies favors the production of radicals and causes the substance to polymerize at ordinary temperatures.[107] Polymerization also can be induced by the introduction of free radicals into a monomer; hydrogen atoms and methyl radicals, for example, bring about the polymerization of vinyl acetate.[108] Substances such as oxygen and peroxides also are found to catalyze polymerization and are known as *sensitizers;* peroxides, for example, readily produce radicals by breaking the $O-O$ bond. In such cases the propagation and termination steps are of the same character, and we will consider a general scheme of reactions, without the nature of the initiating reaction being at first specified.

The type of propagation reaction that is involved is the addition of a radical to a double bond:

$$R + R'CH = CH_2 \rightarrow R'CH - CH_2 R$$

This produces another radical, which in turn can add on to a double bond. This process can continue with the formation of large radicals, which finally react with another radical to give a molecule.

The general reaction scheme for this type of polymerization reaction can be represented as follows:

$$(1) \quad ? \rightarrow R_1 \qquad\qquad \text{Initiation}$$

$$(2) \quad R_1 + M \rightarrow R_2$$

$$(3) \quad R_2 + M \rightarrow R_3 \qquad\qquad \text{Chain propagation}$$
$$\vdots$$
$$(4) \quad R_{n-1} + M \rightarrow R_n$$

$$(5) \quad R_n + R_m \rightarrow M_{n+m} \qquad \text{Termination}$$

Here M represents a monomer molecule, R_1 a free radical produced in the initiation step, and R_n a radical consisting of R_1 added to a chain of $(n-1)$ monomer molecules. The rate of the initial formation of R_1 is written v_1, and the rate constants for the propagation reactions, all assumed to be the same, are written k_p. Termination can involve any two radicals (including identical ones), and the rate constant for termination is written k_t.

Application of the steady-state treatment leads to a set of equations. The first of these, applying to R_1, is

$$v_1 - k_p[R_1][M] - k_t[R_1]([R_1] + [R_2] + \cdots) = 0 \qquad (8.168)$$

The final term is for the removal of radical R_1 by reaction with R_1, R_2, R_3, and so on. The equation may be written as

$$v_1 - k_p[R_1][M] - k_t[R_1] \sum [R_n] = 0 \qquad (8.169)$$

Similarly, for R_2,

$$k_p[R_1][M] - k_p[R_2][M] - k_t[R_2] \sum [R_n] = 0 \qquad (8.170)$$

In general, for R_n,

$$k_p[R_{n-1}][M] - k_p[R_n][M] - k_t[R_n] \sum [R_n] = 0 \qquad (8.171)$$

There is an infinite number of such equations, and the sum of all the equations is§

$$v_1 - k_t(\sum [R_n])^2 = 0 \qquad (8.172)$$

Therefore
$$\sum [R_n] = \left(\frac{v_1}{k_t}\right)^{1/2} \qquad (8.173)$$

The rate of consumption of monomer is

$$-\frac{d[M]}{dt} = k_p[M] \sum [R_n] \qquad (8.174)$$

$$= k_p\left(\frac{v_1}{k_t}\right)^{1/2} [M] \qquad (8.175)$$

Various special cases of this equation, corresponding to different mechanisms of initiation, are considered now.

§ This equation simply states that the rate of initiation is equal to the sum of the rates of all the termination processes; this must be true if steady-state conditions apply.

In a purely thermal polymerization the initiation process may be a second-order reaction between monomer molecules; in this case

$$v_1 = k_1[M]^2 \tag{8.176}$$

Equation (8.175) therefore becomes

$$-\frac{d[M]}{dt} = k_p\left(\frac{k_1}{k_t}\right)^{1/2}[M]^2 \tag{8.177}$$

so that the overall order is the second. This law is obeyed in the polymerization of styrene in the gas phase[109] and in various solvents.[110]

If initiation involves a second-order reaction between a catalyst C and the monomer, the rate of initiation is

$$v_1 = k_1[C][M] \tag{8.178}$$

Introduction of this into Eq. (8.175) gives, for the rate of consumption of the monomer,

$$-\frac{d[M]}{dt} = k_p\left(\frac{k_1}{k_t}\right)^{1/2}[M]^{3/2}[C]^{1/2} \tag{8.179}$$

This law is obeyed accurately in the polymerization of styrene[111] and vinyl acetate,[112] catalyzed by benzoyl peroxide.

Another possibility with respect to a catalytic or sensitized initiation is that the initiating reaction is first order with respect to the catalyst and zero order with respect to the monomer; that is, the initiation rate is

$$v_1 = k_1[C] \tag{8.180}$$

The rate of polymerization is then

$$-\frac{d[M]}{dt} = k_p\left(\frac{k_1}{k_t}\right)^{1/2}[M][C]^{1/2} \tag{8.181}$$

This law is followed accurately by the polymerization of d-sec-butyl α-chloroacrylate, sensitized by benzoyl peroxide.[113]

In photochemical initiation the rate of the initiating step is

$$v_1 = I \tag{8.182}$$

where I is the intensity of light absorbed. The rate of polymerization is then

$$-\frac{d[M]}{dt} = k_p\left(\frac{I}{k_t}\right)^{1/2}[M] \tag{8.183}$$

Such a law is obeyed in the polymerization of ethylene[114] induced by wavelengths of about 186 nm.

It is important for an understanding of the mechanisms of polymerizations to have knowledge of the rate constants of the various elementary processes. This presents little difficulty as far as the initiation reaction is concerned, the procedure being somewhat similar to that used by Bodenstein (Section 8.4.1) for the hydrogen–bromine reaction. The quantities that have to be measured are the rate of the overall process and the value of the kinetic chain length, γ, defined as the ratio of the overall rate to the rate of the initial step:

$$\gamma = \frac{-d[\text{M}]/dt}{v_1} \tag{8.184}$$

The kinetic chain length, for the above mechanism, may be identified with the number of repeating units in the polymer chain, a quantity that can be determined in various ways. The rate v_1 of the initiating reaction and hence its rate constant therefore can be obtained.

To obtain the rate constants for the propagation and termination processes, a procedure devised by Melville[115] often is employed. The overall rate constant for a free-radical polymerization is equal to $k_p(k_1/k_t)^{1/2}$. The evaluation of k_1, described previously, therefore gives the value of $k_p/k_t^{1/2}$ but does not allow a separation of the constants. Melville pointed out that a further relationship can be obtained if a measurement is made of the total steady-state concentration of the free radicals R_n which are concerned in the chain-propagating steps. Thus, the overall rate of reaction is

$$-\frac{d[\text{M}]}{dt} = k_p[\text{M}] \sum [\text{R}_n] \tag{8.185}$$

so that, if $\sum [\text{R}_n]$ is known, k_p can be determined from the overall rate. The method used for determining this concentration is to measure the mean lifetime τ of the active radicals. The stationary-state expression for radicals is

Rate of formation of radicals = rate of removal of radicals

$$= \sum \frac{[\text{R}_n]}{\tau} \tag{8.186}$$

so that if τ and the rate of formation of radicals are known, the concentration $\sum [\text{R}_n]$ can be calculated. The rate of formation of radicals can be determined in various ways: one method involves the use of inhibitors, and the rate also can be determined from the number of quanta absorbed in a photochemical polymerization, assuming that each absorbed photon starts a chain. The mean life τ is measured using the sector method, described in Section 9.1.5.

8.7.3 Cationic Polymerization

Polymerizations in the liquid phase and in solution may occur by mechanisms that do not take place readily in the gas phase. These are the ionic mechanisms, in which the propagation steps involve ions rather than free radicals. The evidence for the occurrence of such mechanisms is that the reactions are catalyzed not by free radicals but by substances that are either acidic or basic, in the general Lewis sense. Also, the rates of such reactions vary with the dielectric constant of the solvent in the manner expected of ionic processes. The ionic mechanisms of polymerization may be classified as *cationic* or *anionic,* according to whether the catalysis is brought about by cationic or anionic species.

Many acidic substances have been found to be effective catalysts for the polymerization of unsaturated substances such as isobutene: they include

1. Hydrogen acids such as HCl, H_2SO_4, and H_3PO_4.
2. Lewis acids, such as $AlCl_3$ and BF_3.
3. Cation-forming substances, such as I_2 and $AgClO_4$.

All these substances are electron acceptors and are therefore acids in the Lewis sense.

To account for polymerizations brought about by such catalysts, Price[116] proposed the following polar-bond mechanism, the acidic catalyst being represented as MX:

(1) $\quad MX + CH_2{=}CHR \rightleftharpoons X{-}M^-{-}CH_2{-}C^+HR$

(2) $\quad X{-}M^-{-}CH_2{-}C^+HR + CH_2{=}CHR \rightarrow$

$\qquad\qquad X{-}M^-{-}CH_2{-}CHR{-}CH_2{-}C^+HR$

Initiation

(3) $\quad X{-}M^-{-}CH_2{-}CHR{-}CH_2{-}C^+HR + CH_2{=}CHR \rightarrow$

$\qquad X{-}M^-{-}CH_2{-}CHR{-}CH_2{-}CHR{-}CH_2{-}C^+HR$

$\qquad\qquad\qquad\qquad \vdots$

(4) $\quad X{-}M^-{-}(CH_2{-}CHR)_n{-}CH_2{-}C^+HR + CH_2{=}CHR \rightarrow$

$\qquad\qquad X{-}M^-{-}(CH_2{-}CHR)_{n+1}{-}CH_2{-}C^+HR$

Propagation

(5) $\quad X{-}M^-{-}(CH_2{-}CHR)_n{-}CH_2{-}C^+HR \rightarrow$

$\qquad\qquad M{-}(CH_2{-}CHR)_n{-}CH{-}CR + HX$

Termination

Reactions (1) and (2) may be regarded as constituting the initiation step; they are written in this way to avoid postulating a third-order initiation. The rate of initiation is

$$v_1 = k_2[X{-}M^-{-}CH_2{-}C^+HR][CH_2{=}CHR] \qquad (8.187)$$

$$= k_2 K_1[MX][CH_2{=}CHR]^2 \qquad (8.188)$$

where K_1 is the equilibrium constant for reaction (1), and k_2 is the rate constant for reaction (2).

A series of steady-state equations can be written: the first is

$$k_i[MX][CH_2{=}CHR]^2 - k_p[X{-}M^-{-}CH_2{-}CHR{-}CH_2{-}C^+HR][CH_2{=}CHR]$$

$$- k_t[X{-}M^-{-}CH_2{-}CHR{-}CH_2{-}C^+HR] = 0 \qquad (8.189)$$

Here k_i has been written for $k_2 K_1$, and k_p is the rate constant for the chain-propagating step (3). All these steps are assumed to have the same rate constant. The general steady-state equation is

$$k_p[X{-}M^-{-}(CH_2{-}CHR)_n{-}CH_2{-}C^+HR][CH_2{=}CHR]$$

$$- k_p[X{-}M^-{-}(CH_2{-}CHR)_{n+1}{-}CH_2{-}C^+HR][CH_2{=}CHR]$$

$$- k_t[X{-}M^-{-}(CH_2{-}CHR)_{n+1}{-}CH_2{-}C^+HR] = 0 \qquad (8.190)$$

The sum of all the equations is

$$k_i[MX][CH_2{=}CHR]^2 - k_t \sum [X{-}M^-{-}(CH_2{-}CHR)_n{-}CH_2{-}C^+HR] = 0$$

$$(8.191)$$

and therefore

$$\sum [X-M^- -(CH_2-CHR)_n-CH_2-C^+HR] = \frac{k_i}{k_t}[MX][CH_2=CHR]^2 \qquad (8.192)$$

The rate of consumption of monomer is

$$-\frac{d[CH_2=CHR]}{dt} = k_p[CH_2=CHR]$$

$$\times \sum [X-M^- -(CH_2-CHR)_n-CH_2-C^+HR] \qquad (8.193)$$

$$= \frac{k_i k_p}{k_t}[MX][CH_2=CHR]^3 \qquad (8.194)$$

This rate equation is in agreement with experiment for a number of ionic polymerizations.

It appears that this polar-bond mechanism is applicable to certain polymerizations, such as the polymerization of styrene catalyzed by aluminum chloride in carbon tetrachloride solution.[117] This type of mechanism, however, cannot be the correct one in a number of cases of cation polymerization. An objection to it as a general mechanism is that the zwitterions postulated as the chain carriers are of low stability when the charges are separated by a long saturated chain.

In many cases of cationic polymerization it is found that a cocataylst must be present in addition to the acidic catalyst. Thus, the polymerization of isobutene catalyzed by boron trifluoride occurs only in the presence of a small amount of water, which is believed to act as a cocatalyst.[118] The polar-bond mechanism provides no interpretation of the necessity of a cocatalyst. To explain this type of behavior Evans and Polanyi[119] suggested that the acidic catalyst reacts with the cocatalyst to form ions and that the cation adds on to the monomer to form a carbonium ion. In catalysis by BF_3 in the presence of moisture, for example, the ions are believed to be formed by the process

$$(1) \quad BF_3 + H_2O \rightleftharpoons HOBF_3^- + H^+$$

and to remain together as an ion pair, $HOBF_3^- H^+$. This is followed by addition of the proton to the monomer; in isobutene polymerization, for example,

$$(2) \quad HOBF_3^- H^+ + CH_2=C(CH_3)_2 \rightarrow C(CH_3)_3^+ + HOBF_3^-$$

In isobutene polymerization this is followed by a series of propagation reactions in which a carbonium ion is added to the monomer:

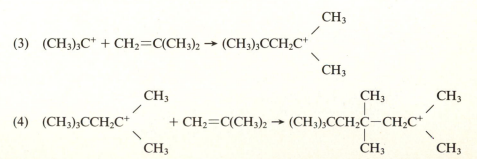

and so on. Termination occurs by one of the polymer cations splitting off a hydrogen ion,

$$(5) \quad \text{$\sim$CH}_2\text{C}^+ \begin{array}{c} \diagup \text{CH}_3 \\ \diagdown \text{CH}_3 \end{array} \longrightarrow \text{$\sim$CHC(CH}_3)_2 + \text{H}^+$$

A number of different kinetic equations can arise from this mechanism, according to the assumption made about the rate-determining initiation step. Suppose, for example, that the equilibrium (1) is rapid and that the slow initiation process is (2); the concentration of ion pairs is given by

$$[\text{HOBF}_3^-\text{H}^+] = K[\text{BF}_3][\text{H}_2\text{O}] \tag{8.195}$$

where K is the equilibrium constant. The rate of initiation is therefore

$$v_i = k_2 K[\text{BF}_3][\text{H}_2\text{O}][\text{monomer}] \tag{8.196}$$

In the steady state this is equal to the sum of the rates of all the termination steps:

$$k_2 K[\text{BF}_3][\text{H}_2\text{O}][\text{monomer}] = k_t \sum [\text{P}_n^+] \tag{8.197}$$

where P_n^+ is the carbonium ion containing n molecules of monomer. It follows that

$$\sum [\text{P}_n^+] = \frac{k_2}{k_t} K[\text{BF}_3][\text{H}_2\text{O}][\text{monomer}] \tag{8.198}$$

and the overall rate of polymerization is

$$v = \frac{k_1 k_p}{k_t} K[\text{BF}_3][\text{H}_2\text{O}][\text{monomer}]^2 \tag{8.199}$$

In many cases either the catalyst or the cocatalyst is present in excess, so that its concentration does not appear explicitly. A number of systems have been found to conform to this type of kinetic equation. Note that if the individual ion HOBF_3^-, rather than the ion pair, were to react with the monomer in reaction (2), the rate equation would involve the catalyst and cocatalyst concentrations to their half powers. Square-root laws also would be found if termination involved two ions. Square-root laws are not generally observed, so that it appears that the ion pairs are the reactive species and that termination involves the elimination of an ion in a unimolecular reaction.

8.7.4 Anionic Polymerization

The polymerization of olefins can also be brought about by basic catalysts such as aqueous alkalis, the alkali metals, sodamide, and the metal alkyls. A kinetic study by Higginson and Wooding[120] of the polymerization of styrene catalyzed by potassamide in liquid ammonia showed that the rate was given by

$$v = k[\text{NH}_2^-][\text{styrene}]^2 \tag{8.200}$$

These results were explained in terms of the following mechanism:

$$NH_2^- + CH_2{=}CHC_6H_5 \rightarrow NH_2{-}CH_2C^-HC_6H_5 \qquad \text{Initiation}$$

$$\left.\begin{array}{l} NH_2{-}CH_2{-}[CHC_6H_5{-}CH_2]_nC^-HC_6H_5 + CH_2{=}CHC_6H_5 \rightarrow \\[2mm] \qquad\qquad NH_2{-}CH_2{-}[CHC_6H_5{-}CH_2]_{n+1}C^-HC_6H_5 \end{array}\right\} \text{Propagation}$$

$$\text{\textasciitilde}C^-HC_6H_5 + NH_3 \rightarrow \text{\textasciitilde}CH_2C_6H_5 + NH_2^- \qquad \text{Termination}$$

With weaker bases, such as the metal alkyls, it appears[121] that these frequently do not act as true catalysts but are incorporated into the polymer. The polymerization of butadiene in the presence of a metal alkyl MR, for example, is believed to occur by the mechanism

$$MR + C_4H_6 \rightarrow R{-}C_4H_6{-}M$$

$$R{-}C_4H_6{-}M + C_4H_6 \rightarrow R{-}[C_4H_6]_2M \quad \text{and so on}$$

Such polymerizations are kinetically analogous to those occurring by a condensation mechanism. The chains are terminated by certain impurities which act as "interceptors."

8.7.5 Emulsion Polymerization

During recent years there has been considerable interest in the kinetics and mechanisms of polymerizations occurring in emulsions. The substituted ethylenes, for example, can polymerize in aqueous emulsions. A typical procedure that has been used with monomers like methyl methacrylate,

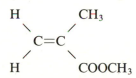

is to form an aqueous emulsion with the monomer, using an emulsifying agent such as cetyltrimethylammonium bromide. Addition of Fenton's reagent (ferrous ions and hydrogen peroxide) may be used to initiate polymerization; this reagent generates radicals, and the polymerization occurs by a free-radical mechanism. The practical advantage of emulsion polymerization is that the reactions proceed much more rapidly than in bulk systems.

Only a brief account is given here of the kinetics of emulsion polymerizations. Emulsions formed by soaps were shown by McBain[122] and Harkins[123] to contain micelles consisting of layers of oriented soap molecules. When monomers such as methyl methacrylate are added to such a soap solution, they are solubilized, which means that some of the monomer penetrates the micelles. The remainder of the monomer exists in the form of suspended droplets in the aqueous phase.

It was suggested by Harkins[124] that in emulsion systems there are two principal loci where polymerization takes place. Initially, when there is not much polymer present, the micelles largely consist of monomer. The radii of the micelles are much smaller than those of the droplets, and the micelles present a much larger interfacial area. As a result, most of the polymerization takes place in the micelles. The free radicals generated by the initiating medium (e.g., Fenton's reagent) in the aqueous phase are captured by the micelles, and at the same time monomer is transferred from the droplets to the micelles, where they react with the radicals. The micelles grow larger and soon consist mainly of polymer particles with soap adsorbed on them. After 2–3% of the monomer has been converted into polymer, the micelles have turned largely into polymer particles. From this point on, the polymerization takes place mainly in the polymer particles.

A quantitative formulation of this theory was put forward by Smith and Ewart.[125] Its application to the second phase of the polymerization, when the polymerization is occurring mainly in the polymer particles, is in brief as follows. The number of polymer particles remains approximately constant and varies little from one system to another; on the average there are about 10^{14} particles per cubic centimeter. Suppose that the initiator concentration is such that about 10^{13} free radicals are produced per cubic centimeter per second. If all these enter the polymer particles, a particle will acquire, on average, a radical once in about every 10 s. Once a radical has entered a particle it adds on to monomer units at a rate equal to $k_p[M]$, where k_p is the rate constant for the propagation process and $[M]$ is the concentration of monomer in the particles.

If a radical enters a particle that already contains a radical, it combines with it at once, and it follows that, at any time, one-half of the polymer particles contain one radical and the other half no radical. The rate of polymerization therefore is equal to

$$v = \tfrac{1}{2}k_p[M]N \tag{8.201}$$

when N is the total number of polymer particles. This relationship has been confirmed for a number of systems,[126] and the k_p values derived are in reasonable agreement with those obtained in other ways. Relationships between the number of particles N and the concentrations of emulsifier and initiator have been derived by Smith and Ewart[125] and by Medvedev,[127] whose model is a little different from that outlined above.

PROBLEMS

8.1. A reaction of stoichiometry

$$A + B \rightarrow Y + Z$$

is second order in A and zero order in B. Suggest a mechanism that is consistent with this behavior.

8.2. The rate of formation of the product of a reaction gives a nonlinear Arrhenius plot, the line being convex to the $1/T$ axis (i.e., the activation energy is higher at higher temperatures). Suggest a reason for this type of behavior.

8.3. An Arrhenius plot is concave to the $1/T$ axis (i.e., a lower activation energy at higher temperatures). Suggest a reason for this type of behavior.

8.4. Confirm that Eq. (8.94) follows from the mechanism given in Section 8.4.6.

8.5. In Section 5.2 the reaction $2NO + O_2 \rightarrow 2NO_2$ was treated as if it were an elementary third-order reaction: the fact is that the reaction is second-order in NO and first-order in oxygen. The reaction, however, may occur by the alternative mechanism

$$2NO \xrightarrow{k_1} N_2O_2$$

$$N_2O_2 \xrightarrow{k_{-1}} 2NO$$

$$N_2O_2 + O_2 \xrightarrow{k_2} 2NO_2$$

Assume N_2O_2 to be in a steady state and derive the rate equation. Under what conditions is the rate equation $v = k[NO]^2[O_2]$? Is it justifiable, if this is the mechanism, to apply CTST as was done in Section 5.2?

8.6. Confirm that the mechanism

$$(1) \quad Cl_2 \underset{k_{-1}}{\overset{k_1}{\rightleftharpoons}} 2Cl$$

$$(2) \quad Cl + Cl_2 \underset{k_{-2}}{\overset{k_2}{\rightleftharpoons}} Cl_3$$

$$(3) \quad Cl_3 + CO \underset{k_{-3}}{\overset{k_3}{\rightleftharpoons}} COCl_2 + Cl$$

gives the correct form of the rate equations (8.76) and (8.77) for the reaction in the forward and reverse directions. Reactions (1) and (2) can be taken to be rapid in both directions.

8.7. The gas-phase reaction

$$Cl_2 + CH_4 \rightarrow CH_3Cl + HCl$$

proceeds by a free-radical chain reaction in which the chain propagators are Cl and CH_3 (but not H), and the chain-ending step is $2Cl \rightarrow Cl_2$. Write the mechanism, identify the initiation reaction and the chain-propagating steps, and obtain an expression for the rate of the overall reaction.

8.8. Confirm that the mechanism given in Section 8.5.5 for the rate of the NO-inhibited ethane decomposition leads to Eq. (8.141).

8.9. Use the activation energies listed in Table 8.2 to estimate the overall activation energy of the thermal hydrogen–bromine reaction under the following conditions:
(a) At the beginning of the reaction.
(b) In the presence of a large excess of HBr.
Also, estimate the overall activation energy for the photochemical reaction, under the same conditions.

8.10. Confirm that the mechanism in Section 8.5.4 for the photochemical decomposition of acetaldehyde leads to Eq. (8.140). Also, obtain an expression for the quantum yield.

8.11. The photochemical chlorination of chloroform, $CHCl_3 + Cl_2 \rightarrow CCl_4 + HCl$, follows the rate equation

$$\frac{d[CCl_4]}{dt} = kI^{1/2}[CHCl_3]$$

where I is the intensity of light absorbed. Devise a reaction mechanism that is consistent with this result.

8.12. Show that the mechanism

$$I_2 \underset{k_{-1}}{\overset{k_1}{\rightleftharpoons}} 2I \quad (fast)$$

$$I + H_2 \underset{k_{-2}}{\overset{k_2}{\rightleftharpoons}} H_2I \quad \text{(fast)}$$

$$H_2I + I \overset{k_3}{\rightarrow} 2HI \quad \text{(slow)}$$

leads to the result that the reaction between hydrogen and iodine obeys the rate equation $v = k[H_2][I_2]$.

8.13. Obtain the rate equation for the H_2–I_2 reaction corresponding to the mechanism

$$I_2 \underset{k_{-1}}{\overset{k_1}{\rightleftharpoons}} 2I$$

$$I + H_2 \overset{k_2}{\rightarrow} HI + H$$

$$H + I_2 \overset{k_3}{\rightarrow} HI + I$$

REFERENCES

1. C. N. Hinshelwood and T. E. Green, *J. Chem. Soc.,* **129,** 730 (1926).
2. W. D. Bonner, W. L. Gore, and D. M. Yost, *J. Am. Chem. Soc.,* **57,** 2723 (1935).
3. C. Capellos and B. H. J. Bielski, *Kinetic Systems: Mathematical Description of Chemical Kinetics in Solution,* Wiley-Interscience, New York, 1972.
4. A. V. Harcourt and W. Esson, *Proc. R. Soc. London,* **14,** 470 (1865); *Philos. Trans.,* **156,** 193 (1866); **157,** 117 (1867).
5. E. Rutherford and F. Soddy, *J. Chem. Soc.,* **81,** 321, 837 (1902).
6. G. R. Branton and H. M. Frey, *J. Chem. Soc.,* 1342 (1966).
7. D. L. Chapman and L. K. Underhill, *Trans. Chem. Soc.,* 496 (1913).
8. A. Skrabal, *Ann. Phys.,* **82,** 138 (1927).
9. M. Bodenstein, *Ann. Phys.,* **82,** 836 (1927). See also D. A. Frank-Kamenetsky, *J. Phys. Chem. USSR,* **14,** 695 (1940).
10. S. Glasstone, K. J. Laidler, and H. Eyring, *The Theory of Rate Processes,* pp. 99–100, McGraw-Hill, New York, 1941.
11. F. R. Cruikshank, A. J. Hyde, and D. Pugh, *J. Chem. Educ.,* **54,** 288 (1977).
12. J. R. Murdoch, *J. Chem. Educ.,* **58,** 32 (1981).
13. This problem is discussed in further detail by R. M. Krupka, H. Kaplan, and K. J. Laidler, *Trans. Faraday Soc.,* **62,** 2754 (1966).
14. W. Nernst, *Z. Elektrochem.,* **24,** 335 (1918).
15. M. Bodenstein and W. Dux, *Z. Phys. Chem.,* **85,** 297 (1913).
16. M. Bodenstein, *Z. Elektrochem.,* **85,** 329 (1913).
17. M. Bodenstein, *Z. Elektrochem.,* **22,** 53 (1916).
18. H. S. Taylor, *Trans. Faraday Soc.,* **21,** 5601 (1926).
19. F. O. Rice, W. R. Johnston, and B. L. Evering, *J. Am. Chem. Soc.,* **54,** 3559 (1932); F. O. Rice and A. L. Glasebrook, *J. Am. Chem. Soc.,* **55,** 4329 (1933); **56,** 2381, 2472 (1934).
20. F. Paneth and W. Hofeditz, *Ber. D. Chem. Ges.,* **B62,** 1335 (1929).
21. H. S. Taylor, *Ann. Rev. Phys Chem.,* **13,** 477 (1962).
22. F. O. Rice and K. F. Herzfeld, *J. Am. Chem. Soc.,* **56,** 284 (1934).
23. M. Bodenstein and S. C. Lind, *Z. Phys. Chem.,* **57,** 108 (1907).
24. J. A. Christiansen, *K. Dan. Vidensk. Selsk. Mat.-Fys. Medd.,* **1,** 14 (1919).
25. K. F. Herzfeld, *Z. Elektrochem.,* **25,** 301 (1919); *Ann. Phys.,* **59,** 635 (1919).
26. M. Polanyi, *Z. Elektrochem.,* **26,** 50 (1920).

27. M. Bodenstein and H. Lutkemeyer, *Z. Phys. Chem.,* **114,** 208 (1924).

28. C. H. Burgess and D. L. Chapman, *J. Chem. Soc.,* **89,** 1399 (1906).

29. J. G. A. Griffiths and R. G. W. Norrish, *Proc. R. Soc. London A,* **135,** 69 (1932); **147,** 140 (1934).

30. D. L. Chapman and P. S. MacMahon, *J. Chem. Soc.,* **95,** 959 (1909); M. Bodenstein and W. Dux, *Z. Phys. Chem.,* **85,** 297 (1913).

31. M. Bodenstein and W. Unger, *Z. Phys. Chem. B,* **11,** 253 (1930). See also N. Thon, *Z. Phys. Chem.,* **124,** 327 (1926); E. Cremer, *Z. Phys. Chem.,* **128,** 285 (1927).

32. W. Jost and H. Schweizer, *Z. Phys. Chem. B,* **13,** 373 (1931).

33. R. Gohring, *Z. Elektrochem.,* **27,** 511 (1921).

34. H. Sirk, *Z. Phys. Chem.,* **61,** 545 (1908); K. H. A. Melander, *Arkiv. Kemi.,* **5,** No. 12 (1913–1915); J. A. Christiansen, *Z. Phys. Chem.,* **82,** 405 (1929); R. N. Pease, *J. Am. Chem. Soc.,* **56,** 2388 (1934); G. Kornfeld and S. Khodschaian, *Z. Phys. Chem. B,* **35,** 403 (1937).

35. R. N. Pease, *J. Am. Chem. Soc.,* **56,** 2388 (1934).

36. M. Bodenstein, *Z. Phys. Chem.,* **13,** 56 (1894); **22,** 1, 23, (1897); **29,** 295 (1899).

37. W. C. McC. Lewis, *J. Chem. Soc.,* **113,** 471 (1918).

38. E. J. Rosenbaum and T. R. Hogness, *J. Chem. Phys.,* **2,** 267 (1934).

39. J. W. Benson and R. Srinivasan, *J. Chem. Phys.,* **23,** 200 (1955).

40. J. H. Sullivan, *J. Chem. Phys.,* **30,** 1291, 1577 (1959); **36,** 1925 (1962); **39,** 300 (1963).

41. N. N. Semenov, *Some Problems in Chemical Kinetics and Reactivity,* vol. 2, pp. 73–74, Princeton University Press, Princeton, NJ, 1959.

42. J. H. Sullivan, *J. Chem. Phys.,* **46,** 73 (1967).

43. J. B. Anderson, *J. Chem. Phys.,* **61,** 3390 (1974); R. L. Jaffe, J. M. Henry, and J. B. Anderson, *J. Am. Chem. Soc.,* **98,** 1140 (1976).

44. M. Ritchie and R. G. W. Norrish, *Proc. R. Soc. London A,* **140,** 112, 713 (1933); J. C. Potts and G. K. Rollefson, *J. Am. Chem. Soc.,* **57,** 1027 (1935); W. J. Kramer and L. A. Moignard, *Trans. Faraday Soc.,* **45,** 903 (1949).

45. J. A. Christiansen, *Z. Phys. Chem.* **103,** 99 (1922).

46. M. Bodenstein and H. Plaut, *Z. Phys. Chem.,* **110,** 399 (1924).

47. M. Bodenstein, S. Lenher, and C. Wagner, *Z. Phys. Chem. B,* **3,** 459 (1959); M. Bodenstein, W. Brenschade, and H. J. Schumacher, *Z. Phys. Chem. B,* **28,** 81 (1935); **40,** 120 (1938).

48. F. Daniels and E. H. Johnston, *J. Am. Chem. Soc.,* **43,** 53, 72 (1921).

49. J. K. Hunt and F. Daniels, *J. Am. Chem. Soc.,* **47,** 1602 (1925).

50. H. C. Ramsperger, M. E. Nordheim, and R. C. Tolman, *Proc. Natl. Acad. Sci. USA,* **15,** 453 (1929); H. C. Ramsperger and R. C. Tolman, *Proc. Natl. Acad. Sci. USA,* **16,** 6 (1930).

51. H. J. Schumacher and G. Sprenger, *Z. Phys. Chem. A,* **140,** 281 (1929); J. H. Hodges and E. F. Linhorst, *Proc. Natl. Acad. Sci. USA,* **17,** 28 (1931); *J. Am. Chem. Soc.,* **56,** 836 (1934).

52. W. F. Busse and F. Daniels, *J. Am. Chem. Soc.,* **49,** 1257 (1927).

53. R. A. Ogg, *J. Chem. Phys.,* **15,** 337 (1947); **18,** 572 (1950).

54. R. L. Mills and H. S. Johnston, *J. Am. Chem. Soc.,* **73,** 938 (1951).

55. R. A. Ogg, *J. Chem. Phys.,* **18,** 573 (1950).

56. D. L. Chapman and H. E. Jones, *J. Chem. Soc.,* **97,** 2463 (1910).

57. J. H. van't Hoff, *Etudes de dynamique chimique,* Muller, Amsterdam, 1884.

58. E. H. Riesenfeld and W. Bohnholtzer, *Z. Phys. Chem.,* **130,** 241 (1927); E. H. Riesenfeld and H. J. Schumacher, *Z. Phys. Chem.* **138,** 268 (1928).

59. D. L. Chapman and H. E. Clarke, *J. Chem. Soc.,* **93,** 1638 (1908).

60. S. Jahn, *Z. Anorg. Chem.,* **48,** 260 (1906). See also E. P. Perman and R. H. Greaves, *Proc. R. Soc. London A,* **80,** 353 (1908).

61. S. W. Benson and A. E. Axworthy, *J. Chem. Phys.,* **26,** 1718 (1957).

62. A. Glissman and H. J. Schumacher, *Z. Phys. Chem. B,* **21,** 323 (1933).

63. H. J. Schumacher, *J. Chem. Phys.,* **33,** 938 (1960).

64. S. W. Benson, *J. Chem. Phys.,* **33,** 939 (1960).

65. D. J. McKenney and K. J. Laidler, *Can. J. Chem.,* **40,** 539 (1962).

66. W. D. McGrath and R. G. W. Norrish, *Proc. R. Soc. London A,* **254,** 317 (1960).

67. K. F. Bonhoeffer and P. Harteck, *Z. Phys. Chem. B,* **4,** 119 (1929); A. Farkas, *Z. Elektrochem.,* **36,** 782 (1930); *Z. Phys. Chem. B,* **10,** 419 (1930); A. Farkas and L. Farkas, *Proc. R. Soc. London A,* **162,** 124 (1935).

68. F. O. Rice and K. F. Herzfeld, *J. Am. Chem. Soc.,* **56,** 284 (1934).

69. P. Goldfinger, M. Letort, and M. Niclause, *Contribution à l'étude de la structure moléculaire,* Victor Henri Commemorative Volume, p. 283, Desoer, Liège, 1948.

70. L. A. K. Staveley and C. N. Hinshelwood, *Proc. R. Soc. London A,* **154,** 335 (1936); *J. Chem. Soc.,* 1568 (1937); L. A. K. Staveley, *Proc. R. Soc. London A,* **162,** 557 (1936).

71. V. A. Poltorak and V. V. Voevodsky, *Dokl. Acad. Nauk SSSR,* **91,** 589 (1953).

72. F. O. Rice and R. E. Varnerin, *J. Am. Chem. Soc.,* **76,** 324 (1954).

73. A. Kuppermann and L. G. Larson, *J. Chem. Phys.,* **33,** 1264 (1960).

74. B. W. Wojciechowski and K. J. Laidler, *Can. J. Chem.,* **41,** 1993 (1963).

75. M. C. Lin and M. H. Back, *Can. J. Chem.,* **44,** 505, 2357 (1965).

76. M. C. Lin and K. J. Laidler, *Trans. Faraday Soc.,* **64,** 79 (1968).

77. C. P. Quinn, *Proc. R. Soc. London A,* **275,** 190 (1963), *Trans. Faraday Soc.,* **59,** 2543 (1963). See also P. D. Pacey and J. H. Purnell, *J. Chem. Soc. Faraday I,* **68,** 1462 (1972); P. D. Pacey and J. H. Wimalasena, *Chem. Phys. Lett.,* **53,** 593 (1976); **76,** 433 (1980).

78. F. O. Rice and K. F. Herzfeld, *J. Am. Chem. Soc.,* **56,** 284 (1934).

79. L. Küchler and H. Theile, *Z. Phys. Chem. B,* **42,** 359 (1939).

80. H. Okabe and J. R. McNesby, *J. Chem. Phys.,* **34,** 668 (1961).

81. R. Gorden and P. Ausloos, *J. Chem. Phys.,* **46,** 4823 (1967).

82. M. Letort, Thesis, University of Paris, 1937; *J. Chim. Phys.,* **34,** 206 (1937).

83. A. Boyer, M. Niclause, and M. Letort, *J. Chim. Phys.,* **49,** 337 (1952); M. Eusuf and K. J. Laidler, *Can. J. Chem.,* **42,** 1851 (1964); M. T. H. Liu and K. J. Laidler, *Can. J. Chem.,* **46,** 479 (1968).

84. F. O. Rice, W. R. Johnston, and B. L. Evering, *J. Am. Chem. Soc.,* **54,** 3529 (1932).

85. M. Burton, J. E. Ricci, and T. W. Davis, *J. Am. Chem. Soc.,* **62,** 265 (1940).

86. A. B. Trenwith, *J. Chem. Soc.,* 4426 (1963); R. W. Dexter and A. B. Trenwith, *J. Chem. Soc.,* 5459 (1964); K. J. Laidler and M. T. H. Liu, *Proc. Roy. Soc. London A,* **297,** 365 (1967); M. T. H. Liu and K. J. Laidler, *Can. J. Chem.,* **46,** 479 (1968).

87. M. T. H. Liu and K. J. Laidler, *Can. J. Chem.,* **46,** 479 (1968).

88. K. J. Laidler and M. T. H. Liu, *Proc. R. Soc. London A,* **297,** 365 (1967).

89. P. A. Leighton and F. E. Blacet, *J. Am. Chem. Soc.,* **55,** 1766 (1933).

90. For a review see T. Bercés, in C. H. Bamford and C. F. H. Tipper (Eds.), *Comprehensive Chemical Kinetics,* vol. 5, chap. 3, Elsevier, New York, 1972.

91. M. A. Khan, R. G. W. Norrish, and G. Porter, *Nature,* **171,** 513 (1953); *Proc. R. Soc. London A,* **219,** 312 (1953).

92. B. W. Wojciechowski and K. J. Laidler, *Can. J. Chem.,* **38,** 1027 (1960); *Trans. Faraday Soc.,* **59,** 369 (1963); M. H. Back and K. J. Laidler, *Can. J. Chem.,* **44,** 215 (1966).

93. J. L. Bancroft, J. M. Hollas, and D. A. Ramsay, *Can. J. Phys.,* **40,** 322L (1962).

94. K. J. Laidler and B. W. Wojciechowski, *Proc. R. Soc. London A,* **259,** 257 (1960).

95. J. H. van't Hoff, *Studies in Chemical Dynamics* (translation by T. Ewan of *Studieren zur chemischen Dynamik*), Williams and Norgate, London, 1896.

96. H. B. Dixon, *Philos. Trans.,* **175,** 617 (1884); **184,** 97 (1893); **200,** 315 (1903).

97. D. L. Chapman, *Philos. Mag.,* **47**(5), 90 (1899). Later work of a similar kind was done by E. Jouguet, *Comptes Rendus,* **138,** 1685 (1904); **139,** 121 (1904); **144,** 415, 530 (1907).

98. C. N. Hinshelwood, *Proc. R. Soc. London A,* **188,** 1 (1946).

99. N. Semenov, *Z. Phys.,* **46,** 109 (1927); *Chemical Kinetics and Chain Reactions,* Clarendon, Oxford, 1935.

100. C. N. Hinshelwood and H. W. Thompson, *Proc. R. Soc. London A,* **118,** 171 (1928); C. N. Hinshelwood, *Kinetics of Chemical Change in Gaseous Systems,* Clarendon, Oxford, 1929 (2nd ed.), 1933 (3rd ed.).

101. See, for example, A. G. Gaydon, *Q. Rev. Chem. Soc.,* **4,** 1 (1950).

102. D. T. A. Townend, *Chem. Rev.,* **21,** 259 (1937); M. Maccormac and D. T. A. Townend, *J. Chem. Soc.,* 238 (1938).

103. N. N. Semenov, *Z. Phys. Chem. B,* **11,** 464 (1930); N. Semenov, *Chemical Kinetics and Chain Reactions,* Clarendon, Oxford, 1935.

104. N. N. Semenov, *Some Problems of Chemical Kinetics and Reactivity,* Pergamon, Oxford, 1959.

105. P. J. Flory, *J. Am. Chem. Soc.,* **61,** 3334 (1939).

106. H. Dostal and R. Raff, *Monash. Chem.,* **68,** 188 (1936); *Z. Phys. Chem. B,* **32,** 11 (1936).

107. H. W. Melville, *J. Chem. Soc.,* 274 (1947).

108. T. T. Jones and H. W. Melville, *Proc. R. Soc. London A,* **187,** 19 (1946). See also G. V. Schultz and G. Wittig, *Naturwissenschaften,* **27,** 387, 659 (1939).

109. H. M. Hulburt, R. A. Harman, A. V. Tobolsky, and H. Eyring, *Ann. N.Y. Acad. Sci.,* **44,** 371 (1943).

110. G. V. Schultz and E. Husemann, *Z. Phys. Chem. B,* **34,** 187 (1936); **43,** 385 (1939); H. Suess, K. Pilch, and R. Rudorfer, *Z. Phys. Chem. A,* **179,** 361 (1937); H. Suess and A. Springer, *Z. Phys. Chem. A,* **181,** 81 (1937); G. V. Schultz, *Z. Elektrochem.,* **47,** 265 (1941).

111. G. V. Schultz and E. Husemann, *Z. Phys. Chem. B,* **43,** 385 (1939); W. Simpson and R. N. Haward, *Trans. Faraday Soc.,* **47,** 226 (1951).

112. A. C. Cuthbertson, G. Gee, and E. K. Rideal, *Proc. R. Soc. London A,* **170,** 300 (1939).

113. C. C. Price and R. W. Kell, *J. Am. Chem. Soc.,* **63,** 2798 (1941).

114. H. S. Taylor and H. J. Emeleus, *J. Am. Chem. Soc.,* **53,** 562, 3370 (1931); R. D. McDonald and R. G. W. Norrish, *Proc. R. Soc. London A,* **157,** 480 (1936).

115. H. W. Melville, *J. Chem. Soc.,* 274 (1947).

116. C. C. Price, *Ann. N.Y. Acad. Sci.,* **44,** 351 (1943); *Reactions at Carbon–Carbon Double Bonds,* Interscience, New York, 1946.

117. D. M. Clark, in *Cationic Polymerization and Related Complexes,* p. 99 (Ed. P. H. Plesch), Heffer, Cambridge, 1953. D. O. Jordan and A. R. Mathieson, ibid., p. 90; *J. Chem. Soc.,* 2354, 2358, 2363 (1952).

118. A. G. Evans and G. W. Meadows, *Trans. Faraday Soc.,* **46,** 327 (1950).

119. A. G. Evans and M. Polanyi, *J. Chem. Soc.,* 252 (1947).

120. W. C. E. Higginson and N. S. Wooding, *J. Chem. Soc.,* 760, 778, 1178 (1952).

121. K. Ziegler, *Ann. Chem.,* **567,** 43 (1950).

122. J. W. McBain, *Adv. Colloid Chem.,* **1,** 124 (1942).

123. W. D. Harkins, *J. Chem. Phys.,* **13,** 381 (1945); W. D. Harkins and R. S. Stearns, *J. Chem. Phys.,* **14,** 215 (1946).

124. W. D. Harkins, *J. Am. Chem. Soc.,* **69,** 1428 (1947).

125. W. V. Smith and R. H. Ewart, *J. Chem. Phys.,* **16,** 592 (1948).

126. W. V. Smith, *J. Am. Chem. Soc.,* **70,** 3695 (1948); **71,** 4077 (1949); M. Morton, P. P.

Sanatiello, and H. Landfield, *J. Polym. Sci.,* **8,** 111, 215, 279 (1952); J. G. Brodyan, J. A. Cala, T. Konen, and E. L. Kelley, *J. Colloid Sci.,* **18,** 73 (1963); K. G. McCurdy and K. J. Laidler, *Can. J. Chem.,* **42,** 825 (1964).

127. S. S. Medvedev, *International Symposium on Macromolecular Chemistry,* p. 174, Pergamon, New York, 1959.

BIBLIOGRAPHY

For a useful compilation of mathematical solutions for various composite mechanisms see:

C. Capellos and B. H. J. Bielski, *Kinetic Systems,* Wiley-Interscience, New York, 1972.

For general treatments of composite reactions see:

L. A. Albright, B. L. Crynes, and W. H. Corcoran (Eds.), *Pyrolysis: Theory and Industrial Practice,* Academic, New York, 1983.

C. H. Bamford and C. F. H. Tipper (Eds.), Decomposition and Isomerization of Organic Compounds, *Comprehensive Chemical Kinetics,* Vol. 5, Elsevier, 1972.

F. S. Dainton, *Chain Reactions: An Introduction,* Methuen, London, 1956.

V. N. Kondratiev, Chain Reactions, in C. H. Bamford and C. F. H. Tipper (Eds.) *Comprehensive Chemical Kinetics,* Vol. 2, Elsevier, 1969.

V. N. Kondratiev, *Kinetics of Chemical Gas Reactions,* Moscow, 1958 [Two English translations have appeared, one published by the U.S. Atomic Energy Commission, the other (edited by N. B. Slater) by Pergamon, Oxford, 1964].

W. A. Pryor (Ed.), *Frontiers of Free Radical Chemistry,* Academic, New York, 1983.

E. W. R. Steacie, *Atomic and Free Radical Reactions,* Reinhold, New York, 1954.

For discussions of inhibition mechanisms see:

B. W. Wojciechowski and K. J. Laidler, Inhibition of Organic Decompositions, *Trans. Faraday Soc.,* **59,** 369 (1963).

M. H. Back and K. J. Laidler, Theories of Inhibition by Nitric Oxide, *Can. J. Chem.,* **44,** 215 (1966).

For accounts of the earlier work on gas-phase and liquid-phase oxidation see:

C. N. Hinshelwood and A. T. Williamson, *The Reaction between Hydrogen and Oxygen,* Clarendon, Oxford, 1934.

W. Jost, *Explosion and Combustion Processes in Gases,* McGraw-Hill, New York, 1939–1946.

B. Lewis and G. von Elbe, *Combustion, Flames, and Explosions of Gases,* Cambridge University Press, 1938; Academic, New York, 1961.

G. J. Minkoff and C. F. H. Tipper, *Chemistry of Combustion Reactions,* Butterworth, London, 1962.

N. N. Semenov, *Chemical Kinetics and Chain Reactions,* Clarendon, Oxford, 1935.

N. N. Semenov, *Some Problems of Chemical Kinetics and Reactivity,* Pergamon, Oxford, 1959.

N. N. Semenov, Modern Concepts of the Mechanism of Hydrocarbon Oxidation in the Gas Phase, in *Photochemistry and Reaction Kinetics* (P. G. Ashmore, F. S. Dainton and T. M. Sugden, Eds.), Cambridge University Press, 1967, Chapter 9 (pp. 229–249).

V. Y. Shtern, *The Gas-Phase Oxidation of Hydrocarbons,* Pergamon, London, 1964.

For more recent accounts of oxidation processes see:

C. H. Bamford and C. F. H. Tipper (Eds.), Gas-Phase Combustion, *Comprehensive Chemical Kinetics,* Vol. 17, (1977).

W. C. Gardiner and D. B. Olson, Chemical Kinetics of High-Temperature Combustion, *Ann. Rev. Phys. Chem.,* **31,** 377 (1980).

R. W. Getzinger and G. L. Schott, Shock Tube Studies of the Hydrogen–Oxygen Reaction System, in *Physical Chemistry of Fast Reactions* (B. P. Levitt, Ed.), Plenum, New York, 1973, Vol. 1.

For accounts of the kinetics of reactions occurring in flames see:

J. N. Bradley, *Flame and Combustion Phenomena,* Methuen, London, 1969.

C. P. Fenimore, *Chemistry in Premixed Flames,* Pergamon, New York, 1964.

R. M. Fristrom and A. A. Westenberg, *Flame Structure,* McGraw-Hill, New York, 1965.

B. Lewis and G. von Elbe, *Combustion, Flames and Explosions of Gases,* Academic, New York, 1961.

F. M. Page, Chemical Reaction and Ionization in Flames, in *Physical Chemistry of Fast Reactions* (B. P. Levitt, Ed.), Plenum, New York, 1973, Vol. 1, Chapter 3 (pp. 161–244).

For accounts of polymerization mechanisms see:

C. H. Bamford, W. G. Barb, A. D. Jenkins, and P. F. Onyon, *The Kinetics of Vinyl Polymerization by Radical Mechanisms,* Butterworth, London, 1958.

C. H. Bamford and C. F. H. Tipper (Eds.), Nonradical Polymerization, *Comprehensive Chemical Kinetics,* Vol. 15, Elsevier, 1976.

C. H. Bamford and C. F. H. Tipper (Eds.), Free-Radical Polymerization, *Comprehensive Chemical Kinetics,* Vol. 14A, Elsevier, 1976.

J. C. Bevington, *Radical Polymerization,* Academic, New York, 1961.

G. M. Burnett, *Mechanism of Polymer Reactions,* Interscience, New York, 1954.

P. J. Flory, *Principles of Polymer Chemistry,* Cornell University Press, Ithaca, 1953.

A. D. Jenkins and A. Ladwith (Eds.), *Reactivity, Mechanism, and Structure in Polymer Chemistry,* Wiley, New York, 1974.

D. C. Pepper, Ionic Polymerization, *Quart. Rev.,* **8,** 88 (1954).

P. H. Plesch (Ed.), *Cationic Polymerization and Related Complexes,* Heffer, Cambridge, 1953.

Chapter 9

Photochemical and Radiation-Chemical Reactions

Radiation is of two kinds, electromagnetic and particle, and examples of each are as follows:

Electromagnetic radiation	Particle radiation
Infrared radiation	α particles (He nuclei)
Visible light	β particles (electrons)
Ultraviolet radiation	Cathode rays (electrons)
X rays, γ rays	Beams of protons, deuterons, and so on, produced in particle-accelerating machines

In some of its effects electromagnetic radiation acts as a beam of particles, known as light quanta or photons.§

Radiation of both types can induce chemical reaction. It has proved convenient to make a distinction, although not a sharp one, between photochemical and radiation-chemical¶ reactions. Sometimes the distinction is made on the basis of whether ions are produced in the reaction. With visible electromagnetic radiation and ultraviolet radiation in which the frequency is not too high, ions are not produced and the resulting chemical reaction is referred to as *photochemical*. On the other hand, with ultraviolet radiation of higher frequencies, X rays, γ rays, and high-energy particle radiation, ions are usually formed, and the process is described then as *radiation-chemical*.

Another distinction between the two types of reaction is based on *specificity*. When lower-energy radiation is used, reaction occurs with fairly simple stoichiometry. Higher-energy radiation, on the other hand, can break up molecules into a number

§ The name "photon," now generally used, was suggested by G. N. Lewis, *Nature,* **118,** 874 (1926).

¶ The word "radiochemical" should not be used in this context; radiochemistry is concerned with the use of radioactive materials in the study of chemical systems.

of fragments, which recombine to give a variety of products. In such nonspecific situations the stoichiometry is by no means straightforward, and the term radiation-chemical is more appropriate.

9.1 PHOTOCHEMICAL REACTIONS

That chemical reaction can be induced by electromagnetic radiation has been known since the 18th century. A number of important investigations, carried out over a period of many years, led to our present understanding of the nature of photochemical processes. One of these was that of von Grotthuss,[1] who in 1819 showed that for radiation to bring about chemical change it must first be absorbed. This conclusion was rediscovered in 1841 by Draper[2] and often is referred to as the *Grotthuss–Draper law.*

Important quantitative studies on the photochemical reaction between hydrogen and chlorine were made by Dalton[3] and later by Wittwer.[4] These and many subsequent investigations, such as those of Bunsen and Roscoe[5] on the same reaction, showed that the rate is proportional to the intensity of the light absorbed by the system [compare Eq. (8.63)]. With other reactions the dependence on the intensity of light absorbed is different. With the hydrogen–bromine reaction, for example, the rate is proportional to the square root of the intensity of light absorbed [Eq. (8.57)]. In all cases it is the light absorbed, and not the incident intensity, that directly controls the rate.

With the advent of the quantum theory and particularly after Einstein's[6] realization that light can be regarded as a beam of particles (later to be called photons), it became possible to relate the number of reactant molecules transformed in a photochemical reaction to the number of photons absorbed. Stark[7] in 1911 and independently Einstein[8] in 1912 proposed that one photon of radiation is absorbed by one molecule. This conclusion, usually referred to as *Einstein's law of photochemical equivalence,* is discussed later in more detail.

9.1.1 Photochemical Primary Process

The process in which electromagnetic radiation is absorbed by a molecule, with the formation of species that undergo further reaction, is known as the photochemical primary process. Absorption of radiation leads initially to the formation of an excited species, which may in a subsequent step produce atoms and free radicals; the ways in which this can occur are discussed later. In photochemistry it is usual to consider the formation of the excited species and the subsequent formation of atoms and free radicals as together constituting the primary process. The production of atoms and free radicals in the primary process is referred to as *photolysis* (Greek *photo,* meaning light, and *lysis,* meaning splitting). These atoms and free radicals may be involved in subsequent reactions, which may involve chains; such reactions are referred to as secondary processes. The term photolysis is applied also to an overall photochemical process in which a larger molecule is broken down into smaller ones.

Reactions in which alkyl radicals are produced from organic molecules by the action of ultraviolet radiation are examples of photochemical primary processes. Acetone, for example, absorbs radiation of wavelengths in the neighborhood of 280 nm to give CH_3CO and CH_3, and at higher temperatures the acetyl radical further splits into CH_3 and CO. The overall primary process is then

$$CH_3COCH_3 + h\nu \rightarrow 2CH_3 + CO$$

where $h\nu$ represents a photon. The photolysis of the mercury alkyls also has been used as a source of free radicals:

$$Hg(CH_3)_2 + h\nu \rightarrow Hg + 2CH_3$$

Methylene is produced conveniently by the photolysis of ketene,

$$CH_2CO + h\nu \rightarrow CH_2 + CO$$

and of diazomethane,

$$CH_2N_2 + h\nu \rightarrow CH_2 + N_2$$

A number of important principles relate to the absorption of electromagnetic radiation with the formation of excited molecules and of atoms and free radicals. In the first place, as already noted, the light must be absorbed in order to have any effect; however, not all the absorbed radiation leads to chemical change.

The probabilities of transitions when light is absorbed are governed by the *Franck–Condon principle.* This was first expressed qualitatively by Franck,[9] who based his conclusions on the relative speeds of electronic transitions and of vibrational motions. The absorption of radiation may occur in 10^{-15}–10^{-18} s and is so rapid compared to vibrational motion ($\approx 10^{-13}$ s) that immediately after an electronic transition has occurred the internuclear distances in a molecule are much the same as before. Figure 9.1 shows a number of possible arrangements of potential-energy curves for a diatomic molecule, and Franck argued that the most probable transitions are those that can be represented by more or less vertical lines in the diagrams. Radiation of frequency ν can be absorbed by a molecule only if the energy $h\nu$ is equal to the energy difference between a quantized energy level in the excited state and a quantized level in the ground state. The probability of a transition is small if the two states correspond to two very different interatomic distances, since a considerable change in the molecule would have to accompany absorption. The probability also is small if certain selection rules are not satisfied; this is the case, for example, if the two states correspond to different spectroscopic multiplicities, for example, if a singlet $\rightarrow$ triplet transition is involved.

Franck's ideas were put on a more rigorous quantum-mechanical basis by Condon.[10] An important matter relates to the most probable internuclear distances in vibrating molecules. According to classical mechanics, the distance is changing more slowly near the turning points of the motion: for example, a vibrating diatomic molecule spends more time in its fully extended or fully compressed configurations than in intermediate configurations. In quantum mechanics, on the other hand, the situation is different. For the lowest (zero-point) vibrational state the probability is a maximum at the *center* of the vibration. In the diagrams in Fig. 9.1 the connecting arrows therefore start at the *centers* of the lines representing the zero-point levels for the ground electronic states. For other vibrational states the maximum is slightly inside each end of the classical vibrational amplitude, and for such states the arrows in Fig. 9.1 end close to

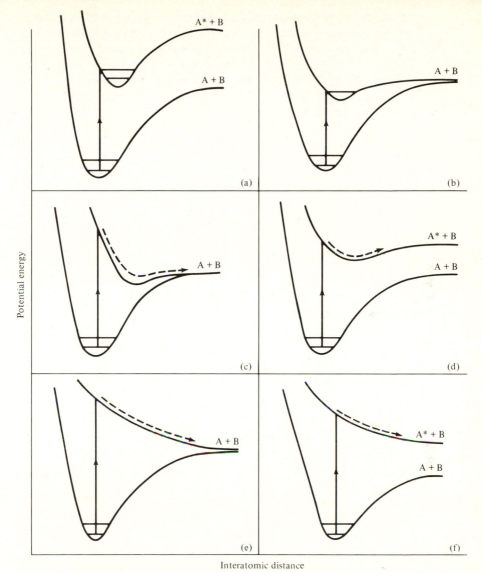

Figure 9.1 Potential-energy curves for a diatomic molecule AB, showing a number of processes that can occur when radiation is absorbed by AB in its ground state.

the extremities of the lines representing the vibrational levels. Of course, this procedure is an oversimplification, because there is a smaller probability of transitions occurring from and to other positions.

In diagrams a and b of Fig. 9.1 the absorption of the radiation results in an excited molecule that is stable with respect to dissociation; it does not give rise to atoms. In diagram c excitation results in a species which at once dissociates to give the atoms in their ground electronic states. In diagram d there is again dissociation, but the atom A is now in an excited electronic state. In diagrams e and f the excited species are repulsive ones, which immediately dissociate into atoms.

Another situation, corresponding to what is called *predissociation,* is illustrated in Fig. 9.2. This phenomenon was discovered by Henri[11] with reference to the absorption spectrum of S_2. It was observed that the spectral bands corresponding to certain transitions were diffuse, in the sense that the rotational fine structure is blurred. Henri's explanation is illustrated by Fig. 9.2. The transition from $v'' = 0$ to $v' = 2$ produces an AB* molecule that has insufficient energy to dissociate, and therefore the rotational fine structure is observed. The diagram shows a repulsive state of AB* that crosses the upper potential-energy curve at a point corresponding to $v' = 4$. It follows that the transition from $v'' = 0$ to $v' = 4$ produces a species that reaches point a in the first vibration; the system then undergoes a transition to the repulsive state and dissociates into the atoms A and B. Transitions to a vibrational level such as $v' = 7$, on the other hand, may produce a species that does not dissociate easily, since when it vibrates it moves rapidly past the crossing point so that the transition probability is reduced. Therefore, the $v' = 7$ state of AB* may have a long enough life for rotations to occur, and there will be a rotational fine structure corresponding to this $v'' = 0 \rightarrow v' = 7$ transition.

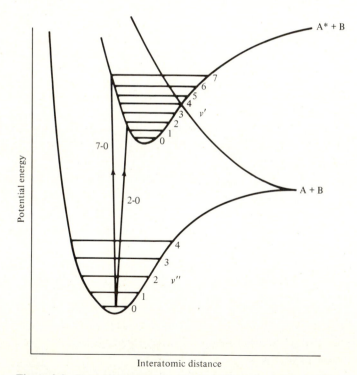

Figure 9.2 Potential-energy curves for a diatomic molecule AB. The $v'' = 0 \rightarrow v' = 4$ transition corresponds to predissociation, with no rotational fine structure.

The characteristic feature of a predissociation spectrum is that there is a diffuse region, corresponding to a transition to the repulsive state followed by rapid dissociation, and that at both longer and shorter wavelengths the spectrum shows a rotational fine structure. This is a different situation from that found with ordinary dissociation (e.g., Fig. 9.1c), where the rotational fine structure can appear only at the lower-energy (longer-wavelength) side of the diffuse bands.

Figures 9.1 and 9.2 relate to a diatomic molecule AB, but the principles are somewhat the same for more complex molecules. The curves in Figs. 9.1 and 9.2 may be regarded as sections through the multidimensional potential-energy curves for polyatomic molecules. However, some additional factors must be taken into account for molecules having more than two atoms. If a *diatomic* molecule has enough energy to undergo dissociation, it must do so within the period of its first vibration (i.e., within 10^{-13}–10^{-14} s). On the other hand, as discussed with reference to unimolecular reactions (Section 5.3), a *polyatomic* molecule having sufficient energy may not dissociate in the first vibration. This is because the energy may not be distributed suitably among the normal modes for bond rupture to occur. As the molecule vibrates, the energy is redistributed constantly among the normal modes, and eventually a vibration leads to dissociation. As a result, a polyatomic molecule with sufficient energy to dissociate may survive for 10^{-8} s or more, and during that time a number of processes may occur, including the emission of radiation.

9.1.2 Reactions of Electronically Excited Species

After an electronically excited molecule has been produced by the absorption of radiation, it can undergo a number of processes, some of which have already been mentioned. The most important of these processes are:

1. *Fluorescence.* The excited molecule may emit radiation and pass into a lower state which may be the ground state or a lower excited state. This process is referred to as *fluorescence.* Usually, this term is restricted to processes in which the molecule does not change its multiplicity, in contrast to *phosphorescence* where there is a multiplicity change (e.g., triplet → singlet) and in which the emission is slower.
2. *Molecular dissociation.* This has been considered already with reference to Figs. 9.1 and 9.2.

In the case of fairly simple gas molecules, these two processes are generally the only ones that can occur. With more complex molecules, however, the dissociative processes are much slower, for reasons just discussed. In the liquid phase, collisions may occur within 10^{-12}–10^{-14} s, which is much shorter than the time required for fluorescence or molecular dissociation to take place. Other processes therefore may be important, such as the following:

3. *Vibrational relaxation.* During this process the molecule passes into lower vibrational states within the same electronic state.

4. *Collisional quenching* or *external conversion.* Upon collision with other molecules, such as solvent molecules, the electronically excited molecule may pass into a lower electronic state, which is often the ground state. The electronic energy is converted in this process into translational and internal energy (vibrational and rotational) and eventually is dissipated as heat.

In addition, a complex molecule in an electronically excited state may undergo *internal quenching* or *internal conversion.* These processes do not involve collisions; the electronic energy is converted into vibrational energy of the molecule itself, which therefore acts as a self-quencher.

Some of the processes that may occur with an electronically excited molecule are summarized in Fig. 9.3, which is known as a *Jablonski diagram.* All these possible processes have to be considered in a detailed photochemical study. Often the processes leading to atoms and free radicals are regarded together as constituting the primary photochemical act, and subsequent processes are referred to as secondary processes. However, it is to be noted that with this definition the primary act itself may be composite. It is particularly important to appreciate that the excited species formed by the absorption of radiation may undergo processes other than the formation of

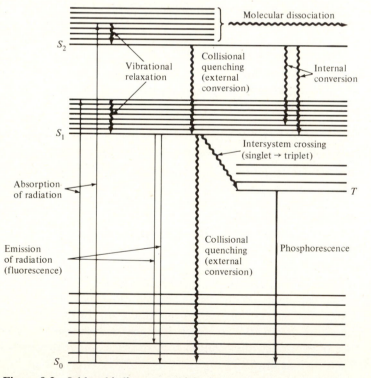

Figure 9.3 Jablonski diagram, which shows the ground state S_0, a singlet state, and also two excited singlet states (S_1 and S_2) and a triplet state T. Various possible processes are indicated: solid lines are radiative, wavy lines are radiationless.

atoms and free radicals (e.g., fluorescence or quenching) and that this must be taken into account in considering the applicability of the law of photochemical equivalence.

9.1.3 Photochemical Thresholds

Since for most of the radiation employed in photochemistry the energy of the photon is considerably greater than the average thermal energy of the absorbing molecule, there is little variation in the effectiveness of a photon from molecule to molecule in the reacting system. At ordinary temperatures, most reacting molecules are in their ground vibrational states. On changing the frequency of the incident radiation there is a fairly sharp transition from a region in which there is no absorption and no chemical reaction, to one in which a considerable amount of reaction occurs. Many examples of such photochemical *thresholds* have been observed. Reaction occurs on the higher-frequency, or lower-wavelength, side of the threshold.

9.1.4 Law of Photochemical Equivalence

Since the lifetime of an electronically excited species is short, usually less than about 10^{-8} s, with ordinary radiation it is unlikely for a molecule that has absorbed one photon to absorb another before it has become deactivated. Thus there is usually a one-to-one relationship between the number of photons absorbed by the system and the number of excited molecules produced. For example, if light is absorbed with the production of an excited species that does nothing but dissociate into two radicals, as in the process

$$CH_3COCH_3 + h\nu \rightarrow CH_3COCH_3^* \rightarrow CO + 2CH_3$$

the rate of production of methyl radicals is twice the rate of absorption of photons. This is the basis of the law of photochemical equivalence, due independently to Stark[7] and Einstein.[8] As is discussed in Section 9.2, exceptions may be found when laser radiation is used: then, the intensity of the radiation can be so great that two or more photons can be absorbed by a molecule.

The principle of photochemical equivalence is of great value in photochemical studies, since it enables the rates of formation of radicals to be calculated from the results of optical measurements. A mole of photons is called an *einstein:*[12] the rate of formation of methyl radicals in the photolysis of acetone, in mol dm^{-3} s^{-1}, is twice the number of einsteins absorbed per cubic decimeter per second.

For ordinary radiation the law of photochemical equivalence is accepted as applying to the formation of excited species in primary processes. For many years, however, the situation was by no means clear; even as late as 1925 there was much dissension.[13] This was due in part to failure to realize that an excited species may emit radiation before it can give rise to atoms and free radicals. Also, in the early days there was failure to recognize the difference between primary and secondary processes, although as early as 1913 Bodenstein[14] had made a clear distinction. Reactions like that between hydrogen and chlorine, in which the *quantum yield* (the number of reactant molecules transformed for each photon absorbed) was often as high as 10^6, were sometimes considered to represent a failure of the law of photochemical equivalence. In reality, one photon interacts with only one chlorine molecule, and the high quantum yield is due to the chain processes.

A simpler example, involving a nonchain reaction, is provided by the photochemical decomposition of hydrogen iodide, for which Warburg[15] found that *two* molecules of HI are transformed into $H_2 + I_2$ by one photon. Warburg showed that this is not a violation of the law of photochemical equivalence, since the mechanism is

$$HI + h\nu \rightarrow H + I$$

$$H + HI \rightarrow H_2 + I$$

$$\underline{I + I \rightarrow I_2}$$

Overall $2HI + h\nu \rightarrow H_2 + I_2$

Thus, although one photon interacts with one HI molecule, in accordance with the law, two molecules are transformed. Later work on many other reactions has confirmed that the law is valid and that apparent deviations from it can be explained in terms of the occurrence of secondary reactions.

9.1.5 Rotating-Sector Technique

In the study of atom and free-radical reactions it is important to be able to measure the rates of radical-combination reactions such as

$$CH_3 + CH_3 \rightarrow C_2H_6$$

and of disproportionation reactions such as

$$C_2H_5 + C_2H_5 \rightarrow C_2H_6 + C_2H_4$$

A useful method for doing this involves generating the radicals photochemically in an experiment in which the radiation is applied intermittently. This is done by interrupting the light beam by means of a rotating slotted disk. This method, usually known as the *rotating sector method,* was devised and first used by Briers, Chapman, and Walter[16] and later was developed by Melville and coworkers[17] with particular reference to polymerization reactions. A theoretical treatment of the method had been given by Shepp[18] and has proved useful in the interpretation of experimental data.

The principle of the method is as follows. Suppose first that the system is illuminated continuously; the concentration of free radicals rises until eventually a steady state is reached, as shown in Fig. 9.4a. If the system is represented as

$$R_2 + h\nu \rightleftharpoons 2R$$

the steady-state equation is

$$I - k_{-1}[R]^2 = 0 \tag{9.1}$$

where I is the intensity of the light absorbed. The radical concentration is thus

$$[R] = \left(\frac{I}{k_{-1}}\right)^{1/2} \tag{9.2}$$

and the rate of the overall reaction is proportional to $I^{1/2}$.

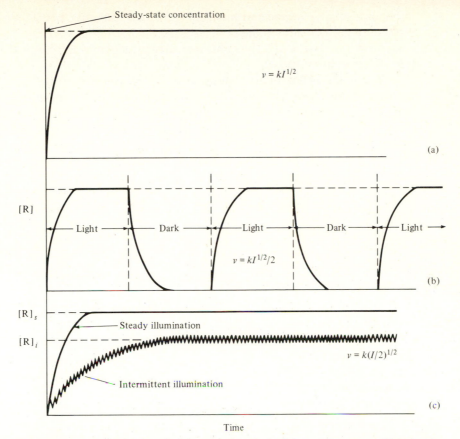

Figure 9.4 Radical concentrations as a function of time: (a) constant illumination; (b) intermittent illumination with equal periods of light and darkness that are long compared with the half-life of the radicals; and (c) intermittent illumination with light and dark periods that are much shorter than the half-life.

Suppose now that the slotted disk is constructed in such a way that the periods of light and darkness are equal. Consider first the case in which the rotation is so slow that the periods of light and dark are much longer than the half-life of the radicals. The variation of radical concentration is now as shown in Fig. 9.4b. During the period of illumination the radical concentration rapidly attains its steady-state value, and during the dark period it falls to zero. Thus, the steady-state concentration of radicals is maintained for only half the period of the experiment, and the rate therefore is one-half of the value for constant illumination. Thus, if the rate with constant illumination is $kI^{1/2}$, that with slow intermittent illumination, with equal light and dark periods, is $kI^{1/2}/2$.

The situation, however, is different if the rate of rotation is so rapid that the periods of light and dark (again assumed equal) are much smaller than the radical half-life. The concentration of radicals now varies as shown in Fig. 9.4c. During the periods of darkness the concentrations decay only to a small extent, and eventually a steady state is reached at which the radical concentration is less than that which corresponds to steady illumination with light of the same intensity. In the limit of infinitely rapid rotation, the effect is the same as if the light intensity were $I/2$ and the rate of reaction in this case is $k(I/2)^{1/2}$. This is greater by a factor of $2^{1/2} = 1.41$ than for the case of low rotation speeds.

Therefore, as the rotation speed is varied, the overall reaction rate changes. The transition from the higher rates with rapid rotation to lower rates with slower rotation varies over a range of speeds that are related to the half-life of the radicals. In a photochemical experiment, the rate of formation of the radicals is known (from the rate of absorption of photons), and this rate is equal to the rate of the combination reaction. From this rate and the half-life of the radicals, the concentration of radicals and the rate constant can be calculated.

Shepp's[18] theoretical treatment led to the result that if the ratio of darkness to light is r, the rate for slow rotation is the fraction

$$\frac{1}{1 + r}$$

of the rate for steady illumination. For rapid rotation the ratio is

$$\frac{1}{(1 + r)^{1/2}}$$

For equal intervals $r = 1$ and the ratios are $\frac{1}{2}$ and $(\frac{1}{2})^{1/2}$, as already seen. If the sector design is such that the period of darkness is three times as long as the period of illumination ($r = 3$), the ratios are $\frac{1}{4}$ and $\frac{1}{2}$, respectively; the rate with rapid rotation is twice that with slow rotation. This case is illustrated in Fig. 9.5, which is based on Shepp's analysis. Curves such as this are displaced to one side or the other if the half-

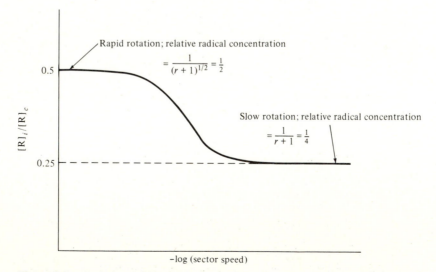

Figure 9.5 Variation of $[R]_i/[R]_c$, the ratio of the steady-state concentrations for intermittent and constant illuminations, with sector speed. The ratio r of dark to light intervals is 3.

life is varied. Diagrams of this type can be fitted to the experimental results, and then the half-life of the radical can be obtained.

This method has been applied in a number of investigations to the methyl radical recombination. Shepp's analysis of earlier work led to a value of 2.2×10^{10} dm^3 mol^{-1} s^{-1} for the rate constant over the temperature range 250–420 K. This is in good agreement with the value of 2.5×10^{10} dm^3 mol^{-1} s^{-1} which Baulch and Duxbury[19] have recommended as the most reliable value; this value is based on their evaluation of the results obtained by a number of methods.

9.1.6 Flash Photolysis

In the technique of flash photolysis an intense flash of light of very short duration initiates reaction by producing excited species and free radicals. Fast photometric methods are used to follow the subsequent reactions that occur. The first use of the method was described in 1950 by Porter.[20] Later, Porter, Norrish, and many others made extensive use of the technique.

Figure 9.6 shows a schematic representation of a conventional flash photolysis apparatus. The flash is produced by the discharge of a capacitor which produces a high voltage between electrodes sealed into an evacuated tube. The flash tube is parallel

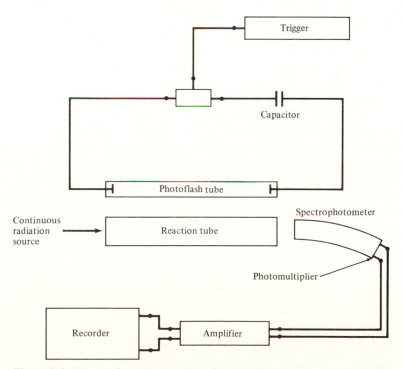

Figure 9.6 Schematic representation of a typical flash-photolysis apparatus. To study the spectra of intermediates formed, the continuous radiation source can be replaced by a subsidiary flashtube.

to the reaction vessel, and a reflector increases the amount of light that impinges on the system. Sometimes two or more flash tubes are used. The duration of a flash is typically a few microseconds.

After the initiating flash has been produced, the spectrum of the resulting system can be recorded by the use of a subsidiary flash lamp which is fired at a specified time after the primary flash. A series of spectra taken after various intervals provides important information about the course of reaction. Alternatively, the rates of reactions occurring subsequent to the initial flash can be measured spectrophotometrically, by choosing a wavelength at which there is substantial absorption by a particular species. This may be done (Fig. 9.6) by passing light from a tungsten-filament lamp or xenon arc to a monochromator and then to a photomultiplier, the output of which is amplified and recorded.

The flash-photolysis method has been applied to many reaction systems, in the gas phase and in solution. Reference has already been made, in Section 5.4.1, to the study by this technique of the combination of iodine atoms.

9.2 LASER PHOTOCHEMISTRY

Since the 1950s the fields of spectroscopy and photochemistry have been enriched greatly by the introduction and development of lasers. The word *laser* is an acronym for "light amplification by stimulated emission of radiation," and the laser was a development of the *maser,* the latter word being an acronym for "microwave amplification by stimulated emission of radiation." The work with masers in the microwave region was extended to other regions of the spectrum, including the visible and ultraviolet.

To produce a laser beam, excitation energy is supplied to a substance in such a way as to produce a *population inversion,* in which there are more atoms in certain excited states than in the ground state. Such excited species can release energy spontaneously, with the emission of photons, and also do so under the influence of an impinging photon; the latter process is referred to as *stimulated emission.* What occurs in a laser beam is that a photon released by one atom or molecule interacts with another atom or molecule having a population inversion, and stimulates the release of another photon of the same wavelength and traveling in the same direction as the original photon. Thus, as a beam of photons travels through the laser medium, its intensity greatly increases. A common device is to reflect the initial beam by means of a mirror, so that it passes again through the medium and its intensity is further amplified.

A commonly used laser is the *ruby laser,* in which the medium is a rod of ruby. This is aluminum oxide, Al_2O_3, in which one out of every 10^2 or 10^3 Al^{3+} ions has been replaced by a Cr^{3+} ion; this gives the crystal its characteristic red color. The population inversion is usually brought about by means of xenon flashtubes. The ruby laser emits red light of wavelength 694.3 nm. Some lasers operate continuously and are called *continuous-wave* (cw) lasers. In pulsed lasers, on the other hand, the exciting light is applied only for short periods, which in modern devices can be less than a picosecond (10^{-12} s). The power developed in pulsed lasers may be many times greater than in conventional light beams.

There are three special features of laser radiation. In the first place, it is highly monochromatic, covering only a very narrow range of wavelengths. Also, with a suitable geometry for the laser medium (e.g., the ruby), the beam can be extremely narrow and is said to be *coherent;* the beam divergence can be of the order of a milliradian. Finally, for reasons already explained, the highly coherent and monochromatic beams can be of very high intensity.

These special features of laser beams are of great value in both spectroscopy and photochemistry. The fact that laser beams are highly monochromatic but are nevertheless coherent and of high energy means that atoms and molecules can be excited selectively to an extent that was not possible with ordinary radiation. Also, with laser radiation an atom or molecule can become excited in stages, by the stepwise absorption of a number of photons.

9.2.1 Pulsed Lasers

In pulsed lasers the radiation is emitted for periods of very short duration. An important and useful example of a pulsed laser is the *atomic iodine laser,* first used by Kasper and Pimentel.[21] This is an example of a *chemical laser,* the effective species being iodine atoms produced photolytically. They are often produced in an excited state by the irradiation of heptafluoropropyl iodide, C_3F_7I, with ultraviolet radiation:

$$C_3H_7I + h\nu \rightarrow C_3F_7 + I^*$$

The transition between these excited iodine atoms and ground-state iodine atoms corresponds to a wavelength of 1315 nm, which is in the infrared. Irradiation of the system with such radiation therefore induces laser action, with the production of a powerful coherent beam of 1315 nm wavelength. The photolysis of the C_3F_7I can lead to almost complete dissociation into $C_3F_7 + I^*$. It is possible to produce a large population inversion with this system and hence to obtain a laser beam of very high intensity.

Various techniques have been used to shorten the duration of the pulse emitted by the iodine laser.[22] In the "free-running" mode the duration is from 2 to 6 μs. There is also a "Q-switched" mode, involving a fast shutter, which reduces the duration to 13–30 ns (1 ns = 10^{-9} s), while a "mode-locked oscillator" system produces pulses of 0.2–2 ns duration.

Pulses of such short durations are important in allowing the fastest of chemical reactions to be studied. One application of the iodine laser is an investigation,[23] using the temperature-jump technique, of the rate of the dissociation of H_2O into H^+ and OH^- ions, and the reverse recombination:

$$H_2O \rightleftharpoons H^+ + OH^-$$

The temperature jump was brought about by the absorption of the iodine laser radiation, which rapidly produces vibrations and rotations in the water molecules. The T-jump occurs as fast as the emission time of the laser, so that the temperature changes have been brought about in as short a time as 0.2 ns (2×10^{-10} s). The reaction was followed by observing the conductivity of the solution, the work being done over a range of temperatures. Some of the results of this investigation were referred to in Section 6.7.3.

Pulsed lasers have now been developed in which the duration of the pulse is less than a picosecond (10^{-12} s). Such instruments have been used to study molecular relaxation processes.[24]

9.2.2 Multiphoton Excitation

In conventional photochemical experiments there is only a low probability that a molecule excited by the absorption of one photon can absorb an additional one; this is because of the very short lifetimes of excited molecules and because of the relatively small number of photons in an ordinary beam of electromagnetic radiation. The unlikelihood of multiple photon absorption is, in fact, the basis of the law of photochemical equivalence.

With lasers, on the other hand, the situation is very different, because of the much higher intensities of laser beams. There is now a substantial probability that a molecule may absorb successively a number of photons and pass in a stepwise fashion to higher excited states. In fact, in a laser beam even infrared radiation, in which each photon carries only small amounts of energy and can bring about only vibrational and rotational excitation, is capable of causing molecular dissociation as a result of the absorption of a number of photons. With infrared radiation of wavelengths suitable for bringing about vibrational excitation, 20–40 photons are usually sufficient to bring about the breaking of a chemical bond.

The process of infrared multiphoton excitation is illustrated schematically in Fig. 9.7. Three energy regions are to be distinguished. For the first few vibrational states (typically from $v = 0$ to $v = 6$) the vibrational levels are discrete, but for higher levels the states overlap owing to anharmonicity and form the so-called "quasicontinuum." Beyond the level for dissociation the states form a true continuum. Once the molecule has absorbed photons that take it into the quasicontinuum, further excitation takes place more readily because of the availability of suitable energy spacings. Then the system can easily pass into the region of the true continuum, where dissociation can occur.

Much work is being done in this field, and it has led to important contributions to the understanding of unimolecular reactions (see also Section 5.3.9). For an account of some of the kinetic studies on infrared multiple photon excitation a review by Ashfold and Hancock[25] should be consulted.

9.3 PHOTOSENSITIZATION

The dissociation of molecular hydrogen into two hydrogen atoms requires energy of 432 kJ mol^{-1}, which corresponds to a wavelength of 277.6 nm. However, as seen from the potential-energy curves in Fig. 9.8, hydrogen does not absorb at all in this region of the ultraviolet, but only at much shorter wavelengths corresponding to energies of more than 1000 kJ mol^{-1}. Working at such short wavelengths is inconvenient, so that alternative methods are needed. A useful technique is *photosensitization,* in which the radiation is absorbed by another species, which transfers energy to a hydrogen molecule on collision.

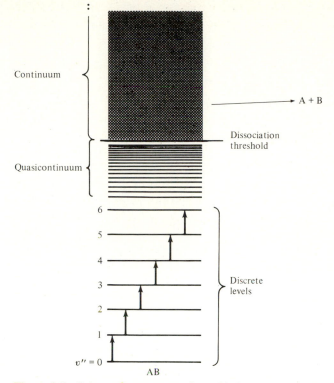

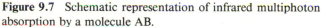

Figure 9.7 Schematic representation of infrared multiphoton absorption by a molecule AB.

Mercury vapor has been used widely as a sensitizer for the production of hydrogen atoms from H_2 and for other processes. Mercury atoms are normally in a $6\ ^1S_0$ state and are excited to the $6\ ^3P_1$ level by illumination with the mercury resonance line at 253.7 nm, which corresponds to an energy of 469.4 kJ mol^{-1}. Upon collision with a hydrogen molecule, the process

$$Hg(6\ ^3P_1) + H_2 \rightarrow Hg(6\ ^1S_0) + 2H$$

occurs very efficiently. The experimental technique is very simple: the hydrogen gas is brought into contact with liquid mercury and rapidly becomes saturated with mercury vapor. Illumination of the mixture with ultraviolet radiation of wavelength 253.7 nm, produced by a mercury resonance lamp, then gives rise to hydrogen atoms. Many bonds having strengths of less than 469.4 kJ mol^{-1} can be split in this way.

Photosensitization was first discovered by Cario and Franck,[26] who demonstrated the formation of hydrogen atoms from hydrogen molecules and excited mercury, detecting them by their chemical reactivity. They described a collision between the excited mercury atom and the molecule as a "Stoss zweiter Art" ("collision of the second kind"). Later, Taylor and his coworkers[27] used the method to investigate the reactions of atomic hydrogen, which they caused to react with carbon monoxide, nitrous oxide, hydrocarbons, and other substances.

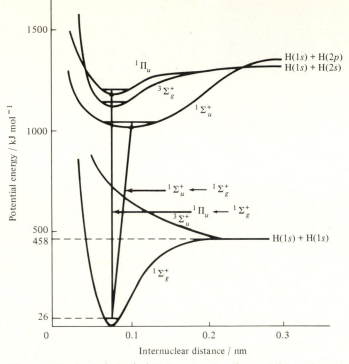

Figure 9.8 Potential-energy diagram for the neutral hydrogen molecule, showing possible transitions from the ground state of H_2.

The mechanism of the reaction between excited mercury and hydrogen has been the subject of some investigation. The reaction is sometimes written in two stages,

$$(1) \quad Hg(^3P_1) + H_2 \rightarrow HgH + H$$

$$(2) \quad HgH \rightarrow Hg(^1S_0) + H$$

and the first reaction is more exothermic than the process

$$(3) \quad Hg(^3P_1) + H_2 \rightarrow 2H + Hg(^1S_0)$$

by the dissociation energy of HgH, which is 35.6 kJ mol^{-1}. The presence of HgH in these systems was investigated spectroscopically by Olsen,[28] who found that it was not present in the normal state, but only in an excited state, presumably having been produced by secondary processes. It is concluded therefore that either process (3) occurs in one stage[29] or the HgH produced in reaction (1) is so excited vibrationally that it dissociates at once.[30] The purely quenching type of process,

$$Hg(^3P_1) + H_2 \rightarrow Hg(^1S_0) + H_2$$

in which electronic energy is converted into translational, rotational, and vibrational energy, cannot occur readily, since it violates the spin-conservation rule.

TABLE 9.1   AMOUNTS OF ENERGY LIBERATED BY EXCITED ATOMS

Metal	Transition	Wavelength/nm	Energy of excited atoms/kJ mol^{-1}	Energy liberated if hydride is formed/kJ mol^{-1}
Mercury	$6\ ^3P_1$–$6\ ^1S_0$	253.7	469.4	505.0
	$6\ ^1P_1$–$6\ ^1S_0$	184.9	643.9	679.5
Cadmium	$5\ ^3P_1$–$5\ ^1S_0$	326.1	365.3	430.1
	$5\ ^1P_1$–$5\ ^1S_0$	228.8	520.5	585.3
Zinc	$4\ ^3P_1$–$4\ ^1S_0$	307.6	387.0	483.7
	$4\ ^1P_1$–$4\ ^1S_0$	213.9	558.1	654.8

Excited mercury also has been caused to react directly with other substances, such as hydrocarbons. With the saturated hydrocarbons the main primary process that takes place is

$$\text{Hg}(^3P_1) + \text{HR} \rightarrow \text{H} + \text{R} + \text{Hg}(^1S_0)$$

where R is a free radical, such as ethyl in the case of ethane. With unsaturated hydrocarbons the same type of process may occur, but in addition an excited molecule can be formed and can undergo subsequent reactions.

Work also has been done using mercury in its singlet-excited $6\ ^1P_1$ state, which corresponds to a wavelength of 184.9 nm and an energy of 643.9 kJ mol^{-1}. Zinc and cadmium, in both singlet- and triplet-excited states, have been used in photosensitization experiments. The amounts of energy liberated in each case are given in Table 9.1, the values with and without the formation of the hydride being included. By the use of these and other excited states the energy available for transfer to the reacting molecule can be varied over a convenient range. Sometimes there is ambiguity as to whether the hydride is formed or not, although it appears that when the reacting molecule requires the additional energy, the hydride will be formed in order to make it available. This may be illustrated by comparing the reactions of triplet mercury and cadmium. Triplet mercury has sufficient energy to split the H—H bond without hydride formation and, as has been seen, the hydride is not formed as a final product. With triplet cadmium, on the other hand, the energy available is only just about sufficient even if the hydride is formed. The fact that the reaction proceeds readily therefore requires that the hydride be a final product, and this has been found spectroscopically to be the case by Bender.[31] The overall process is therefore

$$\text{Cd}(^3P_1) + \text{H}_2 \rightarrow \text{CdH} + \text{H}$$

and other reactions proceed in the same way. For example, Steacie and LeRoy[32] detected CdH in the cadmium-photosensitized reactions of propane.

9.4 RADIATION-CHEMICAL REACTIONS

The field of radiation chemistry involves the study of chemical reactions brought about by radiation in which the photons or other particles are of higher energy than in photochemistry. As previously noted, such radiations bring about some ionization in reaction systems and are less specific in their action than is the case with photochemical processes. The study of radiation-chemical reactions began in about 1935, partly as a result of the construction of particle-accelerating machines at about that time. An additional reason for the interest in the processes was the realization[33] that atoms and free radicals are important products of the primary radiation-chemical step and that these frequently are involved in chain processes. Thus, the field of radiation chemistry is related closely to that of photochemistry and to the study of thermal reactions.

A matter of importance in an experimental study of a reaction induced by ionizing radiations is the rate of absorption of energy by the reaction system: this quantity is commonly expressed in electronvolts (eV) per dm^3 per second (1 eV corresponds to 96.47 kJ mol^{-1}). The energy absorbed may be measured directly by determining the absorption coefficient for the reaction system, although this procedure is somewhat difficult. The method is simplified greatly if it is possible to arrange matters in such a way that all the radiation is absorbed. A second method, more widely used, is to measure the amount of ionization produced by the radiation when it passes through the reaction vessel containing a gas such as air. The rate of production of ions is determined by applying to the system a sufficiently high electrical potential that the current reaches a limiting value (the *saturation current*). When this is done, the ions are swept out of the system as rapidly as they are produced by the radiation, and the rate of production of the ions can be calculated from the saturation current. To convert the rate of ion production into the energy absorbed per second, it is necessary to know what energy is dissipated in creating an ion pair; for air the value usually employed is 32.5 eV per ion pair for electrons, and 35 eV per ion pair for other types of radiation.

A third method for determining the rate of energy absorption, and the one most commonly used, involves allowing the radiation to pass into a reaction vessel that contains reactants, the reaction rate of which can be followed and related in a known manner to the radiation absorbed. Sometimes the oxidation of ferrous sulfate in 0.4 M sulfuric acid is used for this purpose.

In the study of overall reactions initiated by ionizing radiations, two quantities are quoted frequently and lead to important information about mechanisms. The first is the *ion-pair yield* or *ionic yield,* which is denoted as M/N: this is the ratio of the rate of production of product molecules to the rate of formation of ion pairs. The second is the *G value,* which is the number of product molecules produced per 100 eV input. Both these quantities give some idea of chain lengths.

9.4.1 Radiation-Chemical Primary Process

A process in which high-energy radiation interacts with a chemical substance with the formation of atoms and free radicals, as well as ions, is known as a *radiation-chemical primary process*. The species produced subsequently undergo secondary processes, which may involve chains. In a photochemical reaction, the primary process usually occurs in only two steps, the formation of an electronically excited species and its subsequent dissociation into atoms. In a radiation-chemical reaction, on the other hand, the primary step involves the formation of ions together with the formation of excited species, occurring in parallel. The ions and the excited species then undergo subsequent processes which give rise to atoms and free radicals. Radiation-chemical primary processes are therefore more complex than the photochemical ones. The modern convention is to write the primary process using an arrow of the form $\rightsquigarrow$ and to indicate the type of radiation as a superscript above the arrow. Thus, when α radiation interacts with hydrogen, the primary process results in the production of hydrogen atoms, and the process is written as

$$H_2 \overset{\alpha}{\rightsquigarrow} 2H$$

The processes that occur when ionizing radiations interact with matter vary considerably with the type of radiation and with the substance involved. The situation is complicated, and the details have been worked out in only a few cases. As already noted, the radiation produces atoms and free radicals via excited species and also produces ions that are converted later by neutralization into atoms and radicals. One indication that this is so is that the energy W used in the production of an ion pair is always considerably larger than the ionization potential: W is usually about 35 eV, and only about half of this may be required for ionization. Another argument is that *ion-pair yields* (the number of molecules transformed for each ion pair produced) are frequently greater than photochemical quantum yields. Since an ion pair in general cannot produce more than a pair of radicals, this fact can be explained only if atoms and free radicals are produced as well as ions.

Further evidence for the same point of view was provided by the work of Essex and his collaborators,[34] who measured rates of radiation-chemical reactions when saturation currents were passed through reaction systems. If the atoms and free radicals were all produced from the ions formed in the primary process, the saturation current, by completely sweeping out the ions, would reduce the reaction rate to zero. In fact, the rates were reduced by a factor of about 2 by passing the saturation current. This suggests that about half of the reaction is brought about by atoms and free radicals that are produced directly and not from the neutralization of ions.

When a high-energy electron interacts with a molecule, an important process is frequently the production of a positive ion,

$$e^- + M \rightarrow M^+ + 2e^-$$

Since there is an increase in the number of electrons, a single initial electron may bring about the formation of a large number of positive ions. Eventually, however, an electron is slowed down to a speed at which it can no longer eject an electron from a molecule, and then one of several things may happen to it. In the case of some molecules, slow electrons may become attached to them with the formation of the negative ion,

$$e^- + M \rightarrow M^-$$

A second possible process is electron capture associated with the dissociation of the molecule,

$$e^- + R_2 \rightarrow R^- + R$$

A third possibility is that the electron may neutralize a positive ion,

$$e^- + M^+ \rightarrow M$$

Frequently, this neutralization is accompanied by dissociation

$$e^- + R_2^+ \rightarrow R + R$$

A few of these processes have been interpreted in terms of potential-energy curves. The formation of positive ions from the hydrogen molecule, for example, is explained[35] in terms of the curves shown in Fig. 9.9. As with photons, the process of excitation by electrons obeys the Franck–Condon principle, the most probable transitions being those in which there is little change in internuclear separation. It is to be seen from the figure that electrons with energy of about 16 eV may bring about excitation to the lower, $^2\Sigma_g^+$, state of the H_2^+ ion. The most probable transitions are to vibrationally excited states. There is also a certain probability of a transition to a point on the left-hand limb of the $^2\Sigma_g^+$ curve that corresponds to an energy in excess of the dissociation energy: in this case, dissociation occurs within the period of the first vibration. The overall process can be written as

$$e^- + H_2 \rightarrow H + H^+ + 2e^-$$

Electrons having an energy of about 26 eV can also produce molecules in the repulsive $^2\Sigma_u^+$ state of H_2^+: the result is the production of $H + H^+$ having high kinetic energies.

Other processes occurring with hydrogen and the various primary radiolytic reactions of oxygen,[36] water,[37] and methane[38] have been considered from a similar point of view.

Positively charged ions, such as α particles, also interact with molecules, mainly by ejecting electrons; for example,

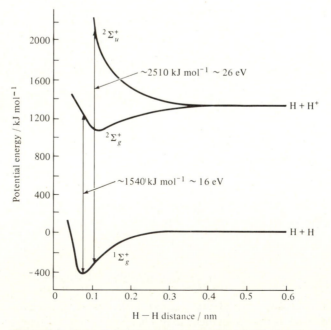

Figure 9.9 Potential-energy curves for H_2 and H_2^+, showing how positive ions are formed by electron impact.

$$\alpha^{2+} + M \rightarrow M^+ + e^- + \alpha^{2+}$$

The positive ion continues on its way with little deflection and ionizes other molecules. The ejected electron often has sufficient energy to bring about ionization of additional molecules, by the mechanisms described above.

Photons differ from the material particles in that they are absorbed and annihilated usually in a single elementary act, instead of undergoing a stepwise loss of energy. Three distinct processes are of importance in reactions brought about by photons:

1. Formation of an excited molecule, which may dissociate into atoms and radicals:

$$M + h\nu \rightarrow M^* \rightarrow 2R$$

   This is the type of process occurring in photochemical initiation.

2. Electromagnetic radiations having energy sufficient to cause ionization but not greater than about 0.5 meV (5×10^5 eV) lead to the ejection of an electron with the formation of the positive ion:

$$M + h\nu \rightarrow M^+ + e^-$$

   There is complete transfer of the energy of the radiation to the electron, which therefore is ejected with an energy equal to that of the photon minus the ionization energy.

3. At energies between about 10 keV (10^4 eV) and 1 meV (10^6 eV) there is Compton scattering of the photon with ejection of the electron and the formation of a positive ion:

$$M + h\nu \rightarrow M^+ + e^- + h\nu'$$

   The difference between this process and that of mechanism 2 is that the photon is not annihilated but continues on its way with a different frequency. The scattered photon is liable to be absorbed by subsequent molecules in its path.

In spite of the fact that the details of radiolytic initiation processes are somewhat complex, the overall result is usually fairly simple. In hydrogen, for example, the initiation process with most types of radiation is largely the production of atoms,

$$H_2 \rightsquigarrow 2H$$

Similarly, with hydrogen iodide the main process is

$$HI \rightsquigarrow H + I$$

In all cases, however, the production of ions also must be taken into consideration.

Reference has already been made, in Section 5.1.1 (see Fig. 5.1), to the technique of producing hydrogen atoms in an electric discharge. In such a discharge a stream of electrons is emitted by the cathode, and the mechanism of atom formation is therefore as previously discussed with reference to Fig. 9.9.

9.4.2 Pulse Radiolysis

In radiation chemistry the analogue of flash photolysis is *pulse radiolysis:* instead of the light flash there is a pulse of high-energy radiation, usually a beam of electrons. For example, linear electron accelerators have been used to give pulses of 1–10 MeV with durations usually of a few microseconds; much shorter pulses, of the order of a nanosecond, also have been obtained. These pulses have been applied to a considerable number of substances, and the results have led to valuable information about the existence of transient intermediates and to the rate constants for the elementary reactions that occur.

9.4.3 Hydrated Electrons

There has been much interest in the radiolysis of liquid water. With various types of ionizing radiation, the first step is the ejection of an electron:

$$H_2O \rightarrow H_2O^+ + e^-$$

If the electron produced were to remain free, it rapidly would undergo processes such as the following:

$$e^- + H_2O \rightarrow H + OH^-$$

and

$$e^- + H_2O^+ \rightarrow H_2O^* \rightarrow H + OH$$

However, in 1953 it was suggested on theoretical grounds by Platzman[39] that an electron might become somewhat stabilized by being hydrated. He suggested that the hydrated electron, given the symbol e_{aq}^-, would have a blue color and that the enthalpy of hydration would be about 200 kJ mol^{-1} (≈ 2 eV). Being stabilized in this way the hydrated electron would have a much longer life than a free electron and would survive long enough to react with solutes, bringing about their reduction.

Indirect support for the existence of the hydrated electron was provided by kinetic studies. For example, the reduction of certain substances in solution showed an ionic-strength effect (Section 6.3.4) that indicated the reducing species to have a negative charge:[40] it therefore could not be the H atom. Thus, it appeared that the reducing species are electrons, which must be hydrated since free electrons would react rapidly with water and would not survive long enough to undergo reaction with solutes. Another application of the ionic-strength method has been to the reaction between the hydrated electron and lysozyme.[41] The reaction was found to be diffusion controlled, and the ionic-strength effect indicated that the hydrated electron was interacting with various ionized states of the lysozyme molecule; as the pH is increased the molecule is deprotonated.

Direct evidence for the existence of the hydrated electron was provided by pulse radiolysis studies of liquid water, in which the transient spectrum of the species was observed.[42] The spectrum showed a maximum absorption at about 700 nm, and this is consistent with the blue color predicted by Platzman.

Since the 1960s many investigations have been made of the properties of the hydrated electron. Some rate constants for reactions involving it are listed in Table 9.2.

TABLE 9.2   RATE CONSTANTS FOR REACTIONS OF THE HYDRATED ELECTRON[a]

Reaction	$k/dm^3\ mol^{-1}\ s^{-1}$	Comments
$e_{aq}^- + e_{aq}^- \rightarrow H_2 + 2OH^-$	5×10^9	Important only at relatively high concentrations of e_{aq}^-
$e_{aq}^- + H_2O \rightarrow H + OH^-$	16	Predominant only at higher pH values (>8); at lower values $e_{aq}^- + H_3O^+$ is more important
$e_{aq}^- + H_3O^+ \rightarrow H + H_2O$	2.06×10^{10}	See above

[a] Additional values of rate constants for the hydrated electron are to be found in M. Anbar, M. Bambenek, and A. B. Ross, *Selected Specific Rates of Reactions of Transients from Water in Aqueous Solution. 1. Hydrated Electron*, NSRDS–NBS Report No. 43, National Bureau of Standards, Washington DC, 1975.

9.5   CHEMILUMINESCENCE

The term *chemiluminescence* is used to refer to the radiation emitted as the result of a chemical reaction. Early investigations of this effect were carried out by Polanyi,[43] who studied gas-phase reactions such as that between sodium vapor and chlorine. He found that the first step in this reaction is

$$Na + Cl_2 \rightarrow NaCl + Cl$$

and that the resulting chlorine atoms react with Na_2 to give NaCl in a vibrationally excited state:

$$Na_2 + Cl \rightarrow Na + NaCl'$$

Vibrationally excited NaCl' molecules may collide with sodium atoms to give electronically excited Na* atoms which emit the yellow sodium *D* radiation:

$$NaCl' + Na \rightarrow NaCl + Na*$$

Similar results were obtained in reactions of sodium with bromine and iodine and of potassium with chlorine, bromine, and iodine. The mechanisms of these processes are considered further in Section 12.2.

The radiation emitted by flames is to some extent blackbody radiation, emitted as a result of the high temperature of the burnt gas. Spectroscopic studies of flames have revealed, however, that some of the radiation is emitted specifically by species that have been formed in excited states. A well-known example is the yellow radiation emitted when sodium is present in a flame. The phenomenon of cool flames has been referred to in Section 8.6.2. These are usually too cool to emit blackbody radiation in the visible; reaction products or intermediates, such as formaldehyde, are formed in electronically excited states and emit radiation.

In the liquid phase, chemiluminescence is less likely than in the gas phase, because of a much greater probability of the deactivation of excited species by collisions. There are, however, some striking examples. One of these is the oxidation by ferricyanide of *o*-aminophthalic cyclic hydrazide in alkaline solution in the presence of a small amount of hydrogen peroxide. The reaction produces blue light with high efficiency, and the radiation emitted is sufficiently intense to enable one to read (for a few seconds!) in a darkened room. A chemiluminescent reaction also occurs in the firefly.

PROBLEMS

9.1. Calculate the maximum wavelength of the radiation that will bring about dissociation of a diatomic molecule having a dissociation energy of 390.4 kJ mol^{-1}.

9.2. Hydrogen iodide undergoes decomposition into $H_2 + I_2$ when irradiated with radiation having a wavelength of 207 nm. When 1 J of energy is absorbed, 440 μg of HI is decomposed. How many molecules of HI are decomposed by one photon of radiation of this wavelength? Suggest a mechanism that is consistent with this result.

9.3. A 100 W mercury-vapor lamp emits radiation of 253.7 nm wavelength and may be assumed to operate with 100% efficiency. If all the light emitted is absorbed by a substance that is decomposed with a quantum yield of unity, how long will it take for 0.01 mol to be decomposed?

9.4. Suppose that the radiation emitted by the lamp in Problem 9.3 is all absorbed by ethylene, which decomposes into $C_2H_2 + H_2$ with a quantum yield of unity. How much acetylene will be produced per hour?

9.5. The enthalpies of formation of CH_4, C_2H_6 and CH_2 are as follows:

$$\Delta_f H^\circ_{298K}(CH_4) = -74.9 \text{ kJ mol}^{-1}$$

$$\Delta_f H^\circ_{298K}(C_2H_6) = -84.5 \text{ kJ mol}^{-1}$$

$$\Delta_f H^\circ_{298K}(CH_2) = 398 \text{ kJ mol}^{-1}$$

Calculate, in electron volts, the minimum energy that an electron must have to produce CH_2 radicals by the reaction

$$C_2H_6 \overset{e^-}{\rightsquigarrow} CH_4 + CH_2$$

9.6. When water vapor is irradiated with a beam of high-energy electrons, various ions such as H^+ and O^- appear. Calculate the minimum energies required for the formation of these ions, given the following thermochemical data:

$$H_2O(g) \rightarrow H(g) + OH(g) \qquad \Delta H^\circ = 498.7 \text{ kJ mol}^{-1}$$

$$OH(g) \rightarrow H(g) + O(g) \qquad \Delta H^\circ = 428.2 \text{ kJ mol}^{-1}$$

$$H(g) \rightarrow H^+(g) + e^-(g) \qquad \Delta H^\circ = 1312.2 \text{ kJ mol}^{-1}$$

$$O(g) + e^-(g) \rightarrow O^-(g) \qquad \Delta H^\circ = -213.4 \text{ kJ mol}^{-1}$$

Are the results you obtain consistent with the experimental appearance potentials of 19.5 eV for H^+ and 7.5 eV for O^-?

9.7. By pulse radiolysis, electrons are generated in aqueous solution of pH 9 at a concentration of 10^{-8} M. Use the data in Table 9.2 to confirm that the reaction $e^-_{aq} + H_2O \rightarrow H + OH^-$ is predominant and calculate the half-life of the hydrated electrons.

9.8. Suppose that the conditions are the same as in Problem 9.7 but that the pH is 3. Which process is now most important, and what is the half-life?

9.9. The following are rate constants at zero ionic strength for reactions of the hydrated electron:

$$e^-_{aq} + Cr^{2+} \qquad 4.2 \times 10^{10} \text{ dm}^3 \text{ mol}^{-1} \text{ s}^{-1}$$

$$e^-_{aq} + Fe^{2+} \qquad 1.2 \times 10^8 \text{ dm}^3 \text{ mol}^{-1} \text{ s}^{-1}$$

$$e^-_{aq} + Fe(CN)_6^{3-} \qquad 3.0 \times 10^9 \text{ dm}^3 \text{ mol}^{-1} \text{ s}^{-1}$$

Estimate the corresponding rate constants at an ionic strength of 0.01 M.

9.10. Confirm that the rate constant given in the previous problem for the reaction between e_{aq}^- and $Fe(CN)_6^{3-}$ is consistent with a diffusion-controlled reaction. Use the following data:

	Radius/nm	Diffusion coefficient/cm^2 s^{-1}
e_{aq}^-	0.27	4.8×10^{-5}
$Fe(CN)_6^{3-}$	0.28	9.0×10^{-6}

REFERENCES

1. C. J. D. von Grotthuss, *Ann. Phys.,* **61,** 50 (1819).
2. J. W. Draper, *Philos. Mag.,* **19,** 195 (1841).
3. J. Dalton, *A New System of Chemical Philosophy,* vol. I, Part ii, p. 300, Bickerstaff, London (1810).
4. W. C. Wittwer, *Ann. Phys.,* **94,** 597 (1855).
5. R. W. Bunsen and H. E. Roscoe, *Pogg. Ann.,* **96,** 373 (1855); **100,** 32, 481 (1857); **101,** 193 (1859).
6. A. Einstein, *Ann. Phys.,* **17,** 549 (1905); **19,** 371 (1906).
7. J. Stark, *Phys. Z.,* **9,** 889, 894 (1908). For a discussion of Stark's priority over Einstein see J. A. Allmand, *Trans. Faraday Soc.,* **21,** 438 (1926).
8. A. Einstein, *Ann. Phys.,* **37**(4), 832 (1912); **38,** 881 (1912); *J. Phys.,* **3,** 227 (1913).
9. J. Franck, *Trans. Faraday Soc.,* **21,** 536 (1925).
10. E. V. Condon, *Phys. Rev.,* **32,** 858 (1928).
11. V. Henri, *Comptes Rendus,* **177,** 1037 (1923); V. Henri and M. C. Teves, *Comptes Rendus,* **178,** 894 (1924).
12. The term "einstein" was proposed by M. Bodenstein and G. Wagner, *Z. Phys. Chem. B,* **3,** 456 (1929).
13. For example, in the 1925 Faraday Society Discussion; see *Trans. Faraday Soc.,* **21,** 437–609 (1926).
14. M. Bodenstein, *Z. Phys. Chem. B,* **5,** 329 (1913); M. Bodenstein and M. Dux, *Z. Phys. Chem. B,* **5,** 297 (1913).
15. E. Warburg, *Sitzungsber. Preuss. Akad. Wiss. Phys.-Math Kl.,* 314 (1916).
16. E. Briers, D. L. Chapman, and E. Walter, *J. Chem. Soc.,* 502 (1926).
17. See, for example, H. W. Melville, *J. Chem. Soc.,* 274 (1947); G. M. Burnett and H. W. Melville, *Nature,* **156,** 661 (1945); H. W. Melville, *Proc. R. Soc. London A,* **237,** 149 (1956).
18. A. Shepp, *J. Chem. Phys.,* **24,** 939 (1956).
19. D. L. Baulch and J. Duxbury, *Combust. Flame,* **37,** 313 (1980).
20. G. Porter, *Proc. R. Soc. London A,* **200,** 284 (1950); *Discuss. Faraday Soc.,* **9,** 60 (1950).
21. J. V. V. Kasper and G. C. Pimentel, *Chem. Phys. Lett.,* **5,** 231 (1964).
22. For a review see J. J. Bannister, J. Gormally, J. F. Holzwarth, and T. A. King, *Chem. Br.,* **20,** 227 (1984).
23. W. C. Natzie, C. B. Moore, D. M. Goodall, W. Frisch, and J. F. Holzwarth, *J. Phys. Chem.,* **85,** 2882 (1981).
24. For a review see P. M. Rentzepis, *Adv. Chem. Phys.,* **23,** 189 (1973).
25. M. N. R. Ashfold and G. Hancock, in P. G. Ashmore and R. J. Donovan (Eds.), *Gas Kinetics and Energy Transfer,* vol. 4, p. 73, Royal Society of Chemistry, London, 1980.
26. G. Cario and J. Franck, *Z. Phys.,* **11,** 161 (1922).
27. H. S. Taylor and J. R. Bates, *Proc. Natl. Acad. Sci. USA,* **12,** 714 (1926); H. S. Taylor and D. G. Hill, *J. Am. Chem. Soc.,* **51,** 2927 (1929); *Z. Phys. Chem. B,* **2,** 449 (1929); H. S. Taylor and A. L. Marshall, *J. Phys. Chem.,* **29,** 1140 (1925).
28. L. O. Olsen, *J. Chem. Phys.,* **6,** 307 (1938).

29. J. Franck and H. Sponer, *Nachr. Ges. Wiss. Goett.,* 241 (1928); H. Beutler and E. Rabinowitch, *Z. Phys. Chem. B,* **8,** 403 (1930).
30. K. J. Laidler, *J. Chem. Phys.,* **10,** 43 (1942).
31. P. Bender, *Phys. Rev.,* **36,** 1535 (1930).
32. E. W. R. Steacie and D. J. LeRoy, *J. Chem. Phys.,* **12,** 34 (1944).
33. H. Eyring, J. O. Hirschfelder, and H. S. Taylor, *J. Chem. Phys.,* **4,** 479, 570 (1937).
34. C. Smith and H. Essex, *J. Chem. Phys.,* **6,** 188 (1938); A. D. Kolumban and H. Essex, *J. Chem. Phys.,* **8,** 450 (1940); N. T. Williams and H. Essex, *J. Chem. Phys.,* **16,** 1153 (1948); **17,** 995 (1949).
35. J. L. Magee and M. Burton, *J. Am. Chem. Soc.,* **72,** 1965 (1950).
36. K. J. Laidler and E. K. Gill, *Trans. Faraday Soc.,* **54,** 633 (1958); D. C. Frost and C. A. McDowell, *J. Am. Chem. Soc.,* **80,** 6183 (1958).
37. K. J. Laidler, *J. Chem. Phys.,* **22,** 1740 (1954); F. Fiquet-Fayard, *J. Chim. Phys.,* 274 (1957).
38. C. A. McDowell, *Trans. Faraday Soc.,* **50,** 423 (1954).
39. R. L. Platzman, in *Basic Mechanism in Radiobiology,* National Research Council Publication 305, Washington, DC, 1953.
40. E. Collinson, F. S. Dainton, D. R. Smith, and S. Tanuke, *Proc. Chem. Soc.,* 140 (1962).
41. M. Z. Hofmann and E. Nayon, *J. Phys. Chem.,* **79,** 1362 (1975).
42. E. J. Hart and J. W. Boag, *J. Am. Chem. Soc.,* **84,** 4090 (1962); J. W. Boag and E. J. Hart, *Nature,* **197,** 45 (1963).
43. M. Polanyi, *Atomic Reactions,* Williams and Norgate, London, 1932.

BIBLIOGRAPHY

The following older books on photochemistry are still useful:

E. J. Bowen, *The Chemical Aspects of Light,* Clarendon, Oxford, 1946.

W. A. Noyes and P. A. Leighton, *The Photochemistry of Gases,* Reinhold, New York, 1941; Dover, New York, 1966.

G. K. Rollefson and M. Burton, *Photochemistry and the Mechanism of Chemical Reactions,* Prentice-Hall, Englewood Cliffs, NJ, 1939.

For more recent general treatments of photochemistry see:

J. C. Calvert and J. N. Pitts, *Photochemistry,* Wiley, New York, 1966.

H. Okabe, *Photochemistry of Small Molecules,* Wiley-Interscience, New York, 1966.

N. J. Turro, *Modern Molecular Photochemistry,* Benjamin-Cummings, Menlo Park, CA, 1978.

General principles relating to photochemical and radiation-chemical reactions are considered in:

K. J. Laidler, *The Chemical Kinetics of Excited States,* Clarendon, Oxford, 1955.

For accounts of the principles and applications of flash photolysis see:

R. G. W. Norrish, Some Fast Reactions in Gases Studied by Flash Photolysis and Kinetic Spectroscopy, in *Les Prix Nobel en 1967,* p. 181, Norstedt and Soner, Stockholm, 1967.

R. G. W. Norrish, Fast Reactions and Primary Processes in Chemical Kinetics, *Nobel Symp.,* **5,** 33 (1967).

R. G. W. Norrish, Some Reactions of Organic Photochemistry as Revealed by Flash Photolysis and Kinetic Spectroscopy, *Mem. Soc. R. Sci. Liège,* **6**(I), 33 (1971).

R. G. W. Norrish and G. Porter, The Application of Flash Techniques to the Study of Fast Reactions, *Discuss. Faraday Soc.,* **17,** 40 (1955).

R. G. W. Norrish and B. A. Thrush, Flash Photolysis and Kinetic Spectroscopy, *Q. Rev. Chem. Soc.,* **10,** 149 (1956).

G. Porter, Flash Photolysis, in P. G. Ashmore, F. S. Dainton, and T. M. Sugden (Eds.), *Photochemistry and Reaction Kinetics,* pp. 93–111, Cambridge University Press, New York, 1967.

G. Porter, Flash Photolysis and Some of Its Applications, *Science,* **160,** 1299 (1968).

For accounts of kinetic studies using laser beams see the following publications:

M. N. R. Ashfold and G. Hancock, in P. G. Ashmore and R. J. Donovan (Eds.), *Gas Kinetics and Energy Transfer,* vol. 4, p. 83, Royal Society of Chemistry, London, 1980.

J. J. Bannister, J. Gormally, J. F. Holzwarth, and T. A. King, The Iodine Laser and Fast Reactions, *Chem. Br.,* **20,** 227 (1984).

M. J. Beezley, *Lasers and their Applications,* Halsted, New York, 1976.

A. Ben-Shaul, Y. Haas, K. L. Komba, and R. D. Levine, *Lasers and Chemical Change,* Springer, Berlin, 1981.

G. Braderlow, E. Fill, and K. J. Witte, *The High Power Iodine Laser,* Springer, New York, 1983.

D. M. Goodall and C. G. Cureton, Vibrational Photochemistry with Lasers, *Chem. Br.,* **19,** 493 (1983).

S. F. Jacobs, M. Sargent, M. O. Scully, and C. T. Walker, *Laser Photochemistry, Tunable Lasers, and Other Topics,* Addison-Wesley, Reading, MA., 1976.

M. C. Lin and J. R. McDonald, Production and Detection of Reactive Species with Lasers in Static Systems, in D. W. Setzer (Ed.), *Reaction Intermediates in the Gas Phase: Generation and Monitoring,* Academic, New York, 1979.

C. B. Moore (Ed.), *Chemical and Biochemical Applications of Lasers,* Academic, New York, 1974.

J. I. Steinfeld (Ed.), *Laser-Induced Chemical Processes,* Plenum, New York, 1981.

H. Walther (Ed.), *Laser Spectroscopy of Atoms and Molecules,* Springer, Berlin, 1976.

For general accounts of radiation chemistry see:

G. Hughes, *Radiation Chemistry,* Clarendon, Oxford, 1973.

J. H. O'Donnell and D. F. Sangster, *Principles of Radiation Chemistry,* Arnold, London, 1970.

J. W. T. Spinks and R. J. Wood, *An Introduction to Radiation Chemistry,* 2nd ed., Wiley, New York, 1976.

A. J. Swallow, *Radiation Chemistry: An Introduction,* Longman, London, 1973.

Special aspects of radiation chemistry are treated in the following publications:

A. O. Allen, *The Radiation Chemistry of Water and Aqueous Solutions,* Van Nostrand, New York, 1961.

F. S. Dainton, The Chemistry of the Electron, in S. Claesson (Ed.), *Fast Reactions and Primary Processes in Chemical Kinetics,* Proceedings of the Fifth Nobel Symposium, Interscience, New York, 1967.

L. M. Dorfman and M. S. Matheson, Pulse Radiolysis, in G. Porter (Ed.), *Progress in Reaction Kinetics,* vol. 3, p. 237, Pergamon Press, Oxford, 1965.

I. G. Draganic and Z. D. Draganic, *The Radiation Chemistry of Water,* Academic, New York, 1971.

M. Ebert, J. P. Keene, A. J. Swallow, and J. H. Baxendale, *Pulse Radiolysis,* Academic, London, 1965.

E. J. Hart and M. Anbar, *The Hydrated Electron,* Wiley, New York, 1970.

S. C. Lind, *Radiation Chemistry of Gases,* Reinhold, New York, 1961.

M. S. Matheson and L. M. Dorfman, *Pulse Radiolysis,* MIT Press, Cambridge, MA, 1969.

P. M. Rentzepis, Picosecond Spectroscopy and Molecular Relaxation, *Adv. Chem. Phys.,* **23,** 189 (1973).

A. J. Swallow, *Radiation Chemistry of Organic Compounds,* Pergamon, New York, 1960.

For an account of chemiluminescence in biological systems see:

F. McCapra, Mechanism and Function of Bioluminescent Systems, *Biochem. Soc. Trans.,* **7,** 1239 (1979).

E. N. Harvey, *Bioluminescence,* Academic, New York, 1952.

P. J. Herring (Ed.), *Bioluminescence in Action,* Academic, New York, 1978.

Chapter 10

Homogeneous Catalysis

One special type of catalysis, *heterogeneous catalysis,* has already been treated in Chapter 7. The present chapter is concerned with various types of catalysis that occur in a single phase and which therefore are referred to as *homogeneous catalysis.* Important examples are catalysis by acids, bases and enzymes.

The concept of catalysis was first recognized clearly by Berzelius[1] who, however, regarded the effect as due to a special type of force that the catalyst exerted on the reaction system. More understanding of catalysis was provided by a series of investigations carried out by Ostwald between 1888 and the turn of the century. In the course of his work, Ostwald proposed several useful definitions of a catalyst. One of them is that a catalyst is[2]

> any substance that alters the velocity of a chemical reaction without modification of the energy factors of the reaction.

A later definition is that[3]

> a catalyst is any substance that alters the velocity of a chemical reaction without appearing in the end product of the reaction.

Another definition, perhaps the best known, is that a catalyst is[4]

> a substance that changes the velocity of a reaction without itself being changed by the process.

All these definitions are quite acceptable and are essentially equivalent to one another. They are consistent with the definition approved in 1981 by the International Union of Pure and Applied Chemistry:[5]

A catalyst is a substance that increases the rate of reaction without modifying the overall standard Gibbs energy change in the reaction; the process is called *catalysis,* and a reaction in which a catalyst is involved is known as a *catalyzed reaction.*

The restriction that a catalyst must emerge unchanged from the reaction system is a rather rigid one, and sometimes a broader definition has been used. For example, a substance may behave like a catalyst in most respects but it may undergo some change, such as decomposition, that is quite independent of the main reaction; in such cases use of the term "catalyst" may be acceptable, although it is well to point out that the changes are taking place. Another situation where a broader definition often is used is exemplified by the hydrolysis of esters under the influence of bases: the acid formed reacts with base to form its anion, so that the base is consumed. However, the base usually is described as a catalyst for this reaction, since in other respects its action is catalytic.

Towards the end of his studies of catalysis Ostward suggested a useful classification of catalyzed reactions as follows:[6]

1. *Crystallization from supersaturated solutions.* For example, crystallization from such solutions often is brought about by a particle of dust or a crystal. Today this effect usually is not referred to as catalysis.
2. *Catalysis in homogeneous systems.* This is the subject of the present chapter.
3. *Catalysis in heterogeneous systems.* Important examples of this are reactions catalyzed at a gas–solid interface, a matter dealt with in Chapter 7.
4. *Catalysis by enzymes.* Except for special applications this often is considered to be homogeneous catalysis and is considered briefly in Section 10.3.

Ostwald himself did little work on heterogeneous catalysis, but he did much on homogeneous catalysis. He summarized his contributions in a number of review articles.[7]

In spite of his important contributions, Ostwald was never able to develop a satisfactory theory of catalysis. He recognized[8] that a catalyzed reaction proceeds by an alternative reaction pathway made possible by the introduction of the catalyst. From this he drew the correct conclusion that a substance that lowers the rate of a reaction (now known as an *inhibitor*) cannot do so by introducing a less efficient pathway, since the reaction would continue to occur at the original rate by following the path for the uncatalyzed process. Ostwald's failure to develop a suitable theory of catalysis was inevitable at the time, since then practically nothing was known about reaction mechanisms.

As knowledge has developed, it has become clear that in a catalyzed reaction the catalyst always is involved temporarily in some kind of chemical transformation and finally is regenerated. In surface catalysis and enzyme catalysis the catalyst usually forms an addition complex with the substrate. In acid and base catalysis there is nearly always proton transfer between the catalyst and substrate molecules, as is considered in more detail in Section 10.2.

Since by its strict definition a catalyst is unchanged at the end of the reaction, it gives no energy to the system; therefore, it can have no influence on the position of equilibrium. It follows that since at equilibrium the equilibrium constant is the ratio of the rate constants for the reaction in the forward and reverse directions (Section 8.2.4), a catalyst must influence the forward and reverse rates in exactly the same way. This conclusion has been verified experimentally in a number of investigations.[9]

An extremely small amount of a catalyst frequently brings about a considerable increase in the rate of a reaction. For example, a solution of hydrogen peroxide decomposes very slowly at ordinary temperatures, but the reaction occurs at a much greater rate under the action of colloidal platinum, even at a concentration as low as 10^{-8} mol dm^{-3}. However, there is nothing in the definition of a catalyst that says that it must act at very low concentrations. Some idea of the effectiveness of a catalyst has sometimes been provided by expressing the *turnover number,* defined as the number of molecules transformed per minute by one molecule of the catalyst. For example, under certain conditions the enzyme catalase has a turnover number of more than 10^6 for the decomposition of hydrogen peroxide. However, the turnover number varies with the temperature, the concentration of substrate (a substance undergoing reaction), and other conditions. Therefore, it is not a useful quantity in kinetic work, for which it is better to use the rate constant and other kinetic parameters.

The rates of catalyzed reactions are sometimes directly proportional to the concentration of the catalyst:

$$v = F[C] \tag{10.1}$$

where [C] is the catalyst concentration and F is a function of the substrate concentration and other factors. This equation implies that the reaction does not occur to a measurable extent in the absence of catalyst, and this is sometimes the case. There are many reactions, however, for which there is significant reaction in the absence of any catalyst, and then an equation such as

$$v = F[C] + F' \tag{10.2}$$

often applies.

10.1 GENERAL CATALYTIC MECHANISMS

Catalyzed reactions occur by a wide variety of mechanisms. There is, however, one pattern that applies to a number of single-substrate reactions catalyzed by surfaces, enzymes, acids, and bases. It is useful to consider this scheme of reactions first so as to appreciate the similarities that exist between certain reactions that are catalyzed by different types of catalyst.

The reaction scheme[10] is

$$(1) \quad C + S \ \rightleftharpoons X + Y$$

$$(2) \quad X + W \rightarrow P + Z$$

Here C represents the catalyst and S the substrate; X and Y are intermediates, the first of which undergoes a second reaction with a species W to give the final product or products P together with an additional substance Z. This scheme shows only the kinetically significant reactions; the substances Y and Z undergo other processes that do not have any effect on the kinetic behavior. To simplify the treatment it is assumed that the second reaction does not occur in the reverse direction; this can be ensured if the product P is removed as fast as it is formed.

In surface catalysis X is an adsorption complex, and Y and W are nonexistent. The constants k_{-1} and k_2 in this case are first-order rate constants, while k_1 is a second-order constant. In catalysis by acids and bases, however, Y and W do play important roles. Thus, if C is an acid catalyst, reaction 1 involves the transfer of a proton to S, so that Y is the base conjugate to the acid C. In acid catalysis the intermediate X is the protonated substrate SH^+, and in reaction 2 a proton is transferred to the species W. This species W therefore has basic (or amphoteric) properties, and it may be a molecule of solvent or of solute; for example, it may be the species Y formed in the first step. We will see that the kinetic behavior depends in an important way on whether the intermediate X (i.e., SH^+) transfers its proton to a solvent molecule or to a solute molecule.

Conversely, in base catalysis Y is the acid conjugate to the base C. The intermediate X is the substrate molecule minus a proton, and in reaction 2 it accepts a proton from W. Again, W may be a solvent molecule or a solute molecule.

In some situations the rate with which the intermediate X undergoes reaction 2 may be sufficiently slow that the first reaction may be regarded as being at equilibrium: the exact condition for this is $k_2[X][W] \ll k_{-1}[X][Y]$. Since this case corresponds to Arrhenius's concept of an intermediate in equilibrium with the reactants, such intermediates have been called *Arrhenius intermediates*.

The converse case is that the condition $k_{-1}[X][Y] \ll k_2[X][W]$ applies. The concentration of X is then small, and the steady-state treatment may be applied to it. Intermediates of this kind have been called *van't Hoff intermediates*.

If neither of these extreme conditions applies, the kinetic situation is more complicated, and the appropriate differential equations have to be solved. Only the equilibrium and steady-state treatments are considered here.

10.1.1 Equilibrium Treatment: Arrhenius Intermediates

In this case the equation

$$\frac{[X][Y]}{[C][S]} = \frac{k_1}{k_{-1}} = K \tag{10.3}$$

applies. However, the concentrations of C and S do not correspond necessarily to the initial concentrations $[C]_0$ and $[S]_0$, since appreciable amounts of C and S have been used to form the intermediate X. These initial concentrations may be expressed as

$$[C]_0 = [C] + [X] \tag{10.4}$$

$$[S]_0 = [S] + [X] \tag{10.5}$$

as long as attention is confined to initial rates (at later times some S has formed product). Equation (10.3) then becomes

$$\frac{[X][Y]}{([C]_0 - [X])([S]_0 - [X])} = K \tag{10.6}$$

This is a quadratic in [X] and can be solved for [X]. Then an expression for the rate, equal to $k_2[X][W]$, can be written down. However, it is more useful to consider two special cases.

Case 1 If the initial concentration of the substrate is much greater than that of the catalyst, that is, if $[S]_0 \gg [C]_0$, it follows that $[S]_0 - [X]$ is very close to $[S]_0$, since [X] cannot exceed $[C]_0$. Equation (10.6) therefore reduces to

$$\frac{[X][Y]}{([C]_0 - [X])[S]_0} = K \tag{10.7}$$

and thus

$$[X] = \frac{K[C]_0[S]_0}{K[S]_0 + [Y]} \tag{10.8}$$

The rate of reaction is therefore

$$v = k_2[X][W] = \frac{k_2 K[C]_0[S]_0[W]}{K[S]_0 + [Y]} \tag{10.9}$$

This rate equation corresponds to a variation of rate of the type represented in Fig. 10.1a. At lower substrate concentrations, when $K[S]_0 \ll [Y]$, the rate varies linearly with $[S]_0$, whereas at higher concentrations, when $K[S]_0 \gg [Y]$, the rate becomes independent of $[S]_0$. As long as the condition $[S]_0 \gg [C]_0$ holds, however, the rate varies linearly with $[C]_0$.

This type of behavior is characteristic of single-substrate reactions on surfaces and of enzyme reactions. For both of these the species Y and W are nonexistent and Eq. (10.9) becomes

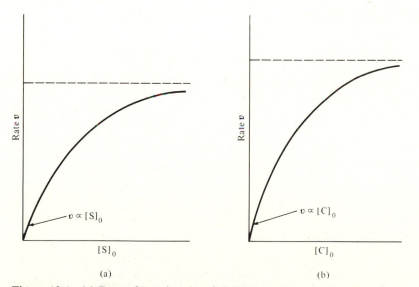

Rate v

$v \propto [S]_0$

$[S]_0$

(a)

Rate v

$v \propto [C]_0$

$[C]_0$

(b)

Figure 10.1 (a) Rate of reaction as a function of substrate concentration for the case in which $[S]_0 \gg [C]_0$. (b) Rate of reaction as a function of catalyst concentration for the case $[C]_0 \gg [S]_0$.

$$v = \frac{k_2 K [C]_0 [S]_0}{K[S]_0 + 1} \tag{10.10}$$

For surface reactions this is equivalent to Eq. (7.40). The equation is also equivalent to the Michaelis–Menten equation, Eq. (10.93), for enzyme reactions. This equation usually is written as

$$v = \frac{k_2 [C]_0 [S]_0}{K_m + [S]_0} \tag{10.11}$$

where K_m, the Michaelis constant, is equal to $1/K$ in the present treatment.

In reactions catalyzed by acids and bases it will be seen that the rate remains linear with the substrate concentration; this is because of the special types of equilibrium that are rapidly established in the solution.

Case 2 On the other hand, if the catalyst is greatly in excess of the substrate, that is, $[C]_0 \gg [S]_0$, Eq. (10.6) reduces to

$$\frac{[X][Y]}{[C]_0([S]_0 - [X])} = K \tag{10.12}$$

and the rate of reaction is

$$v = \frac{k_2 K [C]_0 [S]_0 [W]}{K[C]_0 + [Y]} \tag{10.13}$$

The rate now varies linearly with the concentration of substrate (as long as the condition $[C]_0 \gg [S]_0$ holds), but the variation with the catalyst concentration is as shown in Fig. 10.1b.

10.1.2 Steady-State Treatment: van't Hoff Intermediates

If the condition $k_2[W] \gg k_{-1}[Y]$ applies, the concentration of X is small and the steady-state treatment is applicable. The steady-state equation is

$$\frac{d[X]}{dt} = k_1[C][S] - k_{-1}[X][Y] - k_2[X][W] = 0 \tag{10.14}$$

Substitution of $[C]_0 - [X]$ for $[C]$ and of $[S]_0 - [X]$ for $[S]$ gives

$$k_1([C]_0 - [X])([S]_0 - [X]) - k_{-1}[X][Y] - k_2[X][W] = 0 \tag{10.15}$$

Since $[X]$ is small the term in $[X]^2$ can be neglected; with this approximation Eq. (10.15) gives

$$[X] = \frac{k_1 [C]_0 [S]_0}{k_1([C_0] + [S]_0) + k_{-1}[Y] + k_2[W]} \tag{10.16}$$

The rate is therefore

$$v = \frac{k_1 k_2 [C]_0 [S]_0 [W]}{k_1([C]_0 + [S]_0) + k_{-1}[Y] + k_2[W]} \tag{10.17}$$

This equation again indicates that at low concentrations of either catalyst or substrate the rate is proportional to either $[C]_0$ or $[S]_0$; at a higher concentration of either the rate becomes independent of that concentration.

In catalysis by surfaces and enzymes, W and Y are nonexistent and the rate equation becomes

$$v = \frac{k_1 k_2 [C]_0 [S]_0}{k_1 ([C]_0 + [S]_0) + k_{-1} + k_2} \tag{10.18}$$

An equation of essentially this form was first derived by Briggs and Haldane[11] for enzyme reactions and is considered further in Section 10.3.

10.1.3 Activation Energies for Catalyzed Reactions

Depending on the system and the conditions, there are a number of different relationships between the overall activation energy and the values for the individual steps. Only a few cases need be considered. Suppose, for example, that Eq. (10.10) applies: the intermediate is in equilibrium with the reactant and the catalyst. At low substrate concentrations the rate is $k_2 K [C]_0 [S]_0$, and since $K = k_1/k_{-1}$ the rate constant is $k_1 k_2/k_{-1}$. The overall activation energy is therefore

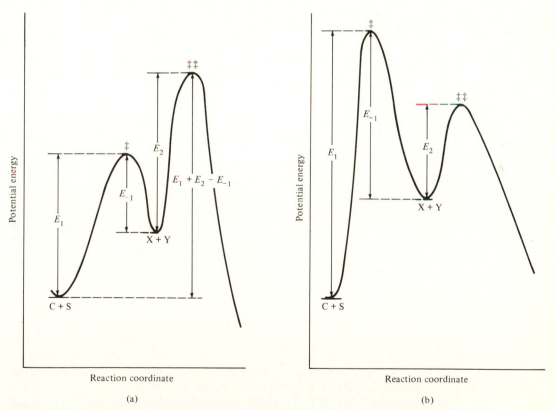

(a)   (b)

Figure 10.2 Potential-energy diagrams for catalyzed reactions: (a) Arrhenius intermediate ($k_{-1} \gg k_2$); the rate-controlling step is the crossing of the second barrier. (b) van't Hoff intermediate, with $k_2 \gg k_{-1}$; the crossing of the first barrier is now the rate-controlling step.

$$E_{\text{low}} = E_1 + E_2 - E_{-1} \qquad (10.19)$$

At high substrate concentrations, on the other hand, the rate is $k_2[C]_0$, and the activation energy is then

$$E_{\text{high}} = E_2 \qquad (10.20)$$

These two relationships, illustrated in Fig. 10.2a, are equivalent to those represented in Fig. 7.5.

In the steady-state case, the rate equation is given by Eq. (10.18). At high substrate concentrations the activation energy is given again by Eq. (10.20). At low substrate and catalyst concentrations the rate is

$$v_{\text{low}} = \frac{k_1 k_2}{k_{-1} + k_2} [C]_0 [S]_0 \qquad (10.21)$$

In general, the Arrhenius equation does not apply, but it does so in two special cases. If $k_2 \gg k_{-1}$, Eq. (10.21) reduces to

$$v_{\text{low}} = k_1 [C]_0 [S] \qquad (10.22)$$

and the activation energy is

$$E_{\text{low}} = E_1 \qquad (10.23)$$

This is illustrated in Fig. 10.2b. On the other hand, if $k_{-1} \gg k_2$, Eq. (10.21) reduces to

$$v_{\text{low}} = \frac{k_1 k_2}{k_{-1}} [C]_0 [S]_0 \qquad (10.24)$$

and E_{low} is given again by Eq. (10.19). For this case of $k_{-1} \gg k_2$ the steady-state treatment has reduced to the equilibrium treatment.

10.2 ACID–BASE CATALYSIS

The study of catalysis by acids and bases played a very important part in the development of physical chemistry, particularly at the end of the 19th century and the beginning of the 20th century. Reference has been made already to the studies of Ostwald on catalysis, including acid–base catalysis, and these were important in the development of chemical kinetics. These and other studies also contributed much to the early theories of solutions of electrolytes.

It was realized by Ostwald[12] and Arrhenius[13] that the ability of an acid to catalyze certain reactions is independent of the nature of the anion but depends on the electrical conductivity of the solution. They concluded that the conductivity is a measure of the strength of the acid and of the concentration of the hydrogen ions, which they assumed to be the sole effective acid catalysts. Similarly, it was shown[14] that for catalysis by bases the rate depends on the conductivity but not on the nature of the cation, indicating that the catalysis is brought about by the hydroxide ions.

As will be seen, the idea that the sole catalyzing species in catalysis by acids and bases are the hydrogen and hydroxide ions has proved to require modification. It is convenient to discuss first the kinetic equations that apply when these ions are the only effective catalysts, as is sometimes the case.

If a reaction is carried out in water in a fairly acid solution (i.e., at low pH), the concentration of hydroxide ions may be reduced to such an extent that they do not have any appreciable catalytic action. The hydrogen ions then may be the only effective catalysts, and the rate expression may be of the form

$$v = k_0[S] + k_{H^+}[S][H^+] \tag{10.25}$$

Here k_0 is the rate constant for the uncatalyzed reaction; $[H^+]$ is the concentration of hydrogen ions§ and k_{H^+} is the rate constant for catalysis by hydrogen ions.

However, if the circumstances are such that there is also catalysis by hydroxide ions, the rate equation may be

$$v = k_0[S] + k_{H^+}[S][H^+] + k_{OH^-}[S][OH^-] \tag{10.26}$$

We can define a first-order rate constant for the reaction by

$$k \equiv \frac{v}{[S]} \tag{10.27}$$

and therefore

$$k = k_0 + k_{H^+}[H^+] + k_{OH^-}[OH^-] \tag{10.28}$$

The constants k_{H^+} and k_{OH^-} are known as the *catalytic constants* for the hydrogen and hydroxide ions, respectively.

For aqueous systems the concentrations of H^+ and OH^- are related approximately by

$$[H^+][OH^-] = K_w \tag{10.29}$$

and Eq. (10.28) can be written as

$$k = k_0 + k_{H^+}[H^+] + \frac{k_{OH^-}K_w}{[H^+]} \tag{10.30}$$

In many solutions one of these concentration terms is negligible compared with the other. For example, in 0.1 M hydrochloric acid the second term in Eq. (10.30) is $10^{-1}k_{H^+}$ mol dm^{-3}, while the third term is $10^{-13}k_{OH^-}$ mol dm^{-3} (since at 25°C $K_w \approx 10^{-14}$ mol^2 dm^{-6}). Consequently, unless k_{OH^-} is greater than k_{H^+} by many orders of magnitude, the third term is negligible compared with the second. At this acid concentration, catalysis by hydroxide ions is expected to be unimportant; similarly, in 0.1 M NaOH, catalysis by hydrogen ions is usually unimportant.

In general, there is an upper range of hydrogen-ion concentrations in which catalysis by hydroxide ions is unimportant, and within this range the rate is linear in $[H^+]$. Within this range the catalytic constant k_{H^+} can be determined easily. Similarly, in solutions of high pH, the catalysis by hydrogen ions is insignificant; the rate varies linearly with $[OH^-]$ and the catalytic constant k_{OH^-} can be determined. The hydrolysis of ethyl acetate was investigated in this manner by Wijs,[15] who obtained the k_{H^+} and k_{OH^-} values and also obtained a value for K_w. This he did by making use of the fact that the rate passes through a minimum when the second and third terms on the right-hand side of Eq. (10.30) are equal. This gives rise to

§ For simplicity the symbol H^+ is used here for the solvated hydrogen ion. In water the concentration of bare protons is negligible; the ion is present as H_3O^+ which itself is strongly hydrated. A similar situation exists in other ionizing solvents.

$$[H^+]_{min} = \left(\frac{k_{OH^-}K_w}{k_{H^+}}\right)^{1/2} \qquad (10.31)$$

so that from the measured values of $[H^+]_{min}$, k_{H^+}, and k_{OH^-} the ionic product K_w can be calculated.

Various possibilities that may arise in reactions of this type have been classified by Skrabal,[16] in terms of plots of the common logarithm of the first-order rate constant against the pH of the solution (see Fig. 10.3). The most general type of behavior is shown in curve I, which shows a region of catalysis by hydrogen ions, one of catalysis by hydroxide ions, and an intermediate region where the catalysis is unimportant. When the catalysis is largely by hydrogen ions (in the low-pH region), the rate constant k is essentially equal to $k_{H^+}[H^+]$ so that

$$\log_{10} k = \log_{10} k_{H^+} + \log_{10} [H^+] \qquad (10.32)$$

$$= \log_{10} k_{H^+} - pH \qquad (10.33)$$

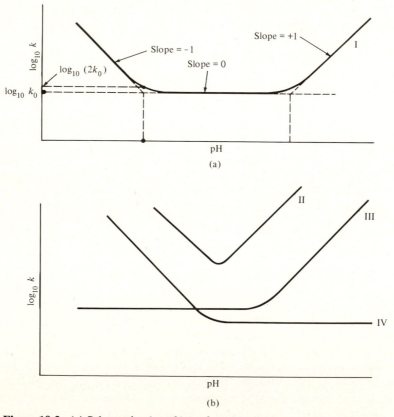

Figure 10.3 (a) Schematic plot of $\log_{10} k$ versus pH for a reaction catalyzed by both hydrogen and hydroxide ions, and for which the uncatalyzed reaction occurs at an appreciable rate. (b) Variants of the plot shown in (a).

The slope of the left-hand limb in curve I is therefore -1. The slope of the right-hand limb is similarly $+1$, since from Eq. (10.30)

$$\log_{10} k = \log_{10} k_{OH^-} + \log_{10} K_w - \log_{10} [H^+] \qquad (10.34)$$

$$= \log_{10} k_{OH^-} + \log_{10} K_w + pH \qquad (10.35)$$

The rate constant in the intermediate region is k_0, and the slope is zero.

The lines in curve I do not meet sharply; instead, there is a smooth transition. The equation for the left-hand limb in curve I is $\log_{10} k = \log_{10} k_{H^+}[H^+]$, while that for the horizontal line is $\log_{10} k = \log_{10} k_0$. If the two straight lines are continued to a sharp intersection point, as shown by the dashed lines in Fig. 10.3a, the point of intersection corresponds to $[H^+] = k_0/k_{H^+}$. At that point the observed rate constant is not k_0 but is

$$k = k_0 + k_{H^+}[H^+] = 2k_0 \qquad (10.36)$$

This is shown in Fig. 10.3a, and the $\log_{10} k$ value is therefore raised by $\log_{10} 2 = 0.301$.

If the rate of the spontaneous reaction is small [specifically, if $k_0 \ll (K_w k_{H^+} k_{OH})^{1/2}$], the horizontal portion of the curve is not found, the two limbs meeting as shown in curve II (Fig. 10.3b). If k_{H^+} is negligible, the left-hand limb is missing (curve III); if k_{OH^-} is negligible, the pattern is that of curve IV. Examples of all these cases are known.

10.2.1 General Acid–Base Catalysis

Early in this century a number of suggestions were made that acid and base catalysis can be brought about by species other than H^+ and OH^- ions. It was not until 1913, however, that the first reliable experimental evidence was presented; this was done by Dawson and Powis,[17] who made careful measurements on the reaction between acetone and iodine. Their results showed convincingly that with acetic acid–acetate buffers there is catalysis not only by H^+ and OH^- ions but also by undissociated acetic acid and acetate ions. Their rate constant for the process could be expressed as

$$k = k_0 + k_{H^+}[H^+] + k_{OH^-}[OH^-] + k_{HA}[HA] + k_A[A^-] \qquad (10.37)$$

Here HA represents acetic acid, A^- represents the acetate ion, and k_{HA} and k_A are the corresponding catalytic constants.

When catalysis is brought about by other species in addition to H^+ and OH^-, one speaks of *general acid–base catalysis*. The term *specific acid–base catalysis* refers to catalysis where species other than H^+ and OH^- have not been observed to have any significant effect. It will be seen later that certain catalytic mechanisms lead to specific catalysis and others to general catalysis. The theory that species other than H^+ and OH^- can be catalysts has been referred to as the *dual theory of catalysis* and as the *multiple theory of catalysis*.

The idea that general catalysis is possible is related to modern theories of acids and bases, which were proposed particularly by Brønsted and Lewis. The definition of acids and bases proposed by Brønsted[18] is that an acid has a tendency to give up a proton, whereas a base has a tendency to accept a proton. According to this definition the ammonium ion, for example, is an acid since it can undergo processes such as§

$$NH_4^+ + H_2O \rightleftharpoons NH_3 + H_3O^+$$

The anions of acids are bases since they accept protons in reactions such as

$$H_3O^+ + CH_3COO^- \rightarrow H_2O + CH_3COOH$$

In all processes of this type an acid reacts with a base to give a new acid and a new base:

$$Acid_1 + base_1 \rightleftharpoons base_2 + acid_2$$

The species $base_2$ is $acid_1$ from which a proton has been removed and is said to be the *conjugate base* to $acid_1$. Similarly, $acid_2$ is the *conjugate acid* to $base_1$.

Brønsted's definition is applicable particularly to aqueous solutions. Lewis's definition[19] of acids and bases is more general; according to him acids are substances that accept, and bases are substances that donate, electron pairs. This definition has the advantage of including as acids substances such as aluminum chloride which do not contain protons but which exhibit typical acidic properties. Similarly, BF_3 is regarded as an acid since it forms complexes such as $H_3N \rightarrow BF_3$.

In most mechanisms for acid–base catalysis there is a transfer of a proton to or from the substrate. It is not surprising, therefore, that species such as NH_4^+, which can donate protons, are capable of bringing about acid catalysis, or that species such as CH_3COO^- can act as base catalysts by accepting protons. In base catalysis, however, there is another mechanism: a base always possesses unshared electron pairs and therefore may act by attaching itself to an electrophilic center rather than by abstracting a proton. For example, in the hydrolysis of esters catalyzed by hydroxide ions the initial process may be a nucleophilic attack, to give the intermediate

$$
\begin{array}{c}
O^- \\
| \\
R-C-OR' \\
| \\
OH
\end{array}
$$

Catalysis occurring by this mechanism is known as *nucleophilic catalysis*. In individual cases it is often difficult to determine whether a base is acting by abstracting a proton or by making a nucleophilic attack on the substrate molecule.

10.2.2 Mechanisms of Acid–Base Catalysis

The present subsection deals with certain mechanisms of acid catalysis that appear to apply to a number of reactions. It must be emphasized that other types of mechanism are possible. The mechanisms to be considered involve, in the first step, proton transfers to or from the substrate molecule. One alternative type of mechanism, occurring in

§ Here it is more convenient to write the hydrogen ion as H_3O^+.

TABLE 10.1 NATURE OF REACTING SPECIES IN ACID–BASE CATALYSIS IN AQUEOUS SOLUTION

Type of catalysis	C	Y	W	Z
Acid	HA	A^-	H_2O	H_3O^+
Acid	H_2O	OH^-	B	BH^+
Base	B	BH^+	H_2O	OH^-
Base	H_2O	H_3O^+	HA	A^-

nucleophilic catalysis, has just been referred to. Another mechanism, the *concerted*§ *mechanism,* was suggested originally by Lowry[20] and also may be involved in some systems. According to this mechanism, the acid and base species arrive *simultaneously* at the substrate molecule, the former donating a proton and the latter removing one or making a nucleophilic attack. Space does not permit a discussion of the evidence with regard to this mechanism.[21]

Only acid catalysis is treated here; the rate equations for base catalysis can be obtained using the same methods, and the conclusions for both acid and base catalyses are summarized later in Table 10.2.

In Section 10.2 catalysis was discussed in terms of the general mechanism

$$(1) \quad C + S \ \rightleftharpoons X + Y$$

$$(2) \quad X + W \rightarrow P + Z$$

If there is acid catalysis, X can be written as SH^+, and Y is C minus a proton. In the second step, X (SH^+) donates a proton to a species W and gives the product or products P. The species W may be a solvent molecule, in which case the step has been referred to as *protolytic* (Greek *lysis,* meaning splitting). Alternatively, W may be a solute species, such as the conjugate base of the catalyst C; in that case the term *prototropic* (Greek *tropos,* meaning turning) has been used. These possibilities are summarized in Table 10.1 which applies to aqueous solutions and includes base catalysis also. The table shows four combinations. Associated with each are two extreme cases: the intermediate X may be in equilibrium with C + S ($k_2[W] \ll k_{-1}[Y]$) or there may be a steady state[22] ($k_{-1}[Y] \ll k_2[W]$). This scheme thus leads to eight cases; only the four corresponding to acid catalysis are treated here.

Consider first the mechanism in which in the second stage the proton is transferred to a water molecule. If the acidic catalyst is written as BH^+ the mechanism can be written as

$$(1) \quad BH^+ + S \ \rightleftharpoons SH^+ + B$$

$$(2) \quad SH^+ + H_2O \rightarrow P + H_3O^+$$

The acidic catalyst is in equilibrium with its conjugate base B, this equilibrium being established extremely rapidly:

---

§ The terms *ternary, synchronous,* and *push–pull* also have been used for this type of mechanism.

$$BH^+ + H_2O \rightleftharpoons B + H_3O^+$$

and therefore

$$\frac{[B][H^+]}{[BH^+]} = K \tag{10.38}$$

Two extreme cases are now of special interest, according to whether $k_2 \ll k_{-1}[B]$ or $k_{-1}[B] \ll k_2$.

Case 1 If $k_2 \ll k_{-1}[B]$ the first reaction is essentially at equilibrium, and therefore

$$\frac{[SH^+][B]}{[BH^+][S]} = \frac{k_1}{k_{-1}} \tag{10.39}$$

The concentrations $[BH^+]$ and $[S]$ that appear in Eqs. (10.38) and (10.39) are not necessarily the initial concentrations, since some of the substances have formed the intermediate SH^+. At the beginning of the reaction, before any substrate has been transformed into product,

$$[BH^+]_0 = [BH^+] + [SH^+] \tag{10.40}$$

and

$$[S]_0 = [S] + [SH^+] \tag{10.41}$$

These equations could be used to eliminate $[BH^+]$ and $[S]$ in Eqs. (10.38) and (10.39); then the solution for $[SH^+]$ would be a quadratic equation. The situation is simpler if either BH^+ or S is in excess of the other; the first possibility is considered now. Equation (10.41) must be used, but $[BH^+]$ can be equated to $[BH^+]_0$ since the amount of BH^+ that gives rise to SH^+ is negligible.

Eqs. (10.38) and (10.39) can be written as

$$\frac{[B][H^+]}{[BH^+]_0} = K \tag{10.42}$$

and

$$\frac{[SH^+][B]}{[BH^+]_0([S]_0 - [SH^+])} = \frac{k_1}{k_{-1}} \tag{10.43}$$

The latter equation gives

$$[SH^+] = \frac{k_1[BH^+]_0[S]_0}{k_1[BH^+]_0 + k_{-1}[B]} \tag{10.44}$$

Use of Eq. (10.42) leads to

$$[SH^+] = \frac{k_1[H^+][S]_0}{k_1[H^+] + k_{-1}K} \tag{10.45}$$

The rate, equal to $k_2[SH^+]$, is

$$v = \frac{k_1 k_2[H^+][S]_0}{k_1[H^+] + k_{-1}K} \tag{10.46}$$

Thus, although the mechanism is such that in the first step the proton can be transferred from any acidic species present, there is *specific hydrogen-ion catalysis:* catalysis by other species could not be detected. The mechanism also requires that the rate will not increase indefinitely as the hydrogen-ion concentration is increased but will attain a limiting value equal to $k_2[S]_0$.

Case 2 If $k_{-1}[B] \ll k_2$, the steady-state treatment must be applied. This is justifiable even if the catalyst is in excess, since the concentration of the intermediate SH^+ must be small in comparison with that of BH^+.

The steady-state equation for SH^+ is

$$k_1[BH^+][S] - k_{-1}[SH^+][B] - k_2[SH^+] = 0 \tag{10.47}$$

Replacement of $[BH^+]$ by $[BH^+]_0$ and of $[S]$ by $[S]_0 - [SH^+]$ then gives

$$k_1[BH^+]_0([S]_0 - [SH^+]) - k_{-1}[SH^+][B] - k_2[SH^+] = 0 \tag{10.48}$$

Thus,
$$[SH^+] = \frac{k_1[BH^+]_0[S]_0}{k_1[BH^+]_0 + k_{-1}[B] + k_2} \tag{10.49}$$

$$= \frac{k_1[BH^+]_0[S]_0}{k_1[BH^+]_0 + k_2} \tag{10.50}$$

Since the concentration $[B]$ no longer appears, Eq. (10.42) cannot be used to eliminate $[BH^+]_0$ as was done in Case 1. The rate is

$$v = \frac{k_1 k_2 [BH^+]_0 [S]_0}{k_1[BH^+]_0 + k_2} \tag{10.51}$$

The rate now depends on the concentration of any acid catalyst BH^+ that is present in the solution; in other words, there is *general acid catalysis*. There may be a limiting rate at high acid concentrations; however, it may be difficult to observe this experimentally, since the rate constant k_1 for the first step is usually small. If $k_2 \gg k_1[BH^+]_0$ at all acid concentrations used, Eq. (10.51) reduces to

$$v = k_1[BH^+]_0[S]_0 \tag{10.52}$$

which means that the rate is controlled entirely by the first step.

Consider now the mechanisms in which the intermediate transfers its proton to a basic species rather than to the solvent:

$$BH^+ + S \rightleftharpoons SH^+ + B$$

$$SH^+ + B \rightarrow P + BH^+$$

Two special cases are $k_2 \ll k_{-1}$, in which the first reaction is at equilibrium, and $k_{-1} \ll k_2$, in which case the steady-state treatment applies. The treatment again is given for the situation in which the catalyst is in excess.

Case 3 If the first reaction is at equilibrium, Eqs. (10.40)–(10.44) again apply. The rate, however, is now $k_2[SH^+][B]$; that is,

$$v = \frac{k_1 k_2 [BH^+]_0 [S]_0}{k_1[BH^+]_0/[B] + k_{-1}} \tag{10.53}$$

$$= \frac{k_1 k_2 [BH^+]_0 [S]_0}{k_1 [H^+]/K + k_{-1}} \tag{10.54}$$

This equation shows that there is *general acid catalysis*. Also, if the concentration of an acidic species BH^+ is increased with the pH held constant, the rate increases indefinitely and does not level off at a limiting value.

Case 4 If $k_{-1} \ll k_2$, the steady-state treatment is applicable. The equation is

$$k_1 [BH^+][S] - (k_{-1} + k_2)[SH^+][B] = 0 \tag{10.55}$$

or

$$k_1 [BH^+]_0 ([S]_0 - [SH^+]) - (k_{-1} + k_2)[SH^+][B] = 0 \tag{10.56}$$

Therefore, with k_{-1} neglected in comparison with k_2,

$$[SH^+] = \frac{k_1 [BH^+]_0 [S]_0}{k_1 [BH^+]_0 + k_2 [B]} \tag{10.57}$$

and the rate is

$$v = \frac{k_1 k_2 [BH^+]_0 [S]_0}{k_1 [BH^+]_0 / [B] + k_2} \tag{10.58}$$

$$= \frac{k_1 k_2 [BH^+]_0 [S]_0}{k_1 [H^+]/K + k_2} \tag{10.59}$$

As in Case 3, there is general acid catalysis and no leveling off of the rate if $[BH^+]$ is increased at constant pH.

These conclusions are summarized in Table 10.2, which also includes the conclusions for base-catalyzed reactions. Intermediate situations are possible, and the resulting equations are correspondingly more complicated. Note in particular that with acid catalysis a limiting rate with increasing concentration of H^+ is found only in Case 1, in which the proton is transferred to the solvent and $k_2 \ll k_{-1}[B]$, which means that the first reaction is at equilibrium. For base catalysis, specific OH^- ion catalysis and a limiting rate are found only if there is equilibrium ($k_2 \ll k_{-1}[B]$) and the proton is transferred to the intermediate from the solvent. This behavior has been observed with the base-catalyzed hydrolysis of $C_6H_5NHCOCH_3$,[23] and it is concluded therefore that this mechanism applies to it. The corresponding behavior for an acid-catalyzed reaction has been found with the hydrolysis of amides.[24]

A well investigated reaction[25] that is catalyzed by both acids and bases is the reaction between acetone and iodine in aqueous solution:

$$CH_3COCH_3 + I_2 \rightarrow CH_3COCH_2I + HI$$

There is general acid–base catalysis and no limiting rate has been found. The rate is linear in the acetone concentration and that of any acid or basic species present in solution, but the rate is independent of the concentration of iodine. Indeed, if the iodine is replaced by bromine, the corresponding bromination reaction proceeds at the same rate. This suggests that the iodine or bromine is involved in a rapid step that has no effect on the overall reaction rate. The evidence is that in acid catalysis the

TABLE 10.2 SUMMARY OF CONCLUSIONS FOR VARIOUS MECHANISMS OF ACID–BASE CATALYSIS

Acid catalysis	Arrhenius intermediates	van't Hoff intermediates
Proton transferred to solvent: $BH^+ + S \rightleftharpoons SH^+ + B$ $SH^+ + H_2O \rightarrow P + H_3O^+$	$k_2 \ll k_{-1}[B]$ Specific H^+ catalysis; limiting rate	$k_2 \gg k_{-1}[B]$ and $k_2 \gg k_1[BH^+]$ General catalysis; no limiting rate
Proton transferred to solute: $BH^+ + S \rightleftharpoons SH^+ + B$ $SH^+ + B \rightarrow P + BH^+$	$k_2 \ll k_{-1}$ General catalysis; no limiting rate	$k_2 \gg k_{-1}$ General catalysis; no limiting rate

Base catalysis	Arrhenius intermediates	van't Hoff intermediates
Proton transferred from solvent: $B + SH \rightleftharpoons S^- + BH^+$ $S^- + H_2O \rightarrow P + OH^-$	$k_2 \ll k_{-1}[BH^+]$ Specific OH^- catalysis; limiting rate	$k_2 \gg k_{-1}[BH^+]$ and $k_2 \gg k_1[B]$ General catalysis; no limiting rate
Proton transferred from solute: $B + SH \rightleftharpoons S^- + BH^+$ $S^- + BH^+ \rightarrow P + B$	$k_2 \ll k_{-1}$ General catalysis; no limiting rate	$k_2 \gg k_{-1}$ General catalysis; no limiting rate

rate-determining step is the conversion of the ordinary keto form of acetone into its enol form:

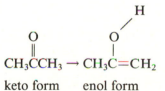

keto form   enol form

The enol form is then rapidly iodinated or brominated:

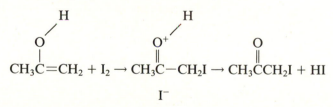

The way in which acids catalyze the conversion of the keto form into the enol form is believed to be as follows. First, the acidic species HA transfers a proton to the oxygen atom on the acetone molecule in a pre-equilibrium:

$$HA + CH_3-\underset{\overset{\|}{O}}{C}-CH_3 \rightleftharpoons CH_3-\underset{\overset{\|}{{}^+O}}{C}-CH_3 + A^-$$

The transferred proton is bound to the oxygen atom by one of the oxygen's lone pairs of electrons. The protonated acetone then gives up one of its other hydrogen atoms to some base B present (which may be water), at the same time forming the enol form of acetone;

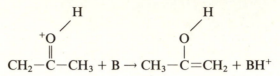

There is evidence[26] that the rate-determining step in the reaction is this latter process.

In basic solution the first step involves the transfer of a proton from the acetone to a basic species:

$$CH_3COCH_3 + B \rightleftharpoons \left\{ \begin{matrix} CH_3COCH_2^- \\ \updownarrow \\ CH_3C{=}CH_2 \\ | \\ O^- \end{matrix} \right\} + BH^+$$

The negative ion formed exists in a resonance state involving the two structures shown and can react rapidly with the halogen:

$$\underset{\underset{O^-}{|}}{CH_3C{=}CH_2} + I_2 \rightarrow \underset{\underset{O}{\|}}{CH_3C{-}CH_2I} + I^-$$

Note that the halogen attacks the anion and that the enol itself need not be formed. The halogen reacts more rapidly with the anion than with the neutral molecule.[27]

10.2.3 Catalytic Activity and Acid–Base Strength

Since catalysis by acids and bases usually involves the transfer of a proton from or to the catalyst, it is natural to seek a correlation between the effectiveness of a catalyst and its strength as an acid or base. The most satisfactory relationship was that suggested by Brønsted and Pedersen,[28] according to whom the catalytic constant k_a of an acid is related to the acid dissociation constant K_a by

$$k_a = G_a K_a^\alpha \tag{10.60}$$

where G_a and α are constants, the latter always being less than unity. Similarly, for a base catalyst it was suggested that

$$k_b = G_b K_b^\beta \tag{10.61}$$

$$= G_b' \left(\frac{1}{K_a} \right)^\beta \tag{10.62}$$

where K_b is the dissociation constant for the base and K_a that for its conjugate acid; β is a constant that is again less than unity.

The equations require modification if they are to be applied to an acid that has more than one ionizable proton or to a base that can accept more than one proton. Thus, if a long-chain fatty acid, $CH_3(CH_2)_nCOOH$, is compared with the dibasic acid $HOOCCH_2(CH_2)_nCOOH$, and if there is negligible interaction between the two carboxyl groups, the dissociation constant of the dicarboxylic acid is twice that of the mono-carboxylic acid, since in the former molecule the ion can be formed by the loss of either of the two protons. For the same reason the catalytic constant for the dicarboxylic acid is twice that of the monocarboxylic acid. These two facts, however, are inconsistent with the Brønsted relationship (10.60) as it stands, since a ratio of 2 in the acid strengths leads to a ratio of 2^α in the catalytic constants, and this is generally less than 2. This anomaly is avoided if both the acid strength and the catalytic constant are divided by the number of protons involved in the dissociation; in the example given, the modified catalytic constants of the two acids are equal, as are the modified acid strengths, and the relationship (10.60) is therefore obeyed. This, however, is still not completely satisfactory. Thus, in the case of the two acids $HOOC(CH_2)_nCOOR$ and $HOOC(CH_2)_nCOO^-$, both of which have one dissociable proton, the catalytic strengths are the same; on the other hand, the acid strength of the second is only one-half that of the first since the ion $^-OOC(CH_2)_nCOO^-$ into which the second dissociates has two points at which a proton may be added, whereas the ion of the first acid has only one. These results are again inconsistent with Eq. (10.60), and to remove the difficulty it is necessary to multiply the dissociation constant of $HOOC(CH_2)_nCOO^-$ by 2 before inserting it into the equation.

The conclusions may be generalized by means of the relationships[29]

$$\frac{k_a}{p} = G_a \left(\frac{qK_a}{p} \right)^\alpha \qquad (10.63)$$

and

$$\frac{k_b}{q} = G'_b \left(\frac{p}{qK_a} \right)^\beta \qquad (10.64)$$

In Eq. (10.63) p is the number of dissociable protons bound equally strongly in the acid, while q is the number of equivalent positions in the conjugate base to which a proton may be attached. Similarly, in Eq. (10.64) q is the number of positions in the catalyzing base to which a proton may be attached, while p is the number of equivalent dissociable protons in the conjugate acid. The ratio q/p in Eq. (10.63) can be identified with the ratio σ_S/σ_{SH^+} of symmetry numbers,[30] where SH^+ is the catalyzing acid, and S is its conjugate base. The statistical factor p is the same as the factors used in Section 4.5.4.

The Brønsted relationships are special cases of the linear Gibbs energy relationships that were discussed in Section 6.6. Consider a reaction catalyzed by a series of homologous acids. The Hammett equation

$$\log k = \log k_0 + \delta\rho \qquad (10.65)$$

may be expected to apply therefore, and the corresponding equation for the dissociation is

$$\log K = \log K_0 + \delta\rho' \qquad (10.66)$$

The substituent constants δ are the same in both equations but the reaction constants ρ and ρ' are different. Equations (10.65) and (10.66) can be written as

$$\frac{1}{\rho} \log k = \frac{1}{\rho} \log k_0 + \delta \tag{10.67}$$

and

$$\frac{1}{\rho'} \log K = \frac{1}{\rho'} \log K_0 + \delta \tag{10.68}$$

Subtraction leads to

$$\frac{1}{\rho} \log k - \frac{1}{\rho'} \log K = \text{const} \tag{10.69}$$

and therefore

$$\log \frac{k^{1/\rho}}{K^{1/\rho'}} = \text{const} \tag{10.70}$$

This may be written as

$$\log \frac{k}{K^{\rho/\rho'}} = \text{const} \tag{10.71}$$

or

$$k = GK^{\alpha} \tag{10.72}$$

where G and α are constants, the latter being equal to the ratio ρ/ρ'.

On the basis of this argument the Brønsted relationships may be expected to apply to catalysis by a series of homologous acids or bases. They also apply, although not so exactly, to acids and bases that do not belong to homologous series.

10.2.4 Salt Effects in Acid–Base Catalysis

The influence of neutral salts on the rates of reactions in solution was discussed in Section 6.3.4. This matter is of particular importance in connection with acid–base catalysis, since in addition to its possible catalytic action a salt may exert an influence by changing the activity coefficients of other species in solution. For this reason much of the early evidence regarding catalysis by species other than hydrogen and hydroxide ions must be discounted, the apparent catalytic action being in reality a salt effect.

In the study of acid–base catalysis it is convenient to classify salt effects as primary and secondary. The *primary effects* arise from the fact that the rate of a bimolecular interaction in solution between A and B is given by Eq. (6.39):

$$v = k_0[\text{A}][\text{B}] \frac{y_\text{A} y_\text{B}}{y_\ddagger} \tag{10.73}$$

where the y's are the activity coefficients. Since salts influence the y's, they affect the rate v and the rate constant, equal to $k_0 y_\text{A} y_\text{B}/y_\ddagger$, as discussed in Section 6.3.4.

The *secondary salt effects* are not concerned with the direct influence of a salt on the rate of reaction but with its influence on the concentrations of the reactants. The equilibrium constant for the dissociation of a weak acid HA can be written as

$$K_a = \frac{[\text{H}^+][\text{A}^-]}{[\text{HA}]} \frac{y_{\text{H}^+} y_{\text{A}^-}}{y_{\text{HA}}} \tag{10.74}$$

where K_a is a true constant at a given temperature and in a given solvent. Since salts influence the activity coefficients, they also influence the concentrations of H^+, A^-, and HA, and in this way indirectly influence the rate of a reaction involving one of these species. The activity coefficient y_{HA} for a neutral molecule is only slightly influenced by the ionic strength, while at low ionic strengths both y_{H^+} and y_{A^-} are decreased by increasing the ionic strength according to the Debye–Hückel equation (6.40):

$$-\log_{10} y_i = B z_i^2 I^{1/2} \tag{10.75}$$

The net result is therefore an increase in the degree of dissociation of HA, that is, an increase in $[H^+]$ and $[A^-]$ and a decrease in $[HA]$. Of course, salt effects of this kind are produced not only by foreign salts but also by salts that are involved directly in the reaction.

The fundamental work on both primary and secondary salt effects in chemical kinetics was done largely by Brønsted and his coworkers.[31] Positive secondary salt effects have been observed in acid-catalyzed reactions involving weak uncharged acids, since salts, as has already been described, increase the degree of dissociation and hence the hydrogen-ion concentration. The effect is similar to a purely catalytic action, with which it is apt to be confused. The same type of result is found in hydroxide-ion catalysis in the presence of weak uncharged bases, the concentration of hydroxide ions being increased by the addition of the salt. Different results are found, however, if the acid or base is charged. With strong acids and bases the secondary salt effects are small, since the species are dissociated almost completely under all conditions.

10.2.5 Acidity Functions

As has been discussed, particularly with reference to Fig. 10.3, there is often a relationship between a rate constant and the pH, which is an acidity function of the simplest kind. In concentrated acid solutions these relationships break down, and attempts have been made to obtain improvements by defining certain other acidity functions. Each of these is related to a type of equilibrium, and a correlation between a rate constant and an acidity function has been taken as evidence for a particular reaction mechanism. The acidity functions usually become identical to the pH as the solution is made more dilute.

The first acidity function proposed, other than the pH, is due to Hammett and Deyrup[32] and is based on equilibria of the type

$$(1) \quad B + H^+ \rightleftharpoons BH^+$$

An example of such an equilibrium is found in solutions of anilines,

$$(2) \quad C_6H_5NH_2 + H^+ \rightleftharpoons C_6H_5NH_3^+$$

The equilibrium constant for such reactions is

$$K = \frac{[BH^+]}{[B][H^+]} \frac{y_{BH^+}}{y_B y_{H^+}} \tag{10.76}$$

where the y's are the activity coefficients. If BH^+ and B are species that can be distinguished spectrophotometrically, it is possible to measure $[BH^+]/[B]$, and K also can be measured in more dilute solutions. Equation (10.76) may be written

$$\log_{10} K - \log_{10} \frac{[\text{BH}^+]}{[\text{B}]} = -\log_{10}\left(a_{\text{H}^+} \frac{y_\text{B}}{y_{\text{BH}^+}}\right) \tag{10.77}$$

where a_{H^+}, equal to $[\text{H}^+]y_{\text{H}^+}$, is the activity of the hydrogen ion. In Eq. (10.77) the quantities on the left-hand side are experimentally observable; the function on the right-hand side therefore can be measured in any acidic solution by introducing a suitable indicator and measuring the concentrations of the two species. The quantity on the right-hand side of Eq. (10.77) is defined as the *acidity function* H_0:

$$H_0 \equiv -\log_{10}\left(a_{\text{H}^+} \frac{y_\text{B}}{y_{\text{BH}^+}}\right) \tag{10.78}$$

The usefulness of the function is that there may be a correlation with rate constants for processes occurring by a mechanism in which there is a pre-equilibrium of the same type as reaction (1) above. Suppose, for example, that a reaction occurs by the mechanism

$$S + \text{H}^+ \rightleftharpoons \text{SH}^+ \qquad \text{(rapid)}$$

$$\text{SH}^+ \rightarrow \text{products} \quad \text{(slow)}$$

If the second step is slow and rate-controlling, the overall rate is proportional to the concentration of activated complexes, $(\text{SH}^+)^\ddagger$, corresponding to this step:

$$v = k^\ddagger[(\text{SH}^+)^\ddagger] \tag{10.79}$$

These activated complexes are in equilibrium with the species SH^+,

$$\frac{a_\ddagger}{a_{\text{SH}^+}} = \frac{[(\text{SH}^+)^\ddagger]}{[\text{SH}^+]} \frac{y_\ddagger}{y_{\text{SH}^+}} = K_\text{S}^\ddagger \tag{10.80}$$

where $a_\ddagger$ and $y_\ddagger$ are the activity and activity coefficient, respectively, of the activated complex. The rate therefore is given by

$$v = k^\ddagger K_\text{S}^\ddagger[\text{SH}^+] \frac{y_{\text{SH}^+}}{y_\ddagger} \tag{10.81}$$

For the pre-equilibrium,

$$\frac{a_{\text{SH}^+}}{a_\text{S} a_{\text{H}^+}} = \frac{[\text{SH}^+]}{[\text{S}]} \frac{y_{\text{SH}^+}}{y_\text{S} a_{\text{H}^+}} = K_\text{S} \tag{10.82}$$

o that the rate is

$$v = k^\ddagger K_\text{S}^\ddagger K_\text{S} \frac{y_\text{S}}{y_\ddagger} a_{\text{H}^+}[\text{S}] \tag{10.83}$$

The first-order rate constant, defined by $v/[\text{S}]$, is therefore

$$k = k^\ddagger K_\text{S}^\ddagger K_\text{S} \frac{y_\text{S}}{y_\ddagger} a_{\text{H}^+} \tag{10.84}$$

and thus

$$\log k = \log(k^\ddagger K_\text{S}^\ddagger K_\text{S}) + \log\left(a_{\text{H}^+} \frac{y_\text{S}}{y_\ddagger}\right) \tag{10.85}$$

Comparison of Eqs. (10.78) and (10.85) shows that the acidity function and the rate constant contain similar ratios, namely,

$$a_{H^+} \frac{y_B}{y_{BH^+}} \quad \text{and} \quad a_{H^+} \frac{y_S}{y_{\ddagger}}$$

The ratios y_B/y_{BH^+} and $y_S/y_{\ddagger}$ may vary in a manner somewhat similar to each other, since in both cases the species in the denominator contains one more proton than that in the numerator. Therefore, a correlation between $\log_{10} k$ and H_0 might be expected; a plot of one against the other might give a line of slope equal to -1. Such relationships have been observed in a number of instances in which there is independent evidence for this type of kinetic mechanism.

An alternative type of kinetic mechanism that undoubtedly occurs in some cases is

$$S + H^+ \rightleftharpoons SH^+ \quad \text{(rapid)}$$

$$SH^+ + H_2O \rightarrow \text{products} \quad \text{(slow)}$$

This mechanism differs from the previous one in that the activated complex in the slow process consists of the substrate S to which a proton and a water molecule have been added; in the previous case the activated complex consisted simply of the substrate molecule plus a proton. Application of the same treatment to this second case leads to

$$\log k = \log (k^{\ddagger} K_S^{\ddagger} K_S) + \log \left(a_{H^+} a_{H_2O} \frac{y_S}{y_{\ddagger}} \right) \tag{10.86}$$

This differs from Eq. (10.85) in containing the additional factor a_{H_2O}.

It was suggested rather tentatively by Hammett and Zucker[33] that if a reaction occurs by this type of mechanism, there might be a correlation between the logarithm of the rate constant and the pH rather than H_0. The suggestion was that it might be possible to decide between the two mechanisms by seeing whether the logarithm of the rate constant showed a better correlation with pH or with H_0. Unfortunately, this "Hammett–Zucker hypothesis" did not prove reliable.

A number of other acidity functions have been suggested, and much work has been done to test their reliability for distinguishing between reaction mechanisms. Conclusions about mechanisms derived from such functions are at the best very tentative, and there is still much confusion about the theory of reaction rates in concentrated solutions of acids and bases. Some books and articles dealing with this topic are listed in the bibliography at the end of this chapter.

10.3 CATALYSIS BY ENZYMES

Catalysis by enzymes, the biological catalysts, is much more specific than that by acids and bases. Some enzymes show absolute specificity; an example is urease, which only catalyzes the hydrolysis of urea,

$$CO(NH_2)_2 + H_2O \rightarrow CO_2 + 2NH_3$$

A lower degree of specificity is shown by such enzymes as the proteolytic enzymes, which catalyze the hydrolysis of the peptide linkage provided that certain structural conditions are satisfied in the neighborhood of the linkage; this is known as *group specificity*. Many enzymes exhibit *stereochemical specificity,* in that they catalyze the reactions of one stereochemical form and not the other; the proteolytic enzymes, for example, only catalyze the hydrolysis of peptides made up from amino acids in the *L* configuration.

The enzymes are proteins, but they may be associated with nonprotein substances (known as coenzymes or prosthetic groups) that are essential to the action of the enzyme. Some enzymes are catalytically inactive in the absence of certain metal ions. For a number of enzymes the evidence is that the catalytic activity is due to a relatively small region of the protein molecule; this region is usually referred to as the *active center.*

The action of enzymes shows some resemblance to the catalytic action of acids and bases but is more complicated. The present treatment of enzyme kinetics is confined to the influence of concentration, pH, and temperature and to some brief comments about enzyme mechanisms.

10.3.1 Influence of Substrate Concentration

The simplest case is that of an enzyme-catalyzed reaction where there is a single substrate (aside from water); an example is the hydrolysis of an ester. The dependence on substrate concentration in such cases is frequently as shown in Fig. 10.4. The rate varies linearly with the substrate concentration at low concentrations (first-order kinetics) and becomes independent of substrate concentration (zero-order kinetics) at high concentrations. This type of behavior, which is reminiscent of that found with unimolecular surface reactions (Section 7.5), was first explained by Michaelis and Menten[34] in terms of the mechanism

$$E + S \underset{k_{-1}}{\overset{k_1}{\rightleftharpoons}} ES$$

$$ES \overset{k_2}{\rightarrow} E + P$$

Here E and S are the enzyme and substrate, P is the product, and ES is an addition complex. The steady-state treatment was first applied to this mechanism by Briggs and Haldane.[35] The steady-state equation is

$$k_1[E][S] - k_{-1}[ES] - k_2[ES] = 0 \tag{10.87}$$

In studies of enzyme reactions the molar concentration of substrate is usually very much greater than that of the enzyme; only a small proportion of the substrate therefore is bound to the enzyme. The total concentration of enzyme, $[E]_0$, is equal to the concentration of free enzyme, $[E]$, plus the concentration of complex, $[ES]$:

$$[E]_0 = [E] + [ES] \tag{10.88}$$

Elimination of $[E]$ between these two equations gives

$$k_1([E]_0 - [ES])[S] - (k_{-1} + k_2)[ES] = 0 \tag{10.89}$$

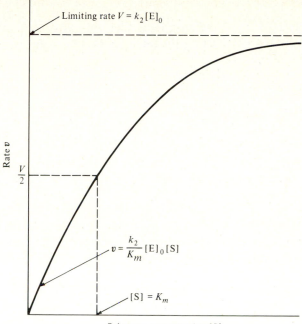

Figure 10.4 Variation of rate with substrate concentration for an enzyme-catalyzed reaction obeying the Michaelis–Menten equation.

and therefore

$$[ES] = \frac{k_1[E]_0[S]}{k_{-1} + k_2 + k_1[S]} \qquad (10.90)$$

The rate of reaction is

$$v = k_2[ES] = \frac{k_1k_2[E]_0[S]}{k_{-1} + k_2 + k_1[S]} \qquad (10.91)$$

$$= \frac{k_2[E]_0[S]}{(k_{-1} + k_2)/k_1 + [S]} \qquad (10.92)$$

$$= \frac{k_2[E]_0[S]}{K_m + [S]} \qquad (10.93)$$

In this equation K_m, equal to $(k_{-1} + k_2)/k_1$, is known as the *Michaelis constant,* and an equation of the form of Eq. (10.93) is referred to as a *Michaelis–Menten equation.*

When [S] is sufficiently small, it may be neglected in the denominator in comparison with K_m,

$$v = \frac{k_2}{K_m} [E]_0[S] \qquad (10.94)$$

so that the kinetics are first order in substrate concentration. When, on the other hand, $[S] \gg K_m$,

$$v = k_2[\text{E}]_0 \tag{10.95}$$

and the kinetics are zero order; the enzyme is then saturated with substrate, and a further increase in [S] has no effect on the rate. Thus, Eq. (10.93) is consistent with the behavior shown in Fig. 10.4.

Equation (10.95) can be rewritten as

$$v = \frac{V[\text{S}]}{K_m + [\text{S}]} \tag{10.96}$$

where V, equal to $k_2[\text{E}]_0$, is the limiting rate at high substrate concentrations; it is often known as the *maximal velocity*. When [S] is equal to K_m, Eq. (10.96) becomes

$$v = \frac{V[\text{S}]}{[\text{S}] + [\text{S}]} = \frac{V}{2} \tag{10.97}$$

This relationship is illustrated in Fig. 10.4. The Michaelis constant K_m can be determined from a plot of v versus [S], by finding the concentration of substrate that gives one-half of the limiting rate. In practice, however, this procedure does not provide a very reliable value.

To see whether experimental data are consistent with Eq. (10.96), the equation can be recast into a form that gives a linear plot. If reciprocals of Eq. (10.96) are taken the result is

$$\frac{1}{v} = \frac{K_m}{V[\text{S}]} + \frac{1}{V} \tag{10.98}$$

Therefore, a plot of $1/v$ versus $1/[\text{S}]$ gives a straight line if Eq. (10.96) applies. This type of plot, suggested by Lineweaver and Burk,[36] is shown schematically in Fig. 10.5a, which gives the intercepts and the slope. The parameters V and K_m can be derived from such a plot. If the enzyme concentration $[\text{E}]_0$ is known, k_2 also can be calculated, since according to this simple mechanism $V = k_2[\text{E}]_0$. However, the individual constants k_1 and k_{-1} cannot be obtained from studies of rate as a function of substrate concentration; the way they can be obtained from transient-phase studies is considered later.

Alternatively, the Michaelis–Menten equation (10.96) can be put into the form

$$vK_m + v[\text{S}] = V[\text{S}] \tag{10.99}$$

or

$$\frac{v}{[\text{S}]} K_m + v = V \tag{10.100}$$

A plot can be made of $v/[\text{S}]$ versus v; Fig. 10.5b shows the slopes and intercepts. A plot of this kind, first suggested by Eadie,[37] has the advantage that it tends to spread out the points to a greater extent than the Lineweaver–Burk plot.

Many enzyme-catalyzed reactions obey the Michaelis–Menten equation. It is important to note, however, that adherence to the empirical equation does not establish the simple mechanism, since more complicated mechanisms can give exactly the same behavior. An example is the mechanism

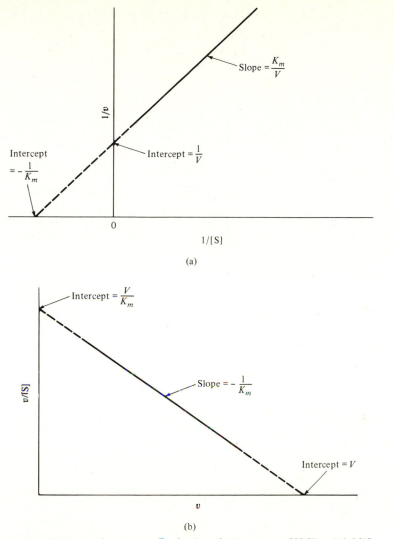

Figure 10.5 (a) Lineweaver–Burk plot of $1/v$ versus $1/[S]$ [Eq. (10.98)]. (b) Eadie plot of $v/[S]$ versus v [Eq. (10.100)].

$$E + S \underset{k_{-1}}{\overset{k_1}{\rightleftharpoons}} ES \overset{k_2}{\rightarrow} ES' \overset{k_3}{\rightarrow} E + Z$$
$$\searrow$$
$$Y$$

for which there is considerable evidence with a number of enzyme systems. Here there are two products of reaction; the first one, Y, is formed by the breakdown of the Michaelis complex ES, and this breakdown also gives rise to a second intermediate

ES′. This second intermediate breaks down in a subsequent stage into enzyme plus the second product Z.

Application of the steady-state treatment to this mechanism leads to the result that the rates of formation of both Y and Z are given by

$$v = \frac{\dfrac{k_2 k_3}{k_2 + k_3} [E]_0 [S]}{\left(\dfrac{k_{-1} + k_2}{k_1}\right)\left(\dfrac{k_3}{k_2 + k_3}\right) + [S]} \tag{10.101}$$

This is of the same form as Eq. (10.96) with

$$V = \frac{k_2 k_3}{k_2 + k_3} [E]_0 \tag{10.102}$$

and

$$K_m = \left(\frac{k_{-1} + k_2}{k_1}\right)\left(\frac{k_3}{k_2 + k_3}\right) \tag{10.103}$$

In the event that k_3 is much greater than k_2, Eq. (10.101) reduces to Eq. (10.92). In this case the intermediate ES′ exists only at very low concentrations and can be neglected so that the mechanism reduces to the simple one-intermediate mechanism.

Because the same type of behavior can arise from a number of different mechanisms, it is customary to write the rate equation as

$$v = \frac{k_c [E]_0 [S]}{K_m + [S]} \tag{10.104}$$

where k_c (also often written k_{cat}) is known as the *catalytic constant*. This equation applies to mechanisms having any number of sequential intermediates ES → ES′ → ES″ → and so on.

When reaction occurs between two solute species, the steady-state equations take a more complicated form than that for the single-substrate reactions. Only three of the possible mechanisms, and the corresponding rate equations, will be considered here.

One mechanism that may apply to a two-substrate reaction is the *random ternary-complex mechanism*. In this mechanism, represented in Fig. 10.6a, the enzyme E forms binary complexes EA and EB with the two substrates A and B; it also forms the ternary complex EAB, with no restriction on the order in which A and B are attached. The steady-state treatment of this mechanism leads to an extremely complicated expression.[38] Considerable simplification results if the assumption is made that the breakdown of EAB is sufficiently slow that EA, EB, and EAB can be regarded as at equilibrium. The rate equation then reduces to the form

$$v = \frac{V[A][B]}{K_{1A} K_{mB} + K_{mB}[A] + K_{mA}[B] + [A][B]} \tag{10.105}$$

where K_{1A}, K_{mA}, and K_{mB} are constants.

If the concentration of B is held constant, the variation of v with [A] is of the Michaelis–Menten form. This may be shown for the special case in which there is an excess of B. Thus, if [B] is sufficiently large the first two terms in the denominator of Eq. (10.105) may be neglected, and the result is

$$v = \frac{V[A]}{K_{mA} + [A]}$$

(10.106)

which is of the Michaelis–Menten form; K_{mA} is thus the Michaelis constant for A in the presence of excess of B. Similarly, if there is excess of A,

$$v = \frac{V[B]}{K_{mB} + [B]}$$

(10.107)

Another mechanism that may apply to a two-substrate reaction is the *ordered ternary-complex mechanism,* shown in Fig. 10.6b. For example, the complex EAB can be formed from EA, by addition of B, but not from EB by addition of A; the substrates must become attached in a particular order. This mechanism also leads to a rate equation of the form of (10.105), but the significance of the constants is different.

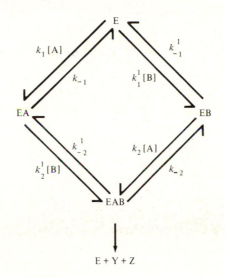

(a)

(b) $E \underset{k_{-1}}{\overset{k_1[A]}{\rightleftharpoons}} EA \underset{k_{-2}}{\overset{k_2[B]}{\rightleftharpoons}} EAB \overset{k_3}{\longrightarrow} E + Y + Z$

(c) $E \underset{k_{-1}}{\overset{k_1[A]}{\rightleftharpoons}} EA \overset{k_2}{\longrightarrow} EA' \overset{k_3[B]}{\longrightarrow} EA'B \overset{k_4}{\longrightarrow} E + Z$

Figure 10.6 Three possible mechanisms for an enzyme-catalyzed reaction involving two substrates: (a) random ternary-complex mechanism; (b) ordered ternary-complex mechanism; and (c) ping-pong bi-bi mechanism.

A third possible mechanism for reaction between two substrates is shown in Fig. 10.6c. In this mechanism one substrate, A, first adds on to the enzyme to form a complex EA, which then eliminates the first product Y *before* there is any reaction with B; B then reacts with the complex EA′ to form the second product Z. Experimentally, this mechanism sometimes can be distinguished from the others by observing the production of Y from A in the complete absence of the second substrate B. This mechanism is called the *ping-pong bi-bi* mechanism, the first "bi" indicating that there are two reactants for the reaction from left to right, the second "bi" that there are two products.

Application of the steady-state treatment to this mechanism leads to an equation of the form

$$v = \frac{V[A][B]}{K_{mB}[A] + K_{mA}[B] + [A][B]} \tag{10.108}$$

This is similar to Eq. (10.105) but the first term in the denominator is missing.

Discrimination between these and other possible mechanisms for two-substrate enzyme-catalyzed reactions is a matter of some difficulty but has been accomplished for a number of systems. Details can be found in references cited in the bibliography at the end of this chapter.

10.3.2 Influence of pH

The pH of the solution usually has a very marked effect on the rate of an enzyme reaction. In most cases the rates of enzyme reactions pass through a maximum as the pH is varied, as shown in Fig. 10.7a. The pH corresponding to the maximum rate is known as the *optimum pH;* its value varies with the nature of the substrate and with the substrate concentration.

Effects of pH are irreversible if the acidity or basicity becomes too high, since the tertiary structure of the protein is destroyed. Reversible pH changes occur when the pH is not taken too far from the pH optimum; within a certain pH range the pH can be changed back and forth without any permanent effects ensuing. This behavior was first explained by Michaelis and his coworkers[39] in terms of the ionizations of groups on the protein. It is necessary to postulate at least two ionizing groups as playing an important role at the active center. If these groups are $-NH_3^+$ and $-COOH$ the ionizations at the active center may be represented as

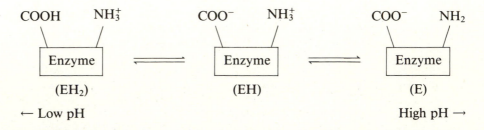

The pH behavior can be explained by postulating that the intermediate, zwitterion, form is enzymically active, but that the species to the left and right are inactive. The

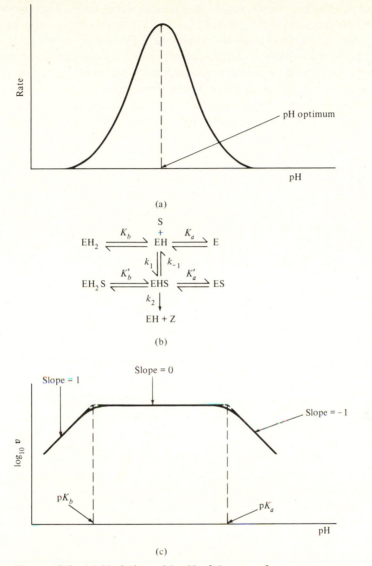

Figure 10.7 (a) Variation with pH of the rate of an enzyme-catalyzed reaction. Contrast Fig. 10.3 for a reaction catalyzed by acids and bases. (b) Simple reaction scheme which interprets pH dependence in some cases. (c) Schematic plot of $\log_{10} v$ versus pH for a system to which Eq. (10.110) applies.

concentration of the intermediate form goes through a maximum as the pH is varied, so that the rate passes through a maximum.

The detailed treatment of pH effects is quite complicated, because various possibilities have to be taken into account. A simple situation is when the reaction follows the original Michaelis–Menten mechanism in which the enzyme and substrate form an addition complex, which breaks down in a single stage. The enzyme–substrate complex also may exist in three states of ionization, and perhaps only the intermediate form is capable of giving rise to products. This simple case[40] is represented in Fig. 10.7b. At low substrate concentrations, when the enzyme exists mainly in the free form, the pH behavior is controlled by the ionization of the free enzyme. Analysis of the experimental pH dependence at low substrate concentrations therefore allows a determination of the acid dissociation constants K_a and K_b for the free enzyme. If, on the other hand, the enzyme is saturated with substrate, analysis of the pH behavior gives the values of K'_a and K'_b, which relate to the ionization of the enzyme-substrate complex.

Application of the steady-state treatment to the mechanism shown in Fig. 10.7b leads to the equation[41]

$$v = \frac{k_2[E]_0[S]}{K_m(1 + K_a/[H]^+ + [H^+]/K_b) + [S](1 + K'_a/[H^+] + [H^+]/K'_b)} \quad (10.109)$$

At sufficiently low substrate concentrations

$$v = \frac{k_2[E]_0[S]}{K_m(1 + K_a/[H^+] + [H^+]/K_b)} \quad (10.110)$$

while at high concentrations of substrate

$$v = \frac{k_2[E]_0}{1 + K'_a/[H^+] + [H^+]/K'_b} \quad (10.111)$$

Suppose that the low-concentration rates have been measured over a range of pH values and that the logarithms of these rates are plotted against the pH (Fig. 10.7c). In a sufficiently acid solution (low pH), the term $[H^+]/K_b$ predominates in the denominator of Eq. (10.110), so that

$$v = \frac{k_2[E]_0[S]K_b}{K_m[H^+]} \quad (10.112)$$

or $\qquad\qquad\qquad\qquad \log_{10} v = \text{constant} - \log_{10}[H^+] \qquad\qquad (10.113)$

$$= \text{constant} + \text{pH} \quad (10.114)$$

The plot of $\log_{10} v$ versus pH therefore has a slope of $+1$ at these low pH values. At intermediate pH values the first term predominates in the denominator of Eq. (10.110), and therefore

$$v = \frac{k_2[E]_0[S]}{K_m} \quad (10.115)$$

The plot of $\log_{10} v$ versus pH has zero slope in this region. These lines at low pH and intermediate pH intersect when the right-hand sides of Eqs. (10.112) and (10.115) are equal, and this occurs when

$$[H^+] = K_b \quad (10.116)$$

that is, when

$$\text{pH} = pK_b \quad (10.117)$$

Thus, the value of pK_b can be determined, as shown schematically in Fig. 10.7c. Of course, the lines do not meet sharply; there is a rounding off, as shown in the diagram.

Similarly, the right-hand inflection point corresponds to pK_a. At high pH values the term $K_a/[H^+]$ predominates in the denominator of Eq. (10.110), so that

$$v = \frac{k_2[E]_0[S][H^+]}{K_m K_a} \tag{10.118}$$

and therefore

$$\log_{10} v = \text{constant} + \log_{10} [H^+] \tag{10.119}$$

$$= \text{constant} - pH \tag{10.120}$$

The slope is now -1 (Fig. 10.7c), and by equating the right-hand sides of (10.115) and (10.118) it follows that, at the inflection point,

$$pH = pK_a \tag{10.121}$$

However, few enzyme systems conform to this simple mechanism. Many single-substrate systems involve additional intermediates, which may exist also in various ionization states, and this will influence the behavior. Additional complexities exist for reactions involving more than one substrate. In spite of these complications it has proved possible to obtain pK values for active groups on a number of enzymes, and this information is important in leading to an understanding of enzyme mechanisms.

10.3.3 Influence of Temperature

Valuable information about mechanisms also has been provided by studies of the influence of temperature on the rates of enzyme reactions. Account must be taken of the fact that the enzymes themselves undergo a deactivation process which has a very high activation energy and also a very high pre-exponential factor. At temperatures of 35°C or higher (depending on the particular enzyme) the enzyme may undergo rapid deactivation during the course of a kinetic experiment, and then a low rate of transformation of the substrate is observed. As a result, the rates of enzyme-catalyzed reactions frequently pass through a maximum as the temperature is raised. The temperature at which the rate is a maximum often is referred to as the *optimum temperature,* its value depending on the conditions of the experiment. The kinetics of enzyme inactivation, which is due to the denaturation of the protein, have been investigated in some detail.

By working at sufficiently low temperatures, at which no appreciable inactivation occurs, or by making a correction for the inactivation, it is possible to determine the effect of temperature on the enzyme-catalyzed reaction itself. The analysis of the results must take account of the fact that the rate equation may be, for example,

$$v = \frac{k_2[E]_0[S]}{K_m + [S]} \tag{10.122}$$

so that a simple dependence of the rate on the temperature is not to be expected. However, at sufficiently high substrate concentrations,

$$v = k_2[\text{E}]_0 \tag{10.123}$$

and, since k_2 is expected to vary with temperature according to the Arrhenius equation, the same temperature dependence is found for v. Under these conditions a plot of $\log_{10} v$ versus $1/T$ should be linear, and this has been found to be the case for a number of enzyme systems.

At sufficiently low enzyme concentration the rate equation becomes

$$v = \frac{k_2}{K_m} [\text{E}]_0[\text{S}] \tag{10.124}$$

For a reaction occurring by the simple Michaelis–Menten mechanism, for which $K_m = (k_{-1} + k_2)/k_1$, this equation becomes

$$v = \frac{k_1 k_2}{k_{-1} + k_2} [\text{E}]_0[\text{S}] \tag{10.125}$$

In general, a simple dependence of rate on temperature is not to be expected under these conditions of low substrate concentrations. In the special case that k_2 is much greater than k_{-1}, Eq. (10.125) becomes

$$v = k_1[\text{E}]_0[\text{S}] \tag{10.126}$$

A plot of $\log v$ versus $1/T$ is a straight line, and the activation energy calculated from its slope corresponds to k_1; it therefore applies to the reaction between enzyme and substrate with the formation of the enzyme–substrate complex. If, on the other hand, k_{-1} is much greater than k_2, the rate equation becomes

$$v = \frac{k_1 k_2}{k_{-1}} [\text{E}]_0[\text{S}] \tag{10.127}$$

The activation energy obtained from a plot of $\log v$ versus $1/T$ is now equal to $E_1 + E_2 - E_{-1}$, where these are the activation energies corresponding to the three elementary reactions. The above considerations are equivalent to those discussed previously with reference to Fig. 10.2.

The activation energies for enzyme-catalyzed reactions are usually much smaller than for the same reaction occurring under the influence of other catalysts. For example, the activation energy for the reaction

$$2\text{H}_2\text{O}_2 \rightarrow 2\text{H}_2\text{O} + \text{O}_2$$

catalyzed by Fe^{2+} ions[42] is 42 kJ mol^{-1}, whereas that catalyzed by the enzyme catalase[43] is 7 kJ mol^{-1}.

10.3.4 Transient-Phase Kinetics

Most kinetic studies on enzyme reactions have been concerned with the behavior after the steady state has become established. Valuable information also has been provided by measurements made during the pre-steady-state period, which is referred to as the *transient phase*. Usually, such investigations must be made within a few milliseconds of the commencement of reaction, and it is necessary to use special techniques such as the stopped-flow and relaxation methods (Section 2.8). Only a very brief account will be given here.

If solutions of enzyme and substrate are brought together, at zero time no enzyme–substrate complex exists. The concentration of this complex rises from zero to its steady-state value in the manner shown in curve *a* of Fig. 10.8. If the simple Michaelis–Menten mechanism applies, the complex is converted directly into product and the rate of formation of product is proportional to the concentration of complex. As shown in curve *b* of Fig. 10.8, the rate of formation of product (the slope of the concentration–time curve) is zero at zero time and rises until it reaches its steady-state value.

The form of the concentration–time curve for the product is such that there appears to be an induction period during which little reaction occurs. As shown in Fig. 10.8, this induction period is determined by extrapolating curve *b* back to zero product concentration. For the particular case of the simple Michaelis–Menten mechanism the induction period is given by[44]

$$\tau = \frac{1}{k_{-1} + k_2 + k_1[A]_0} \tag{10.128}$$

By determining how τ varies with $[A]_0$, the initial concentration of substrate, it is

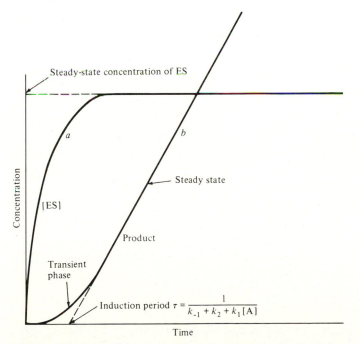

Figure 10.8 Initial variations of the concentration of enzyme-substrate complex (curve *a*) and of the product of reaction (curve *b*) for a reaction occurring by the simple Michaelis–Menten mechanism.

possible to determine k_1 and $k_{-1} + k_2$ separately. The steady-state measurements for a reaction proceeding by the simple mechanism yield k_2 and $K_m \, [=(k_{-1} + k_2)/k_1]$ but do not allow the constants k_1, k_{-1}, and k_2 to be determined separately. The transient-phase studies provide the additional information needed to obtain the three individual rate constants.

For more complex enzyme mechanisms the separation of rate constants is less straightforward, but valuable additional information is provided by transient-phase studies.

10.3.5 Enzyme Mechanisms

Enzyme-kinetic studies, together with other types of investigation, have led to insight into the way in which enzymes exert their catalytic action.

Two aspects of this are of special interest. The specificity of enzymes is explained in terms of an elaboration of Fischer's "lock and key" hypothesis,[45] which is concerned with the way in which an enzyme and a substrate fit together in forming the enzyme–substrate complex and in undergoing subsequent reaction. Recent X-ray studies have allowed the specificity of certain enzymes to be interpreted in some detail.

The second aspect is the very high effectiveness of enzymes in comparison with other catalysts. For example, for the decomposition of hydrogen peroxide,

$$2H_2O_2 \rightarrow 2H_2O + O_2$$

the enzyme catalase is more effective than any other catalyst by many powers of ten.[46] This high effectiveness of catalysts almost always is associated with a low energy of activation for the reaction. In some cases the effect has been attributed to the fact that the enzyme is acting as a *bifunctional catalyst,* in that two catalytic groups are present side by side at the active center. In the case of ester hydrolyses, for example, there is evidence that one group is donating electrons to the substrate and that at the same time another group is donating a proton. This type of behavior, for which there is evidence in nonenzyme systems, is referred to as *bifunctional catalysis.*[47]

10.4 CATALYSIS IN GASEOUS SYSTEMS

Many gas reactions are catalyzed by added gases as well as by solid surfaces, and a number of different types of mechanism have been shown to occur. A much-investigated reaction of great technical importance is the oxidation of sulfur dioxide to sulfur trioxide,

$$2SO_2 + O_2 \rightarrow 2SO_3$$

This reaction is slow in the absence of a catalyst but is accelerated greatly by the addition of nitric oxide. As early as 1806 Désormes and Clément[48] had given the essentially correct explanation that nitric oxide first is oxidized and then transfers oxygen to sulfur dioxide. Later, Davy[49] found that the reaction does not occur in the absence of water vapor, and he identified in the reaction system certain crystalline hydrates which react with water to form sulfuric acid. Much later, Reynolds and Taylor[50] proposed the mechanism

$$2NO + O_2 \rightarrow 2NO_2$$

$$NO_2 + SO_2 + H_2O \rightarrow H_2SO_4 + NO$$

which is probably oversimplified but not far from the truth. As seen in Section 5.2 (see Fig. 5.4) the first of these reactions has zero activation energy and occurs rapidly; the second reaction is also rapid.

10.5 CHAIN MECHANISMS

Chain reactions often are catalyzed, and then there are two possibilities; the catalyst may increase the rate of initiation of chains by introducing an additional process, or it may give rise to new chain-propagating steps. There are many examples of *heterogeneous* catalysts that aid in the initiation of chains. For example, at ordinary temperatures the hydrogen–oxygen reaction (Section 8.6) is extremely slow in the absence of a catalyst, and an important reason for this is the slowness of the initiation process. The rate is increased greatly by the addition of powdered platinum or palladium, which catalyzes the production of hydrogen atoms from hydrogen molecules. The addition of these substances can transform an exceedingly slow reaction into an explosion. Other oxidation reactions are catalyzed in the same way.

Substances that homogeneously catalyze initiation processes are not as common. Initiation rates often are increased by added substances, but usually these are not regenerated and are better called *initiators* rather than catalysts. For example, azomethane frequently increases the rates of organic decompositions, doing so by decomposing with the production of methyl radicals,

$$CH_3N_2CH_3 \rightarrow N_2 + 2CH_3$$

These methyl radicals then contribute to the chain-propagation processes.

Added substances frequently catalyze reactions by taking part in chain-propagation processes. For example, the decomposition of a substance M may involve chain-propagation reactions of the following type, where β is a radical involving a bimolecular reaction and μ is one undergoing a unimolecular reaction:

$$\beta + M \rightarrow \mu + \text{product}$$

$$\mu \rightarrow \beta + \text{product}$$

An added substance HX, which, for example, might be HCl, HBr, or H_2S, may introduce the following processes:

$$\beta + HX \rightarrow X + \text{product}$$

$$X + M \rightarrow \mu + HX$$

It therefore increases the rate of reaction, and since it is regenerated it acts as a true catalyst. Examples that follow essentially this reaction pattern are the decomposition of dimethyl ether catalyzed by hydrogen sulfide[51] and of acetaldehyde catalyzed by hydrogen halides.[52]

   An example of a reaction in which a catalyst both aids in the initiation process and contributes to the chain-propagating processes is the iodine-catalyzed decomposition of acetaldehyde. The mechanism in the absence of catalyst has been discussed in Section 8.5.4. In the presence of iodine the mechanism is approximately as follows:[53]

$$I_2 \xrightarrow{k_1} 2I$$

$$I + CH_3CHO \xrightarrow{k_2} HI + CH_3CO$$

$$CH_3CO \xrightarrow{k_3} CH_3 + CO$$

$$CH_3 + I_2 \xrightarrow{k_4} CH_3I + I$$

$$CH_3 + HI \xrightarrow{k_5} CH_4 + I$$

$$CH_3I + HI \xrightarrow{k_6} CH_4 + I_2$$

$$2I \xrightarrow{k_{-1}} I_2$$

Eventually, the iodine is regenerated, and the overall process is mainly the formation of methane and carbon monoxide, as in the uncatalyzed reaction. Application of the steady-state treatment to this mechanism leads to the result that the rate is

$$v = k_2 \left(\frac{k_1}{k_{-1}}\right)^{1/2} [I_2]^{1/2}[CH_3CHO] \tag{10.129}$$

The activation energy for the uncatalyzed reaction is about 200 kJ mol^{-1} while that for the iodine-catalyzed reaction is about 135 kJ mol^{-1}.

10.6 CATALYSIS BY IONS OF VARIABLE VALENCY

There are many examples of reactions in solution that are catalyzed by ions that can exist in two or more valency states. An example is the reaction

$$Tl^+ + 2Ce^{4+} \rightarrow Tl^{3+} + 2Ce^{3+}$$

which is catalyzed by silver ions. The mechanism appears to be[54]

$$Ce^{4+} + Ag^+ \rightleftharpoons Ce^{3+} + Ag^{2+} \quad \text{(fast)}$$

$$Ag^{2+} + Tl^+ \rightarrow Tl^{2+} + Ag^+ \quad \text{(slow)}$$

$$Tl^{2+} + Ce^{4+} \rightarrow Tl^{3+} + Ce^{3+} \quad \text{(fast)}$$

In each of these three elementary steps only a single electron is transferred; such single-electron transfers are more rapid than those involving more than one electron. Much work has been done on the detailed mechanisms of such processes, particularly by Marcus;[55] for reviews see the bibliography at the end of this chapter.

Another reaction catalyzed by ions of variable valency is the decomposition of hydrogen peroxide in aqueous solution:

$$2H_2O_2 \rightarrow 2H_2O + O_2$$

There are a number of catalysts for this reaction. In glass vessels there is a certain amount of catalysis at the surface, and there is also some catalysis brought about by particles liberated from the glass walls.[56] Little catalysis occurs if quartz vessels are used, and the activation energy is then about 70 kJ mol^{-1}. The reaction is catalyzed strongly by Fe^{2+} ions, the activation energy then being reduced to 42 kJ mol^{-1}. The mechanism suggested[57] is

$$Fe^{2+} + H_2O_2 \rightarrow Fe^{3+} + OH^- + OH$$

$$Fe^{2+} + OH \rightarrow Fe^{3+} + OH^-$$

$$OH + H_2O_2 \rightarrow H_2O + HO_2$$

$$Fe^{2+} + HO_2 \rightarrow Fe^{3+} + HO_2^-$$

$$Fe^{3+} + HO_2 \rightarrow Fe^{2+} + O_2 + H^+$$

This is a chain reaction, and some of the individual rate constants and activation energies have been determined. The hydrogen peroxide decomposition is also catalyzed by the enzyme catalase, and the kinetics and mechanism have been studied in great detail.

10.7 ACTIVATION OF MOLECULAR HYDROGEN

One type of process that has been investigated very extensively is the homogeneous catalytic activation of molecular hydrogen. Many reactions of great technical importance involve hydrogenation, and catalysts that bring about a splitting of the hydrogen molecule can greatly aid these processes.

The first successful catalytic activation by hydrogen was carried out in 1938 by Calvin,[58] who found that in quinoline solution copper(I) salts, notably copper(I) acetate, catalyzed various reductions by molecular hydrogen. The reactions were second order in Cu(I), indicating that two Cu$^+$ ions are involved.

Later, Halpern and his coworkers[59] discovered a number of catalysts for the homogeneous activation of hydrogen in aqueous solution. The first reaction that demonstrated this was the reduction by H$_2$ of copper(II) acetate, with the formation of cuprous oxide:

$$2Cu(OAc)_2 + H_2 + H_2O \rightarrow Cu_2O + 4HOAc$$

A kinetic study of this process[60] showed that the initial, rate-determining process is an interaction between H$_2$ and Cu^{2+}, this process being first order in both H$_2$ and Cu^{2+}. This result contrasts with that obtained by Calvin and indicates that a single Cu^{2+} ion is involved in the activation process. Halpern and his coworkers later demonstrated that the activation of hydrogen can be brought about by a number of other ions, such as Ag(I), Hg(I), Hg(II), MnO_4^-, Pd(II), Rh(III), Ru(II), and Ru(III).

The kinetic results obtained for such processes in which hydrogen brings about the reduction of a substrate molecule showed that the rate was independent of the concentration of the substrate. This indicates that the catalyst ion first brings about the activation of H_2 in a rate-determining step which produces some reaction intermediate; in a subsequent step this intermediate rapidly reacts with the oxidizing substrate. There is evidence for different types of intermediate. There may be a heterolytic dissociation of H_2, as in the process

$$Cu^{2+} + H_2 \rightleftharpoons CuH^+ + H^+$$

For example, in the Cu^{2+} catalyzed oxidation of H_2 by $Cr_2O_7^{2-}$, the kinetic equation[61] is of the form

$$v = \frac{k[H_2][Cu^{2+}]^2}{[H^+] + k'[Cu^{2+}]} \tag{10.130}$$

The inhibiting effect of H^+ suggests the mechanism

$$Cu^{2+} + H_2 \underset{k_{-1}}{\overset{k_1}{\rightleftharpoons}} CuH^+ + H^+$$

$$CuH^+ + Cu^{2+} \overset{k_2}{\rightarrow} 2Cu^+ + H^+$$

followed by rapid reaction between Cu^+ and $Cr_2O_7^{2-}$. The steady-state equation for CuH^+ is

$$k_1[Cu^{2+}][H_2] - k_{-1}[CuH^+][H^+] - k_2[CuH^+][Cu^{2+}] = 0 \tag{10.131}$$

so that

$$[CuH^+] = \frac{k_1[Cu^{2+}][H_2]}{k_{-1}[H^+] + k_2[Cu^{2+}]} \tag{10.132}$$

The rate of formation of Cu^+ is thus

$$v = \frac{k_1 k_2[H_2][Cu^{2+}]^2}{k_{-1}[H^+] + k_2[Cu^{2+}]} \tag{10.133}$$

and if Cu^+ reacts rapidly with the substrate, this is the overall rate.

Evidence that a homolytic splitting of H_2 also can occur is provided, for example, by the kinetic equation for the Ag^+-catalyzed reduction of $Cr_2O_7^{2-}$ ions:[62]

$$v = k[H_2][Ag^+]^2 + \frac{k'[H_2][Ag^+]^2}{[H^+] + k''[Ag^+]} \tag{10.134}$$

The first term suggests that there is a homolytic splitting of hydrogen:

$$2Ag^+ + H_2 \overset{k}{\rightarrow} 2AgH^+$$

This process is followed by the rapid reduction of the dichromate ion by AgH^+. The second term, analogous to that found with Cu^{2+}, is attributed to an initial heterolytic splitting of hydrogen:

$$Ag^+ + H_2 \underset{k_{-1}}{\overset{k_1}{\rightleftharpoons}} AgH + H^+$$

$$AgH + Ag^+ \overset{k_2}{\rightarrow} 2Ag + H^+ \quad \text{or} \quad AgH^+ + Ag$$

Either Ag or AgH^+ then reacts rapidly with the dichromate ion. The steady-state equation for the intermediate AgH is

$$k_1[Ag^+][H_2] - k_{-1}[AgH][H^+] - k_2[AgH][Ag^+] = 0 \tag{10.135}$$

and therefore

$$[AgH] = \frac{k_1[Ag^+][H_2]}{k_{-1}[H^+] + k_2[Ag^+]} \tag{10.136}$$

The rate by this mechanism is thus

$$v = k_2[AgH][Ag^+] = \frac{k_1 k_2[H_2][Ag^+]^2}{k_{-1}[H^+] + k_2[Ag^+]} \tag{10.137}$$

This accounts for the second term in the empirical equation (10.134).

In the above homolytic dissociation of hydrogen, two Ag^+ ions were involved. There also has been identified, for certain catalysts, a third mode of activation, where there is homolytic splitting at a single metal center.[63] This type of process is found most commonly with four-coordinate complexes of Rh(I) and Ir(I), the product being a six-coordinate complex in which two hydrogen atoms are attached to the central metal atom; an example is

$$[Rh^I Cl(PPh_3)_3] + H_2 \rightarrow [Rh^{III}H_2Cl(PPh_3)_3]$$

The catalysts already referred to, although capable of activating molecular hydrogen, proved to be ineffective for the hydrogenation of substrates such as olefins. These are difficult to reduce, and it appears that the hydrides formed as intermediates undergo competing reactions before they can hydrogenate the olefins. This difficulty was overcome[64] by the use of metal complexes containing a coordinated olefin molecule. For example, if maleic acid is attached to ruthenium(II) chloride in aqueous solution, molecular hydrogen brings about the reduction of the acid. Mechanistic studies showed that the two hydrogen atoms are added in separate stages.

The work on the activation of H_2 by ions and their complexes has led to important conclusions about the action of the enzyme hydrogenase. Earlier work by Rittenberg and Krasna,[65] on the basis of studies of the H_2–D_2O exchange reaction, indicated that the splitting of H_2 by this enzyme was heterolytic rather than homolytic. The enzyme contains iron, and in 1975 the active center was shown[66] to be a coordinated iron atom. Previously, however, Halpern and coworkers[67] had suggested a mechanism for hydrogenase action that is based on the results for the catalytic activation of H_2 by ruthenium chloride. This mechanism proved to be consistent with the nature of the active center and is probably the correct one.

Of necessity this account of catalytic hydrogenation has covered only the more fundamental aspects and has not gone into details. This important topic continues to expand very rapidly. In 1953 almost nothing was known of the subject, but James's book,[68] published 20 years later, cited nearly 2000 references. After another 10 years the number of published papers undoubtedly had doubled, and this exponential rise in activity has continued. Much of the later work has been concerned with extending the range of catalysts and substrates for homogeneous hydrogenation. Further details of reaction mechanisms have been explored,[69] and the results have thrown light on the mechanisms of the action of enzymes.

PROBLEMS

10.1. The hydrolysis of a substance is catalyzed specifically by hydrogen ions, the rate constant being given by

$$k/s^{-1} = 4.70 \times 10^{-2}[H^+]/mol\ dm^{-3}$$

When the substance is dissolved in a 1.00×10^{-3} M solution of an acid HA, the rate constant is 3.20×10^{-5} s^{-1}. Calculate the dissociation constant of the acid HA.

10.2. The following are some acid dissociation constants (for the conjugate acid) and catalytic constants for the decomposition of nitramide, for a number of basic ions:

Ion	K_a/mol dm^{-3}	k/s^{-1}
Hydroxide	1.8×10^{-16}	1.67×10^4
Trimethylacetate	9.4×10^{-6}	1.37×10^{-2}
Propionate	1.3×10^{-5}	1.08×10^{-2}
Acetate	1.8×10^{-5}	8.40×10^{-3}
Phenyl acetate	5.3×10^{-5}	3.87×10^{-3}
Benzoate	6.5×10^{-5}	3.15×10^{-3}
Formate	2.1×10^{-4}	1.37×10^{-3}
Monochloroacetate	1.4×10^{-3}	2.67×10^{-4}
o-Nitrobenzoate	7.3×10^{-3}	7.00×10^{-5}
Dichloroacetate	5.0×10^{-2}	1.17×10^{-5}

Test the applicability of the Brønsted relationship by making a suitable plot and determine the constant β.

10.3. A solution is 0.2 M with respect to acetic acid and 0.1 M with respect to sodium acetate. If the dissociation constant of acetic acid is 1.8×10^{-5} mol dm^{-3} and the exponent α in the Brønsted equation is 0.8 for a particular acid-catalyzed reaction, estimate the percentage reaction due to hydrogen ions, acetic acid, and water. [Note: It is important to use K_a values for H^+ and H_2O that are consistent with that used for CH_3COOH.]

10.4. In the presence of monochloroacetic acid the iodination of acetone is catalyzed mainly by hydrogen ions and undissociated $CH_2ClCOOH$:

$$k = k_{H^+}[H^+] + k_{CH_2ClCOOH}[CH_2ClCOOH]$$

The dissociation constant of the acid is 1.55×10^{-3} mol dm^{-3}, and the following rate constants were obtained at various total concentrations c_a of acid:

c_a/mol dm^{-3}	0.05	0.10	0.20	0.50	1.00
$k/10^{-8}$ s^{-1}	7.67	12.7	19.8	39.7	66.8

Determine the catalytic constants k_{H^+} and $k_{CH_2ClCOOH}$ from these results.

10.5. Suppose that a reaction undergoes acid catalysis by the following mechanism:

$$BH^+ + S \underset{k_{-1}}{\overset{k_1}{\rightleftharpoons}} SH^+ + B$$

$$SH^+ + H_2O \overset{k_2}{\rightarrow} P + H_3O^+$$

Assuming that the substrate is in great excess of the catalyst and that $k_{-1}[B] \gg k_2$, derive the rate equation. Is there specific or general catalysis?

10.6. Suppose that the mechanism of an acid-catalyzed reaction is

$$BH^+ + S \underset{k_{-1}}{\overset{k_1}{\rightleftharpoons}} SH^+ + B$$

$$SH^+ + B \overset{k_2}{\rightarrow} P + BH^+$$

Derive the rate equation for $k_{-1} \gg k_2$ and with the substrate in great excess. Is there specific or general catalysis?

10.7. The following results have been obtained by Dahlberg and Long[70] for the base-catalyzed enolization of 3-methyl acetone:

Catalyst	ClCH$_2$COO$^-$	CH$_3$COO$^-$	HPO$_4^{2-}$
K_a/mol dm^{-3}	1.39×10^{-3}	1.80×10^{-5}	6.25×10^{-8}
k/dm^3 mol^{-1} s^{-1}	1.41×10^{-3}	1.34×10^{-2}	0.26

Estimate the Brønsted coefficient β.

10.8. Confirm that Eq. (10.31),

$$[H^+]_{min} = \left(\frac{k_{OH^-} K_w}{K_{H^+}} \right)^{1/2}$$

follows from Eq. (10.30),

$$k = k_0 + k_{H^+}[H^+] + \frac{k_{OH^-} K_w}{[H^+]}$$

10.9. The rates of some single-substrate reactions catalyzed by an enzyme pass through a maximum as the substrate concentration is increased, rather than reaching a limiting rate. Suggest a mechanism that accounts for this behavior and derive an expression for the rate.

10.10. Derive an equation of the form of Eq. (10.105) for the random ternary-complex mechanism, taking the intermediates EA, EB, and EAB to be in equilibrium with E, A, and B.

10.11. Derive the rate equation for the ordered ternary-complex mechanism:

$$E + A \underset{k_{-1}}{\overset{k_1}{\rightleftharpoons}} EA$$

$$EA + B \underset{k_{-2}}{\overset{k_2}{\rightleftharpoons}} EAB \overset{k_3}{\rightarrow} E + Y + Z$$

10.12. Derive the rate equation for the following special case of the ping-pong bi-bi mechanism:

$$E + A \underset{k_{-1}}{\overset{k_1}{\rightleftharpoons}} EA$$

$$EA \overset{k_2}{\rightarrow} EA' + Y$$

$$EA' + B \overset{k_3}{\rightarrow} E + Z$$

where Y and Z are the reaction products.

10.13. The following mechanism has been proposed by Theorell and Chance:[71]

$$E + A \underset{k_{-1}}{\overset{k_1}{\rightleftharpoons}} EA$$

$$EA + B \overset{k_2}{\longrightarrow} EZ + Y$$

$$EZ \overset{k_3}{\longrightarrow} E + Z$$

Derive the rate equation.

10.14. Confirm that Eq. (10.129),

$$v = k_2 \left(\frac{k_1}{k_{-1}} \right)^{1/2} [I_2]^{1/2} [CH_3CHO]$$

results from the mechanism in Section 10.5 for the iodine-catalyzed decomposition of acetaldehyde.

10.15. The following mechanism has been proposed for the alkaline hydrolysis of $Co(NH_3)_5Cl^{2+}$:

$$Co(NH_3)_5Cl^{2+} + OH^- \underset{k_{-1}}{\overset{k_1}{\rightleftharpoons}} Co(NH_3)_4(NH_2)Cl^+ + H_2O$$

$$Co(NH_3)_4(NH_2)Cl^+ \overset{k_2}{\longrightarrow} Co(NH_3)_4(NH_2)^{2+} + Cl^-$$

$$Co(NH_3)_4(NH_2)^{2+} + H_2O \overset{k_3}{\longrightarrow} Co(NH_3)_5(OH)^{2+}$$

Assume $Co(NH_3)_4(NH_2)Cl^+$ and $Co(NH_3)_4(NH_2)^{2+}$ to be in the steady state and derive an expression for the rate of reaction.

Experimentally, the rate is proportional to $[Co(NH)_5Cl^{2+}][OH^-]$. Does this fact tell us anything about the relative magnitudes of the rate constants?

10.16. When an inhibitor I is added to a single-substrate enzyme system, the mechanism is sometimes

$$
\begin{array}{c}
S \\
+ \quad {\scriptstyle k_i} \\
I + E \;\rightleftharpoons\; EI \\
{\scriptstyle k_{-1}} \updownarrow {\scriptstyle k_{-i}} \\
ES \\
\downarrow {\scriptstyle k_2} \\
E + Y
\end{array}
$$

This is known as a competitive mechanism, since S and I compete for sites on the enzyme.

(a) Assume that the substrate and inhibitor are present in great excess of the enzyme, apply the steady-state treatment, and obtain the rate equation.

(b) Obtain an expression for the degree of inhibition ε, defined as

$$\varepsilon = \frac{v_0 - v}{v_0}$$

where v is the rate in the presence of inhibitor and v_0 is the rate in its absence.

10.17. Enzyme-catalyzed reactions frequently follow an equation of the form of Eq. (10.104). Suppose that k_c and K_m show the following temperature dependence:

$$k_c = A_c e^{-E_c/RT} \quad \text{and} \quad K_m = B e^{-\Delta H_m/RT}$$

where A_c, B, E_c, and ΔH_m are temperature-independent parameters. Explain under what conditions, with [S] held constant, the rate may pass through a maximum as the temperature is raised.

10.18. **(a)** Show that, for a reaction occurring by the mechanism

$$C + S \underset{k_{-1}}{\overset{k_1}{\rightleftharpoons}} X \overset{k_2}{\rightarrow} C + Y$$

where the total catalyst concentration $[C]_0$ is much less than the substrate concentration $[S]$, the rate equation at low substrate concentrations is

$$v_{low} = \frac{k_1 k_2}{k_{-1} + k_2} [C]_0 [S]$$

(b) Prove that the observed activation energy under these conditions is

$$E_{low} = \frac{k_{-1}}{k_{-1} + k_2} (E_1 + E_2 - E_{-1}) + \frac{k_2}{k_{-1} + k_2} E_1$$

10.19. Confirm that Eq. (10.101) is the steady-state rate equation that follows from the mechanism

$$E + S \underset{k_{-1}}{\overset{k_1}{\rightleftharpoons}} ES \overset{k_2}{\underset{Y}{\searrow}} ES' \overset{k_3}{\rightarrow} E + Z$$

REFERENCES

1. J. J. Berzelius, *Jahresber. Chem.*, **15**, 237 (1836).
2. W. Ostwald, *Chem. Betrachtungen*, Aula No. 1 (1895).
3. W. Ostwald, *Phys. Z.*, **3**, 313 (1902).
4. W. Ostwald, *Lehrbuch der allgemeinen Chemie*, 2nd ed., vol. 2, pt. 2, pp. 248, 262, Akademische Verlagsgesellschaft, Leipzig, 1902.
5. *Manual of Symbols and Terminology for Physicochemical Quantities and Units, Appendix V. Symbolism and Terminology in Chemical Kinetics, Pure App. Chem.*, **53**, 753 (1981).
6. W. Ostwald, *Z. Elektrochem.*, **7**, 995 (1901); *Nature*, **65**, 522 (1902).
7. W. Ostwald, *Verh. Ges. Dtsch. Naturforsch. Aerzte*, **73**, 184 (1901); *Physik. Z.*, **3**, 312 (1902); *Nature, 65*, 522 (1902); *Las Prix Nobel en 1909*, pp. 63–88, Stockholm, 1910.
8. W. Ostwald, *Z. Elektrochem.*, **7**, 995 (1901); *Nature*, **65**, 522 (1902).
9. See, for example, J. H. van't Hoff, *Vorlesungen über theoretische und physikalische Chemie*, Vol. I, *Die chemische Dynamik*, Brunswick, 1898 (English translation by R. A. Lehrfeld, pp. 204, 214, Edward Arnold, London); S. H. Maron and V. K. La Mer, *J. Am. Chem. Soc.*, **61**, 2018 (1939); R. A. Alberty, *Adv. Enzymol.*, **17**, 1 (1955); R. A. Alberty and R. M. Bock, *J. Am. Chem. Soc.*, **75**, 1921 (1953).
10. K. J. Laidler and I. M. Socquet, *J. Phys. Colloid Chem.*, **54**, 519 (1950). The scheme is a slight elaboration of one originally proposed by K. F. Herzfeld, *Z. Phys. Chem.*, **98**, 151 (1921).
11. G. E. Briggs and J. B. S. Haldane, *Biochem. J.*, **19**, 338 (1925).
12. W. Ostwald, *J. Prakt. Chem.*, **30**, 39 (1884).
13. S. Arrhenius, *Z. Phys. Chem.*, **2**, 495 (1888); **4**, 244 (1889); **28**, 317 (1899).
14. L. T. Reicher, *Ann. Chem.*, **228**, 257 (1885).
15. J. J. A. Wijs, *Z. Phys. Chem.*, **11**, 492 (1893); **12**, 415 (1893).
16. A. Skrabal, *Z. Elecktrochem.*, **33**, 322 (1927).
17. H. M. Dawson and F. Powis, *J. Chem. Soc.*, 2135 (1913). For a detailed and critical discussion

of some of the early evidence see R. P. Bell, *Acid–Base Catalysis,* pp. 48–81, Clarendon, Oxford, 1941.

18. J. N. Brønsted, *Rec. Trav. Chim.,* **42,** 718 (1923).

19. G. N. Lewis, *Valency and the Structure of Atoms and Molecules,* Reinhold, New York, 1923.

20. T. M. Lowry, *J. Chem. Soc.,* 2554 (1927).

21. The matter is complicated and the evidence is by no means unequivocal. For a detailed discussion see R. P. Bell, *The Proton in Chemistry,* 2nd ed., pp. 149–158, Chapman and Hall, London, 1973.

22. This possibility in acid–base catalysis was first considered by J. N. Brønsted, *Chem. Rev.,* **5,** 321 (1928).

23. S. S. Biechler and R. W. Taft, *J. Am. Chem. Soc.,* **79,** 4927 (1957).

24. H. von Euler, *Sven. Vet. Akad. Forh.,* No. 4 (1899); *Z. Phys. Chem.,* **28,** 619 (1899); **32,** 384 (1900); **36,** 641 (1901); B. S. Rabinovitch and C. A. Winkler, *Can. J. Res. B,* **20,** 73 (1942); J. T. Edward, H. P. Hutchison, and S. C. R. Meacock, *J. Chem. Soc.,* 2520 (1955); J. T. Edward and S. C. R. Meacock, *J. Chem. Soc.,* 2000 (1957); D. Rosenthal and T. I. Taylor, *J. Am. Chem. Soc.,* **79,** 2684 (1957).

25. R. P. Bell and K. Yates, *J. Chem. Soc.,* 1927 (1962); K. Yates and W. V. Wright, *Can. J. Chem.,* **41,** 2882 (1963).

26. For a discussion of the evidence see R. P. Bell, *The Proton in Chemistry,* p. 177, Chapman and Hall, London, 1973.

27. Further evidence that this is the mechanism for the base-catalyzed reactions is given by R. P. Bell, *The Proton in Chemistry,* p. 46, Chapman and Hall, London, 1973.

28. J. N. Brønsted and K. J. Pedersen, *Z. Phys. Chem.,* **108,** 185 (1923). Similar but less satisfactory relationships (involving exponents of 0.5) had been suggested previously by H. S. Taylor, *Z. Elektrochem.,* **20,** 201 (1914).

29. J. N. Brønsted, *Chem. Rev.,* **5,** 322 (1928).

30. S. W. Benson, *J. Am. Chem. Soc.,* **80,** 5151 (1958); D. M. Bishop and K. J. Laidler, *J. Chem. Phys.,* **42,** 1688 (1965).

31. J. N. Brønsted and K. J. Pedersen, *Z. Phys. Chem.,* **108,** 185 (1923); J. N. Brønsted and C. E. Teeter, *J. Phys. Chem.,* **28,** 579 (1924); J. N. Brønsted and C. V. King, *J. Am. Chem. Soc.,* **47,** 2523 (1925); J. N. Brønsted and W. F. K. Wynne-Jones, *Trans. Faraday Soc.,* **25,** 58 (1929); M. Kilpatrick, *J. Am. Chem. Soc.,* **48,** 2091 (1926).

32. L. P. Hammett and A. J. Deyrup, *J. Am. Chem. Soc.,* **54,** 2721 (1932); L. P. Hammett, *Physical Organic Chemistry,* chap. 9, McGraw-Hill, New York, 1940.

33. L. P. Hammett and L. Zucker, *J. Am. Chem. Soc.,* **61,** 2779, 2785 (1939).

34. L. Michaelis and M. L. Menten, *Biochem. Z.,* **49,** 333 (1913).

35. G. E. Briggs and J. B. S. Haldane, *Biochem. J.,* **19,** 338 (1925).

36. H. Lineweaver and D. Burk, *J. Am. Chem. Soc.,* **56,** 658 (1934).

37. G. S. Eadie, *J. Biol. Chem.,* **146,** 85 (1942). See also B. H. J. Hofstee, *Science,* **116,** 329 (1952); *Nature,* **184,** 1296 (1959).

38. K. J. Laidler, *Trans. Faraday Soc.,* **51,** 528 (1955).

39. L. Michaelis and H. Davidsohn, *Biochem. Z.,* **35,** 386 (1911); L. Michaelis and M. Rothstein, *Biochem. Z.,* **110,** 217 (1920).

40. H. von Euler, K. Josephsohn, and K. Myrbäck, *Z. Physiol. Chem.,* **134,** 39 (1924).

41. S. G. Waley, *Biochem. Biophys. Acta,* **10,** 27 (1953).

42. V. Barb, J. H. Baxendale, P. George, and K. R. Hargrave, *Trans. Faraday Soc.,* **47,** 462 (1951).

43. R. K. Bonnichsen, B. Chance, and H. Theorell, *Acta Chim. Scand.,* **1,** 658 (1947).

44. K. J. Laidler, *Can. J. Chem.,* **33,** 1614 (1955); L. Ouellet and K. J. Laidler, *Can. J. Chem.,* **34,** 146 (1956).

45. E. H. Fischer, *Ber. Dtsch. Chem. Ges.,* **27,** 2987 (1894).

46. For details see, for example, K. J. Laidler and P. S. Bunting, *The Chemical Kinetics of Enzyme Action,* chap. 7, Clarendon, Oxford, 1973.

47. For a critical discussion of the evidence for bifunctional catalysis see R. P. Bell, *The Proton in Chemistry,* 2nd ed., pp. 155–158, Chapman and Hall, London, 1973.

48. J. B. Désormes and F. Clément, *Ann. Chim.,* **59,** 329 (1806).

49. H. Davy, *Philos. Trans.,* **102,** 463 (1812); *Elements of Chemical Philosophy,* p. 276, J. Johnson and Co., London, 1812.

50. W. C. Reynolds and W. H. Taylor, *J. Soc. Chem. Ind.,* **31,** 367 (1912).

51. D. J. McKenney and K. J. Laidler, *Can. J. Chem.,* **41,** 2009 (1963).

52. N. Inai, T. Fuju, and O. Toyama, *Bull. Chem. Soc. Jpn.,* **38,** 410 (1965); N. Inai and O. Toyama, *Bull. Chem. Soc. Jpn.,* **40,** 81 (1967).

53. G. K. Rollefson and R. F. Faull, *J. Am. Chem. Soc.,* **59,** 625 (1937).

54. W. C. E. Higginson, D. R. Rosseinsky, J. B. Stead, and A. G. Sykes, *Discuss. Faraday Soc.,* **29,** 49 (1960).

55. R. A. Marcus, *Ann. Rev. Phys. Chem.,* **15,** 155 (1964); *J. Phys. Chem.,* **72,** 891 (1968).

56. F. O. Rice, *J. Am. Chem. Soc.,* **48,** 2099 (1926); *J. Phys. Chem.,* **31,** 1507 (1927).

57. W. G. Barb, J. H. Baxendale, P. George, and K. R. Hargrave, *Trans. Faraday Soc.,* **47,** 462 (1951). An alternative chain mechanism was proposed earlier by F. Haber and K. Weiss, *Proc. R. Soc. London A,* **147,** 333 (1934).

58. M. Calvin, *Trans. Faraday Soc.,* **34,** 1181 (1938); *J. Am. Chem. Soc.,* **61,** 2230 (1939).

59. For reviews of the earlier work, in which the basic mechanisms were established, see J. Halpern, *Q. Rev.,* **10,** 473 (1956); *Adv. Catalysis,* **9,** 302 (1957); *J. Phys. Chem.,* **63,** 398 (1959); *Adv. Catalysis,* **11,** 301 (1959).

60. J. Halpern and R. G. Dakers, *J. Chem. Phys.,* **22,** 1272 (1954); R. G. Dakers and J. Halpern, *Can. J. Chem.,* **32,** 969 (1954).

61. J. Halpern, E. R. Macgregor, and E. Peters, *J. Phys. Chem.,* **60,** 1455 (1956).

62. A. H. Webster and J. Halpern, *J. Phys. Chem.,* **60,** 280 (1956); **61,** 1239, 1240 (1957). Other examples of homolytic splitting were identified later; for example, N. K. King and M. E. Winfield, *J. Am. Chem. Soc.,* **83,** 3366 (1961); J. Halpern and M. Pribanic, *Inorg. Chem.,* **9,** 2616 (1970).

63. L. Vaska, *Acta Chem. Res.,* **1,** 186 (1968), L. Vaska and J. W. Diluzio, *J. Am. Chem. Soc.,* **84,** 679 (1972). See also J. Halpern, *Acc. Chem. Res.,* **3,** 386 (1970) for additional examples.

64. J. Halpern, J. F. Harrod, and B. R. James, *J. Am. Chem. Soc.,* **83,** 753 (1961); **88,** 5150 (1966).

65. A. I. Krasna and D. Rittenberg, *J. Am. Chem. Soc.,* **76,** 3015 (1954); D. Rittenberg and A. I. Krasna, *Discuss. Faraday Soc.,* **20,** 185 (1955).

66. W. O. Gillum, L. E. Mortensen, J. S. Chen, and R. H. Holm, *J. Am. Chem. Soc.,* **97,** 463 (1975). For a review see R. H. Holm, in A. W. Addison, W. R. Cullen, D. Dolphin, and B. R. James (Eds.), *Biological Aspects of Inorganic Chemistry,* p. 71, Wiley, New York, 1977.

67. J. F. Harrod, S. Ciccone, and J. Halpern, *Can. J. Chem.,* **37,** 1372 (1961); J. Halpern and B. R. James, *Can. J. Chem.,* **44,** 671 (1966).

68. B. R. James, *Homogeneous Hydrogenation,* Wiley, New York, 1973.

69. See, for example, L. Y. Wong and J. Halpern, *J. Am. Chem. Soc.,* **90,** 6665 (1968); H. M. Feder and J. Halpern, *J. Am. Chem. Soc.,* **97,** 7186 (1975); R. L. Sweeney and J. Halpern, *J. Am. Chem. Soc.,* **99,** 8335 (1977); J. Halpern, *Trans. Am. Crystall. Assoc.,* **14,** 59 (1978); *Pure Appl. Chem.,* **51,** 2171 (1979); A. S. C. Chan and J. Halpern, *J. Am. Chem. Soc.,* **102,** 838 (1980).

70. D. B. Dahlberg and F. A. Long, *J. Am. Chem. Soc.,* **95,** 3825 (1973).

71. H. Theorell and B. Chance, *Acta Chem. Scand.,* **5,** 1127 (1951).

BIBLIOGRAPHY

For general treatments of catalysis see the following publications:

P. G. Ashmore, *Catalysis and Inhibition of Chemical Reactions,* Butterworths, London, 1963.

W. P. Jencks, *Catalysis in Chemistry and Enzymology,* McGraw-Hill, New York, 1969.

For reviews of the early work on acids and bases and of acid–base catalysis, see:

R. P. Bell, *Acid–Base Catalysis,* Clarendon, Oxford, 1941.

K. J. Laidler, The Development of Theories of Catalysis, *Arch. Hist. Exact Sci.,* **35,** 345 (1986).

W. F. Luder and S. Zuffanti, *The Electronic Theory of Acids and Bases,* Wiley, New York, 1946.

G. M. Schwab, *Catalysis* (translated with additions by H. S. Taylor and R. Spence), Van Nostrand, New York, 1937.

For a discussion of the contributions of catalytic studies to the early development of the theory of electrolytic dissociation see:

K. J. Laidler, Chemical Kinetics and the Origins of Physical Chemistry, *Arch. Hist. Exact Sci.,* **31,** 43 (1985).

R. D. Root-Bernstein, "The Ionists: Founding Physical Chemistry, 1872–1890," Ph.D. Thesis, Princeton University, 1980.

More recent treatments of acid–base catalysis are to be found in the following publications:

R. P. Bell, *The Proton in Chemistry,* Cornell University Press, Ithaca, NY, 1959; 1973 (2nd ed.).

R. P. Bell, *The Tunnel Effect in Chemistry,* Chapman and Hall, London, 1980.

J. E. Crooks, Proton Transfer to and from Atoms Other Than Carbon, in C. H. Bamford and C. F. H. Tipper (Eds.), *Comprehensive Chemical Kinetics,* vol. 8, pp. 197–250, Elsevier, New York, 1977.

F. Hibbert, Proton Transfer to and from Carbon, in C. H. Bamford and C. F. H. Tipper (Eds.), *Comprehensive Chemical Kinetics,* vol. 8, pp. 97–196, Elsevier, New York, 1977.

A. V. Willi, Homogeneous Catalysis of Organic Reactions (Mainly Acid–Base), in C. H. Bamford and C. F. H. Tipper (Eds.), *Comprehensive Chemical Kinetics,* vol. 8, pp. 1–95, Elsevier, New York, 1977.

Catalysis by enzymes is treated in:

M. L. Bender, *Mechanism of Homogeneous Catalysis from Protons to Proteins,* Wiley-Interscience, New York, 1971.

A. J. Cornish-Bowden, *Principles of Enzyme Kinetics,* Butterworths, London, 1976.

W. P. Jencks, *Catalysis in Chemistry and Enzymology,* McGraw-Hill, New York, 1969.

K. J. Laidler and P. S. Bunting, *The Chemical Kinetics of Enzyme Action,* 2nd ed., Clarendon, Oxford, 1973.

For reviews of acidity functions and related matters see the following publications:

F. A. Long and M. A. Paul, Applications of the H_0 Acidity Function to Kinetics and Mechanisms of Acid Catalysts, *Chem. Rev.,* **57,** 935 (1957).

M. A. Paul and F. A. Long, H_0 and Related Indicator Acidity Functions, *Chem. Rev.,* **57,** 1 (1957).

C. H. Rochester, *Acidity Functions,* Academic, London, 1970.

J. Shorter, *Correlation Analysis in Organic Chemistry,* Clarendon, Oxford, 1973.

J. Shorter, *Correlation Analysis of Organic Reactivity,* Research Studies Press, Chichester, 1982. This book is an expanded and more advanced version of the previous one.

For reviews of the earlier work on the catalytic activation of molecular hydrogen see:

J. Halpern, Homogeneous Reactions of Molecular Hydrogen in Solution. *Q. Rev.,* **10,** 463 (1956).

J. Halpern, The Catalytic Activation of Hydrogen in Homogeneous, Heterogeneous, and Biological Systems, *Adv. Catalysis,* **11,** 301 (1959).

For more recent reviews of the same topic see:

J. Halpern, Homogeneous Catalytic Hydrogenation: A Retrospective Account, *J. Organometallic Chem.,* **200,** 133 (1980).

B. R. James, *Homogeneous Hydrogenation,* Wiley, New York, 1973.

B. R. James, Addition of Hydrogen and Hydrogen Cyanide to Carbon–Carbon Double and Triple Bonds, in G. Wilkinson, F. G. A. Stone, and E. W. Abel (Eds.), *Comprehensive Organometallic Chemistry,* vol. 8, Pergamon, Oxford, 1982.

Catalysis by transition metals and their complexes is treated in the above reviews and also in:

C. Masters, *Homogeneous Transition-Metal Catalysis,* Chapman & Hall, New York, 1981.

G. W. Parshall, *Homogeneous Catalysis,* Wiley, New York, 1980.

R. Pearse and W. R. Patterson, *Catalysis and Chemical Processes,* Wiley, New York, 1981.

A. E. Shilov, *Activation of Saturated Hydrocarbons by Transition Metal Complexes,* Reidel, Boston, 1984.

F. G. A. Stone and R. West (Eds.), *Catalysis and Organic Synthesis,* Vol. 17 of *Advances in Organometallic Chemistry,* Academic, New York, 1979.

G. Wilkinson, F. G. A. Stone, and E. W. Abel (Eds.), *Comprehensive Organometallic Chemistry,* vol. 8, Pergamon, Oxford, 1982. Most of the chapters in this volume deal with catalysis.

For reviews of the theory of electron-transfer processes see:

L. Eberson, Electron-Transfer Reactions in Organic Chemistry, *Adv. Phys. Org. Chem.,* **18,** 79 (1982).

R. A. Marcus, Chemical and Electrochemical Electron-Transfer Theory, *Ann. Rev. Phys. Chem.,* **15,** 155 (1964).

T. W. Newton, The Kinetics of Oxidation–Reduction Reactions, *J. Chem. Educ.,* **45,** 571 (1968).

W. L. Reynolds and R. W. Lumry, *Mechanisms of Electron Transfer,* Ronald, New York, 1966.

H. Taube, *Electron Transfer Reactions of Complex Ions in Solution,* Academic, New York, 1970.

The following serial publications contain articles on catalysis:

Advances in Catalysis, Academic, New York.

Catalysis, Specialist Periodical Reports, Chemical Society, London.

Catalysis Reviews, Marcel Dekker, New York.

Advances in Organometallic Catalysis, Academic, New York.

Metal Complexes in Catalysis, Reidel, Dordrecht.

Chapter 11

Isotope Effects

If an atom in a reactant molecule is replaced by one of its isotopes, both the equilibrium constant of the reaction and the rate constants are altered. The term *equilibrium isotope effects* relates to change in equilibrium constants, and the term *kinetic isotope effects* to changes in rates.

Isotope effects are small if the isotopic substitution is far from the seat of reaction. If a bond involving the substituted atom is broken or formed during the reaction, the effects are much greater. Consequently, measurements of isotope effects are useful for deciding whether or not a particular bond is involved directly in the rate-controlling step of a reaction. If it is, the effect of changing an atom connected by that bond is referred to as a *primary* effect; otherwise it is described as *secondary*.

Isotope effects, being influenced solely by the atomic masses, are greatest when there is a large relative change in the masses. The effects are usually large when an ordinary hydrogen atom (H) is replaced by deuterium (D) or tritium (T), the relative masses then being approximately 1:2:3; they are even larger when it is replaced by muonium, the mass ratio being 8.83:1. Replacement of ^{12}C by ^{13}C gives a much smaller effect, since the ratio is only 12:13. Even very small isotope effects, however, can provide valuable mechanistic information.

The theory of equilibrium isotope effects was first formulated by Urey[1] and by Bigeleisen and Mayer.[2] Later Bigeleisen[3] dealt with kinetic-isotope effects on the basis of conventional transition-state theory. These early treatments essentially contain the theory that is accepted today; later treatments have involved improvements in computational methods and in refinements in the rate theories. There have been attempts to simplify the original treatments, but some of these have not been too successful and care must be taken in applying them to actual reaction systems.

Under most circumstances the predominant factor in an isotope effect is the zero-point energy, which is effected strongly by atomic masses. When an atom is

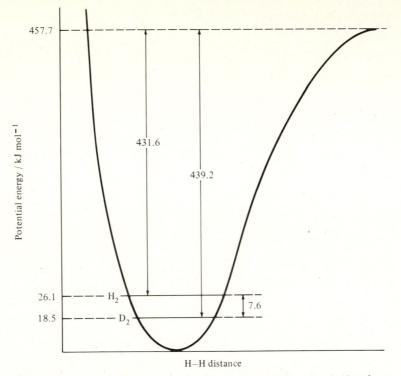

Figure 11.1   Potential-energy curve for the H_2 and D_2 systems, showing the zero-point levels (not to scale).

replaced by an isotope, there is no change in the classical potential-energy surface, but there are changes in vibrational frequencies and therefore in zero-point energies. The situation with H_2 and D_2 is illustrated in Fig. 11.1. The classical dissociation energy is 457.7 kJ mol^{-1}, and the zero-point energy for H_2 is 26.1 kJ mol^{-1}. Replacement of both atoms by D leads to a zero-point energy§ of 18.5 kJ mol^{-1} for D_2. At relatively low temperatures, and this includes temperatures around room temperature, H_2 and D_2 molecules are largely at their zero-point levels. Thus, as far as the initial state of the reaction is concerned, an H_2 molecule, having 7.6 kJ mol^{-1} more zero-point energy than a D_2 molecule, will tend to react more rapidly and its reactions will tend to have a higher equilibrium constant. However, these effects may be offset or even reversed by factors involving the activated state and the products of reaction.

11.1 EQUILIBRIUM ISOTOPE EFFECTS

It is useful to consider first the effects of isotope substitutions on equilibrium constants, for two reasons. First, the theory is much simpler for equilibrium constants, and it

---

§ If vibrations were purely harmonic, the zero-point energy would be inversely proportional to the square root of the reduced mass. In the H_2/D_2 case, the reduced masses are in the ratio of 1/2, and $18.5 = 26.1/\sqrt{2}$. The effect of anharmonicity is to replace $\sqrt{2}$ (=1.41) by a smaller value.

provides a useful basis for understanding the more complicated problem of rate constants. Second, the rates of reactions which occur by composite mechanisms frequently involve equilibrium constants as well as rate constants. Isotope effects must be considered with reference to the rates of elementary reactions and to relevant equilibrium constants and can be understood only if the detailed reaction mechanism is known.

A simple comparison is between the equilibria

$$(1) \quad H_2 \rightleftharpoons 2H$$

and

$$(2) \quad D_2 \rightleftharpoons 2D$$

According to the methods of statistical mechanics (Section 4.5.2), the equilibrium constant for reaction (1) is given by

$$K_H = \frac{q_H^2}{q_{H_2}} e^{-431.6 \text{ kJ mol}^{-1}/RT} \tag{11.1}$$

The energy 431.6 kJ mol^{-1} is the energy required for 1 mol of H_2 to become 2 mol of H (Fig. 11.1). The partition functions are (Table 4.1)

$$q_H = \frac{(2\pi m_H kT)^{3/2}}{h^3} \tag{11.2}$$

and

$$q_{H_2} = \frac{(2\pi m_{H_2} kT)^{3/2}}{h^3} \frac{8\pi^2 IkT}{h^2} q_v \tag{11.3}$$

where I is the moment of inertia of H_2 and q_v is the vibrational partition function. The moment of inertia is given by Eq. (4.69):

$$I = \frac{m_H m_H}{m_H + m_H} d^2 = \tfrac{1}{2} m_H d^2 \tag{11.4}$$

where d is the internuclear distance, which is almost the same in D_2 as in H_2.

In dealing with the isotope effect, one is concerned only with the effect of the masses. From Eqs. (11.1)–(11.4) it follows that, since q_v at ordinary temperatures is close to unity,

$$K_H \propto \frac{(m_H^{3/2})^2}{m_H^{3/2} m_H} e^{-431.6 \text{ kJ mol}^{-1}/RT} \tag{11.5}$$

or

$$K_H \propto m_H^{1/2} e^{-431.6 \text{ kJ mol}^{-1}/RT} \tag{11.6}$$

The corresponding expression for D_2 is

$$K_D \propto m_D^{1/2} e^{-439.2 \text{ kJ mol}^{-1}/RT} \tag{11.7}$$

The proportionality factors are the same in both cases and therefore

$$\frac{K_H}{K_D} = \left(\frac{m_H}{m_D}\right)^{1/2} e^{7.6 \text{ kJ mol}^{-1}/RT} \tag{11.8}$$

Since $m_D = 2m_H$, the pre-exponential factor is $(\tfrac{1}{2})^{1/2} = 0.707$; thus,

$$\frac{K_H}{K_D} = 0.707 e^{7.6 \text{ kJ mol}^{-1}/RT} \tag{11.9}$$

At 300 K

$$\frac{K_H}{K_D} = 0.707 \times 21.1 \tag{11.10}$$

$$= 14.9 \tag{11.11}$$

If only one H atom in H_2 had been replaced by D, to give HD, a similar treatment leads to

$$\frac{K_H}{K_{HD}} \approx 7 \tag{11.12}$$

Another comparison is between the equilibria

$$(1) \quad R-H \rightleftharpoons R + H$$

and

$$(2) \quad R-D \rightleftharpoons R + D$$

where R is an organic radical, assumed to be much heavier than H or D. A typical vibrational frequency for a C—H bond is 2900 cm^{-1}, which corresponds to a zero-point energy ($\frac{1}{2}h\nu$) of 17.4 kJ mol^{-1}. For the C—D bond the vibrational frequency is estimated§ to be 2150 cm^{-1}, which gives a zero-point energy of 12.9 kJ mol^{-1}; the difference is 4.5 kJ mol^{-1}. The equilibrium constant for reaction (1) is then given by

$$K_{RH} = \frac{q_R q_H}{q_{RH}} e^{-17.4 \text{ kJ mol}^{-1}/RT} \tag{11.13}$$

The vibrational factors are again close to unity at ordinary temperatures, so that

$$K_{RH} \approx \frac{m_R^{3/2} I_R m_H^{3/2}}{m_{RH}^{3/2} I_{RH}} e^{-17.4 \text{ kJ mol}^{-1}/RT} \tag{11.14}$$

where I_R is the moment of inertia of R and I_{RH} that of RH. Similarly, for reaction (2),

$$K_{RD} \approx \frac{m_R^{3/2} I_R m_D^{3/2}}{m_{RD}^{3/2} I_{RD}} e^{-12.9 \text{ kJ mol}^{-1}/RT} \tag{11.15}$$

The ratio is thus

$$\frac{K_{RH}}{K_{RD}} \approx \frac{m_{RD}^{3/2} I_{RD} m_H^{3/2}}{m_{RH}^{3/2} I_{RH} m_D^{3/2}} e^{4.5 \text{ kJ mol}^{-1}/RT} \tag{11.16}$$

However, the assumption that R is much heavier than H or D means that

$$m_{RD} \approx m_{RH} \quad \text{and} \quad I_{RD} \approx I_{RH} \tag{11.17}$$

Therefore, since $m_D = 2m_H$,

$$\frac{K_{RH}}{K_{RD}} \approx (\tfrac{1}{2})^{3/2} e^{4.5 \text{ kJ mol}^{-1}/RT}$$

$$= 0.353 \times 6.05 = 2.1 \quad \text{at } 300 \text{ K} \tag{11.18}$$

An interesting situation arises if we compare the abstraction equilibria

$$(1) \quad R' + H-R \rightleftharpoons R'-H + R$$

§ The ratio 2900/2150 = 1.35 differs from $\sqrt{2}$ = 1.41 because of anharmonicity and because the ratio of the reduced masses is slightly greater than 2.

and   (2) $R' + D—R \rightleftharpoons R'—D + R$

The potential-energy profiles in Fig. 11.2 show three different situations for these equilibria. In Fig. 11.2a the potential-energy curves for the H—R and H—R' vibrations have the same shapes, so that the zero-point levels are the same: this means that the zero-point energy differences for H—R and D—R are the same as for H—R' and D—R'. The energies E_D and E_H are therefore equal, and there is no equilibrium isotope effect. This conclusion is independent of whether the reaction is exothermic or endothermic.

Figure 11.2b shows an endothermic reaction (the H—R' level is above that of H—R) in which the H—R' bond is tighter than the H—R bond. In this case the zero-point levels for H—R' and D—R' are further apart than those for H—R and D—R,

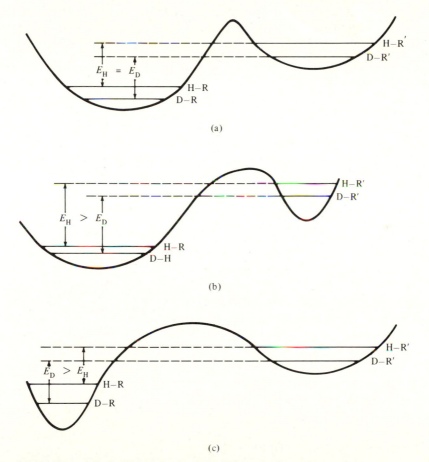

Figure 11.2 Three potential-energy profiles relevant to the equilibria

(1) $R' + H—R \rightleftharpoons R'—H + R$

(2) $R' + D—R \rightleftharpoons R'—D + R$

In (a) the bond strengths are the same; in (b) and (c) they are different on the two sides of the equation. Substitution of H by D always favors the stronger bond.

and the increase in energy E_H is greater than E_D. Because of the endothermicity the equilibrium tends to lie to the left, and since $E_H > E_D$ it lies more to the left for reaction (1). The heavier atom D therefore favors the formation of the stronger bond on the right-hand side of the equation.

Figure 11.2c shows an endothermic reaction in which the bonds on the right are weaker (lower frequencies) than those on the left. Now $E_D > E_H$, and the equilibrium again tends to be to the left, more so for reaction (2) than for reaction (1). Substitution of D for H therefore now causes a shift in equilibrium to the left, that is, in the direction of the stronger bond.

These last two examples illustrate an important principle, namely,

the heavier atom favors the stronger bond

A special case of the same principle was found for the reactions considered with reference to Fig. 11.1. The stronger bond is with the molecules H_2 and D_2, since when there is dissociation there is no bond at all. There is less dissociation with D_2, which means that the heavier atom is favoring the formation of the molecule.

11.1.1 Equilibria in Solution

Similar principles apply to equilibria in solution, but there are additional complications. The possibility of exchange with the solvent must be considered. For example, hydrogen atoms attached to oxygen, nitrogen, and certain other atoms (but not carbon) are exchanged rapidly with an ionizing solvent because of the rapidity of the ionization processes that occur. For example, in water an alcohol undergoes some ionization,

$$-O-H \rightleftharpoons -O^- + H^+$$

the processes occurring rapidly in both directions. Therefore, if a deuterated alcohol ROD is dissolved in ordinary water, the rapid exchange disperses the D atoms in the solvent, which is present in large excess, and the alcohol becomes mainly ROH. Similarly, ROH dissolved in D_2O rapidly becomes ROD. To investigate ROH/ROD isotope effects it is therefore necessary to dissolve the ROH in H_2O and the ROD in D_2O. This gives rise to some difficulties of interpretation, because of the different solvent effects of H_2O and D_2O; for example, there is a little more structure in liquid D_2O than in liquid H_2O.

The relative dissociation constants of an acid HA in H_2O and of DA in D_2O are matters of importance, especially in relation to problems of acid catalysis. The dissociation constant for the process

$$HA \rightleftharpoons H^+ + A^- \quad \text{(in } H_2O\text{)}$$

is usually greater than that for

$$DA \rightleftharpoons D^+ + A^- \quad \text{(in } D_2O\text{)}$$

In many cases the pK difference, pK_D − pK_H, is about 0.6. A somewhat oversimplified explanation for this is as follows.[4] These acid dissociations involve a transfer of H^+ or D^+ to a solvent molecule:

$$HA + H_2O \rightleftharpoons H_3O^+ + A^-$$

$$DA + D_2O \rightleftharpoons D_3O^+ + A^-$$

Usually, HA is a much weaker acid than H_3O^+, the proton being bound more tightly. In terms of Fig. 11.2, replacement of H by D favors the more strongly bonded state DA as compared with D_3O^+, and the dissociation constant is reduced. This explanation is useful for remembering the direction of the isotope effect, but it neglects the numerous other interactions that involve the surrounding solvent molecules. A much more detailed treatment, which takes account of these interactions, has been given by Bunton and Shiner.[5]

An example of how an equilibrium isotope effect can have an important effect on a kinetic problem is provided by studies of the rates of acid-catalyzed reactions in H_2O and D_2O. Although the rates of processes involving H^+ transfers are usually significantly greater than those in which there is a transfer of D^+, the rates of reactions in D_2O catalyzed by D^+ are usually 2–3 times greater than those of reactions in H_2O catalyzed by H^+. This reversal of the expected effect is due to the pre-equilibria that are established. If a substrate S is involved in specific hydrogen-ion catalysis (Section 10.2), the first step is

$$(1) \quad H^+ + S \rightleftharpoons SH^+$$

If the subsequent step

$$(2) \quad SH^+ \rightarrow H^+ + products$$

is rate determining, the overall rate is K_1k_2. Since SH^+ is a stronger acid than SD^+, for reasons just discussed, substitution of H by D causes the pre-equilibrium (1) to lie more over to the right, that is, $K_D > K_H$. Thus, even though k_2 may be reduced by substituting D for H, the effect of the pre-equilibrium usually predominates, with the result that there is an apparent reverse isotope effect.

11.2 PRIMARY KINETIC ISOTOPE EFFECTS

Whereas the treatment of equilibrium isotope effects can be carried out fairly satisfactorily, that of kinetic-isotope effects is much more difficult, for two main reasons:

1. The structure and properties of the activated complexes are involved, and for a detailed knowledge of these it is necessary to calculate a reliable potential-energy surface; as seen in Chapter 3, this presents considerable difficulty for all except the simplest of reactions. It is true that one does not need to know the characteristics of the activated complexes for each of the isotopic forms, but only their differences. However, even this can present a problem.
2. Quantum-mechanical tunneling does not affect equilibrium constants, but it may have a very important influence on the rates of reactions involving H atoms or its ions (H^+ or H^-). As seen in Section 4.9.2, there is still no completely satisfactory treatment of tunneling, and as a result there is uncertainty for some kinetic-isotope effects, especially those involving H, H^+, or H^-.

In spite of these two rather formidable problems, some general principles are now well understood, and treatments have been given which deal satisfactorily with the broader aspects of kinetic-isotope effects.

11.2.1 Semiclassical Treatments

As a first approach to the problem of kinetic-isotope effects it is convenient to neglect quantum-mechanical tunneling; procedures that do so often are referred to as *semi-classical treatments*. Such treatments are reasonably satisfactory for atoms heavier than H. Even when H, H^+, or H^- is involved, neglect of tunneling may not introduce much error at higher temperatures; at temperatures below about 350 K, however, tunneling often becomes significant.

A general and simplified semiclassical discussion of kinetic-isotope effects in hydrogen atom transfer processes is given first. Consider the abstraction reactions

$$R-H + R' \rightarrow (R \cdots H \cdots R')^{\ddagger} \rightarrow R + H-R'$$

and

$$R-D + R' \rightarrow (R \cdots D \cdots R')^{\ddagger} \rightarrow R + D-R'$$

The situation is particularly simple if R and R' are much heavier than either H or D. The partition functions for R—H and R—D are then practically the same, as are the partition functions for the activated complexes $(R \cdots H \cdots R')^{\ddagger}$ and $(R \cdots D \cdots R')^{\ddagger}$. Also, the relevant vibrational frequencies in the activated complexes are almost the same for the two forms. Thus, the frequencies for the symmetric vibrations

$$\leftarrow (R \cdots H \cdots R')^{\ddagger} \rightarrow$$

and

$$\leftarrow (R \cdots D \cdots R')^{\ddagger} \rightarrow$$

are close to one another, since they depend largely on the masses of R and R' and much less on the central masses (of H or D). The same is true of the bending frequencies. The asymmetric vibrational frequencies are different for the two isotopic forms, but these are the ones that are neglected in transition-state theory, since they correspond to passage over the col.

The zero-point energies for the activated complexes are therefore much the same for H and D. This is illustrated in Fig. 11.3a; the energy values relate to a reaction in which H or D is transferred from one carbon atom to another. Thus, since the partition functions for the two isotopic forms are practically the same, the semiclassical ratio k_H^S/k_D^S is determined almost entirely by the zero-point energy difference in the initial state. Thus, it is given by

$$\frac{k_H^S}{k_D^S} \approx e^{4.5 \text{ kJ mol}^{-1}/RT}$$

$$= 6.1 \quad \text{at 298 K} \tag{11.19}$$

The situation is different if one of the radicals R or R' is much lighter than the other. Suppose, for example, that R' is a hydrogen atom, so that we are comparing the reactions

$$R-H + H \rightarrow (R \cdots H \cdots H)^{\ddagger} \rightarrow R + H_2$$

$$R-D + H \rightarrow (R \cdots D \cdots H)^{\ddagger} \rightarrow R + DH$$

In this case, the symmetric vibrational frequencies of the two activated complexes are different from one another, since the symmetric vibration brings about displacement

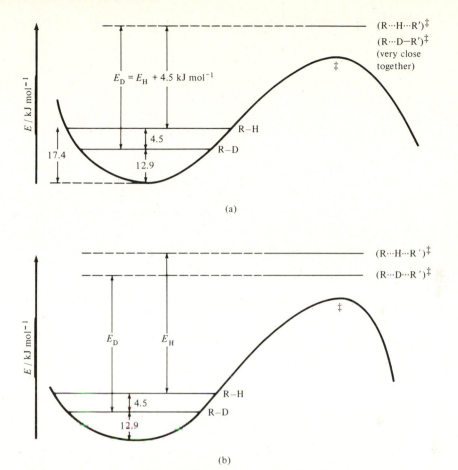

Figure 11.3 Potential-energy profiles for reactions of the type

$$RH + R' \rightarrow R + HR'$$
$$RD + R' \rightarrow R + DR'$$

(a) R and R′ are both much heavier than H or D. (b) Asymmetrical systems in which either R or R′ is lighter than the other. The energy values apply to C—H bonds.

of the central H or D atom. There is now a difference between the zero-point levels in the activated complexes (Fig. 11.3b), and this has the effect of making the k_H^S/k_D^S ratio less than 6.1. Tunneling, of course, increases this ratio.

A similar situation arises for the unsymmetrical systems

$$H—H + R' \rightarrow (H\cdots H\cdots R')^{\ddagger} \rightarrow H + H—R'$$
$$H—D + R' \rightarrow (H\cdots D\cdots R')^{\ddagger} \rightarrow H + D—R'$$

Again there is a difference between the zero-point levels in the two activated complexes, as in Fig. 11.3b, and the k_H^S/k_D^S ratio is reduced.

TABLE 11.1   ESTIMATED MAXIMUM SEMICLASSICAL
k_H^S/k_D^S RATIOS (298 K)

Bond broken	ν/cm^{-1}	$\Delta E_0/\text{kJ mol}^{-1}$	k_H^S/k_D^S
C—H	2900	4.5	6.1
N—H	3100	4.8	7.0
O—H	3300	5.1	7.9
S—H	2500	3.9	4.8

The preceding discussion has been concerned with reactions in which an H or D atom is transferred from a carbon atom. The estimated maximum semiclassical ratios for other bonds are included in Table 11.1. As discussed, these ratios are reduced if there is asymmetry in the activated state, while they are increased if there is quantum-mechanical tunneling.

The discussion so far has been general and considerably simplified; the treatment outlined, however, provides a useful and reliable way of interpreting kinetic-isotope effects and for making approximate predictions. For a complete treatment it is necessary to start with a potential-energy surface since one must know the characteristics of the activated complexes. If the surface is lacking, an approximate semiclassical treatment can be based on the early work of Urey, Bigeleisen, and Mayer, which employed conventional transition-state theory.[1,2] Only a brief account is given here.

The main assumptions made in the Urey–Bigeleisen–Mayer treatment are as follows:

1. Conventional transition-state theory is valid.
2. All vibrations are purely harmonic.
3. Use can be made of the limiting high-temperature expressions for the translational and rotational partition functions (these are given in Table 4.1).
4. The partition functions for types of motion not appreciably affected by the isotopic substitution can be taken to cancel out entirely in the rate expression.
5. Quantum-mechanical tunneling is neglected; that is, the treatment is semiclassical.

Assumptions 1–4 are reasonable as long as the temperature is not too high. They lead to certain *product rules,* long used by spectroscopists, which allow terms in partition functions to cancel out when ratios are written down for two isotopic species. For a reaction of the type XH + Y, where X and Y are heavy compared with H or D, these product rules lead to the following expression for the ratio of rate constants when H is replaced by D:

$$\frac{k_H^S}{k_D^S} = \frac{[m^{\ddagger}(D)/m^{\ddagger}(H)]^{1/2} \prod [u_i^H \operatorname{cosech} (\tfrac{1}{2}u_i^H)]/[u_i^D \operatorname{cosech} (\tfrac{1}{2}u_i^D)]}{\prod [u_i^H \operatorname{cosech} (\tfrac{1}{2}u_i^H)]/[u_i^D \operatorname{cosech} (\tfrac{1}{2}u_i^D)]} \qquad (11.20)$$

In this equation $u_i = h\nu_i/kT$ and $m^{\ddagger}$ is the reduced mass along the reaction coordinate; if X and Y are heavy enough, $m^{\ddagger}(D)/m^{\ddagger}(H)$ is close to $m_D/m_H = 2$. The product in the numerator is taken over all vibrations in the activated state, while that in the denominator is taken over all vibrations of XH or XD. Vibrations that are not much changed by the isotopic substitution are neglected. Sometimes the expression in Eq. (11.20) must be multiplied by a ratio of statistical factors (Section 4.5.4).

Many tests and applications of Eq. (11.20) have been made, particularly by Wolfsberg, Stern, and their collaborators.[6] If as is usual the potential-energy surface is not available, estimates have to be made of the vibrational frequencies ν_1 of the activated complex; sometimes a number of different models for the activated complex are considered, so as to see to what extent the conclusions depend on the model. Calculations made in this way have proved useful, and a correction can be made for quantum-mechanical tunneling.

Calculations on the basis of Eq. (11.20) have led to a number of general conclusions:

1. Only small errors arise if one ignores parts of the reacting molecules that are separated by more than two bonds from the center of reaction. Use of this approximation is referred to as the *cut-off procedure,* and it greatly simplifies the calculations for complicated reacting molecules.

2. If the isotope effect is large, it is sometimes predicted to a useful approximation by the exponential terms alone; in other words, it depends largely on the zero-point-energy differences. This is no longer true for reactions in which no heavy atoms are involved, for example, reactions of the type $H + H_2$; these are considered later. This conclusion—that exponential terms predominate—is by no means true for secondary isotope effects, where the ratios are smaller.

As an example of the applicability of conclusion 2, Schneider and Stern[7] made calculations for a considerable number of reactions over a wide temperature range. They found that the ratios of pre-exponential factors A_H/A_D were usually close to unity and were always between 0.6 and 1.4.

11.2.2 Quantum-Mechanical Tunneling

As far as kinetic-isotope effects are concerned, quantum-mechanical tunneling need be considered only when H or muonium atoms are involved. It has been noted already (Fig. 4.14) that tunneling is most important at lower temperatures. Even at ordinary temperatures k_H/k_D ratios may be substantially greater than the values shown in Table 11.1; a number of values have been compiled in books and articles listed in the bibliography at the end of this chapter.

Some of the lines of evidence pointing to quantum-mechanical tunneling have been mentioned in Section 4.8.3. They include curved Arrhenius plots and abnormally low pre-exponential factors. Studies of kinetic-isotope effects involving H atoms provide additional evidence such as the following:

1. If the pre-exponential factor A_H is small as a result of tunneling, the ratios A_H/A_D and A_H/A_T will be abnormally low, since tunneling will be unimportant with D and T. If there is no tunneling, these ratios are usually not far from unity, and much smaller values indicate that tunneling is occurring with H.

For example, for the base-catalyzed bromination of 2-carbothoxycyclopen-tanone. Bell and coworkers[8] found an A_H/A_D ratio of about 1/24. This is impossible to explain without invoking tunneling.

2. Useful evidence for tunneling is obtained if the ratio k_H/k_T is measured as well as k_H/k_D. In the absence of tunneling there is a predictable relationship between these two ratios, obtained as follows. The isotope effects are now mainly due to zero-point energy differences, which to an approximation vary inversely with the square roots of the reduced masses. If H, D, and T are being transferred from one heavy atom to another, the reduced masses are in the ratio 1:2:3. The logarithms of the rate constants vary approximately linearly with the zero-point energy differences, and it follows that for reactions involving the transfer of H, D, and T from one heavy atom to another, the semiclassical rate constants can be expressed approximately as follows:

$$\ln k_H^S = a + \frac{b}{\sqrt{1}} \tag{11.21}$$

$$\ln k_D^S = a + \frac{b}{\sqrt{2}} \tag{11.22}$$

$$\ln k_T^S = a + \frac{b}{\sqrt{3}} \tag{11.23}$$

where a and b are constants. It then follows that

$$\ln \frac{k_H^S}{k_T^S} = \frac{1 - 1/\sqrt{3}}{1 - 1/\sqrt{2}} \ln \frac{k_H^S}{k_D^S} \tag{11.24}$$

or $$\ln \frac{k_H^S}{k_T^S} = 1.44 \ln \frac{k_H^S}{k_D^S} \tag{11.25}$$

This relationship is represented schematically in Fig. 11.4.

On the other hand, if there is tunneling, $\ln k_H$ is increased but $\ln k_D$ and $\ln k_T$ remain the same. It follows (Fig. 11.4) that the proportionality factor is now less than 1.44. A low value for this ratio is therefore good evidence for tunneling.

11.2.3 Reactions of the Type H + H₂

As discussed in Section 5.1.1, there now exist some reliable experimental data for the rate constants of the four reactions

$$(1) \quad D + H_2 \rightarrow DH + H$$

$$(2) \quad H + H_2 \rightarrow H_2 + H$$

$$(3) \quad D + D_2 \rightarrow D_2 + D$$

$$(4) \quad H + D_2 \rightarrow HD + D$$

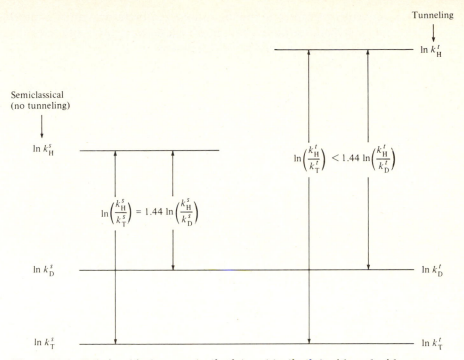

Figure 11.4 Relationship between $\ln (k_H/k_T)$ and $\ln (k_H/k_D)$ with and without tunneling. The superscript s indicates the semiclassical prediction (no tunneling), while the superscript t indicates the value with tunneling.

Some of the results are shown as Arrhenius plots in Fig. 5.2, from which it is seen that the rate constants are in the order

$$k_1 > k_2 > k_3 > k_4$$

Other things being equal, abstraction *of* an H atom is faster than abstraction of D ($k_1 > k_3$ and $k_2 > k_4$). However, abstraction *by* D is faster than abstraction by H ($k_1 > k_2$ and $k_3 > k_4$). The fact that abstraction of H is faster than abstraction of D is understood in terms of initial-state zero-point energies, which are higher for H_2 than for D_2. The differences for abstraction *by* D, however, must be explained in terms of zero-point energies of the activated complexes. Thus, the zero-point energy of $D \cdots H \cdots H^{\ddagger}$ is lower than that of $H \cdots H \cdots H^{\ddagger}$, and therefore the activated state is reached more easily from $D + H_2$ than it is from $H + H_2$. Similarly, $D \cdots D \cdots D^{\ddagger}$ has a lower zero-point energy than $H \cdots D \cdots D^{\ddagger}$. Therefore, the relative rates are easily understood from a qualitative point of view.

Putting the matter on a quantitative basis, however, is more difficult. A number of theoretical treatments have been given on the basis of conventional transition-state theory.[9] The following simple treatment makes use of the activated-complex vibrational frequencies estimated by Shavitt and provides a reasonably reliable interpretation of the ratios at 300 K, where tunneling can be neglected to a good approximation. This is a convenient temperature to use, since the vibrational partition functions are close to unity and therefore do not have much effect on the ratios of pre-exponential factors.

All four reactions can be represented as

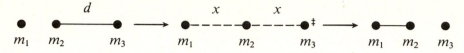

where m_1, m_2, and m_3 are the masses, d is the bond length in the reactant molecule, and x is the distance between neighboring atoms in the activated complex. On the basis of CTST, with the vibrational factors taken as unity, the pre-exponential factor is

$$A = \frac{q_\ddagger}{q_1 q_{23}} \propto \frac{m_\ddagger^{3/2} I_\ddagger}{m_1^{3/2}(m_2 + m_3)^{3/2} I_{23}} \tag{11.26}$$

where the moments of inertia are given by

$$I_{23} = \frac{m_2 m_3}{m_2 + m_3} d^2 \tag{11.27}$$

and

$$I_\ddagger = \frac{m_1 m_2 + 4 m_1 m_3 + m_2 m_3}{m_1 + m_2 + m_3} x^2 \tag{11.28}$$

The distances d and x are the same for all four reactions, and the relative A values calculated on this basis are shown in the last column of Table 11.2. These values do not fall in line with the experimental rates, which must be explained largely in terms of the exponential factors and of the differences between the zero-point energies.

The estimated stretching and bending frequencies for the four activated complexes are shown in columns 2 and 3 of Table 11.3, and column 4 gives the zero-point energies that are calculated from these frequencies. Figure 11.5 shows the relative energy levels for the reactants H_2 and D_2 and the four activated complexes. If the energy difference between H_2 and $D \cdots D \cdots D^\ddagger$ is taken to be x kJ mol^{-1}, the differences for the four reactions are as shown in Fig. 11.5; they also are listed in the fifth column of Table 11.3. The next column of Table 11.3 shows the values of $e^{-E_0/RT}$ at 300 K, normalized to a value of unity for $H + D_2$, the slowest reaction. These values correspond to the correct order of the rates. If the $e^{-E_0/RT}$ values are multiplied by the relative A factors, given in the last column of Table 11.2, one obtains the ratios shown in the last column of Table 11.3, namely,

TABLE 11.2 SEMICLASSICAL RELATIVE A FACTORS FOR REACTIONS OF THE TYPE $H + H_2 \rightarrow H_2 + H$

Reaction	I_{23}	$I_\ddagger$	$m_\ddagger$	Relative A factor
(1) $D + H_2$	1	2.75	4	1.40
(2) $H + H_2$	1	2	3	1.87
(3) $D + D_2$	2	4	6	0.66
(4) $H + D_2$	2	2.8	5	1

TABLE 11.3 CALCULATED ENERGIES AND RELATIVE RATES AT 300 K

Reaction	$\nu_s^{\ddagger}/cm^{-1}$	$\nu_b^{\ddagger}/cm^{-1}$	$E_0^{\ddagger}/kJ\ mol^{-1}$	$E_0/kJ\ mol^{-1}$	$e^{-E_0/RT}$	k_{rel}
(1) $D + H_2 \rightarrow DH + H$	1732	924	21.40	$x + 4.79\,^a$	8.61	12.0
(2) $H + H_2 \rightarrow H_2 + H$	2012	965	23.57	$x + 6.96\,^a$	3.60	6.7
(3) $D + D_2 \rightarrow D_2 + D$	1423	683	16.61	$x + 7.60\,^a$	2.76	1.7
(4) $H + D_2 \rightarrow HD + D$	1730	737	19.17	$x + 10.16\,^a$	$1\,^b$	$1\,^b$

^{a} x kJ mol^{-1} is the energy difference indicated in Fig. 11.6.
^{b} The $e^{-E_0/RT}$ and k_{rel} values are normalized to unity for reaction (4).

$$12.0 : 6.7 : 1.7 : 1$$

The order is correct, although the spread is somewhat greater than found experimentally, which at 300 K is more like 6 : 4 : 2 : 1.

Garrett and Truhlar[10] have carried out detailed treatments of reactions (1) and (4), on the basis of conventional transition-state theory (CTST) and canonical variational transition-state theory (CVTST). Their treatments are based on the ab initio potential-energy surface of Liu and Siegbahn (Section 3.3.2) and on the Marcus–Coltrin theory of tunneling (Section 4.9.2). They arrived at a satisfactory interpretation

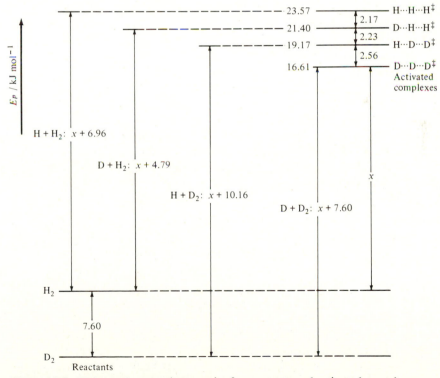

Figure 11.5 Estimated zero-point energies for reactants and activated complexes corresponding to reactions of the type $H + H_2 \rightarrow H_2 + H$.

of the kinetic-isotope effect by locating the dividing surface on a variational basis, even when tunneling is neglected. This, however, they recognize to be somewhat artificial, since there is no doubt that tunneling is important in these reactions. As an alternative procedure they applied CTST but combined it with a quantum treatment of tunneling for a vibrationally adiabatic potential-energy surface; this again gave satisfactory agreement with experiment. Finally, in a more elaborate and detailed treatment they arrived at good agreement with the experimental rates, and therefore with the k_H/k_D ratios for these two reactions, by incorporating the following features:

1. Including anharmonicity more accurately.
2. Using the vibrationally adiabatic model, a least-action tunneling path (following Marcus and Coltrin), and including quantum effects.
3. Including classical recrossing effects.

Garrett and Truhlar have also treated other kinetic-isotope effects along similar lines.[11]

11.2.4 Transfer of H⁺, H, and H⁻

An interesting question that arises for reactions in solution is whether, on the basis of kinetic-isotope ratios, one can tell whether H⁺, H, or H⁻ is being transferred. In some enzyme-catalyzed redox reactions, for example, there is uncertainty as to whether a hydrogen atom is transferred or whether a proton is transferred in the slow step, an electron transfer occurring in a separate step.

From a consideration of the types of activated state that are expected to arise in H⁺, H, and H⁻ transfers, it has been suggested[12] that k_H/k_D ratios will be larger for H⁺ transfer than for H transfer, and larger for H transfer than for H⁻ transfer. The factors invoked in arriving at these conclusions are real ones, but it is realized now that kinetic-isotope effects are considerably more complicated than was originally envisaged. As we have seen, the form of the potential-energy surface and the extent of tunneling are of great importance. For these reasons it now seems doubtful whether one can distinguish between H⁺, H, and H⁻ transfers on the basis of experimental kinetic-isotope ratios.

11.2.5 Reactions of Muonium

In 1960 Hughes and coworkers[13] discovered muonium and recognized it to be an isotope of hydrogen. They later carried out some kinetic studies,[14] and these have been extended to cover a variety of reactions. Such work is of particular value in leading to an understanding of kinetic-isotope effects, since the mass range is greatly extended; a comparison of H and T gives only a threefold range in mass, whereas the H/Mu mass ratio is 8.83 and the T/Mu ratio is 26.4. Table 11.4 summarizes the properties of Mu compared with H, D, and T. The Bohr radii are essentially the same, and the potential-energy surfaces for all the reactions can be taken to be identical.

Of particular interest is some work[15] at high temperatures on the reactions

$$X + H_2 \rightarrow XH + H$$

and

$$X + D_2 \rightarrow XD + D$$

TABLE 11.4 PROPERTIES OF MUONIUM COMPARED WITH H, D, AND T

Atom	Relative mass[a]	Bohr radius/pm	Ionization energy/eV
Mu	208 (1)	0.0532	13.541
^{1}H	1837 (8.83)	0.0529	13.599
^{2}H (D)	3675 (17.67)	0.0529	13.601
^{3}H (T)	5498 (26.4)	0.0529	13.603

[a] Relative to the mass of the electron; in parentheses, relative to the mass of Mu.

where X is Mu or H. The highest temperature used was 850 K, where tunneling is unimportant. Muonium reacted less rapidly than H, the k_{Mu}/k_H ratio at 800 K for the reaction $X + H_2$ being 0.005. This greater reactivity of the heavier isotope H is reminiscent of the fact that D reacts with H_2 and D_2 more rapidly than does H (Table 11.3). The explanation must lie in the zero-point energies of the activated states. Theoretical treatments of this reaction, in terms of variational transition-state theory and reaction dynamics,[16] have given very satisfactory agreement with these experimental results.

Tunneling is expected to be more important at lower temperatures, but these reactions then become too slow for their rates to be measured, in view of the extremely short half-life of muonium (2.2 μs). A number of other reactions have been studied at lower temperatures, and some results are listed in Table 11.5. At the lower temperatures the light isotope reacts more rapidly, and it follows that the influence of the zero-point energy in the activated state is unimportant. The effects must be due to the greater amount of tunneling with the lighter isotopes. For the $X + F_2$ reactions a number of different theoretical treatments have been presented,[17] and they provide useful interpretations of the experimental results.

Some interesting results also have emerged from kinetic-isotope studies of muonium reactions in aqueous solution. Table 11.6 summarizes some results obtained[18] for five different types of reaction involving Mu or H. The rate constants for the first three reactions are those expected for a fully diffusion-controlled reaction (Section 6.7), and the same is true of the activation energies of $\sim$17 kJ mol^{-1}. Furthermore, the temperature dependence of the rate constants divided by the temperature,

TABLE 11.5 KINETIC-ISOTOPE RATIOS (k_{Mu}/k_H AND k_H/k_D) FOR REACTIONS OF Mu ATOMS, H ATOMS, AND D ATOMS

Reaction (X = Mu or H)	Temperature/K	k_{Mu}/k_H	k_H/k_D	Reference
$X + H_2 \rightarrow XH + H$	800	0.055	0.4	_a_
$X + F_2 \rightarrow XF + F$	295	9.2	—	_b_
$X + Cl_2 \rightarrow HCl + Cl$	295	3.5	1.5	_b_
$X + HBr \rightarrow XH + Br$	295	2.9	1.8	_b_
$X + C_2H_4 \rightarrow C_2H_4X$	295	3.8	1.4	_b_

[a] P. M. Garner et al., quoted by D. C. Walker, _Muon and Muonium Chemistry_, p. 101, Cambridge University Press, New York, 1983.

[b] D. G. Fleming, D. M. Garner, and J. R. Mikula, _Hyperfine Interactions_, **8**, 337 (1981).

TABLE 11.6 KINETIC PARAMETERS FOR FIVE TYPES OF MUONIUM REACTION
IN AQUEOUS SOLUTION AND THE k_{Mu}/k_H RATIOS

Reaction	$k/dm^3\ mol^{-1}\ s^{-1}$	$A/dm^3\ mol^{-1}\ s^{-1}$	$E_a/kJ\ mol^{-1}$	k_{Mu}/k_M
Mu + MnO_4^- (reduction)	2.5×10^{10}	3.5×10^{13}	18	1.0
Mu + maleic acid (addition)	1.1×10^{10}	2.3×10^{13}	19	1.4
Mu + Ni^{2+} (spin conversion)	1.7×10^{10}	0.9×10^{13}	16	—
Mu + $HCOO^-$ (abstraction)	8×10^6	0.4×10^{13}	33	0.03
Mu + NO_3^- (?)	1.5×10^9	2×10^{10}	6	155

k/T, is very close to that found for the reciprocal of the viscosity of water. The other two reactions, however, are slower, and the activation energies differ from those associated with diffusion. For these reactions there is less diffusion control and more chemical control.

These results are supported by the observed kinetic-isotope ratios. For the first three reactions, believed to be diffusion controlled, there is essentially no isotope effect. The X + $HCOO^-$ abstraction reaction has a strong inverse kinetic-isotope effect ($k_H > k_{Mu}$), and this is due presumably to the higher zero-point energy in the activated complex when H is replaced by Mu. The large k_{Mu}/k_H ratio for X + NO_3^- must be due to more tunneling with Mu; this is supported by the very low values for both Arrhenius parameters and by a curvature in the Arrhenius plots.

Figure 11.6 summarizes the type of behavior to be expected for reactions in solution where there can be diffusion control or tunneling. The Arrhenius lines are drawn for a common value of the pre-exponential factor. For chemical control the activation energy in the absence of tunneling is greater than for diffusion control, and this is the case for the Mu + $HCOO^-$ reaction. When there is tunneling, however, the Arrhenius plots are curved, and the activation energies at lower temperatures may be lower than those found for diffusion control; this is the case for the Mu + NO_3^- reaction.

It was noted in Section 6.3.4 that at first the nature of muonium was uncertain and that its neutral character was established by ionic-strength studies of its reactions. One reason for this uncertainty was that the kinetic behavior of muonium sometimes closely resembles that of the hydrated electron. This is because many of the reactions of muonium and of the hydrated electron are almost fully diffusion controlled, so that their kinetics depend more on the characteristics of the solvent than on their chemical nature.

11.2.6 Isotope Effects with Heavier Atoms

So far the discussion has been confined to hydrogen and muonium isotope effects, which are the largest because of the large mass ratios. Investigations also have been carried out with some heavier elements. For example, the most abundant isotope of carbon is ^{12}C, and rate ratios with ^{13}C and ^{14}C have been measured. Again, the ratios depend on the differences in zero-point energies, which are very small because of the small mass ratios. Some predicted rate ratios, arrived at on the basis of typical vibrational frequencies as done for the k_H^S/k_D^S ratios, are listed in Table 11.7.

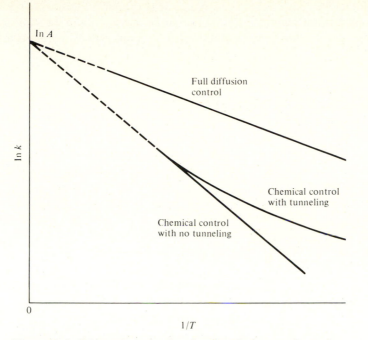

Figure 11.6 Schematic Arrhenius plots for solution reactions where there may be diffusion control or tunneling, for a common pre-exponential factor.

Since the ratios for the heavier elements do not differ greatly from unity, it is unsatisfactory to determine values experimentally by separate experiments with the light and heavy compounds. The best procedure is to allow the processes with the light and heavy reactants to occur together in the same reaction system; this minimizes the experimental errors.

Although the ratios are so small for the heavier elements, the technique is sometimes preferable to studying H/D ratios, since there is no longer the problem of exchange with the solvent. As already noted, the rates of reactions involving deuterated com-

TABLE 11.7 TYPICAL KINETIC ISOTOPE RATIOS AT 25°C

Isotopic forms	Ratios
H, D	6^a
H, T	15^a
^{12}C, ^{13}C	1.04
^{12}C, ^{14}C	1.08
^{14}N, ^{16}N	1.04
^{16}O, ^{18}O	1.04

^{a} These values can be much greater if tunneling occurs.

pounds frequently have to be measured in heavy water, so that a change of solvent is introduced; reactions with heavier atoms can be carried out in the same solvent. Also, the isotope results with heavier atoms are not complicated by quantum-mechanical tunneling.

11.3 SECONDARY KINETIC ISOTOPE EFFECTS

When the isotopic substitution involves an atom that is not at the end of a bond that is broken or formed during reaction, the effect is generally much smaller than when it is a primary effect. Only a brief account of secondary effects is given here, since they are by no means as useful in providing mechanistic information.

Secondary effects are due largely to changes in vibrational frequencies in the reactants and activated complexes, these changes affecting the zero-point levels and hence the activation energies. An example is provided by S_N1 substitution reactions, the rates of which are controlled by an ionization process such as

$$\overset{\diagdown}{\underset{\diagup}{}}C-X \longrightarrow \overset{\diagdown}{\underset{\diagup}{}}C^{\delta+}\cdots X^{\delta-\ddagger} \longrightarrow \overset{|}{\underset{\diagup\diagdown}{}}C^+ + X^-$$

Initially, there is sp^3 hybridization about the carbon atom, and finally, there is sp^2 hybridization; in the activated complex the hybridization is intermediate between

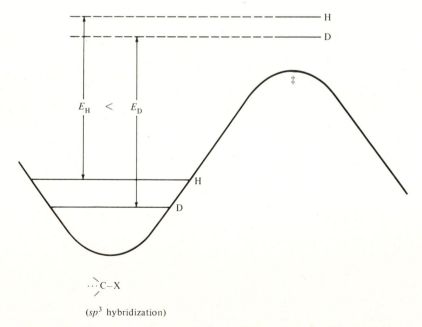

Figure 11.7 Potential-energy profile for an S_N1 reaction, the slow step being the ionization of the C—X bond. A small secondary isotope effect is obtained when H is replaced by D at a position close to the C—X bond.

sp^3 and sp^2. The bending vibration frequencies are lower in sp^2 hybridization than in sp^3. There is therefore a decrease in the frequencies when the activated complex is formed. If an H atom attached to the carbon atom at the reaction center is replaced by D, the zero-point energies in the reactant and in the activated complex are lowered. However, since the vibrational frequencies are lower in the activated state, the H and D levels are closer together than in the initial state (Fig. 11.7). The result is a small positive kinetic-isotope effect; k_H/k_D ratios for this type of system are typically 1.3–1.4.

Sometimes secondary kinetic-isotope effects are interpreted in terms of differing inductive and steric effects for the isotopic forms. For example, D appears to be slightly more electron-donating than H and to occupy somewhat less space; the latter effect arises from the anharmonicity of the vibration and from the fact that the amplitude of the vibration of a C—D bond is less than that of a C—H bond.

REFERENCES

1. H. C. Urey, *J. Chem. Soc.,* 569 (1947).
2. J. Bigeleisen and M. G. Mayer, *J. Chem. Phys.,* **15,** 261 (1947).
3. J. Bigeleisen, *J. Chem. Phys.,* **17,** 675 (1949).
4. R. P. Bell, *The Proton in Chemistry,* 2nd ed., pp. 232–234, Chapman and Hall, London, 1973.
5. C. A. Bunton and V. J. Shiner, *J. Am. Chem. Soc.,* **83,** 42 (1961).
6. M. Wolfsberg and M. J. Stern, *Pure Appl. Chem.,* **8,** 225, 325 (1964); M. J. Stern and M. Wolfsberg, *J. Chem. Phys.,* **39,** 2776 (1963); **45,** 2618 (1966); *J. Pharm. Sci.,* **54,** 849 (1965); M. J. Stern, A. E. Schneider, and P. C. Vogel, *J. Chem. Phys.,* **55,** 4286 (1971); P. C. Vogel and M. J. Stern, *J. Chem. Phys.,* **54,** 779 (1971); M. J. Stern and P. C. Vogel, *J. Am. Chem. Soc.,* **63,** 4664 (1971); M. E. Schneider and M. J. Stern, *J. Am. Chem. Soc.,* **94,** 1517 (1972).
7. M. E. Schneider and M. J. Stern, *J. Am. Chem. Soc.,* **94,** 1517 (1972).
8. R. P. Bell, J. A. Fendley, and J. R. Hulett, *Proc. R. Soc. London A,* **235,** 453 (1956); J. R. Hulett, *Proc. R. Soc. London A,* **251,** 274 (1959).
9. For example, R. E. Weston, *J. Chem. Phys.,* **31,** 892 (1959); D. Rapp and R. E. Weston, *J. Chem. Phys.,* **36,** 2807 (1962); R. E. Weston, *Science,* **158,** 332 (1967); I. Shavitt, *J. Chem. Phys.,* **31,** 1359 (1959).
10. B. C. Garrett and D. G. Truhlar, *J. Chem. Phys.,* **76,** 3460 (1980).
11. B. C. Garrett and D. G. Truhlar, *J. Am. Chem. Soc.,* **102,** 2559 (1980); B. C. Garrett, D. G. Truhlar, R. S. Grey, A. W. Magnuson, and J. N. L. Connor, *J. Chem. Phys.,* **73,** 1721 (1980); B. C. Garrett, D. G. Truhlar, and A. W. Magnuson, *J. Chem. Phys.,* **74,** 1029 (1981); **76,** 2321 (1982); D. C. Clary, B. C. Garrett, and D. G. Truhlar, *J. Chem. Phys.,* **78,** 777 (1983).
12. C. G. Swain, R. A. Wiles, and R. F. W. Bader, *J. Am. Chem. Soc.,* **83,** 1945 (1961).
13. V. W. Hughes, D. W. McColm, K. Kiock, and R. Prepost, *Phys. Rev. Lett.,* **5,** 63 (1960).
14. R. M. Mobley, J. M. Bailey, W. E. Cleland, V. W. Hughes, and F. E. Rothberg, *J. Chem. Phys.,* **44,** 4354 (1966).
15. D. M. Garner et al., quoted in D. C. Walker, *Muon and Muonium Chemistry,* vol. 6, p. 99, Cambridge University Press, New York, 1983.
16. D. K. Bondi, D. C. Clary, J. N. L. Connor, B. C. Garrett, and D. G. Truhlar, *J. Chem. Phys.,* **76,** 4986 (1982); N. C. Blais, D. G. Truhlar, and B. C. Garrett, *J. Chem. Phys.,* **18,** 2363 (1983).

17. J. N. L. Connor, W. Jakubetz, and J. Manz, *Chem. Phys. Lett.,* **45,** 265 (1977); *Chem. Phys.,* **28,** 219 (1978); S. F. Fisher and G. Venzl, *J. Chem. Phys.,* **67,** 1335 (1977); J. N. L. Connor, W. Jakubetz, and A. Lagana, *J. Phys. Chem.,* **83,** 73 (1979); J. N. L. Connor and A. Lagana, *Mol. Phys.,* **38,** 657 (1979); W. Jakubetz, *J. Am. Chem. Soc.,* **101,** 298 (1979); J. N. L. Connor, *Hyperfine Interactions,* **8,** 423 (1981); J. N. L. Connor, A. Lagana, A. F. Turfa, and J. C. Whitehead, *J. Chem. Phys.,* **75,** 3301 (1981).

18. B. W. Ng, Y. C. Jean, Y. Ito, T. Suzuki, J. H. Brewer, D. G. Fleming, and D. C. Walker, *J. Phys. Chem.,* **85,** 454 (1981).

BIBLIOGRAPHY

For general treatments of isotope effects see the following publications:

R. P. Bell, Recent Advances in the Study of Kinetic Hydrogen Isotope Effects, *Chem. Soc. Rev.,* **3,** 513 (1974).

J. Bigeleisen, M. W. Lee, and F. Mandell, Equilibrium Isotope Effects, *Ann. Rev. Phys. Chem.,* **24,** 407 (1973).

E. F. Caldin, *Fast Reactions in Solution,* Blackwell, Oxford, 1964.

W. P. Jencks, *Catalysis in Chemistry and Biochemistry,* chap. 4, McGraw-Hill, New York, 1969.

H. S. Johnston, *Gas-Phase Reaction Rate Theory,* chap. 13, Ronald, New York, 1966.

F. S. Klein, Isotope Effects in Chemical Kinetics, *Ann. Rev. Phys. Chem.,* **26,** 191 (1975).

L. Melander and W. H. Saunders, *Reaction Rates of Isotopic Molecules,* Wiley, New York, 1980.

For treatments of tunneling see:

R. P. Bell, *The Tunnel Effect in Chemistry,* Chapman and Hall, London, 1980.

D. G. Truhlar and B. C. Garrett, Variational Transition State Theory, *Ann. Rev. Phys. Chem.,* **35,** 159 (1984), especially pp. 178–182.

For accounts of kinetic-isotope effects involving muonium see:

J. N. L. Connor, Isotope Effects and Chemical Reaction Dynamics of Muonium in the Gas Phase, *Hyperfine Interactions,* **8,** 423 (1981).

D. G. Fleming, D. M. Garner, L. C. Vaz, D. C. Walker, J. H. Brewer, and K. M. Crowe, Muonium Chemistry—A Review, in H. J. Ashe (Ed.), *Positronium and Muonium Chemistry,* American Chemical Society, Washington, DC, 1979.

P. W. Percival, E. Roduner, and H. Fischer, Radiation Chemistry and Reaction Kinetics of Muonium in Liquids, in H. J. Ashe (Ed.), *Postronium and Muonium Chemistry,* American Chemical Society, Washington, DC, 1979.

D. C. Walker, Muonium: A Light Isotope of Hydrogen, *J. Phys. Chem.,* **85,** 3960 (1981).

D. C. Walker, *Muon and Muonium Chemistry,* chaps. 7 and 8, Cambridge University Press, New York, 1983.

Chapter 12

Reaction Dynamics

In the early years of physical chemistry the word "dynamics" had a much broader meaning than it has today, in that it comprised all of chemical kinetics and also the study of chemical equilibrium. According to van't Hoff,[1] for example,

> dynamics is devoted to the mutual actions of several substances, i.e., to chemical change, affinity, velocity of reaction, and chemical equilibrium.

Today, however, the term "dynamics," or more specifically "reaction dynamics," has a much narrower meaning: it refers to an important and rapidly growing branch of kinetics that is concerned with the details of the elementary act that occurs during the course of chemical change. Since it deals with processes at the molecular, or microscopic, level, the term "molecular dynamics" often is employed. Reaction dynamics deals with the intermolecular and intramolecular motions that occur in the elementary act of chemical change and with the quantum states of the reactant and product molecules.

The importance of reaction dynamics is that it greatly supplements and enriches the work on macroscopic aspects of kinetics. The disadvantage of macroscopic kinetic investigations is that the reactant molecules are present in a range of energy states, so that one gains only indirect information about the individual molecular events that occur during a chemical reaction. Statistical treatments have to be applied to understand the behavior of bulk systems, and such treatments are essential to an understanding of how chemical change occurs under ordinary conditions. Over recent years these treatments have been assisted greatly by both experimental and theoretical investigations in reaction dynamics. However, reaction dynamics can never supersede macroscopic kinetics. As seen throughout the preceding chapters, applications of thermodynamics and statistical mechanics have led to many simple and important generalizations in macroscopic kinetics. Reaction dynamics, with its heavy dependence on computations for individual reaction systems, does not lead as readily to such generalizations, although it may confirm and greatly supplement them.

There are two main reasons why it is important for dynamical studies to be made. In the first place, questions have been raised and considered in Section 4.8 as to the validity of the macroscopic theories, such as transition-state theory in its different versions. There is no doubt that these theories do break down to some extent, especially for very rapid reactions. Experimental and theoretical studies in molecular dynamics provide an alternative approach to the problem. The experimental and theoretical work in reaction dynamics has indicated that transition-state theory is not far from the truth under many circumstances, and has given some indication of the conditions under which it becomes less satisfactory. Some of this work is outlined in Section 12.1, and no doubt further dynamical work will contribute much more to solving this problem.

A second reason for the increasing interest in reaction dynamics is that there are many applications for which it is important to have information about the precise quantum states of reactant and product molecules. For example, the development of lasers, so important in modern technology, owes much to this kind of information. The macroscopic experimental studies and the theories of kinetics, such as transition-state theory, are concerned with Boltzmann distributions among energy states and provide no information about the electronic, vibrational, rotational, and translational states of product molecules; molecular dynamic studies are essential for this purpose.

The subject of reaction dynamics is being approached from a number of different directions, and the techniques are developing rapidly. The ultimate objective of all the work is to investigate what is referred to as *state-to-state kinetics*. As far as possible, reactant molecules are prepared in selected electronic, vibrational, and rotational states, and by the use of spectroscopic techniques the quantum states of the product molecules are determined. Broadly speaking, we can classify the experimental procedures as follows:

1. Experiments in which the energy states of the reaction products are determined by spectroscopic techniques. The reaction itself may occur in a static or a flow system, and the reactant molecules may have been put into particular energy states. The term *chemiluminescence* is used for certain studies of this type.
2. Experiments in which narrow beams of reactant molecules, in preselected states, are brought into contact with each other, often at an angle of 90°. In these *molecular-beam* experiments the primary purpose is to determine the fate of the reactants and products after the beams have come together.

To complement these experimental studies, dynamical calculations are made. These theoretical studies are made in parallel with the experimental investigations and are helpful in interpreting the results of the experiments; at the same time the experimental studies are important in suggesting new lines of theoretical study. It seems expedient first to deal briefly with the theoretical calculations and then to discuss the main experimental results against the background of the theory.

12.1 MOLECULAR-DYNAMICAL CALCULATIONS

The starting point for a dynamical calculation relating to a chemical reaction is a suitable potential-energy surface. As discussed in Chapter 3, for simple molecular systems these can be calculated a priori by the methods of quantum mechanics. For other systems it is necessary to introduce a good deal of empiricism. For certain purposes it is useful to make dynamical calculations using stylized potential-energy surfaces, the shapes of which are suggested by experimental results, particularly those obtained from chemiluminescence or molecular-beam experiments. Calculations of this type have proved useful for determining what features of potential-energy surfaces lead to particular types of energy disposal during a reaction.

After a potential-energy surface has been constructed, dynamical computations involve considering a variety of initial conditions for the reacting molecules and determining how, for each choice of initial conditions, the system moves on the surface. A diagram or mathematical description that describes the motion of a reaction system over a potential-energy surface is known as a *trajectory*. A useful analogy, first pointed out by Hirschfelder,[2] is that if a surface is constructed with suitable skewed coordinates, a trajectory would be represented satisfactorily by the motion of a frictionless particle sliding over the surface. This analogy is useful in visualizing some of the trajectories considered later in this chapter.

Ideally, trajectory calculations would be done quantum mechanically, and some reviews of such calculations are listed in the bibliography at the end of this chapter. In view of the difficulty in making such calculations, however, trajectories often are obtained by classical-mechanical methods, and when this is done certain errors are involved. One of these is that reaction is allowed to proceed along paths that are inconsistent with quantum restrictions; even if the reactant molecules are put into quantized states, the transition species and the products are not necessarily in such states if a classical trajectory is calculated.

Another important source of error in a classical calculation is that there may be a significant amount of tunneling through the potential-energy barrier, and this effect is neglected in the classical calculations. This error is serious if hydrogen atoms are involved at the reaction center. In any case, a correction for tunneling (Sections 4.8 and 4.9) can be applied to the results of the calculations.

In spite of these drawbacks to the classical calculations, there is good reason to believe that they provide a reliable guide to the general features of reaction dynamics. One point of some importance is that, except for certain reactions where tunneling is significant, quantum effects tend to be blurred when collisions are treated, as they should be, in three dimensions; the quantum effects therefore do not have much influence on the behavior that is observed experimentally in reaction-dynamical studies. These conclusions are drawn from quantum-mechanical and classical calculations and the comparison of them with experiment.

Classical calculations of the motion over potential-energy barriers were discussed for the $H + H_2$ system by Eyring and Polanyi,[3] and actual computations were carried out by Hirschfelder, Eyring, and Topley.[4] The results are not considered here since these pioneering calculations, important when they were done, were necessarily not very reliable. Since computers had not been developed, point-by-point computations had to be made of the successive coordinates of the moving system. Because of the great amount of work involved, the calculations could be made only for linear approaches of the H atom to the H_2 molecule, and the number of trajectories calculated was too small to make the results statistically meaningful.

With the development of digital computers, many more detailed calculations have been made. The first computer calculations of reaction dynamics were performed by Wall and coworkers,[5] for the H + H_2 system. These calculations related to the LEP potential-energy surface, which has a basin, and integrations were carried out for the motion over the surface. The earlier calculations related to the linear H$\cdots$H$\cdots$H complex, and the initial conditions were varied systematically. Nonlinear complexes were considered also, but attention was restricted to collisions in which the plane of the H_3 complex does not rotate in space; calculations of this type are referred to as two-dimensional (2D) calculations. The starting conditions (rotational, vibrational, and translational energies, etc.) were chosen by a weighted random method. Unfortunately, in the second series of calculations, very few collisions (6 out of 700) led to reaction, so that the results for the reactive collisions were not statistically significant. Wall and Porter[6] later carried out calculations using an empirical surface which had no basin,[7] but they also considered only linear complexes. Similar calculations were made by Blais and Bunker[8] and by Polanyi and Rosner.[9] They used a basinless LEPS surface, without the linearity restriction, but confined the motion of the atoms to a single plane. These calculations were the first to show that for a basinless potential-energy surface the system passes directly through the activated state; in other words, the reaction is *direct*. Other calculations, such as those of Child and Roach[10] on the K + NaCl system, have shown that when there is a basin the complex performs a number of vibrations before giving rise to products; it is said to occur by an *indirect* or *complex-mode* mechanism.

Subsequent to this work, a very large number of dynamical calculations have been made, and the results have contributed greatly to an understanding of both macroscopic and microscopic kinetics. Many of these results will be referred to in later pages, and the present section merely summarizes some of the main procedures and conclusions.

In carrying out trajectory calculations, the following initial parameters have to be considered:

1. The velocities of the reactant molecules relative to the center of mass of the system. Thus, for a reaction A + BC → AB + C, an initial parameter is the relative velocity of A with respect to the center of mass of BC.
2. The vibrational energy of the reactant molecules.
3. The vibrational phase of the reactant molecules relative to their approach.
4. The rotational energy of the reactants.
5. The rotational phase of the reactants.
6. The impact parameter b. This, as was illustrated in Fig. 4.2, is the closest distance of approach of A to the center of mass of BC if the two molecules continued with their initial velocities without interacting with each other.

If a large number of trajectories are calculated, corresponding to appropriate variations in these initial conditions, valuable information is obtained about the effect of the various factors on the probability of reaction. For example, it is possible to establish, for a given potential-energy surface, the relative importance of vibrational and translation energy on the reaction probability.

To calculate the macroscopic rate constant for a reaction, it is necessary to use probability distribution functions to make an appropriate weighted random selection of the initial parameters. This is done commonly by use of the so-called *Monte Carlo* procedure, the details of which have been explained by Blais and Bunker.[11]

Frequently, the results of dynamical calculations and experiments are quoted in terms of reaction cross sections rather than of rate constants. In Section 4.1.1 the *reaction cross-section* was defined [Eq. (4.19)] as $P_r \pi d^2$ where P_r is the probability of reaction and d (sometimes called the *collision diameter*) is the distance between the centers of the reactant molecules when they collide. More generally, when one does not make the assumption of a well-defined collision diameter, the reaction cross section is defined as $P_r \pi b_{max}^2$, where b_{max} is the maximum value of the impact parameter b that leads to reaction. In practice, the reaction cross section σ is the quantity that has to be postulated in using collision theory to interpret calculated or experimental rates. It can be used for macroscopic rates and also for rates corresponding to specified conditions (e.g., for reactants in particular quantum states).

If simple collision theory were applicable, and there were no activation energy, the reaction cross section σ would be πd^2. Often, the probability of reaction, P_r, is less than unity because of an activation energy or for other reasons; in that case, σ is less than πd^2.

Table 12.1, which relates reaction cross sections and collision diameters, may be helpful in assessing the significance of values quoted for particular reactions. It will be seen in Sections 12.2 and 12.4 that some bimolecular reactions, such as those between alkali metal atoms and halogen molecules, have cross sections of about 10^6 pm^2 (100 Å^2). They have zero activation energies so that P_r is close to unity; reaction thus occurs when the reactant species are over 500 pm (5 Å) apart. Other

TABLE 12.1 REACTION CROSS SECTIONS AND
COLLISION DIAMETERS
FOR $P_r = 1$

Reaction cross section		Collision diameter	
σ/pm^2	σ/Å^2 [a]	d/pm	d/Å[a]
10^6	100	560	5.6
7.9×10^5	79	500	5.0
10^5	10	180	1.8
3.1×10^4	3.1	100	1.0
10^4	1	56	0.56
7.9×10^3	0.79	50	0.5
10^3	0.1	18	0.18
310	0.031	10	0.1

[a] One angstrom (Å) = 10^{-10} m = 100 pm; 1 Å^2 = 10^{-20} m^2 = 10^4 pm^2.

reactions, such as those between K and CH_3I, have cross sections that are less than 10^5 pm^2 (10 Å^2), which means that the collision diameter is somewhat below 200 pm (2 Å). Such collision diameters are close to the sum of the kinetic theory radii, whereas the larger diameters call for a special explanation, to be considered in Section 12.3. When reaction cross sections are substantially less than 10^5 pm^2 (10 Å^2) the reaction probability must be small, probably as a result of an activation energy.

12.1.1 The Reaction H + H$_2$

A large number of dynamical calculations have now been made for the reaction

$$H + H_2 \rightarrow H_2 + H$$

for a variety of potential-energy surfaces. Very significant results were revealed by some calculations made by Karplus and coworkers,[12] on the basis of a semiempirical potential-energy surface calculated by Porter and Karplus.[13] The zero-point level of the H_2 molecule was taken into account, but from then on the reaction was treated classically. This procedure is sometimes called the *quasiclassical trajectory* (QCT) *method*. The rotational and vibrational states, the initial orientation of the H_2 molecule, and other factors were varied systematically to determine their effect on the reaction probability.

An energy diagram for the system is shown in Fig. 12.1. The classical barrier height is 38.2 kJ mol^{-1}, and the calculations showed that the minimum translational energy required to cause the system in its classical ground state to pass over the col was slightly more than this, 39.3 kJ mol^{-1}. The zero-point vibrational energy of H_2 is 25.9 kJ mol^{-1}, and in this level the system therefore requires 12.3 kJ mol^{-1} to reach the barrier from the zero-point level. However, the calculations showed that much more than this, 23.8 kJ mol^{-1}, was required for the system to pass through the col from the zero-point level. Since the calculation was classical, this is not due to the zero-point level for the activated complex. Instead, it shows that of the 25.9 kJ mol^{-1} provided by the zero-point energy, only about two-thirds contributes to reaction. The rest goes into vibrational energy of the activated complex and ends up as the rotational, vibrational, and translational energies of the products. In other words, for purely dynamical reasons, the system cannot pass through the col at its lowest point.

By using suitable weighted averaging procedures, Karplus and coworkers made estimates of the rate constants over the range of temperature from 300 to 1000 K. The results were in good agreement with experiment and were consistent with the Arrhenius equation with an activation energy of 31.1 kJ mol^{-1}. The value corresponding to transition-state theory for the surface they used is somewhat higher, 36.8 kJ mol^{-1}, and this is because the zero-point energy of the activated complex is now involved. Because the assumptions are so different, no precise comparison can be made between these dynamical calculations and transition-state theory, but there are no serious inconsistencies.

The calculations of Karplus and coworkers also provided some very interesting results about the details of collisions in the H + H$_2$ system. Some typical results for reactive collisions are shown in Fig. 12.2. In these collisions and in most of those investigated, the collision complex passes directly into products and does not survive

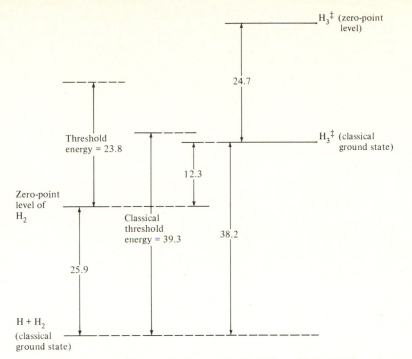

Figure 12.1 Energy diagram for the system $H + H_2 \rightarrow H_2 + H$. The values are energy/kJ mol^{-1} and relate to the calculations of Karplus, Porter, and Sharma.[12]

long enough for any vibrations or rotations to occur; this is shown by the fact that the curves cross one another only once. In a very small fraction of the trajectories, the activated complex did undergo a vibration or two, but this is very atypical behavior that does not affect calculated macroscopic rates to any extent.

In later investigations Karplus and coworkers[14] carried out detailed trajectory calculations designed to test conventional transition-state theory (CTST). To help in the comparison of the different treatments, they defined and calculated three different reaction probabilities, relating to a total reactant energy E:

1. $\bar{P}_r(E)$ is the average reaction probability obtained in the classical dynamical calculations.
2. $\bar{P}_{CTST}(E)$ is that obtained, on the same potential-energy surface, by CTST.
3. $\bar{P}_{trans}(E)$ is the probability of reaching the transition state.

These definitions are useful in leading to conclusions about the reasons for discrepancies between CTST and trajectory values of the reaction probability. If $\bar{P}_{CTST}$ is larger than $\bar{P}_{trans}$, this means that there is an error arising from the CTST assumption that activated complexes are in equilibrium with the reactants: in reality, the concentration of complexes is lower than that corresponding to equilibrium. If $\bar{P}_r$ is lower than $\bar{P}_{trans}$, this means that not all the activated complexes become products; after reaching the tran-

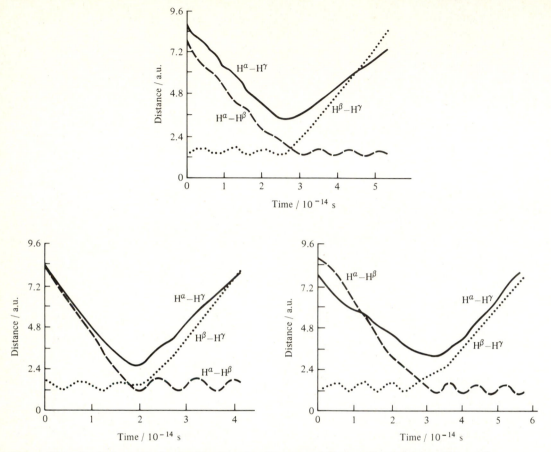

Figure 12.2 Typical results obtained by Karplus, Porter, and Sharma for the system $H^\alpha + H^\beta$ — $H^\gamma \rightarrow H^\alpha - H^\beta + H^\gamma$. The curves show the variations with time for the H—H distances for three reactive collisions. [1 atomic unit (a.u.) = 59.92 pm.]

sition state some of the systems are bounced back by the walls of the potential-energy surface. Thus, it is possible to see whether CTST is in error because of its equilibrium assumption or because of its assumption that the transition state is a state of no return.

Comparisons were made for the reactions $H + H_2$ and $T + H_2$ and also for the hypothetical process $\Delta + H_2$ where Δ is a "hydrogen" atom of mass 15. The surface used was a modified London–Eyring–Polanyi–Sato (LEPS) surface[15] (see Section 3.4). For the $H + H_2$ system, $\bar{P}_{CTST}$ values were in satisfactory agreement with the $\bar{P}_r$ values up to energies of about 25 kJ mol^{-1} in excess of the threshold value (38.2 kJ mol^{-1} for the surface used). In ordinary kinetic studies, the energies are mostly well below this excess value, so that errors in using CTST are small for this reaction.

At higher energies, the $\bar{P}_{CTST}$ values become somewhat higher than the $\bar{P}_r$ values, and this is due largely to the fact that the $\bar{P}_{CTST}$ values are higher than the $\bar{P}_{trans}$ values. This means that there is a nonequilibrium distribution of activated complexes, the concentration of which is less than the equilibrium value assumed in CTST. At these higher energies, the calculated $\bar{P}_r$ values are still close to the $\bar{P}_{trans}$ values, so that systems reaching the activated state almost inevitably become products.

For the systems T + H_2 and Δ + H_2, the agreement was again good at lower energies, but at higher energies the discrepancies were somewhat greater than for H + H_2. Again, CTST predicts too high values, largely because of a nonequilibrium population in the transition state.

Morokuma and Karplus[16] also explored the effects of making certain modifications to the potential-energy surface. They found that placing a small basin at the col or a hill in the entrance valley had little effect. Introducing a hump on the inner wall or shifting the inner wall had more effect on the agreement between $\bar{P}_{CTST}$ and $\bar{P}_r$.

These comparisons show that for these systems CTST is in satisfactory agreement with trajectory calculations in the lower-energy range. This agreement is favored by the symmetry of the potential-energy surface for this H + H_2 system. As is now discussed, for the Br + H_2 system, where the surface is unsymmetrical, the agreement is not as good.

12.1.2 The Reaction Br + H$_2$

Morukuma and Karplus[16] also made a comparison between CTST and trajectory calculations for the endothermic reaction

$$Br + H_2 \rightarrow HBr + H \qquad \Delta H° = 80.3 \text{ kJ mol}^{-1}$$

and for the reverse process. They made use of a fitted LEPS surface obtained by Weston;[17] the classical barrier height is 80.8 kJ mol^{-1}, so that the barrier is very low in the reverse direction. This surface is of the late-barrier type, of the general form shown later in Fig. 12.10; it therefore provides a means of testing the effect of a highly unsymmetrical potential-energy surface.

For the Br + H_2 reaction, there was good agreement between $\bar{P}_{CTST}$ and $\bar{P}_r$ at excess energies up to about 4 kJ mol^{-1}. At higher energies, however, the discrepancies became considerable; at an excess energy of about 40 kJ mol^{-1}, for example, $\bar{P}_{CTST}$ was about 10 times greater than $\bar{P}_r$. At these high energies, $\bar{P}_r$ and $\bar{P}_{trans}$ were still close to each other, so that the error in CTST arises largely from lack of equilibrium at the activated state. What these dynamical calculations reveal is that at higher energies many trajectories are bounced back by the inner walls of the potential-energy surface (cf. Fig. 12.10) and never reach the activated state; those that do so are almost certain to become products. The calculations show that whereas $\bar{P}_{CTST}$ increases steadily with increase in reactant energy, $\bar{P}_r$ passes through a maximum. The decrease at the higher energies results from an increase in the probability of bouncing back.

In spite of these discrepancies at high energies, CTST still leads to fairly satisfactory rates under ordinary circumstances, because of the much greater importance of reaction occurring at low energies. Thus, at 500 K the rate constant calculated by CTST was 1.38 times that calculated dynamically.

The situation with the reverse reaction H + HBr → Br + H_2 is different. There is now good agreement between $\bar{P}_{CTST}$ and $\bar{P}_{trans}$ up to quite high energies. For the reaction proceeding in this direction, the col lies in the entrance valley and is reached

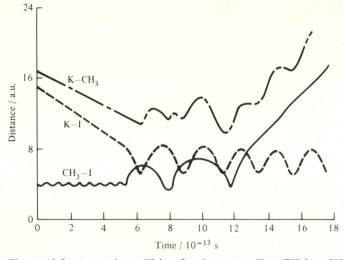

Figure 12.3 A reactive collision for the system $K + CH_3I \rightarrow KI + CH_3$. The complex undergoes some vibrations before splitting into products. [1 atomic unit (a.u.) = 59.92 pm.]

easily; as a result, the activated complexes are essentially in equilibrium with the reactants. However, at energies higher than about 4 kJ mol^{-1} there is a decrease in $\bar{P}_r$, while $\bar{P}_{CTST}$ and $\bar{P}_{trans}$ continue to rise with increasing energy. The reason for the decrease in $\bar{P}_r$ is that systems that have reached the transition state often are bounced back by the inner wall of the potential-energy surface and return to the reactant valley.

The general conclusion from these studies is that CTST is very satisfactory for fairly symmetrical surfaces, but less so when there is much asymmetry in the potential-energy surface.

12.1.3 More Complex Reactions

Many dynamical calculations have been made on reactions involving more than three atoms and are referred to in a number of places later in this chapter. The reaction

$$K + CH_3I \rightarrow KI + CH_3$$

deserves mention since its potential-energy curve has a basin at the col, and its characteristics therefore are different from those proceeding by a direct mechanism. The dynamics have been studied by Karplus and Raff[18] and a typical trajectory is shown in Fig. 12.3. Because of the basin, the collision complex is of sufficiently long life to undergo a few vibrations; as a result, the curves in Fig. 12.3 cross several times. Reactions of this type are said to be *indirect* or to occur by a *complex-mode mechanism.*§ The lifetimes are usually greater than 5×10^{-12} s, which is several times the rotational period of $1–2 \times 10^{-12}$ s.

§ Such reactions sometimes are said to be "complex," but this is apt to give rise to confusion with reactions occurring by more than one elementary reaction.

12.2 CHEMILUMINESCENCE

The bulk experiments in the field of reaction dynamics have their origin in important investigations carried out in the 1920s and 1930s by Michael Polanyi and his collaborators.[19] This work was concerned mainly with the radiation emitted when alkali metals react with halogens or certain halides in the gas phase. Two different techniques were used, according to whether the reaction had zero activation energy, and was therefore fast, or occurred more slowly. The fast reactions were studied by the "method of highly dilute flames," the slower ones by the "method of diffusion flames."

12.2.1 Highly Dilute Flames

In the method of highly dilute flames,[20] the two reactants, such as sodium vapor and chlorine gas, were allowed to flow into an evacuated vessel from opposite ends, the pressure being kept so low that the mean free paths were greater than the tube diameter. Under these conditions, the gases mix by diffusion and the reaction products, for example, sodium chloride, are solids which are deposited on the walls of the vessel. From the distribution of the deposit the rate constant of the reaction could be deduced. At the same time, measurements were made of the amount of radiation emitted.

    The sodium–chlorine reaction is typical of reactions that can be studied by this method. From the results, Polanyi concluded that the first step is

$$(1) \quad Na + Cl_2 \rightarrow NaCl + Cl$$

The energy released in this reaction is 170 kJ mol^{-1}; sodium-D radiation, which is emitted in the reaction, requires an energy of 202 kJ mol^{-1}, so that further reactions must occur. Polanyi concluded that the Cl atoms produced in reaction (1) react with Na_2 molecules,

$$(2) \quad Cl + Na_2 \rightarrow NaCl + Na$$

a process that liberates 335 kJ mol^{-1}. Evidence that this reaction occurs is provided by the fact that the intensity of the luminescence decreases with increasing temperature, a result attributed to the dissociation of the Na_2 molecules. The decrease is exponential, and a plot of the logarithm of the intensity versus $1/T$ gave an energy of 75 kJ mol^{-1}, which is the dissociation energy of Na_2. Further evidence that reaction (2) occurs is that the intensity of the emitted radiation varies with the square of the total sodium pressure, to which the concentration of Na_2 is proportional.

    Since the energy released in reaction (2) is more than sufficient to excite a sodium atom, it might be expected that the sodium atom produced in the reaction would be in the electronically excited ^2P state. Quenching experiments, however, showed that this is not the case. It was concluded that the energy released in reaction (2) does not pass into the sodium atom, but produces a vibrationally excited NaCl molecule, indicated in the equation by a prime:

$$(2') \quad Cl + Na_2 \rightarrow NaCl' + Na$$

The excited NaCl' molecule then transfers its energy to a sodium atom, which becomes electronically excited,

$$(3) \quad NaCl' + Na(^2S) \rightarrow NaCl + Na*(^2P)$$

The $Na*(^2P)$ atom then emits sodium-D radiation,

$$(4) \quad Na*(^2P) \rightarrow Na(^2S) + h\nu$$

As a result of the formation of NaCl′ in reaction (2′), there is more quenching than if only Na* were involved because the NaCl′ molecules have a much longer lifetime than Na*.

Very similar results to these were obtained for reactions of sodium vapor with other halogens and also with reactions of potassium vapor.[21]

12.2.2 Diffusion Flames

The method of highly dilute flames is not suitable for a slower reaction having a significant activation energy, and such reactions were studied by the method of diffusion flames.[22] In this technique, the alkali metal vapor was allowed to diffuse into a large excess of the other reactant gas, which was maintained at a low pressure. The distance of diffusion was measured by illumination with a suitable resonance lamp, and the rate constant then could be calculated.

The investigations by the use of this technique may be exemplified by the work on the reaction between sodium and mercuric chloride. The reaction steps are concluded to be

$$(1) \quad Na + HgCl_2 \rightarrow NaCl + HgCl \qquad \Delta H° = -105 \text{ kJ mol}^{-1}$$

$$(2) \quad HgCl + Na \rightarrow NaCl' + Hg \qquad \Delta H° = -264 \text{ kJ mol}^{-1}$$

$$(3) \quad NaCl' + Na(^2S) \rightarrow NaCl + Na*(^2P)$$

In reaction (2) the HgCl reacts not with Na_2 but with Na to give vibrationally excited NaCl′. The pressure effects found in the sodium–chlorine system are not observed here, showing that it is Na and not Na_2 that is involved in reaction (2). The energy liberated in reaction (2) is more than enough to produce $Na*(^2P)$ which emits the yellow radiation.

Since World War II there have been many important developments in the study of the various types of reaction involved in chemiluminescence. Many of the elementary processes have been studied by the method of molecular beams, and some details are to be found in Section 12.4.

Also, many of the reactions have been investigated by modern spectroscopic techniques, particularly by J. C. Polanyi and his coworkers who have been concerned with the vibrational and rotational states of products in relation to the states of the reactants. These studies have explored the relationship between the excitation of the product of a reaction and the form of its potential-energy surface, and this topic is now discussed in a separate section.

12.3 FEATURES OF POTENTIAL-ENERGY SURFACES

From the results of their chemiluminescence experiments Evans and M. Polanyi[23] drew important conclusions about the nature of the potential-energy surfaces for some of the reactions. An interesting feature of the reactions of the type Na + Cl_2 and Na_2 + Cl is that they occur at high rates that correspond to very large reaction cross sections, of the order of 10^6 pm^2 (100 $Å^2$); this means that reaction occurs when the reactants are several hundred picometers (several angstroms) apart. Evans and Polanyi explained their results in terms of an ionic intermediate, so that the processes can be represented as

$$A + B—C \rightarrow (A^{\delta+}\cdots B^{\delta-}\cdots C)^{\ddagger} \rightarrow A—B + C$$

Polanyi referred to this type of mechanism as an *electron-jump* mechanism and as a *harpooning* mechanism.

12.3.1 Attractive Surfaces for Exothermic Reactions

To explain why an exothermic reaction of this type leads to vibrational energy in the product molecule AB, Evans and Polanyi suggested that the potential-energy surface has the form shown schematically in Fig. 12.4. Such surfaces have been described more recently as *attractive* or *early-downhill surfaces.*[24] It is to be noted, particularly with reference to the energy profile (Fig. 12.4b), that the barrier occurs in the early stage of the reaction path. It is convenient to designate the type of barrier in terms of the relative extensions of the bonds A—B and B—C. If r_{AB} is the A—B distance at any point and r_{AB}^0 is the equilibrium distance in the product molecule AB, a quantity r_{AB}^* can be defined as $r_{AB} - r_{AB}^0$; it is the extension of the bond. Similarly, r_{BC}^* is defined as $r_{BC} - r_{BC}^0$. In the surface shown in Fig. 12.4, the barrier occurs when $r_{AB}^* > r_{BC}^*$ and

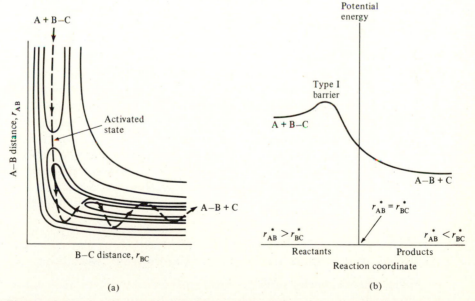

(a)   (b)

Figure 12.4 (a) An attractive (early-downhill) surface for an exothermic reaction A + BC → AB + C. The dashed line shows a schematic trajectory; much of the energy released goes into A—B vibration. (b) The corresponding energy profile. The starred quantity r_{AB}^* is the A—B distance minus the equilibrium value r_{AB}^0; similarly, $r_{BC}^* = r_{BC} - r_{BC}^0$. The barrier, corresponding to $r_{AB}^* > r_{BC}^*$, is called a *Type I barrier*.

is said to be a *Type-I barrier*. The essential feature of this type of surface is that the activated state is reached before there is much extension of the B—C bond. After the activated state has been reached and the system is proceeding down the exit valley, with the liberation of energy, the A$\cdots$B distance decreases, so that a considerable amount of the released energy passes into vibrational energy of the A—B bond.

The location of the covalent–ionic crossing for reactions of this type was calculated by Magee,[25] and a number of later calculations of the same kind have been made.[26] These treatments have given a simple interpretation of the observed high reaction cross sections, in terms of the crossing of potential-energy surfaces for ionic and covalent states of the product molecule AB. For example, Magee's treatment of the reaction K + Br$_2$ showed that as the K and Br$_2$ approach one another there is at first little interaction; then, at a separation of about 700 pm (7 Å), the harpooning electron from the K atom produces Br$_2^-$. The complex K$^+$$\cdots$Br$^-$$\cdots$Br finally dissociates into KBr + Br, with much of the liberated energy passing into vibration of the KBr molecule.

12.3.2 Repulsive Surfaces for Exothermic Reactions

There have been many studies in which the vibrational states of reactants and products have been determined by spectroscopic techniques, and in addition a number of mass-spectrometric studies have been made of a variety of reactions. For example, J. C. Polanyi and coworkers[27] used the chemiluminescence technique to make extensive investigations of the reaction

$$H + Cl_2 \rightarrow HCl + Cl \qquad \Delta H° = -188 \text{ kJ mol}^{-1}$$

Their studies showed that for this reaction a smaller proportion of the released energy passes into vibrational energy of the product HCl, a large part of it going into translational and rotational energies. This behavior is very different from that found with the alkali-metal reactions and was explained in terms of the type of potential-energy surface shown in Fig. 12.5. This type of surface is referred to as a *repulsive* or *late-downhill surface,* and the energy barrier, now corresponding to $r_{BC}^* > r_{AB}^*$, is said to be a *Type-II barrier*. The essential feature of this type of surface for an exothermic reaction is that the energy is released while there is increasing separation of the products AB + C, as further illustrated in the reaction profile shown as Fig. 12.5b. It might be thought that most of the energy released would now go into translational energy of the products, and this is sometimes the case. However, dynamical calculations[28] have shown that for some mass combinations a substantial amount of the released energy may go into product vibration.

12.3.3 Surfaces of Intermediate Types for Exothermic Reactions

The potential-energy surfaces shown schematically in Figs. 12.4 and 12.5 are extreme types, and those for most reaction systems lie between them. Aside from this, with a given potential-energy surface a whole range of trajectories is possible, depending on the initial conditions. Also, as will be discussed later, even for a given type of surface significant variations are found for different combinations of masses in the reacting

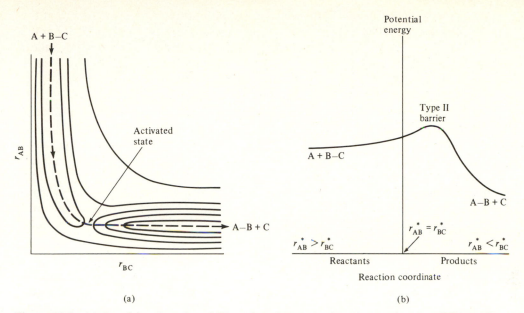

Figure 12.5 (a) Repulsive (late-downhill) surface for an exothermic reaction A + BC → AB + C. The dashed line shows a schematic reaction path; little of the energy released passes into vibration of AB. (b) The corresponding potential-energy profile. This is a *Type II barrier* ($r_{BC}^* > r_{AB}^*$).

species. In spite of these complications the concept of attractive and repulsive surfaces has proved extremely useful in interpreting the results of chemiluminescence and molecular-beam experiments.

This aspect of the problem has been considered in some detail, particularly by Blais and Bunker[29] and by J. C. Polanyi and his coworkers.[30] Much of this work has been concerned with calculating trajectories on various types of surface for a variety of initial conditions and comparing the results with experiment. Special attention has been paid to the influence of the relative masses of the atoms in the reaction system on the distribution of energy in the reaction products.

Polanyi and his coworkers[31] considered a graded series of potential-energy surfaces for an exothermic reaction A + BC → AB + C. Figure 12.6 shows such a series: (a) is an attractive surface equivalent to that shown in Fig. 12.4, and (e) is a repulsive surface (cf. Fig. 12.5), while (b), (c), and (d) are intermediate in form. The results of the trajectory calculations for the case in which A is much lighter than B and C (e.g., in H + Cl_2) and in which there is only translational energy in the reactants A and BC are shown schematically by the dashed lines in the figures. This particular mass combination is designated as **L + HH,** where **L** stands for a light atom and **H** for a heavy one. For the attractive surface (a) much energy passes into vibration, while for the repulsive surface (e) little does so. For the intermediate surfaces (b), (c), and (d), decreasing amounts of energy pass into vibration.

As noted earlier, the rather clear-cut differences in behavior found with the **L + HH** mass combination are not typical of other mass combinations. The fraction

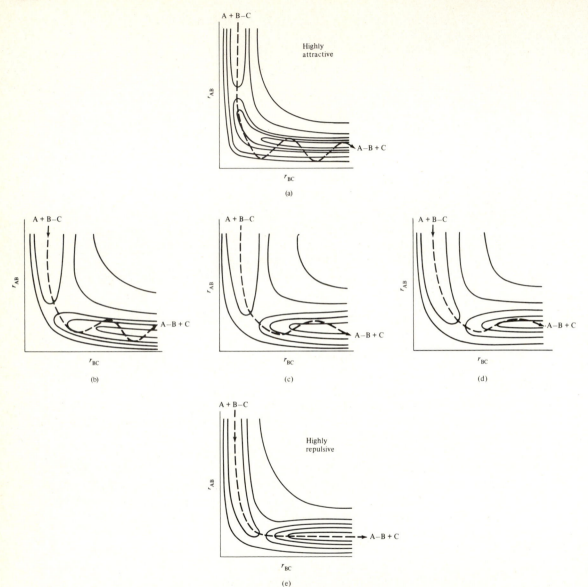

Figure 12.6 Graded series of potential-energy surfaces for an exothermic reaction: (a) is highly attractive, (e) is highly repulsive, while the others are intermediate. The dashed trajectories show purely schematically the conclusions from the calculations for the case in which A is light compared with B and C (**L + HH**).

of released energy that passes into translational and vibrational energies of the product depends to quite a significant extent on the relative masses of the atoms. This mass effect often is referred to as a *kinematic* effect, the word "kinematic" referring to the properties of motion that are independent of the nature of the forces.

It turns out that the combination **L + HH** is rather different from other mass combinations, which show closer similarities to one another; the term *light-atom anomaly* has been applied to the **L + HH** case. If the attacking atom is light, it can approach the molecule BC to something like the normal AB equilibrium separation *before* the repulsion between B and C begins to separate these heavy atoms. As a result, the B—C repulsion will be ineffective in channeling energy into AB vibration.

On the other hand, if the attacking atom A is not light compared with B and C (e.g., for **H + HL**), it approaches B *while* the B—C repulsion is being released. When the A—B bond is extended, the repulsion between B and C is able to cause B to recoil, but A—B does not recoil as a whole since the A—B bond is still being formed. Instead, as shown schematically in Fig. 12.7, there is considerable vibration in the product molecule AB. What is now happening is that the released repulsive energy pushes B towards A and produces AB vibration. This situation has been referred to as *mixed energy release.*[32] The difference between the cases **L + HH** and **H + HL** is further illustrated in Fig. 12.8. Because A is now heavy the A—B distance continues to decrease while the repulsive energy of BC is being released.

A good example of mixed energy release, arising for an **H + LL** reaction on a highly repulsive surface, is in the reaction

$$F + H_2 \rightarrow HF + H$$

The calculated potential-energy surface for this exothermic reaction is very repulsive,[33] but in spite of this the chemiluminescence studies[34] showed that a considerable amount of the released energy passed into vibrational energy of HF.

The classification just described, based on the ideas of attractive and repulsive surfaces and of mixed energy release in some circumstances, has proved very useful. Some insight into the reason that it applies has been provided by an analytical treatment

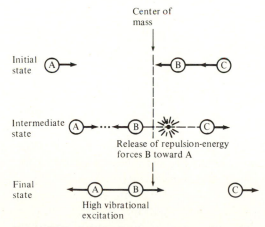

Figure 12.7 Schematic representation of a reaction of the type **H + HL** occurring on a repulsive surface. The release of energy can produce considerable vibrational excitation in the products AB (*mixed energy release*).

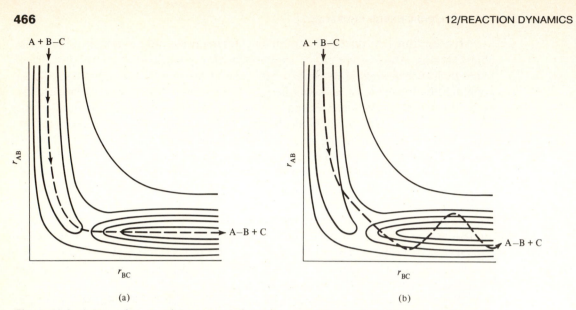

Figure 12.8 Schematic potential-energy surfaces for a repulsive surface. The two surfaces are identical, and (a), corresponding to Fig. 12.5, shows a trajectory for an **L** + **HH** reaction. A trajectory for an **H** + **HL** reaction is shown in (b). Now the AB distance is decreasing as repulsive energy is released, and there is vibrational energy in the product AB; the system "cuts the corner."

based on schematic potential-energy surfaces.[35] However, when a broad range of shapes of potential-energy surfaces is examined, the attractive–mixed–repulsive (A-M-R) classification is not always completely reliable, and additional factors must be considered. The most important of these appears to be the curvature of the reaction path.[36] Duff and Truhlar[37] in particular have made an assessment of the A-M-R model by making trajectory calculations on a variety of surfaces in which they varied the amount of curvature in the neighborhood of the saddle point. They found that difficulties could arise when only the correlations predicted by the A-M-R classification were taken into account, but that consideration of the curvature of the reaction path produced improvement. It is difficult to make predictions on the basis of any simple model.

12.3.4 Selective Enhancement of Reaction

So far the discussion has been confined to exothermic reactions, for which energy disposal is important; the interest has been in whether the released energy goes into translational or vibrational states of the product molecules. Another matter of importance is the kind of energy that is most effective in leading to chemical reaction. This question of *selective enhancement* by different forms of energy is significant not only for exothermic reactions having an activation energy but also for thermoneutral and endothermic reactions.

For an endothermic reaction to occur, the reactant molecules must carry energy at least equal to the endothermicity, and more if there is an activation energy for the reverse reaction. The question of whether this energy should reside in translational or vibrational energy depends on the nature of the potential-energy surface, and the answer is arrived at for a given surface by dynamical calculations and from the results of experiments in which reactants have been put into various energy states.

Endothermic reactions were first treated from this point of view by J. C. Polanyi and coworkers.[38] Figure 12.9 shows a surface of the early-barrier type for an endothermic reaction A + BC → AB + C. The reverse of an endothermic reaction on such a surface is an exothermic reaction occurring on a repulsive surface, and Fig. 12.9 is therefore Fig. 12.5 drawn in reverse. For an exothermic reaction of the type

$$L + HH \rightarrow LH + H$$

occurring on an attractive surface the energy passes mainly into translational energy of the product LH (Fig. 12.5). It follows from the principle of microscopic reversibility (Section 4.10)[39] that for the reverse reaction

$$H + HL \rightarrow HH + L$$

which is endothermic on an early-barrier surface, it is *translational* energy that will be most effective in bringing about reaction. In Fig. 12.9, curve *a* shows a trajectory corresponding to considerable translational energy and leading to reaction. Curve *b* shows a trajectory where there is mainly vibrational energy in the reactant molecules; reaction does not occur.

This, however, is the case of the "light-atom anomaly" for the reaction in the exothermic direction. As previously discussed (Section 12.3.3), other mass combinations will lead to more product vibrational energy in an exothermic reaction (see Figs. 12.7 and 12.8b). This is true, for example, for an exothermic process of the type

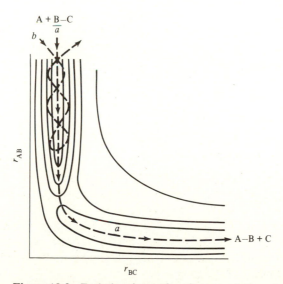

Figure 12.9 Early-barrier surface for an endothermic reaction A + BC → AB + C.

$$H + HL \rightarrow HH + L$$

It follows that for an endothermic reaction of the type

$$L + HH \rightarrow LH + H$$

occurring on a surface having an early barrier, vibrational energy makes more contribution to reaction than it did in the **H + HL** case. The trajectory for **L + HH** would be the reverse of that shown in Fig. 12.8b.

The converse type of surface for an endothermic reaction has a late potential-energy barrier and therefore is the reverse of an attractive surface for an exothermic reaction, which was shown in Fig. 12.4. Such a late-barrier surface for an endothermic reaction is shown schematically in Fig. 12.10. For the surface shown in Fig. 12.4 for an exothermic reaction, the energy passed largely into vibrational energy of the product, and this applies to all mass combinations. It follows from the principle of microscopic reversibility that for a surface such as shown in Fig. 12.10 it is the vibrational energy that will be most effective in leading to reaction. Curve *a* shows a schematic trajectory corresponding to considerable vibration, and reaction occurs; for the curve *b* trajectory, the system bounces back from the wall and there is no reaction.

The majority of endothermic reactions appear to occur on late-barrier surfaces of the type shown in Fig. 12.10. As will be discussed in Section 12.5.1, a state-to-state study[40] of the slightly endothermic reaction

$$K + HCl \rightarrow KCl + H$$

has shown that vibration in the HCl is most effective in leading to reaction. Vibrational energy is also most effective for the substantially endothermic reaction[41]

$$Br + HCl \rightarrow HBr + Cl$$

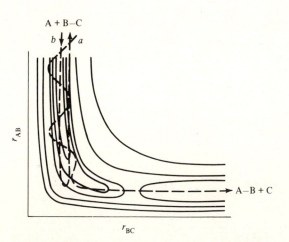

Figure 12.10   Late-barrier surface for an endothermic reaction A + BC → AB + C, showing two schematic trajectories.

and for the ion–atom reaction[42]

$$He + H_2^+ \rightarrow H + HeH^+$$

These reactions appear to occur on surfaces of the type shown in Fig. 12.10. For the endothermic reaction

$$H + HF \rightarrow H_2 + F$$

detailed trajectories have been calculated[43] and they show that vibration is much more effective than translation in leading to reaction.

The situation with thermoneutral reactions is much the same as with endothermic reactions.[44] If there are late barriers, vibrational energy is most effective; otherwise, translational energy is relatively more important.

12.3.5 Disposal of Excess Energy

One matter of special interest, especially in relation to some of the state-to-state studies (Section 12.5), is what happens to energy that is in excess of that required to surmount the energy barrier.

For all types of reaction it appears that there is a tendency for excess translational energy to appear as translational energy in the products, and for the excess vibrational energy to appear as vibrational energy in the products. This effect has been referred to as *adiabaticity*.[45] A number of experimental[46] and theoretical[47] studies have shown this to be the case for exothermic reactions.

Dynamical calculations for an endothermic reaction[48] led to a number of instructive trajectories, one of which is shown in Fig. 12.11. The reactants had about twice as much energy as was required to cross the barrier, and about 96% of the energy was vibrational energy. The excess energy causes the trajectory in the region of the activated state to correspond to a more extended A···B···C configuration than would otherwise be the case. The system thus "cuts the corner" on the potential-energy surface. Relaxation from this stretched configuration gives rise to vibrational excitation in the new bond. Different initial conditions give different trajectories, but the dynamical calculations indicated that the type of behavior shown in Fig. 12.11 is most likely, leading to the conclusion that excess vibrational energy in the reactants leads to excess vibrational energy in the products. A similar set of calculations for excess translational energy indicated that it was most likely to lead to excess translational energy in the products.

A good example of adiabaticity has been found[49] for the reaction

$$Cl + OH \rightarrow HCl + O$$

which is almost thermoneutral. This reaction is considered in further detail in Section 12.5.1; here it may be noted that excess vibrational energy in the hydroxyl radical becomes vibrational energy in the prduct HCl.

12.3.6 Gradual and Sudden Surfaces

Besides the position of the barrier crest, another factor that has an important influence on reaction dynamics is whether the potential-energy change during the course of a reaction is *gradual* or *sudden*.[50] This is illustrated by the surfaces and corresponding profiles for an endothermic reaction occurring on a late-barrier surface (Fig. 12.12). For a sudden surface it is, as expected, mainly vibrational energy that leads to reaction; translational energy makes an insignificant contribution. On a gradual surface, on the other hand, both translational and vibrational energies make significant contributions to crossing the barrier. The reason is that when the approach to the barrier is gradual, the barrier is encountered at an early stage, and the early part of the barrier is surmounted best by translational energy, that is, by motion along the entry valley of the surface.

Potential-energy surfaces have been calculated[51] for the reactions

$$Li + FH \rightarrow LiF + H$$

and

$$Be + FH \rightarrow BeF + H$$

and are found to be of the sudden type. The different dynamical behavior on the gradual and sudden surfaces may be related to the curvatures of the minimal reaction paths.

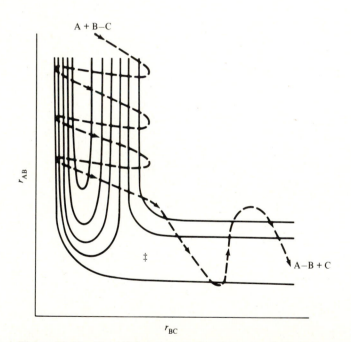

Figure 12.11 Potential-energy surface for an endothermic reaction A + BC → AB + C. A trajectory corresponding to excess reactant vibrational energy (dashed curve) leads to product vibrational energy. This figure is adapted from a diagram based on dynamical calculations given by D. S. Perry, J. C. Polanyi, and C. W. Wilson, *Chem. Phys.,* **3,** 317 (1974). The A—B—C angle was held at 130° which explains why, in the entrance valley, the r_{AB} distance is sometimes decreasing.

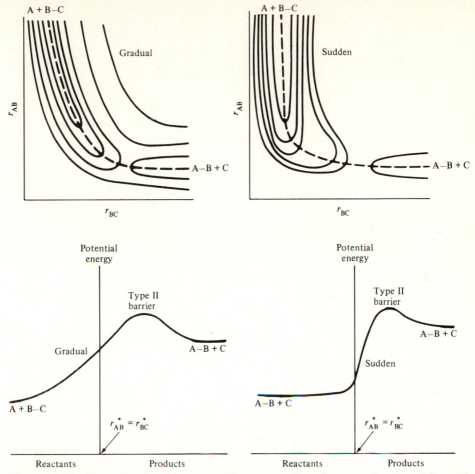

Figure 12.12 Gradual and sudden potential-energy surfaces for a reaction A + BC → AB + C, and the corresponding reaction profiles. The dashed lines show the minimal reaction paths.

When there is excess energy it is found[52] that the degree of adiabaticity is significantly greater on the gradual surfaces than on the sudden surfaces.

12.3.7 Influence of Rotational Energy

So far nothing has been said about rotational energy in chemical reactions. The amount of rotational energy residing in a molecule is usually much less than the vibrational and translational energies, and thus rotational energy is expected to play a less important role. Partly because of this, it has proved more difficult to investigate the importance of rotation. Another complicating factor is that rotation is dissipated rapidly, so that it is not easy to maintain molecules in specified rotational states.

Some progress, however, has been made. In Section 12.5.2 some state-to-state studies[53] are considered for the reactions

$$Na + HF \rightarrow H + NaF$$

$$Na + HCl \rightarrow H + NaCl$$

For a given value of the vibrational quantum number v, the rates at first decreased and then increased as the rotational quantum number J was increased.

A helpful but inevitably oversimplified explanation[54] for the effects of rotational energy is as follows. Different orientations between reacting molecules lead to different probabilities of reaction. As the rotational speeds increase, the time that the system spends in a favorable orientation decreases, and this causes a decrease in reaction probability. This effect may explain the effects found at lower J values. At higher values, however, an increase in reaction rate is observed as J increases, and another explanation is needed. At higher speeds of rotation the preferred orientations are obtained again at short intervals during the approach of the reactants, and this can give rise to enhanced rates as J increases. In addition to this factor, at high rotational speeds there is a significant amount of energy in the form of rotational energy, and rotation therefore contributes in this way to the probability of reaction.

12.4 MOLECULAR BEAMS

The essential feature of a molecular-beam study of a bimolecular reaction is that narrow beams of atoms or molecules are caused to cross one another. Movable detectors determine the directions and speeds of the product molecules and of the unreacted species. Analysis of the experimental results yields detailed information about the distribution of energy and angular momentum among the reaction products, the dependence of the total cross section on the states of the reactant molecules, the quantum states taken up by the products, the lifetime of the collision complex, and a number of other important features of the elementary event involved in the reaction.

Figure 12.13 shows in a very schematic way the type of apparatus used in a molecular-beam experiment. The two essential features of the apparatus are:

1. Devices for producing very narrow beams of the reactants, with some control of the speeds of the molecules and of their rotational, vibrational, and electronic states.
2. Movable devices for detecting the products of reaction at various positions.

In the earlier molecular-beam experiments, the beams were produced by containing the reactants in a vessel having a very small opening and by allowing the atoms or molecules to effuse into an evacuated vessel. The orifice was usually in the form of a rectangular slit, and a second small slit, placed a short distance from it, allowed the passage of those molecules that traveled in the required direction. These early beams had the disadvantage of being of very low intensity, and another disadvantage was that initially there was a thermal spread of velocities. This spread could be reduced by the use of a velocity selector consisting of a series of slotted disks mounted on the same axis; when these were rotated only molecules with velocities within a narrow range were able to pass through. The selected speed could be varied by altering the

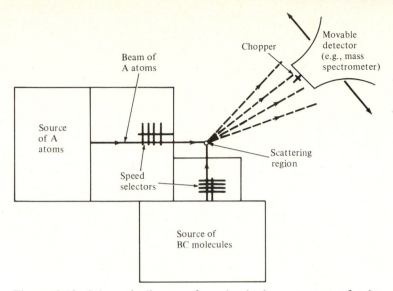

Figure 12.13 Schematic diagram of a molecular-beam apparatus for the study of a reaction A + BC → AB + C.

speed of rotation of the slotted disks. This procedure, however, resulted in drastic reductions in the beam intensity.

In more recent work, these techniques have been replaced by supersonic nozzle sources, which can produce beam intensities greater by 2 to 3 orders of magnitude. A schematic representation of a nozzle beam source is shown in Fig. 12.14. The gas is contained in an oven at a pressure that is several orders of magnitude greater than used for effusive flow. The molecules emerge through the slit in a slightly divergent beam at supersonic velocities. Part of the beam then passes through a specially designed skimmer which selects only the core of the beam, which then travels through one or more evacuated vessels, finally entering the main scattering chamber through a slit. The beams produced in this way have a much narrower range of molecular speeds than the effusive beams, as well as being more intense. By altering the oven temperature the translational energy of the molecules can be varied. For heavy molecules the energy range can be increased by a technique known as *seeding.* This involves mixing the heavy molecules with an excess of a much lighter gas, which emerges from the oven at much higher velocities and sweeps the heavier molecules along with it. The heavy molecules tend to travel along the center of the beam, whereas the light molecules are scattered toward the edges of the beam and are trapped out by the skimmer.

Sometimes the atoms or molecules in molecular beams are put into selected electronic, vibrational, and rotational states; some of the techniques used are briefly referred to in Section 12.5.

When the beams collide, molecular collisions occur, and three effects are possible:

1. *Elastic scattering,* in which there is no transfer of energy among different degrees of freedom.

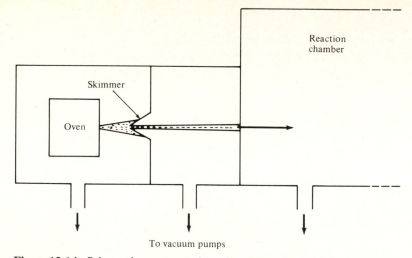

Figure 12.14 Schematic representation of a supersonic nozzle beam source.

2. *Inelastic scattering,* in which there is transfer of energy among degrees of freedom, but in which no chemical reaction occurs.
3. *Reactive scattering,* in which there is chemical change.

Even if the main interest is in chemical change, valuable information is provided by a study of elastic and inelastic scattering, as well as of reactive scattering.

In earlier molecular-beam investigations, the scattered beams were studied by means of differential surface-ionization detectors. This technique is based on the fact that alkali-metal atoms and their salts can be ionized with high efficiency when they come into contact with a heated wire of appropriate work function. By the use of different metals with special pretreatments, it was possible to distinguish between an alkali metal and its salt. This technique, however, is restricted to the reactions of alkali metals.

Later, mass spectrometers were introduced for the analysis of scattered beams; since these can be used for all types of reaction, the range of molecular-beam investigations has been extended greatly. Usually, the velocities of scattered beams are measured by a time-of-flight technique. A chopper is inserted about 1 cm in front of the mass spectrometer and is rotated at high speed. When the product molecules can pass the chopper the time they take to reach the detector depends on their translational velocities, and in this way it is possible to determine the distribution of speeds in the scattered beams.

The first experiment that can be described as a molecular-beam experiment appears to have been that of F. O. Rice, Urey, and Washburne,[55] who in 1928 studied the decomposition of nitrogen pentoxide in a narrow beam so as to reduce the effect of collisions. The first bimolecular reaction to be studied in crossed molecular beams was that between Cs and CCl_4, which was studied by Bull and Moon[56] in 1954. Shortly afterward, Taylor and Datz[57] and later in more detail Greene, Roberts, and Ross[58] investigated the reaction

$$K + HBr \rightarrow KBr + H$$

and obtained information about the collision yields and the energy of activation. Subsequent to these early investigations a large number of reactions have been studied in molecular beams, and the results have contributed greatly to an understanding of elementary chemical reactions.

12.4.1 Stripping and Rebound Mechanisms

A number of interesting patterns of behavior have been revealed by the molecular-beam experiments. Studies have been made, for example, of reactions of the alkali-metal atoms with halogens, such as

$$K + Br_2 \rightarrow KBr + Br$$

and

$$Cs + I_2 \rightarrow CsI + I$$

These reactions were found to have very high reaction cross sections of over 10^6 pm^2 (100 Å^2), so that they occur when the reactants are fairly far apart (500 pm or 5 Å). These results confirm those of the chemiluminescence experiments (Sections 12.2 and 12.3), but the molecular-beam experiments provide the further information that in these reactions the products are scattered forward with respect to the center of gravity of the system. This effect is illustrated in Fig. 12.15, and reactions of this type are said to occur by a *stripping* mechanism. They tend to occur when the surfaces are *attractive*.

A stripping reaction of a rather extreme type is referred to as a *spectator-stripping* reaction, a term borrowed from the theory of nuclear reactions; it was first applied to a chemical reaction by Henglein and Muccini.[59] The characteristic of a spectator-stripping reaction $A + BC \rightarrow AB + C$ is that there is very little interaction between A and the remote atom C, which therefore acts as a "spectator" to the process. The products separate so rapidly that there is no time for much of the momentum of the impact to be transferred to C.

When a stripping process occurs, an observer riding on the center of mass of the $A-B-C$ system first sees A and BC approaching; A strips off B, and A and C are seen to continue almost undisturbed in their original direction, as illustrated in Fig. 12.15b.

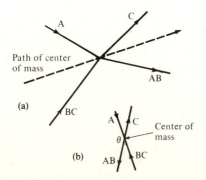

(a)

(b)

Figure 12.15 Directions of motion of reactants and products, for a stripping mechanism. In (a) the actual directions, relative to the laboratory, are shown. The directions relative to the center of mass are shown in (b).

By contrast, certain reactions such as

$$K + CH_3I \rightarrow KI + CH_3$$

have lower cross sections, of about 10^5 pm^2 (10 Å^2) or less, and exhibit *backward* scattering, as illustrated in Fig. 12.16. Now an observer traveling on the center of mass sees the alkali-halide product returning roughly in the direction from which the alkali-metal atom came. Such reactions are said to occur by a *rebound* mechanism. They tend to occur when the surfaces are *repulsive*.

The relationship between the shapes of potential-energy surfaces has been discussed in detail by J. C. Polanyi,[60] with special attention to the mass effects. When a reaction occurs by either a stripping mechanism or a rebound mechanism the lifetime of the activated complex $A \cdots B \cdots C^{\ddagger}$ must be short and the reaction is said to be *direct* or *impulsive;* if it were long enough for rotations to occur, the products AB and C would separate in random directions. The majority of reactions are direct and the results indicate that the lifetimes of the complexes are usually less than 5×10^{-13} s.

For some reactions, however, there is symmetry in the angular distribution of products; in other words, they do not separate in preferred directions. An example[61] of such a reaction is

$$Cs + RbCl \rightarrow CsCl + Rb$$

Such *indirect* or *complex-mode* reactions usually have a basin in the potential-energy surface.

12.5 STATE-TO-STATE KINETICS

The preceding sections have dealt in a rather general way with the extent to which different forms of energy are released in an exothermic reaction or are required to promote reaction. There has also been much work on a more specific aspect of this problem, namely, investigating reactions in which the reactants are in particular vibrational and rotational states. This presents considerable technical difficulties, which are being overcome. One problem is that vibrational and, more particularly, rotational energies are dissipated rapidly, either by internal conversion or by collisions with other

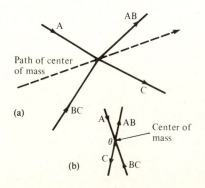

Figure 12.16 Directions of motion of reactants and products in a rebound mechanism. The actual directions are shown in (a), while the directions relative to the center of mass are shown in (b).

molecules. The frequency of collisions can be kept to a minimum, but the problem of internal conversion remains a difficulty for polyatomic reactant molecules.

Two procedures have been used in investigations of state-to-state kinetics. One is the molecular-beam method, first employed in this type of work by Brooks and coworkers.[62] Reactant molecules are put into desired vibrational and rotational states by laser excitation, and the states identified by their fluorescence. In a molecular-beam experiment, as noted in Section 12.4, the translational energy of the reactants can be well controlled. It has been possible also to determine the degree of polarization of the rotational angular momentum of the products and to control the mutual orientation of the reactants.

The second technique, devised by J. C. Polanyi and coworkers,[63] involves bulk experiments. Reactants are put into desired vibrational and rotational states by producing them in an exothermic prereaction, the details of which are well understood as a result of previous chemiluminescence studies. For example, Polanyi and coworkers formed hydroxyl radicals in various vibrationally excited states by means of the reactions

$$H + O_3 \rightarrow OH \ (v = 6 \text{ to } 9) + O_2$$

$$H + NO_2 \rightarrow OH \ (v = 1 \text{ to } 3) + NO$$

They determined the course of the subsequent reactions by two methods. One was to follow product concentrations by monitoring their infrared chemiluminescence. The other was to follow the decrease in the luminescence of the reactant molecules that had been excited vibrationally and rotationally; this is known as the method of *luminescence depletion*. This second method is particularly useful if reactant molecules have been formed in a wide range of vibrationally and rotationally excited states.

The bulk method has the disadvantage, as compared with the molecular-beam method, that it allows no control of translational energy. It has the important advantage, however, that the reactant molecules can be formed in a much wider range of vibrational and rotational states.

12.5.1 Influence of Reactant Vibrational Energy

Investigations of the effects of changing the vibrational quantum states of reactant molecules are of special interest in connection with the shapes of potential-energy surfaces. As discussed in Section 12.3, trajectory calculations for endothermic reactions have led to the conclusion that if there is a late (Type II) barrier, vibrational energy will be more effective than translational energy in leading to reaction (Fig. 12.10). On the other hand, if the barrier crest is an early one, translational energy is more effective (Fig. 12.9). Also, it has been noted[64] that when reactions are substantially endothermic their barriers are usually late ones. On the other hand, when reactions are only slightly endothermic or are exothermic the barriers tend to be earlier. Therefore, substantially endothermic reactions generally are enhanced more by vibrational than by translational energy, while for other types of reactions translational energy tends to be relatively more important.

Evidence in support of these conclusions has been mentioned in previous sections, and further evidence is provided by a number of state-to-state studies.

The first experimental study of the rates of reaction of reactant molecules in different vibrational states was carried out by Brooks and coworkers[65] using the molecular-beam technique. The reaction they investigated was

$$K + HCl(v) \rightarrow KCl + H$$

With the HCl in its ground vibrational state ($v = 0$) the reaction is endothermic by 6.3 kJ mol^{-1}. There was thus no reaction at very low translational energy, but when the translational energy was increased beyond the threshold value of about 6.3 kJ mol^{-1} reaction took place.

The HCl was produced in the $v = 1$ state by laser excitation, when it has an energy of 34.5 kJ mol^{-1}, which is greatly in excess of the threshold energy. There was then a considerable amount of reaction even at very low translational energies, and an increase in translational energy had little effect on the rate. The conclusion from these experiments is that for this endothermic reaction the effect of vibrational energy is much greater, by a factor of about 10, than that of translational energy. Work on rotational enhancement for the same reaction is considered later.

Similar conclusions about vibrational enhancement were drawn by Polanyi and coworkers[66] for the endothermic reaction§

$$Br + HCl(v) \rightarrow HBr + Cl \qquad \Delta H° = 65 \text{ kJ mol}^{-1}$$

The prereaction method was used to put the HCl molecules into various vibrational states, and rate constants were obtained by use of the luminescence depletion method. Some results are shown in Fig. 12.17, which shows the energy levels corresponding to the first four vibrationally excited states and the rate constants. It is to be seen that excitation to the $v = 1$ and $v = 2$ states gives considerable rate increases compared with $v = 0$. When the molecule HCl is in the $v = 2$ state the reaction has become exothermic, and further excitation has a smaller effect.

Similar results have been obtained[67] with the reactions

$$Cl + HF(v) \rightarrow HCl + F \qquad \Delta H° = 142 \text{ kJ mol}^{-1}$$

and
$$Br + HF(v) \rightarrow HBr + F \qquad \Delta H° = 205 \text{ kJ mol}^{-1}$$

For the first reaction, the HF was excited to vibrational levels up to $v = 4$, which corresponds to about 176 kJ mol^{-1} of excess energy. For the second reaction, HF levels up to $v = 6$ were used, this level corresponding to 251 kJ mol^{-1} of excess energy. Both these reactions occurred rapidly when the vibrational energy exceeded the endothermicity, the reaction cross sections being somewhat greater than 10^4 pm^2 (1 Å^2); this corresponds to a collision efficiency of about 0.1.

The reaction

$$OH(v) + Cl \rightarrow HCl(v') + O$$

is almost thermoneutral when the OH and HCl are in their ground vibrational states. An investigation by the luminescence depletion method[68] showed a rate increase by a factor of about 10 as v increased from 0 to 9. Moreover, studies of the vibrational states of the product HCl showed that the vibrational energy of the OH was converted

---

§ Here and elsewhere the $\Delta H°$ values are for the reactants in their lowest vibrational states.

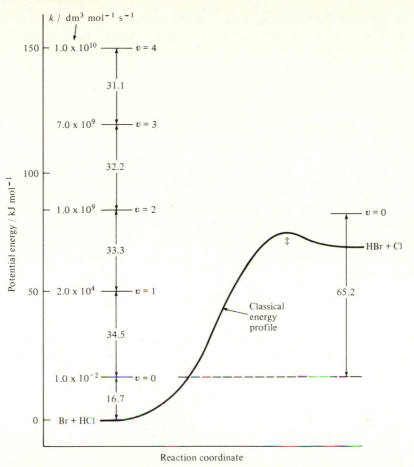

Figure 12.17 Energy diagram for the reaction Br + HCl(v) → HBr + Cl, showing the first four excited vibrational levels. The rate constants, given by R. B. Bernstein, *Chemical Dynamics via Molecular Beam and Laser Techniques,* Clarendon, Oxford, 1982, are based on the experiments of D. J. Douglas, J. C. Polanyi, and J. J. Sloan, *Chem. Phys.* **13**, 15 (1976).

with high efficiency into vibrational energy of the product HCl. The reason for this *adiabaticity* has been discussed in Section 12.3.5 (Fig. 12.11).

Heismann and Loesch[69] have carried out a detailed investigation, using the molecular-beam technique, of the endothermic reactions

$$K + HCl(v) \rightarrow KCl + H \qquad \Delta H° = 4.15 \text{ kJ mol}^{-1}$$

$$K + HF(v) \rightarrow KF + H \qquad \Delta H° = 72.4 \text{ kJ mol}^{-1}$$

They raised the reactant molecules to the $v = 1$ state and covered a wide range of translational energies by seeding the beams of potassium atoms with various carrier gases. For both reactions their results showed that the barrier heights were very close to the endothermicities of the reactions, which means that there is only a small crest in the potential-energy profiles. For the K + HCl reaction, which is only slightly endothermic, they found that at low translational energies the change from $v = 0$ to $v = 1$ increased the rate by a factor of about 10; at translational energies of 7.7 kJ mol^{-1} or more there was little or no vibrational enhancement. For the more endothermic reaction K + HF, the vibrational enhancement was demonstrated very strikingly. At low translational energies the change from $v = 0$ to $v = 1$ was 380 times more effective than the same amount of translational energy. Even with a translational energy of 164 kJ mol^{-1}, which is much greater than the endothermicity, the change from $v = 0$ to $v = 1$ increased the rate by a factor of 70.

These results are in agreement with the conclusion that late barriers give rise to more vibrational enhancement than early ones (Fig. 12.10). Potential-energy surfaces have been calculated[70] for both reactions. For the K + HF reaction, the barrier is a late one, and it was for this reaction that the very large vibrational effects were observed in the experiments of Heismann and Loesch. Much smaller vibrational enhancements were found for the K + HCl reaction, and for this the barrier is much earlier. The effect of rotational energy on these reactions is considered later.

Some work has also been done[71] on the slightly exothermic reaction

$$D + H_2(v) \rightarrow HD + H \qquad \Delta H° = -3.3 \text{ kJ mol}^{-1}$$

in which the H_2 was in its $v = 0$ and $v = 1$ states; H_2 in the $v = 1$ state was formed by collision with HF($v = 1$) formed by laser irradiation. The rate constant with H_2 in the $v = 1$ state was found to be about 4×10^4 times larger than for the $v = 0$ state. The corresponding slightly endothermic reaction

$$H + D_2(v) \rightarrow HD + D \qquad \Delta H° = 4.3 \text{ kJ mol}^{-1}$$

has been investigated in great detail.[72] The hydrogen atoms were produced by photodissociation of HI, and the collision energies could be well defined. The vibrational and rotational quantum states of the product HD molecules and of the unreacted D_2 molecules were determined either by multiple-photon ionization spectroscopy or by coherent anti-Stokes Raman scattering spectroscopy. The results showed that there is a greater population of higher-energy states in the HD produced in reactive collisions than in D_2 molecules that have undergone unreactive collisions. The techniques producing the results of these studies represent a remarkable experimental achievement, and no doubt will reveal important results with other reactions. Blais and Truhlar[73] have carried out trajectory calculations for this system, and the results are in excellent agreement with experiment.

Detailed state-to-state studies, using the molecular-beam technique, also have been made[74] on the reactions

$$F + H_2 \rightarrow HF + H$$

$$F + D_2 \rightarrow DF + D$$

and $\qquad\qquad\qquad$ $F + HD \rightarrow DF + H$

Some important evidence for resonance effects has been obtained. In the case of the F + H_2 reaction, for example, the scattering pattern was very different for the different vibrational states of the product HF. The $v = 1$ and $v = 2$ states were mainly backward scattered, but substantial forward scattering was observed for HF in the $v = 3$ state. This indicates that resonance effects are most prominent for the $v = 3$ state, and attempts are being made to explain this in terms of the potential-energy surface for the system. Lauderdale and Truhlar[75] have discussed resonance effects for the reactions H + H$_2$ and H + FH.

Some state-to-state studies also have been carried out on unimolecular reactions. For example, Reddy and Barry[76] investigated the reaction

$$CH_3NC \rightarrow CH_3CN$$

the energy diagram for which is shown in Fig. 12.18. By laser irradiation the reactant molecules were put into a number of vibrational states. Vibrational excitation was found to give considerable enhancement of rates, and the results were in very good quantitative agreement with the predictions of RRKM theory (Section 5.3.5).

12.5.2 Influence of Reactant Rotational Energy

Some experimental studies also have been made on the influence of rotational quantum states on reactivity, but understanding of these effects is still imperfect. One fundamental reaction studied[77] was

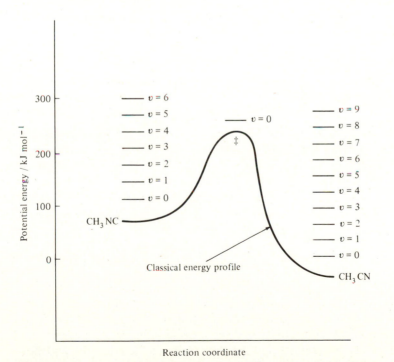

Figure 12.18 Energy diagram for the methyl isocyanide isomerization CH$_3$NC → CH$_3$CN. Vibrational states are shown for the reactant and product.

$$F + H_2 \, (v = 0, J) \rightarrow HF \, (v', J') + H$$

which is exothermic by 133.5 kJ mol^{-1}. By varying the spin isomer of H_2 and the temperature it was possible to produce H_2 in its $v = 0$ state and with various distributions among the rotational levels. The chemiluminescence technique was used to study the detailed state-to-state kinetics from specified $H_2 \, (v = 0, J)$ states to specified states (v', J') of the product HF. In going from $H_2 \, (v = 0, J = 0)$ to $H_2 \, (v = 0, J = 1)$ there was a small decrease in the fraction of total energy that became *vibrational* energy of the product HF. On going from $H_2 \, (v = 0, J = 1)$ to $H_2 \, (v = 0, J = 2)$ there was a small increase in the fraction. Changes in J, however, had little effect on the amount of energy going into *rotation* of the product HF. These results still lack a completely satisfactory explanation.

A somewhat simpler, although still difficult, problem is the effect of the rotational state of the reactant on the rate of reaction. This has been investigated[78] by the luminescence depletion method, for the reactions

$$Na + HF \, (v, J) \rightarrow H + NaF \, (v', J')$$

and
$$Na + HCl \, (v, J) \rightarrow H + NaCl \, (v', J')$$

For the first reaction the HF was put into the vibrational states $v = 2, 3,$ and 4, and the rotational levels were varied from $J = 0$ to 14. For the second reaction the HCl molecules were in the states $v = 1$ to 4 and $J = 0$ to 19. For each reaction, with v held constant, the rates first decreased and then increased as J was increased. A simplified explanation for this behavior was considered in Section 12.3.7.

The influence of rotational quantum states on reaction cross sections also has been investigated[79] for the endothermic reactions

$$K + HCl \, (v, J) \rightarrow KCl + H \qquad \Delta H° = 4.15 \text{ kJ mol}^{-1}$$

$$K + HF \, (v, J) \rightarrow KF + H \qquad \Delta H° = 72.4 \text{ kJ mol}^{-1}$$

For both reactions, increase in the J value from 0 to 4 brought about a considerable decrease in rate. At higher J values the rates remained fairly constant or perhaps showed a small rise as J was increased. This behavior is similar to that found with the Na + HF and Na + HCl reactions.

Some molecular-beam studies also have been carried out, by Bernstein and co-workers,[80] on two reactions that occur through complexes having fairly long lifetimes, that is, by a complex-mode mechanism:

$$CsF(J) + K \rightarrow Cs + KF$$

$$RbF(J) + K \rightarrow Rb + KF$$

The first of these is endothermic, the second exothermic. The results are somewhat complicated but may be summarized by saying that changes in rotational energy have the same effect on reactivity as do changes in translational energy; in other words, the reactivity is determined by the total energy of the reactants and not specifically by particular types of energy. This is understandable when there is a reaction complex of long life, since the different types of energy will have time to exchange before the complex breaks up into products.

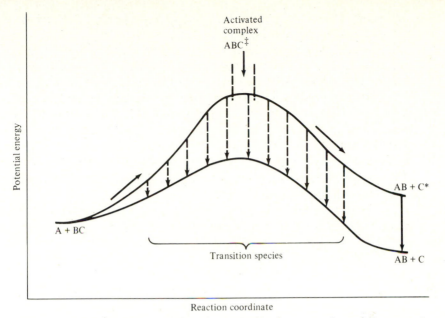

Figure 12.19 Schematic potential-energy profile for a reaction of the type A + BC → AB + C*. The firm vertical line shows the emission C* → C + $h\nu$; the dashed lines show emissions from various transition species.

12.6 SPECTROSCOPY OF TRANSITION SPECIES

An important development has been the spectroscopic detection of transition species, which are defined as molecular entities having configurations intermediate between those of the reactants and products of an elementary reaction. The term "transition species" covers a much broader range of configurations than does "activated complex," which is defined as existing in an arbitrarily small region of phase space.§ The distinction is illustrated by the schematic potential-energy profile, shown in Fig. 12.19, for a chemiluminescent reaction of the type

$$A + BC \rightarrow AB + C*$$

The product C* is in an excited electronic state and emits radiation as it passes into its ground state. There is in addition a much less intense radiation emitted by species corresponding to various configurations intermediate between the reactants and products. These emissions are represented by vertical dashed lines in the figure and correspond to a continuous range of wavelengths.

There is also a possibility of detecting transition species by absorption spectroscopy, and this might be done even for a reaction that does not involve any excited

§ What are here called "transition species" commonly are referred to as "transition states." However, it seems preferable to use the word "state" only as an abstract noun and not to apply it to the molecular entity itself; otherwise, there is a serious danger of confusing a so-called "transition state" with an activated complex.

electronic states. This is represented schematically in Fig. 12.20, which shows not only the ground-state potential-energy profile on which the reaction occurs, but also one of higher energy corresponding to electronic excitation. The detection of transition species by absorption spectroscopy is difficult owing to the low concentrations involved, but the use of laser techniques makes the detection more feasible.

Spectroscopic studies of transition species may be said to be an extension of work on collision damping of spectral lines. Much early work in spectroscopy was devoted to the study of emission from the sun, and by the 19th century it had been observed that the emission lines are broadened. This results from the enormously high pressures involved, and the broadening was explained, particularly by Lorentz,[81] as due to the fact that the emission is from atoms or molecules that are actually in collision with one another.

Spectroscopic evidence for intermediates in *reactive* collisions was not obtained until 1980 when J. C. Polanyi and coworkers[82] investigated the reaction

$$F + Na_2 \rightarrow F \cdots Na \cdots Na \rightarrow NaF + Na^*$$

in crossed molecular beams. The product Na* is in an electronically excited state and emits the familiar yellow *D*-line. On both sides of this line there was "wing" emission, and the evidence indicated that this was due to the transition species $F \cdots Na \cdots Na$. A similar result has been obtained for the $Mg + H_2$ reaction.[83]

Polanyi and his coworkers[84] also have made theoretical studies of the possibility of detecting transition species in emission and absorption. This is an important field that is likely to develop considerably.

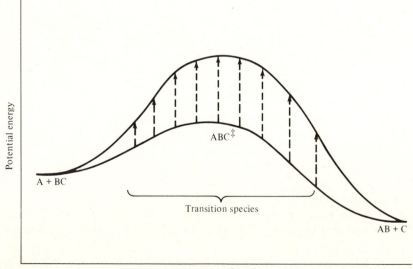

Figure 12.20 Schematic potential-energy surface for a reaction occurring on ground electronic states. The existence of a surface corresponding to excited states makes it possible to detect transition species by absorption spectroscopy.

REFERENCES

1. J. H. van't Hoff, *Lectures on Theoretical and Physical Chemistry,* vol. 3, p. 9, Edward Arnold, London, 1899–1900. W. Ostwald [*J. Chem. Soc.,* **85,** 506 (1904)] similarly defined dynamics as "the theory of the progress of chemical reactions and of chemical equilibrium."
2. J. O. Hirschfelder, Ph.D. dissertation, Princeton University, 1935.
3. H. Eyring and M. Polanyi, *Z. Phys. Chem., B.,* **12,** 279 (1931).
4. J. O. Hirschfelder, H. Eyring, and B. Topley, *J. Chem. Phys.,* **4,** 170 (1936).
5. F. T. Wall, L. A. Hiller, and J. Mazur, *J. Chem. Phys.,* **29,** 255 (1958); **35,** 1284 (1961).
6. F. T. Wall and R. N. Porter, *J. Chem. Phys.,* **39,** 3112 (1963).
7. F. T. Wall and R. N. Porter, *J. Chem. Phys.,* **36,** 3256 (1962).
8. N. C. Blais and D. L. Bunker, *J. Chem. Phys.,* **37,** 2713 (1962).
9. J. C. Polanyi and S. D. Rosner, *J. Chem. Phys.,* **38,** 1028 (1963).
10. M. S. Child and A. Roach, *Mol. Phys.,* **14,** 1 (1968). Compare M. S. Child, *Discuss. Faraday Soc.,* **44,** 68 (1967).
11. N. C. Blais and D. L. Bunker, *J. Chem. Phys.,* **37,** 2713 (1962).
12. M. Karplus, R. N. Porter, and R. D. Sharma, *J. Chem. Phys.,* **34,** 3259 (1965).
13. R. N. Porter and M. Karplus, *J. Chem. Phys.,* **40,** 1105 (1964).
14. K. Morokuma, B. C. Eu, and M. Karplus, *J. Chem. Phys.,* **51,** 5193 (1969); K. Morokuma and M. Karplus, *J. Chem. Phys.,* **55,** 63 (1971).
15. J. C. Polanyi and W. H. Wong, *J. Chem. Phys.,* **51,** 1439 (1969).
16. K. Morokuma and M. Karplus, *J. Chem. Phys.,* **55,** 63 (1971).
17. R. E. Weston, *J. Chem. Phys.,* **31,** 892 (1959).
18. M. Karplus and L. M. Raff, *J. Chem. Phys.,* **41,** 1267 (1964); L. M. Raff and M. Karplus, *J. Chem. Phys.,* **44,** 1212 (1966).
19. M. Polanyi, *Atomic Reactions,* Ernest Benn, London, 1932.
20. H. Beutler, S. von Bogdandy, and M. Polanyi, *Naturwissenschaften,* **14,** 164 (1926); H. Beutler and M. Polanyi, *Naturwissenschaften,* **13,** 711 (1925); *Z. Phys.,* **47,** 379 (1928); S. von Bogdandy and M. Polanyi, *Z. Phys. Chem. B.,* **1,** 21 (1928).
21. H. Otouka, *Z. Phys. Chem. B,* **7,** 422 (1930); M. Krocsak and G. Schay, *Z. Phys. Chem. B,* **19,** 344 (1932); E. Roth and G. Schay, *Z. Phys. Chem. B,* **28,** 323 (1935).
22. H. von Hartel and M. Polanyi, *Z. Phys. Chem. B,* **11,** 97 (1930); C. E. H. Bawn, *Ann. Rept. Chem. Soc.,* **39,** 36 (1942); E. Warhurst, *Q. Rev. (London),* **5,** 44 (1951); D. Garvin and G. B. Kistiakowsky, *J. Chem. Phys.,* **20,** 105 (1952); F. T. Smith and G. B. Kistiakowsky, *J. Chem. Phys.,* **31,** 621 (1959); D. Garvin, P. P. Gwyn, and J. W. Kokowitz, *Can. J. Chem.,* **38,** 1795 (1960).
23. M. G. Evans and M. Polanyi, *Trans. Faraday Soc.,* **31,** 875 (1935); **35,** 178 (1939).
24. J. C. Polanyi, *Acc. Chem. Res.,* **5,** 161 (1972).
25. J. L. Magee, *J. Chem. Phys.,* **8,** 687 (1940).
26. For a review see D. R. Herschbach, in J. Ross (Ed.), *Molecular Beams,* chap. 9, Wiley-Interscience, New York, 1966.
27. J. K. Cashion and J. C. Polanyi, *Proc. R. Soc. London A,* **258,** 570 (1960); P. E. Charters and J. C. Polanyi, *Discuss. Faraday Soc.,* **33,** 107 (1962); F. D. Findlay and J. C. Polanyi, *Can. J. Chem.,* **42,** 2176 (1964); J. C. Airey, R. R. Getty, J. C. Polanyi, and D. R. Snelling, *J. Chem. Phys.,* **41,** 3255 (1964); K. G. Anlauf, P. J. Kuntz, D. H. Maylotte, P. D. Pacey, and J. C. Polanyi, *Discuss. Faraday Soc.,* **44,** 183 (1967); A. M. G. Ding, L. J. Kirsch, D. S. Parry, J. C. Polanyi, and J. L. Schreiber, *Faraday Discuss. Chem. Soc.,* **55,** 73 (1973).
28. For a discussion of this problem see J. C. Polanyi and J. L. Schreiber, in W. Jost (Ed.), *Physical Chemistry: An Advanced Treatise,* chap. 6, Academic, New York, 1974.
29. N. C. Blais and D. L. Bunker, *J. Chem. Phys.,* **37,** 2713 (1962); **39,** 315 (1963); D. L. Bunker and N. C. Blais, *J. Chem. Phys.,* **41,** 2377 (1964).

30. J. C. Polanyi and S. D. Rosner, *J. Chem. Phys.,* **38,** 1028 (1963); J. C. Polanyi, *J. Quant. Spectrosc. Radiat. Transfer,* **3,** 471 (1963); P. J. Kuntz, E. M. Nemeth, J. C. Polanyi, S. D. Rosner, and C. E. Young, *J. Chem. Phys.,* **44,** 1168 (1966); J. C. Polanyi and W. H. Wong, *J. Chem. Phys.,* **51,** 1439 (1969); M. H. Mok and J. C. Polanyi, *J. Chem. Phys.,* **51,** 1451 (1969); A. M. G. Ding, L. J. Kirsch, D. S. Perry, J. C. Polanyi, and J. L. Schreiber, *Faraday Discuss. Chem. Soc.,* **55,** 252 (1973); J. C. Polanyi, J. L. Schreiber, and W. J. Skrlac, *Faraday Discuss. Chem. Soc.,* **67,** 66 (1979).

31. P. J. Kuntz, E. M. Nemeth, J. C. Polanyi, D. S. Rosner, and C. E. Young, *J. Chem. Phys.,* **44,** 1168 (1966).

32. J. C. Polanyi, *J. App. Optics, Supp.* 2, 109 (1965).

33. J. T. Muckerman, *J. Chem. Phys.,* **54,** 1155 (1971); R. L. Jaffe and J. B. Anderson, *J. Chem. Phys.,* **54,** 2224 (1971); C. F. Bender, S. V. O'Neil, P. K. Pearson, and H. F. Schaeffer, *Science,* **176,** 1412 (1972); J. C. Polanyi and J. L. Schreiber, *Chem. Phys. Lett.,* **29,** 319 (1974).

34. J. C. Polanyi and K. B. Woodall, *J. Chem. Phys.,* **57,** 1574 (1972).

35. N. H. Hijazi and K. J. Laidler, *J. Chem. Phys.,* **58,** 349 (1973).

36. D. G. Truhlar and D. A. Dixon, in R. B. Bernstein (Ed.), *Atom–Molecule Collision Theory— A Guide for the Experimentalist.* Plenum, New York, 1979.

37. J. W. Duff and D. G. Truhlar, *J. Chem. Phys.,* **62,** 2477 (1975).

38. D. S. Perry, J. C. Polanyi, and C. W. Wilson, *Chem. Phys.,* **3,** 317 (1974).

39. For a discussion of this principle in relation to the present type of problem, see D. S. Perry, J. C. Polanyi, and C. Woodrow Wilson, *Chem. Phys. Letters,* **24,** 484 (1974).

40. T. J. Odiorne, P. R. Brooks, and J. V. V. Kasper, *J. Chem. Phys.,* **55,** 1980 (1971).

41. D. J. Douglas, J. C. Polanyi, and J. J. Sloan, *Faraday Discuss. Chem. Soc.,* **55,** 310 (1973); *J. Chem. Phys.,* **59,** 6679 (1973).

42. W. A. Chupka, in *Ion–Molecule Reactions* (J. L. Franklin, Ed.), Plenum, New York, 1973, Chapter 3.

43. J. B. Anderson, *J. Chem. Phys.,* **52,** 3849 (1970); R. L. Wilkins, *J. Chem. Phys.,* **58,** 3038 (1973).

44. J. C. Polanyi and W. H. Wong, *J. Chem. Phys.,* **51,** 1439 (1969).

45. D. S. Perry, J. C. Polanyi, and C. W. Wilson, *Chem. Phys.,* **3,** 317 (1974).

46. L. T. Cowley, D. S. Horne, and J. C. Polanyi, *Chem. Phys. Lett.,* **12,** 144 (1971); A. M. Rulis and R. B. Bernstein, *J. Chem. Phys.,* **57,** 5497 (1972); J. D. McDonald, P. R. LeBreton, Y. T. Lee, and D. R. Herschbach, *J. Chem. Phys.,* **56,** 769 (1972); A. G. M. Ding, L. J. Kirsch, D. S. Perry, J. C. Polanyi, and J. L. Schreiber, *Faraday Discuss. Chem. Soc.,* **55,** 252 (1973); D. S. Perry and J. C. Polanyi, *Chem. Phys.,* **12,** 419 (1976); J. C. Polanyi and W. J. Skolac, *Chem. Phys.,* **23,** 167 (1977); D. Brandt and J. C. Polanyi, *Chem. Phys.,* **35,** 23 (1978); D. Brandt, L. W. Dickson, L. N. Y. Dwan, and J. C. Polanyi, *Chem. Phys.,* **39,** 189 (1979); D. Brandt and J. C. Polanyi, *Chem. Phys.,* **45,** 65 (1980); M. A. Nazar and J. C. Polanyi, *Chem. Phys.,* **55,** 229 (1981).

47. C. A. Parr, J. C. Polanyi, and W. H. Wong, *J. Chem. Phys.,* **58,** 5 (1973); A. G. M. Ding *et al., loc. cit.;* D. L. Thompson, *J. Chem. Phys.,* **56,** 3570 (1972); J. C. Polanyi, J. L. Schreiber, and J. J. Sloan, *Chem. Phys.,* **9,** 403 (1975); J. C. Polanyi, J. L. Schreiber, and W. J. Skrlac, *Faraday Discuss. Chem. Soc.,* **67,** 66 (1979).

48. D. S. Perry, J. C. Polanyi, and C. W. Wilson, *Chem. Phys.,* **3,** 317 (1974).

49. B. A. Blackwell, J. C. Polanyi, and J. J. Sloan, *Chem. Phys.,* **24,** 25 (1977).

50. J. C. Polanyi and N. Sathyamurthy, *Chem. Phys.,* **33,** 287 (1978); **37,** 259 (1979); L. Noor-Batcha and N. Sathyamurthy, *J. Chem. Phys.,* **76,** 6447 (1982).

51. G. G. Balint-Kurti and R. N. Yardley, *Faraday Discuss. Chem. Soc.,* **62,** 77 (1977); H. Schor, S. Chapman, S. Green, and R. N. Zare, *J. Chem. Phys.,* **69,** 3790 (1978); M. M. L. Chan and H. F. Schaefer, *J. Chem. Phys.,* **72,** 4376 (1980).

52. J. C. Polanyi and N. Sathyamurthy, *Chem. Phys.,* **37,** 259 (1979).

53. A. M. G. Ding, L. J. Kirsch, D. S. Perry, J. C. Polanyi, and J. L. Schreiber, *Faraday Discuss. Chem. Soc.,* **55,** 252 (1973); B. A. Blackwell, J. C. Polanyi, and J. J. Sloan, *Chem. Phys.,* **30,** 299 (1978).

54. B. A. Hodgson and J. C. Polanyi, *J. Chem. Phys.,* **55,** 4745 (1971); Ding *et al., loc. cit.;* Blackwell *et al., loc. cit.* Related matters are discussed in G. G. Balint-Kunti and R. N. Yardley, *Faraday Discuss. Chem. Soc.,* **62,** 77 (1977); A. C. Roach, *Faraday Discuss. Chem. Soc.,* **62,** 151 (1977); J. C. Polanyi and N. Sathyamurthy, *Chem. Phys.,* **37,** 259 (1979); *J. Phys. Chem.,* **83,** 978 (1979).

55. F. O. Rice, H. C. Urey, and R. N. Washburne, *J. Am. Chem. Soc.,* **50,** 2402 (1928).

56. T. H. Bull and P. B. Moon, *Discuss. Faraday Soc.,* **17,** 54 (1954).

57. E. H. Taylor and S. Datz, *J. Chem. Phys.,* **23,** 1711 (1955).

58. E. F. Greene, R. W. Roberts and J. Ross, *J. Chem. Phys.,* **32,** 940 (1960).

59. A. Henglein and G. A. Muccini, *Z. Naturforsch.,* **17a,** 452 (1962).

60. J. C. Polanyi, *Faraday Discuss. Chem. Soc.,* **55,** 389 (1973).

61. W. H. Miller, S. A. Safron, and D. R. Herschbach, *Discuss. Faraday Soc.,* **44,** 108, 292 (1967).

62. T. J. Odiorne, P. R. Brooks, and J. V. Kasper, *J. Chem. Phys.,* **55,** 1980 (1971); for a review see M. Kneba and J. Wolfrum, *Ann. Rev. Phys. Chem.,* **31,** 47 (1980).

63. D. J. Douglas, J. C. Polanyi, and J. J. Sloan, *J. Chem. Phys.,* **59,** 6679 (1973); *Chem. Phys.,* **13,** 15 (1976); D. J. Douglas and J. C. Polanyi, *Chem. Phys.,* **16,** 1 (1976); B. A. Blackwell, J. C. Polanyi, and J. J. Sloan, *Chem. Phys.,* **24,** 25 (1977); **30,** 299 (1978).

64. J. C. Polanyi, *J. Chem. Phys.,* **31,** 1338 (1959); M. H. Mok and J. C. Polanyi, *J. Chem. Phys.,* **51,** 1451 (1969); M. Shapiro and Y. Zeiri, *J. Chem. Phys.,* **70,** 5264 (1979).

65. T. J. Odiorne, P. R. Brooks, and J. V. Kasper, *J. Chem. Phys.,* **55,** 1980 (1971).

66. D. J. Douglas, J. C. Polanyi, and J. J. Sloan, *J. Chem. Phys.,* **59,** 6679 (1973); *Chem. Phys.,* **13,** 15 (1976); see also D. Arnoldi and J. Wolfrum, *Ber. Bunsenges. Phys. Chem.,* **80,** 892 (1976).

67. D. J. Douglas, J. C. Polanyi, and J. J. Sloan, *Chem. Phys.,* **13,** 15 (1976).

68. B. A. Blackwell, J. C. Polanyi, and J. J. Sloan, *Chem. Phys.,* **24,** 25 (1977).

69. F. Heismann and H. J. Loesch, *Chem. Phys.,* **64,** 43 (1982).

70. M. Shapiro and Y. Zeim, *J. Chem. Phys.,* **70,** 5264 (1979).

71. M. Kneba, W. Wellhausen, and J. Wolfrum, *Ber. Bunsenges. Phys. Chem.,* **83,** 940 (1979).

72. D. P. Gerrity and J. J. Valentini, *J. Chem. Phys.,* **79,** 5202 (1979); **81,** 1298 (1984); **82,** 1323 (1985); **83,** 2207 (1985); E. E. Marinero, C. T. Rettner, and R. N. Zare, *J. Chem. Phys.,* **80,** 4142 (1984); C. T. Rettner, E. E. Marinero, and R. N. Zare, in *Physics of Electronic and Atomic Collisions* (J. Eichler, I. V. Hertel, and N. Stolterfoht, Eds.), North Holland, Amsterdam, 1985.

73. N. C. Blais and D. G. Truhlar, *J. Chem. Phys.,* **83,** 2201 (1985).

74. D. M. Neumark, A. M. Wodte, G. N. Robinson, C. C. Hayden, and Y. T. Lee, in D. G. Truhlar (Ed.), *Resonances, Am. Chem. Soc. Symp. Ser.,* **263,** 479 (1984); *J. Chem. Phys.,* **82,** 3045 (1985); D. M. Neumark, A. M. Wodtke, C. N. Robinson, C. C. Hayden, K. Shobatake, R. K. Sparks, T. P. Schafer, and Y. T. Lee, *J. Chem. Phys.,* **82,** 3067 (1985).

75. J. G. Lauderdale and D. G. Truhlar, *J. Chem. Phys.,* **84,** 192 (1986).

76. K. V. Reddy and M. J. Berry, *Chem. Phys. Lett.,* **52,** 111 (1977).

77. R. D. Coombe and G. C. Pimentel, *J. Chem. Phys.,* **59,** 1535 (1973); F. S. Klein and A. Persky, *J. Chem. Phys.,* **61,** 2472 (1974); D. J. Douglas and J. C. Polanyi, *Chem. Phys.,* **16,** 1 (1976).

78. A. M. G. Ding, L. J. Kirsch, D. S. Perry, J. C. Polanyi, and J. L. Schreiber, *Faraday Discuss. Chem. Soc.,* **55,** 252 (1973); B. A. Blackwell, J. C. Polanyi, and J. J. Sloan, *Chem. Phys.,* **30,** 299 (1978).

79. H. H. Dispert, M. W. Geis, and P. R. Brooks, *J. Chem. Phys.,* **70,** 5317 (1979); F. Heismann and H. J. Loesch, *Chem. Phys.* **64,** 43 (1982).

80. S. Stolte, A. E. Proctor, and R. B. Bernstein, *J. Chem. Phys.,* **65,** 4990 (1976); S. Stolte, A. E. Proctor, W. M. Pope, and R. B. Bernstein, *J. Chem. Phys.,* **66,** 3468 (1977); L. Zandee and R. B. Bernstein, *J. Chem. Phys.,* **68,** 3760 (1978).

81. H. A. Lorentz, *Proc. Acad. Sci. Amsterdam,* **18,** 154 (1915).

82. P. Arrowsmith, F. E. Bartoszak, S. H. P. Bly, T. Carrington Jr., P. E. Charters, and J. C. Polanyi, *J. Chem. Phys.,* **73,** 5895 (1980); P. Arrowsmith, S. H. P. Bly, P. E. Charters, and J. C. Polanyi, *J. Chem. Phys.,* **79,** 283 (1983); H. J. Foth, H. R. Mayne, R. A. Poirier, J. C. Polanyi, and H. H. Telle, *Laser Chem.,* **2,** 229 (1983). A report by P. Hering, P. R. Brooks, R. F. Curl, R. S. Judson, and R. S. Lower, *Phys. Rev. Lett.,* **44,** 687 (1980), that they had obtained similar evidence in the reaction K + HgBr$_2$ → KBr + HgBr was not confirmed by P. R. Brooks, R. F. Curl and T. C. Maguire, *Ber. Bunsenges. Phys. Chem.,* **96,** 401 (1982).

83. P. D. Kleiber, A. M. Lyyra, K. M. Sando, S. P. Heneghan, and W. C. Stwalley, *Phys. Rev. Lett.,* **54,** 2003 (1985).

84. J. C. Polanyi, *Faraday Discuss. Chem. Soc.,* **67,** 129 (1979); J. C. Polanyi and R. J. Wolf, *J. Chem. Phys.,* **75,** 5951 (1981); H. R. Mayne, R. A. Poirier, and J. C. Polanyi, *J. Chem. Phys.,* **80,** 4025 (1984); H. R. Mayne, J. C. Polanyi, N. Sathyamurthy, and S. Raynor, *J. Phys. Chem.,* **88,** 4064 (1984).

BIBLIOGRAPHY

For general reviews of reaction dynamics see:

M. Baer, A Review of Quantum-Mechanical Approximate Treatments of Three-Body Reactive Systems, *Adv. Phys. Chem.,* **49,** 191 (1982).

M. Baer (Ed.), *Theory of Chemical Reaction Dynamics,* 4 vols., CRC Press, Boca Raton, FL, 1986.

R. B. Bernstein (Ed.), *Atom–Molecule Collision Theory—A Guide for the Experimentalist,* Plenum, New York, 1979.

R. B. Bernstein, *Chemical Dynamics via Molecular Beam and Laser Techniques,* Clarendon, Oxford, 1982.

R. D. Levine and R. B. Bernstein, *Molecular Reaction Dynamics,* Clarendon, Oxford, 1974.

J. C. Polanyi, Some Concepts in Reaction Dynamics, *Acc. Chem. Res.,* **5,** 161 (1972).

J. C. Polanyi and J. L. Schreiber, The Dynamics of Bimolecular Reactions, in W. Jost (Ed.), *Physical Chemistry: An Advanced Treatise,* vol. VIA, Academic, New York, 1974, Chapter 6 (pp. 383–487).

N. Sathyamurthy and T. Joseph, Potential-Energy Surfaces and Molecular Reaction Dynamics, *J. Chem. Educ.,* **61,** 968 (1984).

D. G. Truhlar (Ed.), *Potential-Energy Surfaces and Dynamics Calculations,* Plenum, New York, 1981.

R. B. Walker and J. C. Light, Reactive Molecular Collisions, *Ann. Rev. Phys. Chem.,* **31,** 401 (1980).

Accounts of dynamical calculations are to be found in some of the above publications, and also in:

J. T. Hynes, Chemical Reaction Dynamics in Solution, *Ann. Rev. Phys. Chem.,* **36,** 573 (1985).

K. J. Laidler and J. C. Polanyi, Theories of the Kinetics of Biomolecular Reactions, in G. Porter (Ed.), *Progress in Reaction Kinetics,* Pergamon, Oxford, vol. 3, 1965, p. 1.

K. J. Laidler, *Theories of Chemical Reaction Rates,* McGraw-Hill, New York, 1969, Chapter 7.

K. R. Lawley (Ed.), *Potential-Energy Surfaces, Adv. Chem. Phys.,* **42** (1980).

I. W. M. Smith, *Kinetics and Dynamics of Elementary Gas Reactions,* Butterworth, London, 1980, Chapter 3.

D. G. Truhlar and R. E. Wyatt, History of H_3 Kinetics, *Ann. Rev. Phys. Chem.,* **27,** 1 (1976).

D. G. Truhlar and R. E. Wyatt, $H + H_2$: Potential-Energy Surfaces and Elastic and Inelastic Scattering, *Adv. Chem. Phys.,* **36,** 141 (1977).

For reviews of quantum-mechanical calculations for reaction systems, see the previous references, and also:

R. G. Gordon, Rational Selection of Methods for Molecular Scattering Calculations, *Faraday Discuss. Chem. Soc.,* **55,** 22 (1973).

R. D. Levine, *Quantum Theory of Molecular Rate Processes,* Clarendon, Oxford, 1964.

J. C. Light, in *Methods in Computational Physics,* Academic, New York, 1971, vol. 10, p. 110.

R. A. Marcus, Introduction to the *Discussion on Molecular Beam Scattering, Faraday Discuss. Chem. Soc.,* **55,** 372 (1973).

C. Schlier (Ed.), *Molecular Beams and Reaction Kinetics,* Academic, New York, 1970; particularly articles by M. Karplus (p. 407) and J. Ross (p. 392).

For accounts of treatments of chemical kinetics that are not based on potential-energy surfaces (nonequilibrium and stochastic treatments) see:

E. W. Montroll and K. E. Shuler, The Application of the Theory of Stochastic Processes to Chemical Kinetics, *Adv. Chem. Phys.,* **1,** 361 (1958).

H. O. Pritchard, *The Quantum Theory of Unimolecular Reactions,* Cambridge University Press, 1984.

B. Widom, Molecular Transitions and Chemical Reaction Rates, *Science,* **148,** 1555 (1965).

B. J. Zwolinski and H. Eyring, The Nonequilibrium Theory of Absolute Reaction Rates, *J. Am. Chem. Soc.,* **69,** 2702 (1947).

For treatments of the relationship between molecular dynamics and macroscopic kinetics see:

K. E. Shuler, Reaction Cross Sections, Rate Coefficients, and Nonequilibrium Kinetics, in H. Hartmann (Ed.) *Chemische Elementarprozesse,* Springer-Verlag, Berlin, 1968, pp. 1–22.

J. C. Light, J. Ross, and K. E. Shuler, in A. R. Hochstein (Ed.), *Kinetic Processes in Gases and Plasmas,* Academic, New York, 1969, p. 281.

General accounts of the study of chemical reactions in molecular beams are to be found in:

E. F. Greene and J. Ross, Molecular Beams and a Chemical Reaction, *Science,* **159,** 587 (1965).

D. R. Herschbach, Molecular Beam Studies of Internal Excitation of Reaction Products, *Appl. Opt.,* Supp. 2 of *Chemical Lasers,* 128 (1965).

Y. T. Lee and Y. R. Shen, Studies with Crossed Laser and Molecular Beams, *Phys. Today,* **33**(11), 52 (1980).

M. R. Levy, in G. Porter (Ed.) *Progress in Reaction Kinetics,* Pergamon, Oxford, 1979, vol. 10, p. 1.

N. F. Ramsay, *Molecular Beams,* Clarendon, Oxford, 1956.

J. Ross (Ed.), *Molecular Beams, Adv. Chem. Phys.,* **10,** 1966.

C. Schlier (Ed.), *Molecular Beams and Reaction Kinetics,* Academic, New York, 1970.

J. P. Toennies, Molecular Beam Studies of Chemical Reactions, in H. Hartmann (Ed.), *Chemische Elementarprozesse,* Springer-Verlag, Berlin 1968, pp. 157–218.

Molecular Beam Scattering, *Faraday Discuss. Far. Soc.,* **55,** 1–410 (1973).

For accounts of state-to-state kinetic investigations see:

R. B. Bernstein, *Chemical Dynamics via Molecular Beam and Laser Techniques,* Clarendon, Oxford, 1982; especially Chapter 7.

S. R. Leone, State-Resolved Molecular Reaction Dynamics, *Ann. Rev. Phys. Chem.,* **35,** 109 (1984).

C. B. Moore and I. W. M. Smith, Vibrational–Rotational Excitation: Chemical Reactions of Vibrationally Excited Molecules, *Faraday Discuss. Chem. Soc.,* **67,** 146 (1979).

R. N. Zare, Polanyi Memorial Lecture, *Faraday Discuss. Chem. Soc.,* **67,** 6 (1979).

For reviews of work on the spectroscopy of transition species see:

T. Carrington, Jr., J. C. Polanyi, and R. J. Wolf, Probing the Transition State in Reactive Collisions, in S. Datz (Ed.) *Physics of Electronic and Atomic Collisions,* North Holland, Amsterdam, 1982.

H. J. Foth, J. C. Polanyi, and H. H. Telle, Emission from Molecules and Reaction Intermediates in the Process of Falling Apart, *J. Phys. Chem.,* **86,** 5027 (1982).

Biographical Sketches

These brief biographical notes are arranged in chronological order of date of birth. Included are those who, in the author's opinion, have made the most important contributions to the development of chemical kinetics, particularly in earlier years. Some of them, such as Nernst and Lindemann, worked mainly in other fields.

An alphabetic listing of the biographical entries is as follows:

Arrhenius, S. A. [9]
Baker, H. B. [10]
Bjerrum, N. J. [18]
Bodenstein, M. [14]
Bonhoeffer, K. F. [35]
Brønsted, J. N. [17]
Chapman, D. L. [12]
Christiansen, J. A. [24]
Daniels, F. [25]
Dixon, H. B. [7]
Esson, W. [4]
Evans, M. G. [39]
Eyring, H. [38]
Harcourt, A. G. V. [3]
Herzfeld, K. F. [30]
Hinshelwood, C. N. [34]
Kramers, H. A. [31]
Langmuir, I. [19]
Lewis, W. C. McC. [21]
Lindemann, F. A. [23]
London, F. W. [36]

Marcelin, R. [22]
Michaelis, L. [16]
Moelwyn-Hughes, E. A. [40]
Nernst, W. [11]
Norrish, R. G. W. [33]
Ostwald, F. W. [8]
Perrin, J. B. [13]
Pfaundler, L. [5]
Polanyi, M. [29]
Rice, F. O. [28]
Rice, J. [15]
Rideal, E. K. [27]
Semenov, N. N. [32]
Slater, N. B. [41]
Steacie, E. W. R. [37]
Taylor, H. S. [26]
Thénard, L. J. [1]
Tolman, R. C. [20]
van't Hoff, J. H. [6]
Wilhelmy, L. F. [2]

A good deal of the information in these sketches was derived from original scientific publications. The references given are to biographical material and commentaries where the reader will find additional information. Brief but useful factual information about most of the scientists included is to be found in *World Who's Who in Science,* A. N. Marquis, Chicago, 1968. Somewhat more detailed information about some of them is to be found in *Modern Scientists and Engineers,* McGraw-Hill, New York, 1980.

In compiling these biographical notes the author has received much help in the form of information, suggestions, photographs, and permission to reproduce photographs in this book. He is particularly grateful to the following: A. R. Allan, Archives, University of Liverpool; Prof. C. J. Ballhausen, University of Copenhagan; Prof. C. H. Bamford, University of Liverpool; Dr. Erik Bohr, Royal Danish Embassy, London; Center for History of Chemistry, Philadelphia; Dr. Regina Herzfeld, Washington, D.C.; Prof. Erwin Hiebert, Harvard University; Prof. K. A. Jensen, University of Copenhagen; Dr. M. Christine King, London; Prof. Y. Molin, Academy of Sciences, U.S.S.R.; Laurent Morelle, Archivist, Académie de Paris; Prof. John Neu, University of Wisconsin; Aubrey Rendell, University of Hull; Prof. F. O. Rice, Mishawaka, Indiana; Royal Society, London; Royal Society of Chemistry, London; Andrea Rudd, Archives, University of Liverpool; Dr. John Shorter, University of Hull.

[1] LOUIS JACQUES THÉNARD (1777–1857)

Thénard was born in La Louptière, Aube, France; after his death the village was renamed La Louptière-Thénard. The son of a peasant, Thénard suffered many privations to obtain an education and was helped in his efforts by the chemist L. N. Vauquelin (1763–1829). Thénard carried out some of his research with financial support from Napoleon and in collaboration with J. L. Gay-Lussac (1778–1850). In 1828 he was elected to the French Chamber of Deputies, and in 1832 he was created a baron.

Thénard's work was mainly on the preparation of new elements and compounds. He and Gay-Lussac were the first to prepare boron; their announcement of its discovery, on June 21, 1808, was nine days before Humphry Davy's announcement of the same isolation. Thénard also developed a color, known as "Thénard's blue," which is as bright as ultramarine but which withstands heat and can be used in the preparation of porcelain.

In 1818 Thénard discovered hydrogen peroxide and measured the rate of its decomposition. This was one of the earliest measurements of the rate of a chemical reaction.   (1, 14, 56)

[2] LUDWIG FERDINAND WILHELMY (1812–1864)

Wilhelmy was born in Stargard, Pomerania (now Poland). He first studied pharmacy and owned and operated a pharmacy, which he sold in 1843 when he began to study chemistry and physics at the Universities of Berlin, Giessen (now the Justus Liebig University), and Heidelberg. He obtained his doctorate from Heidelberg in 1846 and then studied with H. V. Regnault (1810–1878) at the Collège de France in Paris. He was appointed a privatdozent at Heidelberg in 1849 and remained in that position until 1854, when he retired to private life.

During his brief scientific career, Wilhelmy worked on a variety of problems in physics and chemistry, including capillary action, coefficients of expansion, the equivalence of heat and energy, and relationships between different physical properties.

Wilhelmy is remembered particularly for his pioneering work in chemical kinetics, the results of which were published in 1850. By means of a polarimeter he followed the inversion of cane sugar in the presence of different acids. He found that the instantaneous rate of change of the sugar concentration was proportional to the concentrations of both the sugar and the acid, and he set up a differential equation to express the rate. He also proposed, in the same publication, an empirical equation to express the influence of temperature. Wilhelmy's work received little attention until Ostwald [8] called attention to it in 1884.   (46, 56)

[3] AUGUSTUS GEORGE VERNON HARCOURT (1834–1919)

Harcourt was born in London and attended Harrow School and Balliol College, Oxford. Later, he was a pupil and assistant to Sir Benjamin Brodie (1817–1880). In 1859 he was appointed Dr. Lee's Reader in Chemistry, a position he held until his retirement in 1902. He became President of the Chemical Society of London in 1895.

He was an excellent teacher of chemistry and a number of his students, such as H. B. Dixon [7] and D. L. Chapman [12], went on to do distinguished work in kinetics. Some of his research was very practical: he designed the pentane lamp as a standard of brightness, and he developed a chloroform inhaler that was used for many years.

Harcourt is remembered best for his pioneering work, which he began in 1865, on the rate laws for chemical reactions. His very precise kinetic measurements were analyzed by William Esson [4], since Harcourt's knowledge of mathematics was very weak—he knew no calculus. He also worked with Esson on the temperature dependence of rates, but here they were less successful. However, they did establish for the first time a kinetic absolute zero.   (18, 19, 42, 43, 44, 49, 56, 60)

[4] WILLIAM ESSON (1839–1916)

Esson was born in Carnoustie, Forfarshire, Scotland, and at the age of 16 went to St. John's College, Oxford, where he obtained a first-class degree in mathematics and a second-class in classics. He became a Fellow of Merton College, Oxford, where he tutored in mathematics. In 1897 he was appointed Savilian Professor of Geometry at Oxford.

He published little original work in mathematics, his main distinction being his kinetic research in collaboration with Vernon Harcourt [3]. In a series of papers starting in 1865, Harcourt and Esson described the first systematic and precise studies of the dependence of the rate of a reaction on the concentration of the reactants. Harcourt performed the experiments and the mathematical treatment was worked out by Esson, whose procedures were essentially those used today. In 1912, 47 years after this pioneering work, they again collaborated on a paper dealing with the temperature dependence of rates; this work, however, was not fruitful. (22, 56, 66)

[5] LEOPOLD PFAUNDLER (1839–1920)

Pfaundler was born in Innsbruck, and during the period 1857–1865 he studied at the Universities of Innsbruck, Munich, and Paris. He obtained his Ph.D. in physics in 1861 from the University of Innsbruck. In 1867 he was appointed professor of physics at Innsbruck, and in 1891 he became professor at the University of Graz.

Pfaundler's research covered a wide field of physics and chemistry, including organic chemistry. He had a particular interest in teaching.

He published two important papers dealing with chemical equilibrium and the rates of chemical reactions. The first, published in 1867, was concerned mainly with the application of kinetic theory to the dissociation of gases at high temperatures. He made use of Clausius's kinetic theory of gases and took into account the distribution of speeds, realizing that molecules with higher energies were the most likely to dissociate. He recognized, following A. W. Williamson (1824–1904), that equilibrium is dynamic, forward and reverse reactions occurring at the same rate. He was the first to state clearly that reaction occurs only between molecules having at least a certain critical energy. He developed these ideas more fully in a paper that appeared in 1872.

In his classic book of 1884, *Études de dynamique chimique,* van't Hoff [6] acknowledged the debt he owed Pfaundler for clarifying these ideas.   (56, 58)

[6] JACOBUS HENRICUS VAN'T HOFF (1852–1911)

Van't Hoff was born in Rotterdam and studied in The Netherlands, Germany, and France; he later returned to the University of Utrecht where he obtained his Ph.D. degree in 1874. In the meantime, he had published an important paper proposing the tetrahedral carbon atom, but shrewdly he made no mention of this in his Ph.D. thesis, realizing that its controversial nature would do him no good. In 1878 he was appointed professor of chemistry, mineralogy, and geology at the University of Amsterdam, where he embarked on his research in the new field of physical chemistry. In 1896, he moved to the University of Berlin, and in 1901 he was awarded the first Nobel Prize in Chemistry for his work on solutions.

Much of van't Hoff's work was on fundamental aspects of thermodynamics and its application to solutions. He was the first to derive the equations for the effects of temperature and of pressure on the equilibrium constant for a reaction. He worked on the properties of dilute solutions of both electrolytes and nonelectrolytes and showed that such solutions obey equations that are similar to those applicable to gases. He observed anomalously high osmotic pressures of electrolytes, and these were later related to their dissociation into ions.

His work in chemical kinetics covered a wide range. In 1884 his *Études de dynamique chimique* appeared. This was the first monograph on chemical kinetics, and it described many of his own early studies on gas-phase reactions, reactions on surfaces, and reactions in solution. This book gives a number of equations relating rate to reactant concentration and demonstrates their applicability to experimental results. It also discusses several possible equations relating rate constants to temperature, on the basis of van't Hoff's equation for the temperature dependence of equilibrium constants; one of these was taken up later by Arrhenius [9] and is known by his name. In this book van't Hoff introduced for the first time the differential method, still commonly used in the analysis of kinetic results. Through this and other books on physical chemistry and his lectures, van't Hoff exerted a very important influence on the subject. van't Hoff, Ostwald [8], and Arrhenius [9] commonly are regarded as the founders of physical chemistry.   (1, 20, 37, 39, 49, 56, 58, 62, 63, 72)

[7] HAROLD BAILY DIXON (1852–1930)

Dixon was born in London, and in 1871 went to Christ Church, Oxford. Having failed to be successful in classics, he studied chemistry under Vernon Harcourt [3] and obtained a first-class honors degree in 1875. In that year he was elected a Fellow of Trinity College, Oxford, and he also held a position at Balliol College. In 1887 he succeeded Sir Henry Roscoe (1833–1915) as professor of chemistry in Owens College, Victoria University, Manchester (which in 1903 became the University of Manchester). Through his meticulously prepared lectures, he played an important part in the development of physical chemistry. In 1886 women were allowed to attend his lectures at Oxford "but only by special permission in each case and with the accompaniment of some elderly person."

Dixon made important contributions to the understanding of explosions in gaseous systems. Some of his experiments were carried out in a makeshift subterranean laboratory under the dining hall of Balliol College, and his results were described in three long papers in the *Philosophical Transactions of the Royal Society,* appearing in 1884, 1893, and 1903. Together, these three papers constitute a valuable monograph on explosions. In World War I, Dixon carried out important investigations on explosions in coal mines.

Beginning in 1880, Dixon carried out investigations on the influence of water vapor on explosions and other reactions in the gas phase; these studies were extended by his student H. B. Baker [10].   (2, 10, 12, 56)

[8] FRIEDRICH WILHELM OSTWALD (1853–1932)

Ostwald was born in Riga, Latvia, and studied at the University of Dorpat (now Tartu, Estonia). He devoted much time there to literature, music, and painting but managed to obtain his doctorate in chemistry in 1878. He was appointed professor of chemistry at the Riga Polytechnic Institute in 1881 and in 1887 moved to the University of Leipzig. He resigned his professorship there in 1906, moved to a country estate, and for the rest of his active life worked on a variety of topics including energetics, the history and philosophy of science, a world language, pacifism, and color theory. He received many honors, including the Nobel Prize in Chemistry in 1909.

It is impossible in a brief summary to do any kind of justice to Ostwald's extraordinary scientific accomplishments. He made particularly important contributions to the theory of electrolytic solutions and to thermodynamics. He was a man of remarkable personality with an unlimited capacity for work, and as a lucid expositor of science he exerted a powerful influence. His views were controversial and often turned out to be wrong. Following the physicist Ernst Mach (1838–1916), he was a positivist, believing that scientists should confine their attention to directly measurable quantities such as energy changes. Consequently, although his own work had done much to demonstrate the existence of ions, he denied the real existence of atoms and ions for many years, believing them to be convenient fictions. He was convinced finally by the experiments of Perrin [13] on the Brownian movement.

It was Ostwald who first defined, in 1887, the "order of reaction," although the concept had been introduced previously in 1884 by van't Hoff [6], who called it "molecularity." Ostwald's most important contribution to kinetics was undoubtedly his work on catalysis. He was the first to realize that a catalyst acts without altering the energy relationships of the overall reaction and that it speeds up a reaction usually by lowering the energy of activation. His Nobel Prize was awarded to him specifically for his work on catalysis.   (1, 4, 21, 32, 33, 35, 49, 56, 58)

[9] SVANTE AUGUST ARRHENIUS (1859–1927)

Arrhenius was born in Vik, near Uppsala, Sweden, and carried out research in Stockholm on electrolyte conductivities under the direction of the physicist Eric Edlund (1819–1888). His Ph.D. thesis, presented to the University of Uppsala in 1884, made little impression on his examiners, and it was rated as fourth class; his defense of it was rated as third class. During the next five years Arrhenius traveled extensively and worked with Ostwald [8], Friedrich Kohlrausch (1840–1910), Ludwig Boltzmann (1844–1906), and van't Hoff [6]. After lecturing in physical chemistry at the University of Uppsala, he was appointed professor of physics at the Technical High School in Stockholm in 1895. From 1905 until his death, he was director of physical chemistry at the Nobel Institute in Stockholm. He received the 1903 Nobel Prize in Chemistry.

Arrhenius's Ph.D. thesis did not mention dissociation, but his collaboration with Ostwald and van't Hoff led him in 1887 to formulate his theory of electrolytic dissociation. This theory slowly gained acceptance, having been supported strongly by Ostwald and van't Hoff.

Arrhenius worked in a variety of fields, including immunochemistry, cosmology, the origin of life, and the causes of the ice age; he attributed climatic changes to absorption of radiation by carbon dioxide in the atmosphere.

He did not work extensively in chemical kinetics, but what he did was of lasting importance. His interpretation in 1889 of one of the temperature-dependence equations given earlier by van't Hoff led to its being known universally as the "Arrhenius equation."   (1, 35, 48, 49, 51, 56, 58, 62)

[10] HERBERT BRERETON BAKER (1862–1935)

Baker was born in Blackburn, Lancashire, and went to Balliol College, Oxford, where he became a demonstrator in chemistry and an assistant to H. B. Dixon [7]. For a number of years following that he was a schoolmaster at Dulwich College. He later succeeded Vernon Harcourt [3] as Dr. Lee's Reader in Chemistry at Oxford and from 1912 to 1932 was professor of chemistry at Imperial College, London.

Baker is known best for his work on the effect of moisture on chemical change. The elaborate procedures he used for removing water have become known as "Bakerian drying," and he himself was referred to as "Dry Reaction Baker" or as "Dry Baker." He found that in the complete absence of moisture phosphorus and sulfur may be distilled in oxygen without any reaction taking place. He also found that, when completely dry, ammonia and hydrogen chloride do not combine and that ammonium chloride does not decompose.

It is reported that on one important occasion Baker, having predicted that no explosion would occur, applied a spark to a mixture of hydrogen and oxygen that he had been drying for a considerable time. There was a violent explosion, after which Baker blandly remarked that evidently the drying had not been sufficiently prolonged.   (10, 56)

[11] WALTHER NERNST (1864–1941)

Nernst was born in Brieson, West Prussia (now Wabrzezno, Poland), and he attended the University of Würzburg where he obtained his doctorate in 1887. That same year he joined Ostwald [8] in Leipzig as a research assistant, and in 1891 he became professor of physical chemistry at the University of Göttingen. In 1905 he was appointed professor of physical chemistry at the University of Berlin, and he was awarded the Nobel Prize in Chemistry in 1920.

Nernst made many outstanding contributions to physics and physical chemistry. He applied the principles of thermodynamics to electrochemical cells and arrived at an expression for the electric potential as a function of concentrations. He also provided explanations for the behavior of ions in aqueous solution. In 1905 he gave a theoretical treatment of the entropies of substances in the neighborhood of the absolute zero; this, his most important contribution to science, is now know as the Nernst heat theorem or as the third law of thermodynamics.

Nernst's main contribution to kinetics was to explain the photochemical combination of hydrogen and chlorine in terms of a chain reaction involving hydrogen and chlorine atoms. Previously, Bodenstein [14] had given explanations in terms of excited molecules and ions, but these proved unsatisfactory. Nernst's explanation led to important developments in the chemical kinetics of chain reactions.   (1, 34, 35, 36, 53, 56)

[12] DAVID LEONARD CHAPMAN (1869–1958)

Chapman was born in Wells, Norfolk, and was educated at the Manchester Grammar School and Christ Church, Oxford, where he was a student of Vernon Harcourt [3]. He wrote final examinations at Oxford in both chemistry and physics. After a period of school teaching he joined H. B. Dixon [7] in 1897 at Owens College, Victoria University, Manchester. In 1907 he was appointed a Fellow of Jesus College, Oxford, with the responsibility of directing the College chemistry laboratories which had just been founded in honor of Sir Leoline Jenkins; he retained that position until his retirement in 1944.

Compared with other chemists of his time, Chapman had an unusual proficiency in mathematics and physics, and this was reflected in his excellent teaching and research. His first research was on what is now called detonation. Dixon had made measurements of velocities of explosions in gases, and Chapman presented the first sound theoretical treatment of these explosions. Some of his equations were later arrived at independently by Émile Jouguet, and the region immediately behind a detonation wave is still referred to as the "Chapman–Jouguet layer."

In 1913 Chapman published an important treatment of electrocapillarity, in which he worked out the distribution of ions at a charged surface; the mathematical treatment he gave is a two-dimensional version of the Debye–Hückel theory of strong electrolytes, given more than 10 years later. The electric double layer considered in his theory is now known as the "Gouy–Chapman layer."

Most of Chapman's research was on the chemical kinetics of gas reactions, and in it he was ably assisted by his wife, Muriel C. C. Chapman, who herself independently did much important work in the field. Chapman followed van't Hoff [6] in recognizing, in a paper on the ozone decomposition published in 1910, that the relative importance of surface reaction can be established by varying the surface/volume ratio of the reaction vessel. He and his wife also made important studies on the thermal and photochemical reactions between hydrogen and chlorine. At first, however, Chapman favored the view of Bodenstein [14] that excited molecules, rather than atoms, are the chain carriers.

The introduction of the steady-state treatment into kinetics often is attributed to Bodenstein [14], who indeed made much use of it and ably defended it against its critics. However, the first application of the steady-state treatment was by Chapman, in a paper published in 1913.

In 1924 Mrs. Chapman had made studies of the effect of light intensity on the hydrogen–chlorine reaction. Her results led Chapman to develop the rotating-sector technique for measuring the mean lives of chain carriers. Subsequently, this procedure has been used widely for various types of reaction, particularly polymerizations. (8, 26, 29, 49, 56)

[13] JEAN BAPTISTE PERRIN (1870–1942)

Perrin was born in Lille, France, and obtained his doctorate in 1897 at the École Normale Supérieure in Paris. He was appointed professor of physical chemistry at the Université de Paris in 1910 and held that post until 1941. In that year, after the defeat of France by Nazi Germany, he left for the United States, where he lived the remaining year of his life.

During the 1890s Perrin studied cathode rays and showed them to consist of a beam of negative particles; the mass and charge of these, the electrons, were measured later by J. J. Thomson (1856–1940). In 1908 Perrin embarked on a study of the Brownian movement of particles in an electric field and was able to obtain direct evidence for the existence of electrons, atoms, and molecules. This evidence was enough to convince even the sceptical Ostwald [8] who had regarded atoms as a convenient fiction.

Perrin's work in kinetics was concerned mainly with theoretical arguments in favor of the ill-fated radiation hypothesis. Even after the hypothesis was rendered extremely unlikely, Perrin does not seem to have renounced it.   (1, 45, 55, 64)

[14] MAX BODENSTEIN (1871–1942)

Bodenstein was born in Magdeburg and obtained his Ph.D. degree in 1893 at the University of Heidelberg, where his research was directed by Viktor Meyer (1848–1897). He later worked with Ostwald [8] at Leipzig and then carried out research in Berlin. From 1908 he was professor at the University of Hannover and in 1923 he succeeded Nernst [11] as director of the Institut für Physikalische Chemie in Berlin, a position he held until 1936.

Bodenstein's research was entirely in chemical kinetics, in which he excelled in being a remarkably skillful experimentalist. His work was mainly on thermal and photochemical gas-phase reactions and on reactions at surfaces. Prior to the turn of the century he worked extensively on the reaction between hydrogen and iodine, the reverse decomposition of hydrogen iodide, and the equilibrium established between these species; he investigated both the thermal and photochemical reactions.

Beginning in about 1906 Bodenstein investigated the thermal and photochemical reactions between hydrogen and bromine, and later the corresponding reactions between hydrogen and chlorine. In about 1913 he observed very high quantum yields for the photochemical hydrogen–chlorine reaction and proposed the idea of a chain reaction. He first (1913) postulated ions as chain carriers and later (1916) excited chlorine molecules; in 1918 Nernst [11] proposed the now accepted view that the chain reaction involves hydrogen and chlorine atoms. Later, Bodenstein investigated other gas-phase reactions, including oxidations, and proposed chain mechanisms for them.   (1, 49)

[15] JAMES RICE (1874–1936)

Rice was a native of Northern Ireland and went to Queen's College (now Queen's University), Belfast, where he obtained a master's degree in mathematical science in 1900. After teaching physics for some years in a high school in Liverpool, he was appointed lecturer in physics at the University of Liverpool in 1914. In 1924 he was promoted to an associate professorship.

Rice exerted a considerable influence in his time through his writings on physical problems of fundamental importance. His 1923 book *Relativity,* one of the first on that subject in English, was widely acclaimed and was described by A. S. Eddington (1882–1944) as "first-rate." His book *Introduction to Statistical Mechanics for Students of Physics and Physical Chemistry* appeared in 1930, and it included many applications to kinetic problems. He collaborated extensively with his colleague W. C. McC. Lewis [21] and contributed appendices to the second edition of the latter's important book *A System of Physical Chemistry* (1918). He also helped Lewis greatly in his formulation of the collision theory of chemical reactions and in his support for the radiation hypothesis.

At the 1915 meeting of the British Association for the Advancement of Science there was a discussion of the idea of Marcelin [22] of representing chemical reaction as the motion of a point in many-dimensional phase space. Rice felt that Marcelin's definition of the activated state was too vague, and he suggested a more precise formulation which led to an explicit expression for a rate constant; for a time this work was known as the "Marcelin–Rice treatment." Rice also arrived at the important conclusion that the experimental activation energy is the difference between the average energy of the activated complexes and the average energy of all the reactant molecules. Unfortunately, the report of this work is so brief that it is now impossible to know by what reasoning he obtained this result, which was later obtained more rigorously by Tolman [20].   (45, 47)

[16] LEONOR MICHAELIS (1875–1949)

Michaelis was born in Berlin and obtained his M.D. degree in 1896 at the University of Berlin. He then worked for some time under the German bacteriologist Paul Ehrlich (1854–1915) and from 1899 to 1922 was on the staff of the Berlin Municipal Hospital; he was director of its laboratories from 1906 to 1922. He later held positions in the medical school of Nagoya University, Japan, and at Johns Hopkins University in Baltimore, Maryland. From 1929 until his retirement in 1940 he was a member of the Rockefeller Institute for Medical Research in New York, where he died in 1949.

Most of Michaelis' research was on the application of physicochemical principles to reactions catalyzed by enzymes, and much of his work was in the chemical kinetics of enzyme action. Some of his work, carried out between 1910 and 1920, involved clarifying the influence of hydrogen-ion concentration on the rates of these reactions.

The work for which Michaelis is remembered best is his formulation of an equation expressing the dependence of the rate of an enzyme-catalyzed reaction on the concentration of the substrate. This equation, developed in 1913 in collaboration with the Canadian biochemist Maud L. Menten (1879–1960), is known as the Michaelis–Menten equation, and a constant that occurs in it is called the Michaelis constant.   (1)

[17] JOHANNES NICOLAUS BRØNSTED (1879–1947)

Brønsted was born in Varda, Denmark, and in 1908 he obtained his doctorate in chemistry at the University of Copenhagen. He was then appointed professor of physical chemistry at the University, where he remained until his death.

Brønsted's work was mainly on the thermodynamics and kinetics of electrolytic solutions. In 1923, almost simultaneously with the British chemist T. Martin Lowry (1874–1936), he proposed definitions of acids and bases in terms of proton transfer. In an important paper with the American chemist Victor K. La Mer (1895–1966), published in 1924, he used the Debye–Hückel theory to give a relationship between the activity coefficient of an ion and the ionic strength, the latter quantity having been defined in 1921 by the American chemist Gilbert Newton Lewis (1875–1946).

Brønsted did much important work on the kinetics of reactions catalyzed by acids and bases. In 1922 he proposed an expression for the influence of ionic strength on the rates of reactions between ions; this expression, involving a "kinetic activity factor," was later derived more explicitly by Bjerrum [18] and Christiansen [24]. On the basis of his definitions of acids and bases, he further developed the concept of general acid–base catalysis. In 1924 he proposed relationships, now known as Brønsted relationships, between the catalytic activity of an acid or base and its acid or base strength.   (1, 11, 51, 56, 71)

[18] NIELS JANNIKSEN BJERRUM (1879–1958)

Bjerrum was born in Copenhagen, Denmark, and obtained his doctorate in 1908 at the University of Copenhagen; his research was directed by S. M. Jørgensen. In 1914 he was made professor of chemistry at the Royal Veterinary and Agricultural College in Copenhagen, a post he held until his retirement in 1949. During his early years he collaborated with Perrin [13] in Paris and with Nernst [11] in Berlin.

Bjerrum made distinguished contributions in various fields of chemical physics; he worked on kinetic and quantum theory, specific heats, absorption spectra, Brownian movement (with Perrin), and on a variety of other topics. He was the first to suggest, in 1909 and in more detail in 1916, that strong electrolytes are dissociated completely in solution and that interionic forces are important. This idea offended Arrhenius [9] but led in 1923 to the Debye–Hückel quantitative theory. Bjerrum also made important contributions to the theory of ion association.

In 1925 Bjerrum derived in a much more explicit way the equation of Brønsted [17] for the effect of ionic strength on the rates of reactions between ions. He clarified Brønsted's "kinetic activity factor" by postulating a pre-equilibrium involving an intermediate X, the rate being proportional to the *concentration* of X.   (31, 38, 51, 56)

[19] IRVING LANGMUIR (1881–1957)

Langmuir was born in Brooklyn, New York, and in 1903 obtained a degree in metallurgical engineering at Columbia University. He then went to Göttingen University, where he worked under Nernst [11] and obtained a Ph.D. in chemistry in 1906. After teaching chemistry at the Stevens Institute of Technology he joined the General Electric Company laboratories at Schenectady, New York, in 1909, where he remained until his retirement in 1950. He received the Nobel Prize in Chemistry in 1932.

At General Electric, Langmuir first worked on extending the life of tungsten-filament electric light bulbs. This led him to investigate adsorbed films on metals and to make important contributions to the understanding of the nature of the chemical bond in terms of electronic structures.

Langmuir's work on tungsten filaments also led him to do pioneering work on adsorption on surfaces and the mechanisms of chemical reactions at surfaces. He suggested what has become known as the Langmuir adsorption isotherm, and in 1921 he interpreted the orders of a number of surface-catalyzed reactions in terms of this isotherm. He investigated the kinetics of the dissociation of hydrogen molecules at tungsten surfaces. His work did much to discredit the radiation hypothesis in the 1920s. He later confessed to some frustration over the controversy regarding that hypothesis, and he abandoned his kinetic work in about 1922.   (45, 57, 59, 67)

[20] RICHARD CHASE TOLMAN (1881–1948)

Tolman was born in West Newton, Massachusetts, and obtained a bachelor's degree in chemical engineering at the Massachusetts Institute of Technology in 1903. After a year in Germany he returned to M.I.T. where he obtained a Ph.D. degree in physical chemistry in 1910. After periods at the Universities of Michigan, Cincinnati, and California (at Berkeley), he became professor of physical chemistry at the University of Illinois in 1916. In 1919 he became associate director and in 1920 director of the Fixed Nitrogen Research Laboratory of the U.S. War Department. In 1922 he was appointed professor of physical chemistry at the California Institute of Technology, Pasadena, where he remained until the end of his life.

Tolman's research covered a wide range in both physics and chemistry, and it included both theoretical and experimental investigations. His theoretical work was mainly in statistical mechanics, relativity, and cosmology. In 1909 he published an article on relativity with G. N. Lewis (1875–1946), and he followed it by a book, *The Theory of the Relativity of Motion,* which appeared in 1917. His book *Statistical Mechanics,* which was published in 1938, was a classic that was used widely for many years.

One of Tolman's important contributions to chemical kinetics, made in 1921, was to provide an interpretation of the experimental activation energy, in terms of the average energies of colliding molecules and of molecules actually undergoing reaction. During the 1920s Tolman also carried out much experimental work on gas reactions, designed to test the radiation hypothesis. Although he was unable to obtain any evidence that infrared radiation has any effect on reaction rates, he continued to support that hypothesis at least until 1927; he was of the opinion that the negative evidence was not conclusive. In spite of this, however, his gas-phase work was of considerable value, and he made a particularly useful contribution by expressing very clearly the arguments for and against the radiation hypothesis.   (25, 45)

[21] WILLIAM CUDMORE McCULLAGH LEWIS (1885–1956)

Lewis was born in Belfast, Northern Ireland, and received an M.A. degree from the Royal University of Ireland (now Queen's University, Belfast). He then carried out research at Liverpool University under F. G. Donnan (1870–1956), and later spent a year at Heidelberg University with the colloid chemist Georg Bredig (1868–1944). In 1913 he succeeded Donnan as professor of physical chemistry at Liverpool and retired from that position in 1948.

Much of Lewis's research was in colloidal and biological chemistry. He studied the adsorption of solutes at liquid surfaces, the coagulation of sols, the behavior of oil droplets in water, and the electrophoresis of organic and biological materials. Lewis also wrote an important book, *A System of Physical Chemistry*, which was one of the earliest books in English in that field and which exerted a considerable influence; it was first published in 1916.

Lewis was well informed about physical theory and was one of the first to apply quantum theory to chemical problems. In his work in chemical kinetics, in particular, he applied statistical mechanics and quantum theory to reaction rates. He believed that the distribution of molecular energies was maintained by the absorption of infrared radiation, but at the same time he realized correctly that reaction rates are dependent on the frequency of molecular collisions. On this basis he obtained, in 1917, the simple kinetic theory expression for a rate constant and showed it to apply satisfactorily to the reaction between hydrogen and iodine. (5, 45)

[22] RENÉ MARCELIN (1885–1914)

Marcelin was born in Cagny, Seine-et-Oise, France, and in 1904 became a student at the Faculté des Sciences de Paris, where one of his teachers was Jean Perrin [13]. In 1909 he obtained his Diplôme d'Études Supérieures, and from then until 1914, when he obtained his Ph.D. degree, he carried out independent research on thin films and on theories of chemical kinetics. Immediately upon the outbreak of World War I on August 4, 1914, Marcelin enlisted in the French Army and was killed in the following month.

His research for the Ph.D. degree was concerned almost entirely with the theory of the rates of chemical and physical processes. His work followed two different but parallel paths, one thermodynamic and the other statistical. He obtained an equation which, translated into modern terminology, expresses the rate in terms of a Gibbs energy of activation, a concept further developed in transition-state theory.

In a paper published in 1914 he described a reaction for the first time on the basis of the motion of a point in many-dimensional phase space. He then applied statistical mechanics to obtain an expression for the concentration of species "present at a critical surface in this phase space" and showed that his expression led to the correct type of temperature dependence. (50)

[23] FREDERICK ALEXANDER LINDEMANN, 1ST VISCOUNT CHERWELL (1886–1957)

Lindemann was born in Baden-Baden, Germany, of parents who normally resided in England; his father came from Alsace-Lorraine, and his mother came from the United States. He went to school in Scotland and later attended the Hochschule at Darmstadt. In 1910 he obtained a Ph.D. degree from the Institut für Physikalische Chemie, Berlin, his research having been directed by Nernst [11]. In 1919 he was appointed professor of experimental philosophy at Oxford and retained that position until he resigned in 1955. He was a close personal friend of Winston Churchill, becoming Paymaster General in Churchill's Cabinet in 1942–1945 and 1951–1953.

Lindemann worked on a variety of problems in physics, including the theory of specific heats, the ionization of stars, and temperature inversion in the stratosphere. He played an important part in building up the Clarendon Laboratory at Oxford, to which he attracted a number of European scientists.

Lindemann did little work in physical chemistry, but what he did was very effective. He played an important role in the overthrow of the radiation hypothesis, and in so doing he suggested, in 1921, a mechanism for unimolecular reactions that has formed the basis of all subsequent theories.   (7, 27, 45, 65, 70)

[24] JENS ANTON CHRISTIANSEN (1888–1969)

Christiansen was born in Vejle, Denmark, and studied at the Polytekniske Laereanstalt in Copenhagen. In 1911 he became assistant to S. P. L. Sørensen (1868–1939) at the Carlsberg Laboratory in Copenhagen, and in 1915 he transferred to the University of Copenhagen, where he obtained his Ph.D. degree in 1921. He taught at the University of Copenhagen from 1919 until his retirement in 1959. From 1931 to 1948 he was professor of inorganic chemistry, and in 1948 he succeeded Brønsted [17] as professor of physical chemistry.

In 1919, while still a candidate for the Ph.D. degree, Christiansen published an important paper on the reaction between hydrogen and bromine; very similar treatments were published at almost the same time by K. F. Herzfeld [30] and by M. Polanyi [29]. In these three papers, the now accepted chain mechanism for the reaction was proposed, involving the participation of hydrogen and bromine atoms. This work made use of the steady-state treatment, which had first been used by Chapman [12] in 1913.

Christiansen's Ph.D. thesis, submitted in October 1921, used for the first time the expression "chain reaction" (*Koedereaktion* in Danish). It also contained essentially the same treatment of unimolecular reactions that a few days earlier (September 28, 1921) had been suggested verbally by Lindemann [23] at a meeting of the Faraday Society.

During his early years at the University of Copenhagen, Christiansen studied at Bohr's Institute of Theoretical Physics, where he met the Dutch physicist H. A. Kramers [31]. In 1923 Christiansen and Kramers published a paper in which they attempted to explain unimolecular gas reactions in terms of energy chains. Product molecules were supposed to transfer excess energy to reactant molecules; however, this suggestion did not prove fruitful.

Christiansen also directed his attention to solution reactions, and in 1924 he suggested that rates of collision in solution should be comparable with those in the gas phase; he pursued this idea further in later investigations.   (3, 6, 51, 72)

[25] FARRINGTON DANIELS (1889–1972)

Daniels was born in Minneapolis and studied at the University of Minnesota where he obtained an M.Sc. degree in 1911. He then went to Harvard where his research, on the electrochemistry of thallium alloys, was directed by T. W. Richards (1868–1926). After a period of teaching in a polytechnic and of chemical warfare service, Daniels joined the U.S. Bureau of Soils in 1919, working at their Experimental Station in Arlington, Virginia. In 1920 he was appointed to the University of Wisconsin, where he remained in various capacities until his retirement in 1959. He was active in various fields of scientific research until his death.

Daniels carried out important work on many scientific problems, some of great practical significance. He worked on nitrogen fixation, photosynthesis, thermoluminescence, and solar energy. During World War II he directed important research on problems related to the development of nuclear energy.

In a number of ways Daniels had an important influence on the teaching of physical chemistry. In 1931 he collaborated with F. H. Getman on a revision of the latter's *Outlines of Theoretical Chemistry.* This book went through a number of revisions and changes of title; in 1943 it became *Outlines of Physical Chemistry,* and in 1955 a revision with R. A. Alberty was simply *Physical Chemistry.* Daniels also published *Chemical Kinetics,* which appeared in 1938; it was based on the Baker lectures delivered in 1935 at Cornell University.

In 1919 Daniels began a series of investigations on the decomposition of nitrogen peroxide, which he and his students studied mainly in the gas phase but also in a variety of solvents. His early studies were done to test the radiation hypothesis, and by 1928 he had obtained strong evidence against that hypothesis. Later kinetic work was on a number of other homogeneous gas reactions and surface reactions. (16, 45)

[26] HUGH STOTT TAYLOR (1890–1974)

Taylor was born in St. Helens, Lancashire, and attended the University of Liverpool where he obtained an M.Sc. degree in 1910. He then carried out research at Liverpool in collaboration with Henry Bassett and in 1912 went to work with Arrhenius [9] in Stockholm. From 1913 to 1914 he worked with Bodenstein [14] in Hannover. By 1914 he had published nine papers, and on the basis of them he was awarded the D.Sc. degree by Liverpool University. In 1914 he went to Princeton University where, except for some war work in London, he stayed until the end of his life. He remained a British subject and was knighted in 1953.

During his stay in Arrhenius's institute, Taylor worked on acid–base catalysis and was led to suggest that undissociated species as well as hydrogen and hydroxide ions can bring about catalysis. This view displeased Arrhenius, and although Taylor's idea of general catalysis is correct, it later emerged that his evidence for it was not valid. Taylor also proposed a relationship, later improved by Brønsted [17], between catalytic activity and acid or base strength.

In Bodenstein's laboratory, Taylor worked on the hydrogen–chlorine reaction, and he continued this work at Princeton. In 1925 he suggested for the first time that organic free radicals such as ethyl are chain carriers, a view that was confirmed experimentally by F. O. Rice [28] in 1933. He made many studies of reactions involving atoms and free radicals, produced by photochemical or radiation-chemical processes.

Particularly during the 1920s Taylor worked extensively on the kinetics of reactions on surfaces. He was the first to suggest, in 1925, that reactions occur primarily on "active centers" on surfaces. He also proposed and demonstrated that the process of absorption can involve an activation energy, and he distinguished between two types of adsorption—physical adsorption and chemisorption.　(41)

[27] ERIC KEIGHLEY RIDEAL (1890–1974)

Rideal was born in Sydenham and went to Trinity Hall, Cambridge, in 1907. After obtaining his B.A. degree, he went to the University of Bonn where he obtained a Ph.D. degree in 1912. After two years of private consulting and some years of war work, he was appointed visiting professor at the University of Urbana, Illinois, in 1919. In 1921 he became a Fellow of Trinity Hall, Cambridge, and in 1930 was appointed professor of colloidal physics at Cambridge; this title later became professor of colloid science. In 1946 he was appointed director of the Davy–Faraday Laboratory at the Royal Institution, London. However, he found the duties of this position irksome, and from 1950 to 1955 he was a professor of chemistry at King's College, London. He was knighted in 1951.

Rideal worked on a very wide range of topics in physical chemistry, including electrochemistry, colloid and surface chemistry, spectroscopy, and chemical kinetics. His work in electrochemistry included some studies of overvoltage, electrode kinetics, and diffusion currents. His work in colloid and surface chemistry included investigations of surface films of fatty acids, of ionizations of proteins, and of the electrophoretic mobilities of colloidal particles.

Some of Rideal's work in chemical kinetics was on homogeneous gas reactions; for example, he did some work on active nitrogen and on reactions giving rise to the emission of light. He also carried out some studies of polymerization kinetics. Perhaps his most important work in kinetics was on chemisorption and surface kinetics. He studied a variety of reactions, including the addition of hydrogen to ethylene, on a number of different surfaces, and elucidated some of their mechanisms by the use of isotopes.

Rideal wrote a number of books. One of them, *Catalysis in Theory and Practice* (1st ed., 1919), was coauthored by H. S. Taylor [26] and exerted a considerable influence in the early days of the subject.   (23)

[28] FRANCIS OWEN RICE (b. 1890)

Rice was born in Liverpool, England, and attended Liverpool University. He obtained a D.Sc. degree there in 1916, his research having been directed by the photochemist E. C. C. Baly. In 1919 Rice moved to the United States where he was first at New York University and then at Johns Hopkins University. In 1938 he was appointed chairman of the Chemistry Department at the Catholic University of America, Washington, D.C., where he remained until 1959. From then until 1962 he was chairman of the Chemistry Department at Georgetown University, also in Washington, D.C.

Rice's research was mainly in the field of gas-phase kinetics. In 1928, with Harold C. Urey (1893–1981) and R. N. Washburn, he published a significant paper on the nitrogen pentoxide decomposition, in which the reactant was formed in a narrow molecular beam so as to reduce the effect of collisions. This was probably the earliest molecular-beam kinetic study, and it was important in producing strong evidence against the radiation hypothesis.

Most of Rice's work, however, was on organic reactions, particularly decompositions. Following the discovery by F. A. Paneth (1887–1958) in 1929 of the free methyl radical, Rice was able to demonstrate in 1933 that free ethyl radicals are present in decomposing ethane. A year later Rice and K. F. Herzfeld [30] published a pioneering paper on organic chain reactions, in which they proposed specific mechanisms that led to the correct overall kinetics. Although some of their mechanisms inevitably have required revision in detail, the general principles developed by Rice and Herzfeld still are valid today. In the face of evidence that appeared to suggest that organic decompositions occur in part by molecular mechanisms, Rice insisted that they occur almost entirely by chain mechanisms. Further work by Rice and others, involving isotopic mixing, demonstrated that this is indeed the case.

[29] MICHAEL POLANYI (1891–1976)

Polanyi was born in Budapest and obtained an M.D. degree before World War I, in which he served as a medical officer in the Hungarian army. After the war, having developed an interest in thermodynamics and quantum theory, he worked for a Ph.D. degree; for this he had no formal supervision but received much inspiration from Nernst [11] and from Einstein (1879–1955). He worked for some time at the Kaiser Wilhelm Institut für Physikalische Chemie in Berlin and in 1933 was appointed professor of physical chemistry at Manchester University. In 1948 he resigned that position to become professor of social studies at the same University.

Polanyi carried out some early work on X-ray diffraction and on the adsorption of gases at surfaces, but his main scientific work was in chemical kinetics, particularly on fundamental aspects involving the applications of thermodynamics and of quantum theory. In 1931 he and Henry Eyring [38] constructed for the first time a potential-energy surface for the $H + H_2$ reaction and they considered some dynamical calculations using it. During the 1930s Polanyi and his students made some important fundamental studies on elementary gas-phase reactions, including some atomic reactions, and on solution reactions. In 1935 Polanyi and M. G. Evans [39], at almost the same time as Eyring, developed the transition-state theory of chemical reactions, applying it in particular to the influence of pressure on reaction rates.

Polanyi's son John Charles Polanyi (b. 1929), now at the University of Toronto, carries out important theoretical and experimental work in the field of reaction dynamics. He was awarded a 1986 Nobel Prize in chemistry.   (74)

[30] KARL FERDINAND HERZFELD (1892–1978)

Herzfeld was born in Vienna and studied at the Universities of Vienna, Zurich, and Göttingen; he took his Ph.D. degree in physics at the University of Vienna in 1914. After serving in the Austro-Hungarian army during World War I, he became in 1920 Privat-dozent in theoretical physics and chemistry at the University of Munich, where the professor was Arnold Sommerfeld (1868–1951). From 1926 to 1936 he was visiting professor and then professor of physics at the Johns Hopkins University in Baltimore, Maryland. From 1936 to 1961 he was professor and chairman of the department of physics at the Catholic University of America.

Herzfeld's work in theoretical physics and chemistry covered a very wide range. In 1925, with his student W. H. Heitler (b. 1904), he worked on theories of binary solutions of electrolytes, and he further developed this field for a number of years. He also investigated the structure and dynamics of crystalline solids. He worked on the quantum theory of optical dispersion, and on the absorption of ultrasonic waves, with particular reference to problems of energy transfer. With T. A. Litovitz he wrote an important book in 1959, *Absorption and Dispersion of Ultrasonic Waves.*

Herzfeld's work in chemical kinetics included a suggestion, which proved correct, for the mechanism of the hydrogen–bromine reaction in the gas phase; similar proposals were made at the same time (1919) by Christiansen [24] and Polanyi [29]. It was Herzfeld who first pointed out in 1921 that the recombination of atoms and the reverse dissociation of a diatomic molecule require the presence of a third body, a conclusion that was later confirmed experimentally by Bodenstein [14].

In 1931 Herzfeld collaborated with F. A. Paneth (1887–1958) on the chemistry of free organic radicals. Soon afterwards, F. O. Rice [28] showed that such radicals are intermediates in a number of organic decompositions, and in 1934 Rice and Herzfeld published an important paper in which they proposed specific free-radical mechanisms for a number of organic decompositions and showed by application of the steady-state treatment that they led to the correct kinetic behavior. This work formed the basis for many subsequent investigations of reaction mechanisms.

[31] HENDRIK ANTHONY KRAMERS (1894–1952)

Kramers was born in Rotterdam, and from 1919 to 1926 he studied under Niels Bohr (1885–1962) at the Institute of Theoretical Physics of the University of Copenhagen, where he met the Danish physical chemist J. A. Christiansen [24]. In 1926 he returned to Holland to become professor of theoretical physics at the University of Utrecht. In 1934 he was appointed to a similar position at the University of Leiden.

Kramers' main work was in quantum physics, some of it being on the intensities of atomic spectra. In 1925 he and Heisenberg (1901–1976) predicted the Raman effect, three years before its actual discovery.

Kramers made two important contributions to chemical kinetics. In 1923 he and Christiansen [24] suggested an energy-chain mechanism to explain unimolecular gas reactions; products of reaction were supposed to pass on their energy to reactant molecules, which were then enabled to undergo transformation. Although this idea proved incorrect, it contributed to the development of theories of chain reactions.

In a paper published in 1940, Kramers made an important contribution to the theory of reactions in solution. He treated such reactions in terms of the motion of a particle undergoing Brownian motion as a result of interaction with solvent molecules; the particle finally surmounted a potential-energy barrier. Some recent treatments of such reactions by generalized transition-state theory have been based on Kramers' treatment.

[32] NIKOLAI NIKOLAEVICH SEMENOV (1896–1986)

Semenov was born in Saratov, Russia, and entered the University of St. Petersberg in 1913; by the time he graduated in 1917 it had changed its name to the University of Petrograd. He later was director of chemical physics at the Physico-Chemical Institute in Leningrad (the new name for Petrograd), attaining the rank of professor in 1928. In 1944 he became head of the Department of Chemical Kinetics at the Moscow State University and at the same time was director of the Institute for Chemical Physics of the Academy of Sciences in Moscow. In 1956 he shared the Nobel Prize in Chemistry with C. N. Hinshelwood [34].

Semenov's work has been mainly in gas-phase kinetics. Two of his books have appeared in English editions: one was *Some Problems in Chemical Kinetics and Reactivity* which was published in English in 1958.

His other book in English, *Chemical Kinetics and Chain Reactions,* appeared in 1935 and describes various types of chain process, including gaseous explosions. In the 1920s Semenov and Hinshelwood independently worked out the theory of these explosions, introducing in 1927 the concept of branching chain processes in which one chain carrier undergoes a reaction and produces two or more carriers. Semenov also developed the theory of degenerate branching, which accounts for the cool flames frequently found with hydrocarbon oxidations.   (1)

[33] RONALD GEORGE WREYFORD NORRISH (1897–1978)

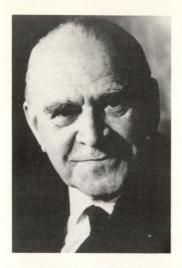

Norrish was born in Cambridge, England, and in 1915 won a scholarship to Emmanuel College, Cambridge; however, he then served in the Royal Field Artillery and only began his undergraduate career in 1919. He obtained his B.A. degree in chemistry in 1921, when he undertook research under E. K. Rideal [27] and obtained his Ph.D. degree in 1924. In that year he was elected a Fellow of Emmanuel College, a position he held for the rest of his life. He was professor of physical chemistry at Cambridge from 1937 to 1965. In 1967 he shared the Nobel Prize in Chemistry with George Porter (b. 1920) and Manfred Eigen (b. 1927).

Most of Norrish's work was in gas-phase kinetics, but he also did research on reactions in solution, particularly oxidations and polymerizations. He studied the kinetics and mechanisms of a number of free-radical and cationic polymerizations. He made important contributions to the understanding of the kinetics of combustion of hydrocarbons and other substances. Some of his work was concerned with the role of the methylene radical in reaction mechanisms.

Norrish also studied the photolysis of a number of compounds, including chlorine, nitrogen dioxide, and a variety of organic compounds, especially those containing the carbonyl group. With George Porter and others he developed in 1950–1952 the technique of flash photolysis, which he and his coworkers applied to a number of reactions. It was these studies in particular that gained Norrish and Porter the Nobel Prize.   (15)

[34] CYRIL NORMAN HINSHELWOOD (1897–1967)

Hinshelwood was born in London and attended Balliol College, Oxford, where, because of World War I, he took a shortened course in chemistry. Immediately after taking his bachelor's degree in 1920, he was made a Fellow of Balliol College, and in 1921 he became Fellow and Tutor of Trinity College, Oxford. In 1937 he was appointed Dr. Lee's professor of chemistry at Oxford, a position he held until his retirement in 1964. He was president of the Chemical Society from 1946 to 1948, president of the Royal Society from 1955 to 1960, and president of the Classical Association in 1959. He was knighted in 1948, received a Nobel Prize in 1956, and was appointed to the Order of Merit in 1960.

Most of Hinshelwood's work was in chemical kinetics, and he made many pioneering contributions. In the 1920s he investigated experimentally a number of gas reactions that were believed to be unimolecular, and on the basis of a suggestion by Lindemann [23] Hinshelwood made an important contribution to the theory of those reactions in 1927. During the same period, Hinshelwood carried out a number of experimental studies of reactions on surfaces, and on the basis of the theory of adsorption proposed by Langmuir [19], he contributed greatly to the understanding of the mechanisms of heterogeneous catalysis.

During the same period, he also investigated gaseous explosions, such as the hydrogen–oxygen reaction, and almost simultaneously with Semenov [32] explained in 1928 their mechanisms in terms of branching chains. Later, Hinshelwood did much work on the mechanisms of other chain reactions and of reactions in solution. He also made important contributions to an understanding of processes occurring in the bacterial cell.

Hinshelwood exerted considerable influence on physical chemistry through his books. One of them, *The Kinetics of Chemical Change in Gaseous Systems,* was first published in 1926 and it was an important new approach to the subject; it underwent several revisions as the field advanced. Hinshelwood's *Structure of Physical Chemistry* presented the subject in an unusual way and provides interesting new insights. (9, 69)

[35] KARL FRIEDRICH BONHOEFFER (1899–1957)

Bonhoeffer was born in Breslau and studied at the Universities of Tübingen and Berlin. His research at Berlin was directed by Nernst [11], and he obtained his doctorate in 1922. From 1923 to 1930 he was assistant to Fritz Haber (1868–1934) in the Kaiser Wilhelm Institut für Physikalische Chemie in Dahlem. Later, he was professor of physical chemistry at the University of Frankfurt and was director of the Institut at Dahlem. In 1949 he became director of the newly named Max Planck Institute für Physikalische Chemie in Göttingen.

Bonhoeffer made distinguished contributions in a variety of branches of physical chemistry. He worked on the thermal dissociation of water vapor and on reactions of the hydroxyl radical. He also studied various aspects of flames and of detonation. In 1924 he improved the method of R. W. Wood (1868–1955) for generating hydrogen atoms in a silent electric discharge. In 1929 he and Paul Harteck (b. 1902) separated hydrogen into its two forms, *para*-hydrogen and *ortho*-hydrogen, and by producing atomic hydrogen by the discharge-tube method they were able to make the first experimental study of the reaction $H + p\text{-}H_2 \rightarrow o\text{-}H_2 + H$. Later, Bonhoeffer and A. Farkas studied the kinetics of the combination of hydrogen atoms at surfaces.

Bonhoeffer also worked on the kinetics of electrode processes, the passivity of metals, and on periodic (oscillating) reactions. He also studied biological reactions and the function of the central nervous system.   (28, 40)

[36] FRITZ WOLFGANG LONDON (1900–1954)

London was born in Breslau, Germany (now Wroclaw, Poland), and received his Ph.D. degree in philosophy in 1921 from the University of Munich. He later turned to theoretical physics and carried out research under Arnold Sommerfeld (1868–1951) and then with Erwin Schrödinger (1887–1961). He left Germany in 1933, worked for some time in the United Kingdom and France, and later became a professor of physics at Duke University, Durham, North Carolina.

In 1927 London and W. H. Heitler (b. 1904) published an important quantum-mechanical treatment of the hydrogen molecule. They were the first to appreciate the significance of exchange energies, and their treatment formed the basis of much further work on molecular structure. In the following year London developed an approximate expression for the potential energy of the triatomic H_3 complex, in terms of the coulombic and exchange energies that apply to the H_2 molecule. His expression, known as the *London equation,* was later used by Eyring [38] and Polanyi [29] to obtain a potential-energy surface for the $H + H_2$ reaction and has been applied in many subsequent treatments of reaction rates: these are referred to as London–Eyring–Polanyi (LEP) treatments.

In 1930 London published a quantum-mechanical treatment of dispersion forces.   (1)

[37] EDGAR WILLIAM RICHARD STEACIE (1900–1962)

Steacie was born in Montreal and was educated at the Royal Military College, Kingston, Ontario, and at McGill University, Montreal, where he obtained a B.Sc. degree in chemical engineering in 1923. He then did research in physical chemistry at McGill under Otto Maass, obtaining his Ph.D. degree in 1926, when he was appointed to the academic staff at McGill. From 1934 to 1935 he worked with K. F. Bonhoeffer [35] at Frankfurt and with A. J. Allmand at King's College, London. In 1939 he was made director of the Division of Chemistry of the National Research Council of Canada, and he became its president in 1952. He was remarkable in being able to carry out important scientific research at the same time as performing heavy administrative duties in which he had a far-reaching influence on the organization and expansion of science throughout the world.

Steacie's research was mainly on the kinetics and mechanisms of organic reactions in the gas phase. He investigated a number of thermal reactions and, more particularly, reactions brought about by visible and ultraviolet radiation. He was interested especially in photosensitization, in which heavy atoms such as mercury become excited by absorption of radiation and transfer their energy by collision to other molecules. He attacked the problem of composite reaction mechanisms from two directions. He studied the overall kinetics of such reactions and also investigated elementary processes brought about by atoms and free radicals produced by photolysis or by photosensitization. In this way he was able to establish the detailed mechanisms of a number of composite reactions and to determine the kinetic parameters for their elementary steps.

His monograph, *Atomic and Free-Radical Reactions,* published in 1946, rapidly became a standard reference work.   (52, 68)

[38] HENRY EYRING (1901–1981)

Eyring was born in Colonia Juarez, Mexico, and attended the University of Arizona where he obtained a bachelor's degree in mining engineering in 1923; in 1924 he obtained a master's degree in metallurgy. In 1925 he went to the University of California at Berkeley where he studied physical chemistry and carried out research under G. E. Gibson (1884–1959), obtaining his Ph.D. degree in 1927. After teaching at the University of Wisconsin where he collaborated on some research with Farrington Daniels [25], Eyring spent a year (1929–1930) working with M. Polanyi [29] at the Kaiser Wilhelm Institut in Berlin. In 1931 he went to Princeton University on the invitation of H. S. Taylor [26] and remained there until 1946 when he moved to the University of Utah, where he remained until the end of his life. He became a naturalized U.S. citizen in 1935.

Eyring's research covered a very wide range. Much of it was in chemical kinetics, but he concerned himself to some extent with the structures of solids and liquids and with physical processes. Although Eyring carried out little experimental research, his theoretical approaches always were related closely to experimental results.

One of Eyring's outstanding contributions to kinetics was his construction, with M. Polanyi [29], of a semiempirical potential-energy surface for the $H + H_2$ reaction. Perhaps of even greater significance for chemical kinetics was his formulation of transition-state theory. In later years Eyring developed his theories in various directions and applied them to a wide variety of chemical and physical processes and to biological systems.

Eyring's book *The Theory of Rate Processes,* coauthored with S. Glasstone and K. J. Laidler in 1941, was for many years the standard monograph on theories of chemical reaction rates. His other books, such as *The Theory of Rate Processes in Biology and Medicine,* written with F. H. Johnston and B. J. Stover in 1974, also have exerted considerable influence.   (30)

[39] MEREDITH GWYNNE EVANS (1904–1952)

Evans was born in Atherton, Lancashire, and attended the University of Manchester. From 1934 to 1935 he worked with H. S. Taylor [26] at Princeton University and afterward became a lecturer at Manchester. In 1939 he was appointed professor of inorganic and physical chemistry at the University of Leeds, where he remained until 1948 when he succeeded M. Polanyi [29] as professor of physical chemistry at the University of Manchester.

Evans worked on a wide range of kinetic problems, from both the theoretical and experimental points of view. During the 1930s he collaborated with Polanyi on studies of sodium atoms with halogens and halides and arrived at important conclusions about the relationship between activation energies and heats of reaction. In 1935 Evans and Polanyi published their paper on transition-state theory, and Evans later developed the theory in various directions.

Evans also carried out work on the kinetics of polymerization reactions and on oxidation–reduction reactions.   (54)

[40] EMYR ALUN MOELWYN-HUGHES (1905–1978)

Moelwyn-Hughes was born in Cardigan, Wales, and attended Liverpool University where he took his Ph.D. degree in 1930. His research director was W. C. McC. Lewis [21], and his work was on the acid hydrolysis of several glycosides. From 1930 to 1933 he worked in Oxford with Hinshelwood [34] on various reactions in solution. After a year with Bonhoeffer [35] in Frankfurt, he spent the rest of his career at Cambridge, later being elected a Fellow of Darwin College.

Moelwyn-Hughes's research was almost entirely on the kinetics of reactions in solution, including some reactions catalyzed by enzymes. His early work with Hinshelwood concerned the magnitudes of the pre-exponential factors of such reactions. In 1924 Christiansen [24] had suggested that collision formulae for gases should apply approximately to solutions, but many reactions in solution showed abnormally low pre-exponential factors. By comparing rates in the gas phase and in solution, Hinshelwood and Moelwyn-Hughes showed that these low factors were often an intrinsic property of the reaction rather than being due to a deactivating effect of the solvent. These concepts were developed by Moelwyn-Hughes during the rest of his research career.

Moelwyn-Hughes's early research was included in his important monograph *The Kinetics of Reactions in Solution,* which appeared in 1933 with a second edition in 1947. He also made a major contribution by his large book, *Physical Chemistry,* which appeared in 1940 and in an enlarged form in 1957. Although written in an austere style that hardly makes it attractive to most students, this book is a mine of information for the serious scholar, since it goes back to many original sources and provides derivations that are more rigorous than those found in most textbooks. His monograph *Chemical Statics and Kinetics in Solutions,* published in 1971, is a valuable treatment of chemical equilibrium and kinetics.   (17, 61)

[41] NOEL BRYAN SLATER (1912–1973)

Slater was born in Blackburn, Lancashire, and attended the University of Edinburgh where in 1933 he obtained an M.A. degree in mathematics and natural philosophy. He then went to Gonville and Caius College, Cambridge, where he was a Wrangler. From 1935 to 1939 he carried out theoretical research under R. H. Fowler (1889–1944) and obtained a Ph.D. degree from Cambridge in 1939. After some war work, he joined the Applied Mathematics Department at Leeds University in 1945, and he was later appointed professor of applied mathematics at the University of Hull, where he remained until his death.

Much of Slater's research was on the "fundamental theory" of Sir Arthur Eddington (1882–1944). He collected and edited Eddington's unpublished writings, producing a book which explained very lucidly the development of Eddington's ideas.

In his work with R. H. Fowler at Cambridge, Slater worked on the theory of collisions in solution and on unimolecular gas reactions. Beginning in 1945 he carried out an extensive series of theoretical investigations on unimolecular processes, giving particular attention to the detailed mathematical treatment of molecular vibrations. Although the particular approach he made was not ultimately successful, his treatments were of great significance in leading to important clarifications.   (13)

REFERENCES

1. I. Asimov, *Biographical Encyclopedia of Science and Technology,* 2nd. rev. ed., Doubleday, Garden City, N.Y., 1982.
2. H. B. Baker and W. A. Bone, *J. Chem. Soc.,* 3349 (1931); *Proc. R. Soc. London, A,* **134,** i (1931–1932).
3. C. J. Ballhausen, Festskrift, University of Copenhagen, p. 237 (1969).
4. W. D. Bancroft, *J. Chem. Educ.,* **10,** 539, 1009 (1933).
5. C. E. H. Bawn, *Biog. Mem. F.R.S.,* **4,** 193 (1958).
6. R. P. Bell, *Chem. Br.,* **6,** 491 (1970).
7. Earl of Birkenhead, *The Prof in Two Worlds: The Official Life of Professor F. A. Lindemann, Viscount Cherwell,* Collins, London, 1961 (U.S. title: *The Professor and the Prime Minister*).
8. E. J. Bowen, *Biog. Mem. F.R.S.,* **4,** 35 (1958).
9. E. J. Bowen, *Chem. Br.,* **3,** 534 (1967).
10. E. J. Bowen, "The Balliol-Trinity Laboratories, Oxford," *Notes Records R. Soc.,* **25,** 227 (1970).
11. J. A. Christiansen, *Overs. Selsk. Virksomhed,* 57–59 (1948–1949).
12. A. B. Costa, in ref. 24.
13. W. H. Cockcroft, *Univ. of Hull Gazette,* **15**(3), 15 (1973).
14. M. P. Crosland, in ref. 24.

15. F. S. Dainton and B. A. Thrush, *Biog. Mem. F.R.S.,* **27,** 379 (1981).

16. Olive Bell Daniels, *Farrington Daniels: Chemist and Prophet of the Solar Age,* privately published by Mrs. Daniels, Madison, Wisconsin, 1978.

17. M. Davies, *Chem. Br.,* **15,** 397 (1979).

18. E. E. Daub, in ref. 24.

19. H. B. Dixon, *Proc. R. Soc. London,* **97,** vii (1920).

20. F. G. Donnan, *Proc. R. Soc. London, A,* **86,** xxxiv (1912).

21. F. G. Donnan, "Ostwald Memorial Lecture," *J. Chem. Soc.,* 316 (1933).

22. "E.B.E.," *Proc. R. Soc. London, A,* **93,** 44 (1917).

23. D. D. Eley, *Biog. Mem. F.R.S.,* **22,** 381 (1976).

24. C. C. Gillispie (editor-in-chief), *Dictionary of Scientific Biography,* in 14 volumes, Charles Scribner, New York, 1970–1976.

25. J. R. Goodstein, in ref. 24.

26. D. L. Hammick, *Proc. Chem. Soc.,* p. 101 (1959).

27. R. H. Harrod, *The Prof: A Personal Memoir of Lord Cherwell,* Macmillan, London, 1959.

28. P. Harteck, *J. Colloid Sci.,* **13,** 1 (1958).

29. Sir Harold Hartley, *Studies in the History of Chemistry,* Clarendon, Oxford, 1971.

30. S. H. Heath, "Henry Eyring, Mormon Scientist," M.A. Thesis, University of Utah, 1980.

31. E. N. Hiebert, in ref. 24.

32. E. N. Hiebert and A. G. Körber, in ref. 24.

33. E. N. Hiebert, "The Energetics Controversy and the New Thermodynamics," in D. H. D. Roller (Ed.), *Perspectives in the History of Science and Technology,* University of Oklahoma Press, Norman, 1971.

34. E. N. Hiebert, "Nernst and Electrochemistry," in G. Dubpernell and J. H. Westbrook (Eds.), *Selected Topics in the History of Electrochemistry,* The Electrochemistry Society, Princeton, N.J., 1978.

35. E. N. Hiebert, "Developments in Physical Chemistry at the Turn of the Century," in C. G. Bernhard, E. Crawford and D. Sörbom (Eds.), *Science, Technology, and Society in the Time of Alfred Nobel,* Pergamon, Oxford, 1982.

36. E. N. Hiebert, "Walther Nernst and the Application of Physics to Chemistry," in R. Aris, H. T. Davis, and R. H. Stuewer (Eds.), *Springs of Scientific Creativity,* University of Minneapolis Press, 1983.

37. A. F. Holleman, *J. Chem. Educ.,* **29,** 379 (1952).

38. A. T. Jensen, *Overs. Selsk. Virksomhed.,* 99 (1958–1959).

39. H. C. Jones, *Proc. Am. Phil. Soc.,* **50,** iii (1911).

40. W. Jost, *Naturwissenschaften,* **24,** 51 (1957).

41. C. Kemball, *Biog. Mem. F.R.S.,* **21,** 517 (1975).

42. M. C. King, "Experiments with Time," *Ambix,* **28,** 70 (1981); **29,** 49 (1982).

43. M. C. King, "The Chemist in Allegory: Augustus Vernon Harcourt and the White Knight," *J. Chem. Educ.,* **60,** 177 (1983).

44. M. C. King, "The Course of Chemical Change: The Life and Times of Augustus G. Vernon Harcourt, 1834–1919," *Ambix,* **31,** 16 (1984).

45. M. C. King and K. J. Laidler, "Chemical Kinetics and the Radiation Hypothesis," *Arch. Hist. Exact. Sci.,* **30,** 45 (1984).

46. S. J. Kopperl, in ref. 24.

47. K. J. Laidler and M. C. King, "The Development of Transition-State Theory," *J. Phys. Chem.,* **87,** 2657 (1983).

48. K. J. Laidler, "The Development of the Arrhenius Equation," *J. Chem. Educ.,* **61,** 494 (1984).

49. K. J. Laidler, "Chemical Kinetics and the Origins of Physical Chemistry," *Arch. Hist. Exact. Sci.,* **32,** 43 (1985).

50. K. J. Laidler, "Réné Marcelin (1885–1914): A Short-Lived Genius of Chemical Kinetics," *J. Chem. Educ.,* **62,** 1012 (1985).

51. K. J. Laidler, "The Development of Theories of Catalysis," *Arch. Hist. Exact. Sci.,* **35,** 345 (1986).

52. L. Marion, *Biog. Mem. F.R.S.,* **10,** 257 (1964).

53. K. Mendelssohn, *The World of Walther Nernst,* Macmillan, London, 1973.

54. H. W. Melville, *Obit. Notices F.R.S.,* **8,** 395 (1953).

55. M. J. Nye, *Molecular Reality: A Perspective on the Scientific Work of Jean Perrin,* Science History Publications, Neale Watson Academic Publications, New York, 1972.

56. J. R. Partington, *A History of Chemistry,* vols. 1–4, Macmillan, London, 1961.

57. E. K. Rideal, *Nature,* **180,** 581 (1957); *Proc. Chem. Soc.,* p. 80 (1959).

58. R. S. Root-Bernstein, "The Ionists: Founding Physical Chemistry, 1872–1890," Ph.D. Thesis, Princeton University, 1980.

59. V. J. Schaffer, *J. Colloid Sci.,* **13,** 3 (1958).

60. J. Shorter, "A. G. Vernon Harcourt: A Founder of Chemical Kinetics and a Friend of 'Lewis Carroll'," *J. Chem. Educ.,* **57,** 411 (1980).

61. J. Shorter, "The British School of Physical Organic Chemistry," *Chem. Tech.* April 1985, p. 252. The title of this article is misleading, having been changed from "The Contribution of British Physical Chemistry to Physical Organic Chemistry" without its author's knowledge or permission.

62. H. P. M. Snelders, in ref. 24.

63. H. P. M. Snelders, *Janus,* **60,** 261 (1973); *J. Chem. Educ.,* **29,** 379 (1952).

64. R. H. Stuewer, in ref. 24.

65. C. P. Snow, *Science and Government,* Harvard University Press, Cambridge, Mass., 1961. This book, comprising lectures at Harvard, is concerned largely with Lindemann's quarrels with Sir Henry Tizard over war strategy. Snow's assessment of Lindemann's character has been criticized and should be treated with some reserve.

66. F. Szadadvary, in ref. 24.

67. H. S. Taylor, *Biog. Mem. F.R.S.,* **4,** 167 (1958).

68. H. G. Thode, *Proc. R. Soc. Can. Ser. 4,* **1,** 113 (1963).

69. H. W. Thompson, *Biog. Mem. F.R.S.,* **19,** 375 (1973).

70. G. P. Thomson, *Biog. Mem. F.R.S.,* **4,** 45 (1958).

71. A. S. Veibel, in ref. 24.

72. A. S. Veibel, *Overs. K. Dan. Vidensk. Selsk. Forh.,* p. 107 (1970).

73. J. Walker, "Van't Hoff Memorial Lecture," *J. Chem. Soc.,* **103,** 1127 (1919).

74. E. P. Wigner and R. A. Hodgkin, *Biog. Mem. F.R.S.,* **23,** 413 (1977).

Answers to Problems

Chapter 1

1.1 (a) 0.10 mol; 0.19 mol; 0.33 mol; (b) 0.195 mol   **1.2** 1.68×10^{-4} mol dm^{-3} s^{-1}; 5.04×10^{-4} mol dm^{-3} s^{-1}; 1.68×10^{-4} mol dm^{-3} s^{-1} **1.3** (a) 3; (b) $k_{H_2O_2} = k_{Br^-} = k = 3.6 \times 10^{-3}$ mol dm^{-3} s^{-1}; (c) 8; 8   **1.4** 1; 1; 6.21×10^{-4} dm^3 mol^{-1} s^{-1}; 3.11×10^{-4} dm^3 mol^{-1} s^{-1} **1.5** 2; 1; $k_A = k_B = k_Z = 1.64 \times 10^{-3}$ dm^6 mol^{-2} s^{-1}; $k_Y = k = 8.2 \times 10^{-4}$ dm^6 mol^{-2} s^{-1}.

Chapter 2

2.1 1.419×10^{-4} s^{-1}; 189 min   **2.2** 1.38×10^{-11} s^{-1}; 3180 y   **2.3** 5.18 h; 0.669   **2.4** 6.64 times as long   **2.5** (a) 77 ps; (b) 7.7 ns   **2.6** (a) 0.539 μg; (b) 0.177 μg   **2.7** 4.32×10^{-4} mol dm^{-3} s^{-1} **2.8** ~ 2 **2.9** (a) 61.6%; (b) 37.9%; (c) 0.78%   **2.10** (a) 2653; (b) 233   **2.14** 290 s   **2.15** (a) 1.042×10^{-3} dm^3 mol^{-1} s^{-1}; 9600 s = 160 min; (c) 2.4×10^4 s = 400 min   **2.16** 51.2 kJ mol^{-1} **2.17** 143.0 kJ mol^{-1} **2.18** (a) 6.68×10^3; (b) 6.62   **2.19** 49.5 kJ mol^{-1} **2.20** 120.7 kJ mol^{-1} **2.21** -1.5

Chapter 3

3.1 At 500 K: (a) 6.62×10^{-5}; (b) 3.57×10^{-11}; (c) 1.28×10^{-21}; (d) 4.55×10^{-32}. At 300 K: (a) 1.08×10^{-7}; (b) 3.87×10^{-18}; (c) 1.50×10^{-35}; (d) 5.88×10^{-53}. At 1000 K: (a) 8.13×10^{-3}; (b) 5.98×10^{-6}; (c) 5.57×10^{-11}; (d) 2.13×10^{-16} **3.2** $E_d \approx 13$ kJ mol^{-1}; $E_c \approx 50$ kJ mol^{-1}

Chapter 4

4.1 409   **4.2** 138.8 kJ mol^{-1}; -76.7 J K^{-1} mol^{-1}; 185.4 kJ mol^{-1}; 1.02×10^{10} dm^3 mol^{-1} s^{-1} **4.3** 84.1 kJ mol^{-1}; 81.6 kJ mol^{-1}; -64.1 J K^{-1} mol^{-1}; 7.48×10^9 s^{-1} **4.4** 8.58×10^{-7} dm^3 mol^{-1} s^{-1}; 5.05×10^{10} dm^3 mol^{-1} s^{-1}; -62.1 J K^{-1} mol^{-1} (standard state: 1 mol dm^{-3})   **4.5** 54.7 kJ mol^{-1}; 2.20×10^5 dm^3 mol^{-1} s^{-1}; 97.2 kJ mol^{-1}; -150.9 J K^{-1} mol^{-1}; 52.2 kJ mol^{-1} **4.6** (a) 2, 1; (b) 4, 1; (c) 2, 1; (d) 2, 1; (e) 1, 1; (f) 4, 4; (g) 3, 2   **4.7** (a) 7.91×10^{32} m^{-3} s^{-1}; (b) 1.049×10^{10} dm^3 mol^{-1} s^{-1}; (c) 6.08×10^{20} s^{-1} **4.8** 4.6; 32.8 kJ mol^{-1} **4.9** (a) 3.6×10^8 dm^3 mol^{-1} s^{-1}; (b) 3.6×10^7 dm^3 mol^{-1} s^{-1}; (c) 3.6×10^{-24} dm^6 mol^{-2} s^{-1}

Chapter 5

5.1 (a) 0.563; (b) 0.25; (c) 0.0625   **5.2** (a) 7.52×10^{-3}; (b) 7.63×10^{-6}; 5.82×10^{-11} **5.3** 2.5×10^{-4} mol dm^{-3}

Chapter 6

6.1 ~2   **6.2** 1.92 dm^3 mol^{-1} s^{-1}; yes   **6.4** -16.3 cm^3 mol^{-1} **6.5** -18.7 cm^3 mol^{-1} **6.6** -14.3 cm^3 mol^{-1} **6.7** 220 pm   **6.8** 1   **6.10** 0.59   **6.11** 1.38×10^{11} dm^3 mol^{-1} s^{-1} **6.12** -4.8 cm^3 mol^{-1}

Chapter 7

7.1 (a) 1.5×10^{-3} mol dm^{-3} s^{-1}; 2.0×10^{-2} s^{-1} (b) 1.5×10^{-5} mol dm^{-3} s^{-1}; 2.0×10^{-4} s^{-1} (c) 1.5×10^{-5} mol dm^{-3} s^{-1}; 2.0×10^{-4} s^{-1}; $k' = kV/S$; m s^{-1} **7.2** (a) 2.5×10^{-2} mol dm^{-3} s^{-1}; 2.5×10^{-2} mol dm^{-3} s^{-1} (b) 2.5×10^{-4} mol dm^{-3} s^{-1}; 2.5×10^{-4} mol dm^{-3} s^{-1} (c) 2.5×10^{-4} mol dm^{-3} s^{-1}; 2.5×10^{-4} mol dm^{-3} s^{-1}; $k' = kV/S$; mol m^{-2} s^{-1} **7.12** 126.7 kJ mol^{-1}; 4.75×10^{10} cm^{-2} s^{-1}

Chapter 8

8.2 Simultaneous reactions   **8.3** Consecutive reactions   **8.9** Thermal: (a) 168 kJ mol^{-1}, (b) 168 kJ mol^{-1}; Photochemical: (a) 72 kJ mol^{-1}, (b) 72 kJ mol^{-1}

Chapter 9

9.1 306 nm   **9.2** 2   **9.3** 47 s   **9.4** 0.76 mol   **9.5** 4.2 eV   **9.6** 18.8 eV; 7.4 eV   **9.7** 7.8×10^{-4} s   **9.8** 3.4×10^{-8} s

Chapter 10

10.1 1.45×10^{-3} mol dm^{-3} **10.2** 0.6   **10.3** 96.5; 3.5; 0   **10.4** 8.2×10^{-6} dm^3 mol^{-1} s^{-1}; 4.0×10^{-7} dm^3 mol^{-1} s^{-1} **10.7** 0.5

Index

A page number in **boldface** print indicates that a physical quantity or a concept is defined and explained. An index to the biographical sketches appears on p. 491.

Absolute-rate theory: *see* Transition-state theory
Abstraction reactions, 140–145
Acetaldehyde decomposition, 6, 11, 310, 319–321
  catalyzed by iodine, 414
  on surfaces, 243, 250
Acetone halogenation, 392–394
Acid-base catalysis, 384–399
  and acid-base strength, 394–396
  general, **387,** 388–394
  mechanisms of, 388–394
  salt effects, 396–397
  specific, **387**–394
Acidity functions, 397–399
Activated adsorption, 230
Activated complex, 54, **60**
Activation energy, 40, **46–47,** 54–77
  calculation of, 64–71
  empirical relationships, 71–74
  of catalyzed reactions, 246–248, 251–252, 383–384, 409–410
  temperature dependence, 46–48
Activation enthalpy and entropy, **113**
Activators, **15**
Active centers, **230, 400**
Active energy, **164**
Activity coefficients, 188–190
Added gases, influence on reactions, 167–168

Adiabatic energy, **164**
Adiabatic treatments, **124**–125
Adiabaticity, **469,** 479
Adsorption, 230–244
  activated, **230**
  active centers, **230**
  associative, 240
  chemisorption, **230**–244
  competitive, 233–234
  with dissociation, 232–233
  dissociative, 240
  ideal, **231**
  isotherms, 230–236
  nonideal, 234
  rates, 259–261
  statistical treatment, 235–236
  thermodynamics, 235
  van der Waals, **230**
Alkali metal reactions, 459–460
Ammonia decomposition on surfaces, 245–246, 263–264
Ammonia–deuterium exchange, 256–257
Analysis of kinetic results, 18–48
Arrhenius equation, **39,** 42
Arrhenius intermediate, **380**–382
Atom combinations, 171–176
  on surfaces, 252–253
Atom formation, on surfaces, 252–253

Atom–molecule complex mechanism, 175–176
Attractive–mixed–repulsive (A–M–R)
    classification, 461–466
Attractive potential-energy surfaces, 461–462
Autocatalysis, **15**, 21

Basin at activated state, 69
Berthelot equation, 40–41
Bimolecular reactions, **13**
    in gas phase, 138–147
    on surfaces, 248–258, 264–266
Bifunctional catalysis, **412**
Bond-energy–bond-order (BEBO) method, 70–71
Branching chains, **323**–328
    degenerate, **329**
Branching factor, **327**
Bromine atoms, reaction with H_2, 291–295,
    457–458
Bromine–hydrogen reaction, 291–295
Brønsted equation, 197–201
Brønsted relationships, 394–396

Cage effect, 187–188
Canonical rate constant, **125**–127
Canonical variational transition-state theory,
    125–127
Catalysis, **14**, 229–268, 377–418
    acid-base, 384–399
    activation energies, 246–248, 251–252, 383–
        384, 409–410
    bifunctional, **412**
    of chain reactions, 413–414
    enzyme, 399–412
    of hydrogenations, 415–418
    heterogeneous, **15**, 229–268
    homogeneous, **15**, 377–418
    by ions, 414–415
    intramolecular, **15**
    of polymerization, 334
    surface, **15**, 229–268
Catalytic constant, **385**, 404
Catalytic hydrogenation, 415–418
Chain-ending step, **14**
Chain length, **289–290**
Chain-propagating step, **14**
Chain reactions, **14**, 288–290
    catalysis of, 413–414
Chaperon, **173**
Chemical flux, **278**
Chemiluminescence, **371**, 450, 459–460
Chemisorption, 230–244
    rates, 259–261
Christiansen's theory of unimolecular reactions,
    152–155

Chlorine, reaction with alkali metals, 459–460
    with hydrogen, 295–298, 300–301
    with nitric oxide, 147–148
Col, **59**
Collision diameter, **84**, **453**
Collision efficiency, **168**
Collision frequency factor, **82**
Collision number, **81**
Collision theory, 81–87
Collisions, strong, **160**
Collisions in solution, 186–188
Combination of atoms and radicals, 171–176
Combustion, 322–330
Competition, 278
Complex-mode mechanism, **458**
Composite mechanisms, **13**, 276–340
    classification, 276, 278
    rate equations, 278–288
Concerted mechanisms, **389**
Consecutive reactions, 2, 278–288
Contact time, **33**
Conventional transition-state theory, 89–115,
    455–457
Cool flames, 328–329
Correlation effects, 211–212
Coulombic energy, **62**
Cross section, reaction, **84**, **453**
Cyclopentane dimerization, 185
Cyclopropane isomerization, 11, 13, 151, 166,
    170–171

Decay constant, **28**
Degenerate branching, **329**
Degrees of freedom, 76–77, 155–162
Density of states, **164**
Desorption rates, 261–262
Detailed balance at equilibrium, **129–130**, 285–
    288
Detonation, **323**
Deuterium–ammonia exchange, 256–257
Deuterium–hydrogen exchange, 253–254
Deuterium–methane exchange, 254–256
Dielectric constant, effect on rates, 191–193,
    203–204
Dielectric saturation, 201–202
Differential method, **18**–21, 28
Diffusion-controlled reactions, 212–222
    ionic reactions, 217–222
Dipole–dipole reactions, 203–206
Direct reaction, **476**
Discharge-tube methods, 139–140, 369
Disproportionation, 172, 317
Distribution of collisions in solution, 186–188
Distribution of energy, 54–58, 74–77
Dividing surface, **91**

Double-sphere model, 192–193
Dynamical treatments, 88–89, 449–484

Early-downhill surfaces, 461–462
Einstein (unit), **355**
Elastic scattering, **473**
Electron, hydrated, **370**
Electron-jump mechanism, **461**
Electrostriction, **195**
Elementary reactions, 5, **13**
Encounters, **187**–188
Energization, 152, **157–158**
Energy of activation: *see* Activation energy
Energy distribution, 54–58, 74–77
Energy flow, 163–164
Energy transfer, 168–174
    intermolecular, 168
    intramolecular, 168–170
    mechanisms, 173–174
Enthalpy of activation, **113**
Entropy of activation, **113**
Enzyme catalysis, 399–412
    influence of pH, 406–409
    influence of temperature, 409–410
    transient-phase kinetics, 410–412
Equilibrium hypothesis, 90–92, 116–119
Equilibrium, statistical mechanics, 93–94
Ester hydrolysis, 204
Ethane decomposition, 316–319
Ethyl radical reactions, 145
Ethylene hydrogenation, 257–258
Exchange energy, **62**
Exchange reactions, on surfaces, 253–257
Excitation function, **58**
Excitation, multiphoton, 362
Explosion limits, 323–326
Extent of reaction, 7–8
External conversion, **354**

Fast reactions, 33–39
Feedback, 278
Flames, cool, 328–329
    diffusion, 460
    highly dilute, 459–460
Flash photolysis, 39, 359–360
Flow systems, 31–33
Fluorescence, **353**
Foreign-gas effect, 167–168
Fowler–Guggenheim theorem, 56–58
Franck–Condon principle, **350**
Franck–Rabinowitch effect, **188**
Free radical combinations, 171–176
Frequency factor: *see* Pre-exponential factor
Freundlich isotherm, 234

G value, **366**
Gas-phase reactions, composite, 276–335
    elementary, 137–176
General acid-base catalysis, **387**, 388–394
Generalized transition-state theory, 123–129
Gibbs energy of activation, **113**
Goldfinger–Letort–Niclause rules, 311–314
Gradual potential energy surfaces, 469–471
Grotthuss–Draper law, 349
Grotthuss mechanism, 221

Half-life, 2, **26–28**
Halides, reaction with alkali metals, 460
Halogenation of acetone, 392–394
Halogens, reactions, with alkali metals, 459–460
    with nitric oxide, 147–148
Hammett relationship, 209–211
Hammett–Zucker hypothesis, 399
Harcourt–Esson equation, 42–43, 46–47
Harpooning mechanism, **461**
Heitler–London treatment, 62
Heterogeneity of surfaces, 230, 237–241, 243–244
    induced, 241, 244
Heterogeneous reactions, comparison with
        homogeneous reactions, 266–268
    See also Surface reactions
Hertz–Knudsen equation, 261
Hinshelwood's treatment of unimolecular
        reactions, 155–157
Hydrated electron, **370**
Hydration shell, 202
Hydrocarbon decompositions, 307–319
Hydrocarbon oxidation, 328–330
Hydrogen atoms, preparation, 139
    reaction with H_2, 61–67, 138–140, 438–442,
        454–457
    reaction with HBr, 109–112
Hydrogen–bromine reaction, 291–295, 300–301
Hydrogen–chlorine reaction, 295–298, 300–301
Hydrogen–deuterium exchange, 253–254
Hydrogen–ethylene reaction, 257–258
Hydrogen iodide decomposition, 298–301
    on surfaces, 263–264
Hydrogen–iodine reaction, 298–300
Hydrogen ion catalysis, 384–399
Hydrogen ions, combination with OH^- ions, 38–39, 220–221
Hydrogen molecules, radiolysis, 368
    reaction with CH_3, 144
    reaction with H, 61–67, 438–442, 454–457
Hydrogen–nitric oxide reaction, 277
Hydrogen–oxygen reaction, 2, 323–328
Hydrogen peroxide, decomposition, 1, 415
    reaction with HI, 2

Impact parameter, 85
Impulsive (direct) reaction, **476**
Inactive energy, **164**
Indirect reaction, **458**
Induction period, 280, 411
Inelastic scattering, **474**
Inert-gas effect, 167–168
Inhibition, **15**
    degree of, **15**
    by nitric oxide, 314–316, 321–322
    of surface reactions, 246
Initiation, **14,** 413
    of oxidation, 325
    of polymerization, 334–335
Integration, method of, **21–25,** 28–29
Interconversion of energy, 168–170
Intermolecular energy transfer, 168
Internal conversion, **354**
Internal pressure, 189–190
Intramolecular energy transfer, 168–170
Iodine–acetone reaction, 392–394
Iodine–atom combination, 174–176
Iodine–hydrogen reaction, 298–300
Ion decomposition, 171
Ion–dipole reactions, 203–206
Ion–molecule reaction, 146–147
Ion–pair (ionic) yield, **366–367**
Ionic reactions, 191–206
    diffusion control, 217–222
Ionic recombinations, 38–39, 220–221
Ionic strength effects, 197–201, 205–206
Isolation method, **26**
Isomerization of cyclopropane, 11, 13
Isotope effects, 427–447
    equilibrium, 428–433
    kinetic, 433–447
Isotope exchange reactions, 253–257

Jablonski diagram, 354

Kinematics, **88, 464**
Kinetic absolute zero, 42
Kinetic activity factor, 198
Kinetic-isotope effects, 433–447
Kinetic parameter, **12**
Kinetic results, analysis of, 18–48
Kinetic theory of collisions, 81–87
Kooij's equation, 42

Langmuir–Hinshelwood mechanisms, **242,** 248–249
Langmuir isotherm, 231–232
Langmuir–Rideal mechanisms, **242,** 250

Laser-induced unimolecular reactions, 170–171
Laser photochemistry, 360–362
LEP (London–Eyring–Polanyi) surfaces, 68–69, 70
LEPS (Sato) surfaces, 69–70
Life, mean, of radicals, 335
Light-atom anomaly, **465**
Lindemann–Christiansen hypothesis, 152–155
Line-of-centers model, **85–87**
Linear Gibbs-energy relationships, 209–211
London equation, 62–64
London–Eyring–Polanyi (LEP) method, 68–69, 70
London–Eyring–Polanyi–Sato (LEPS) method, 69–70
Luminescence depletion, **477**

Macroscopic diffusion control, **212**
Macroscopic kinetics, 3
Marcus–Coltrin path, **128**–129, 140
Marcus (RRKM) theory of unimolecular reactions 164–167
Mass action, 1
Maxwell–Boltzmann distributions, 54–58, 74–77
Mean life of radicals, 335
Menschutkin reactions, 185, 191
Metathetical reactions, **140**
Methane, reaction with H atoms, 144
    exchange with deuterium, 254–256
Methyl iodide, reaction with pyridine, 191
Methyl radical reactions, 141–145
Michaelis–Menten equation, 382, 400–405
Microcanonical rate constant, **125,** 159
Microcanonical variational transition-state theory, **125–127**
Microscopic diffusion control, **212**–222
Microscopic kinetics, 3
Microscopic reversibility at equilibrium, **129–130,** 285–288
Minimal (minimum-energy) path, **59**
Mirror technique, 289
Mixed energy release, **465**
Mixing control, **212**
M/N (ionic yields), **366–367**
Molecular beams, 472–476
Molecularity, **13**
    of surface reactions, **242**
Multiphoton excitation, **362**
Multiple crossings, 116–119
Multiplet hypothesis, 237
Muonium kinetics, 129, 200, 442–444

Nitric oxide
    inhibition by, 314–316, 321–322
    reaction with halogens, 147–148

reaction with hydrogen, 277
reaction with oxygen, 147–149, 265
Nitrogen pentoxide decomposition, 150, 303–305
in solution, 184
Non-RRKM kinetics, 167
Nonuniform surfaces, 230, 237–241, 243–244
Normal modes, **94**

Opposing reactions, 29–31
Order of reaction, **10–11**, 19
with respect to concentration, **19**
with respect to time, **19**
Ordered ternary-complex mechanism, 405
Organic decompositions, 307–322
Ortho–para hydrogen conversion, 139, 307
on surfaces, 252
Oxidation, gas-phase, 322–330
of hydrocarbons, 328–329
of hydrogen, 323–328
Oxygen
reaction with hydrogen, 323–328
reaction with nitric oxide, 147–149, 265
Ozone decomposition, 305–307

Paraffin decompositions, 307–319
Para–ortho hydrogen conversion, 139, 307
on surfaces, 252
Partition functions, **93,** 95 (table)
pH effects in enzyme kinetics, 406–409
Phosgene, formation and decomposition, 301–303
Phosphine decomposition on surfaces, 229, 246
Photochemical equivalence, 290, 349, **355–356**
Photochemical reactions, 348–365
Photochemical threshold, **355**
Photolysis, **349**
of acetaldehyde, 320–321
of ethane, 318–319
flash, 39, 359–360
Photosensitization, 362–365
Ping pong mechanism, 406
Pilot reaction, **143**
Plug flow, 31–33
Poisoning of surfaces, 246
Polymerization, 330–340
addition, **330,** 332–335
anionic, 338–339
catalysis of, 334
cationic, 335–338
condensation, 330–332
emulsion, 339–340
free-radical, 332–335
initiation of, 333–335

ionic, 335–339
molecular, 330–332
Potential-energy profile, **60**
Potential-energy surfaces, 59–64, 460–472
attractive, **461–462**
calculation of, 67–71
empirical, 71–74
gradual and sudden, 469–471
intermediate, 462–466
repulsive, **462**
Predissociation, **352–353**
Pre-exponential factor, **39**
for solution reactions, 191, 194–195, 204–205
temperature dependence, 46–48, 145–146
Pressure, influence on rates, 206–209
internal, 189–190
Pressure jump, 35
Primary recombination, **188**
Probability of reaction, **84, 453**
Protolytic mechanisms, **389**
Prototropic mechanisms, **389**
Pseudocatalysts, **15**
Pulse radiolysis, **370**
Pulsed lasers, 361–362
Pyridine, reaction with methyl iodide, 191
Pyrolysis, 307–322

Quantum effects, 121–123, 127–129
Quantum-mechanical tunneling, 122–123, 433, 437–438
Quantum yield, 288–290, 355
Quasiclassical trajectory method, **454**
Quasiequilibrium, 90–92
Quaternary ammonium salts, formation, 185, 191
Quenching, 354

Radiation–chemical reactions, 348–349, 365–370
Radiation hypothesis, 80, 151–152
Radical combination, 142–143
Radioactive decay, 28, 280
Radiolysis, 366–370
of ethane, 319
of hydrogen, 366
pulse, **370**
Random lifetimes, **160**
Random ternary-complex mechanism, 404
Rate of consumption, **6**
Rate of conversion, **9**
Rate constant, **11–12**
microcanonical, **125**–127
Rate-determining step, 283–285
Rate equations, empirical, 10
Rate of formation, **6**

Rate of reaction, **8**–9
Reaction cross section, **84, 453**
Reaction dynamics, 88–89, 449–484
Reaction function, **58**
Reaction path, **59, 451**
Reaction probability, **84, 453**
Reactive scattering, **479**
Rebound mechanisms, 475–476
Recombination of radicals, 142–143
Reduced mass, **82**
Relaxation methods, **35**–39
Relaxation time, **37**
Relaxation, vibrational, 353–354
Repulsive potential-energy surface, **462**
Rice–Herzfeld mechanisms, 308–311
Rice–Ramsperger–Kassel (RRK) treatment, 157–163
Rice–Ramsperger–Kassel–Marcus (RRKM) treatment, 164–167
Rotating sector method, 356–359
Rotational energy, influence on kinetics, 471–472, 481–483

Saddle point, **59**
Sato (LEPS) method, 69–70
Saturation current, 366
Secondary recombination, **188**
Sector method, 356–359
Seeding, 473
Semiempirical calculations (potential-energy surfaces), 67–71
Separability of motions, 119–121
Shock tubes, 39
Simultaneous reactions, 278, 280
Single-sphere complex, 195–197
Slater's treatment of unimolecular reactions, 163–164
Slygin–Frumkin isotherm, 234
Sodium
    reaction with chlorine, 459–460
    reaction with halides, 460
Solution reactions, 183–222
Solvation, influence on rates, 190–191
Solvent binding, 195
Solvent effects, 183–185
Specific acid-base catalysis, **387**–394
Spectator-stripping reactions, 475
State-to-state kinetics, **449**, 476–483
Statistical factors, **101**–104
Statistical mechanics of equilibrium, 93–94
Steady-state hypothesis, 282–283
Stepwise reactions: *see* Composite reactions
Steric factor, **83**
Sticky collisions, 147

Stimulated emission, **360**
Stirred-flow reactor, 33
Stoichiometry, 4–6
Stoichiometric coefficient, **4**
Stopped-flow method, 34–35
Stripping mechanisms, **475**–476
Strong collisions, **160**
Substituent effects, 209–211
Successive (consecutive) reactions, 2, 278–288
Surface heterogeneity, 230, 237–241, 243–244
Surface reactions, 229–268
    activation energies, 246–248, 251–252
    bimolecular, 248–258, 264–266
    inhibition, 246
    mechanisms, 252–258
    molecularity, **242**
    on nonuniform surfaces, 243–244
    poisoning, 246
    transition-state theory, 258–266
    unimolecular, 244–248, 262–264
Surface structure, 237–241
Surface/volume ratio, 229
Symmetry number, **98**–104
Symmetry restrictions, 104

Temperature, and reaction rates, 39–48
Temperature-jump method, **35**–39
Termination step, **14**
Thermodynamic formulation of rates, 112–115
Third body, **173**
Time-dependent stoichiometry, **4**–5
Time-independent stoichiometry, **4**
Tolman's theorem, **56**–58
Trajectory, **59, 451**
Transfer of energy, 168–170, 173–174
Transient-phase kinetics, 410–412
Transition species, 483–484
Transition state, **54, 61**
Transition-state theory, 89–129
    adiabatic, 124–125
    applications, 106–112, 148–149, 188–189
    assumptions, 89–92, 115–123
    conventional, 89–115, 455–457
    extensions, 123–129
    generalized, 123–129
    quantum-mechanical, 127–129
    reactions in solution, 188–189
    thermodynamic formulation, 112–115
    variational, 117, 124–129
Transmission coefficient, **118**
Triethylamine, reaction with iodides, 185
Trimolecular reactions, 147–150
True order, **19**
Tunneling, 122–123, 127–129

Turnover number, **379**
Type-I barrier, **462**
Type-II barrier, **462**

Unimolecular reactions, **13,** 150–171
    of ions, 171
    laser-induced, 170–171
    on surfaces, 244–248, 262–264

van der Waals adsorption, **230**
van't Hoff equation, 41
van't Hoff intermediate, **380,** 382–383
van't Hoff's treatment of temperature
    dependence, 40–41, 43

Variability of surfaces, 230, 237–240
Variational treatments
    of activation energy, 64–67
    of rate constants, 117, 124–129
Vibrational relaxation, 353–354
Volume of activation, **114,** 206–209
Volume changes, 9–10

Wood–Bonhoeffer method, 139

Zero-order kinetics, 246
Zero-point levels, and isotope effects, 427–428,
    431, 435